LONDON MATHEMATICAL SOCIETY LECTURE NOTE S

Managing Editor: Professor Endre Süli, Mathematical Institute, Universit
Woodstock Road, Oxford OX2 6GG, United Kingdom

The titles below are available from booksellers, or from Cambridge Univers... at
www.cambridge.org/mathematics

376 Permutation patterns, S. LINTON, N. RUŠKUC & V. VATTER (eds)
377 An introduction to Galois cohomology and its applications, G. BERHUY
378 Probability and mathematical genetics, N. H. BINGHAM & C. M. GOLDIE (eds)
379 Finite and algorithmic model theory, J. ESPARZA, C. MICHAUX & C. STEINHORN (eds)
380 Real and complex singularities, M. MANOEL, M.C. ROMERO FUSTER & C.T.C WALL (eds)
381 Symmetries and integrability of difference equations, D. LEVI, P. OLVER, Z. THOMOVA & P. WINTERNITZ (eds)
382 Forcing with random variables and proof complexity, J. KRAJÍČEK
383 Motivic integration and its interactions with model theory and non-Archimedean geometry I, R. CLUCKERS, J. NICAISE & J. SEBAG (eds)
384 Motivic integration and its interactions with model theory and non-Archimedean geometry II, R. CLUCKERS, J. NICAISE & J. SEBAG (eds)
385 Entropy of hidden Markov processes and connections to dynamical systems, B. MARCUS, K. PETERSEN & T. WEISSMAN (eds)
386 Independence-friendly logic, A.L. MANN, G. SANDU & M. SEVENSTER
387 Groups St Andrews 2009 in Bath I, C.M. CAMPBELL *et al* (eds)
388 Groups St Andrews 2009 in Bath II, C.M. CAMPBELL *et al* (eds)
389 Random fields on the sphere, D. MARINUCCI & G. PECCATI
390 Localization in periodic potentials, D.E. PELINOVSKY
391 Fusion systems in algebra and topology, M. ASCHBACHER, R. KESSAR & B. OLIVER
392 Surveys in combinatorics 2011, R. CHAPMAN (ed)
393 Non-abelian fundamental groups and Iwasawa theory, J. COATES *et al* (eds)
394 Variational problems in differential geometry, R. BIELAWSKI, K. HOUSTON & M. SPEIGHT (eds)
395 How groups grow, A. MANN
396 Arithmetic differential operators over the p-adic integers, C.C. RALPH & S.R. SIMANCA
397 Hyperbolic geometry and applications in quantum chaos and cosmology, J. BOLTE & F. STEINER (eds)
398 Mathematical models in contact mechanics, M. SOFONEA & A. MATEI
399 Circuit double cover of graphs, C.-Q. ZHANG
400 Dense sphere packings: a blueprint for formal proofs, T. HALES
401 A double Hall algebra approach to affine quantum Schur–Weyl theory, B. DENG, J. DU & Q. FU
402 Mathematical aspects of fluid mechanics, J.C. ROBINSON, J.L. RODRIGO & W. SADOWSKI (eds)
403 Foundations of computational mathematics, Budapest 2011, F. CUCKER, T. KRICK, A. PINKUS & A. SZANTO (eds)
404 Operator methods for boundary value problems, S. HASSI, H.S.V. DE SNOO & F.H. SZAFRANIEC (eds)
405 Torsors, étale homotopy and applications to rational points, A.N. SKOROBOGATOV (ed)
406 Appalachian set theory, J. CUMMINGS & E. SCHIMMERLING (eds)
407 The maximal subgroups of the low-dimensional finite classical groups, J.N. BRAY, D.F. HOLT & C.M. RONEY-DOUGAL
408 Complexity science: the Warwick master's course, R. BALL, V. KOLOKOLTSOV & R.S. MACKAY (eds)
409 Surveys in combinatorics 2013, S.R. BLACKBURN, S. GERKE & M. WILDON (eds)
410 Representation theory and harmonic analysis of wreath products of finite groups, T. CECCHERINI-SILBERSTEIN, F. SCARABOTTI & F. TOLLI
411 Moduli spaces, L. BRAMBILA-PAZ, O. GARCÍA-PRADA, P. NEWSTEAD & R.P. THOMAS (eds)
412 Automorphisms and equivalence relations in topological dynamics, D.B. ELLIS & R. ELLIS
413 Optimal transportation, Y. OLLIVIER, H. PAJOT & C. VILLANI (eds)
414 Automorphic forms and Galois representations I, F. DIAMOND, P.L. KASSAEI & M. KIM (eds)
415 Automorphic forms and Galois representations II, F. DIAMOND, P.L. KASSAEI & M. KIM (eds)
416 Reversibility in dynamics and group theory, A.G. O'FARRELL & I. SHORT
417 Recent advances in algebraic geometry, C.D. HACON, M. MUSTAŢĂ & M. POPA (eds)
418 The Bloch–Kato conjecture for the Riemann zeta function, J. COATES, A. RAGHURAM, A. SAIKIA & R. SUJATHA (eds)
419 The Cauchy problem for non-Lipschitz semi-linear parabolic partial differential equations, J.C. MEYER & D.J. NEEDHAM
420 Arithmetic and geometry, L. DIEULEFAIT *et al* (eds)
421 O-minimality and Diophantine geometry, G.O. JONES & A.J. WILKIE (eds)
422 Groups St Andrews 2013, C.M. CAMPBELL *et al* (eds)
423 Inequalities for graph eigenvalues, Z. STANIĆ
424 Surveys in combinatorics 2015, A. CZUMAJ *et al* (eds)
425 Geometry, topology and dynamics in negative curvature, C.S. ARAVINDA, F.T. FARRELL & J.-F. LAFONT (eds)

426 Lectures on the theory of water waves, T. BRIDGES, M. GROVES & D. NICHOLLS (eds)
427 Recent advances in Hodge theory, M. KERR & G. PEARLSTEIN (eds)
428 Geometry in a Fréchet context, C.T.J. DODSON, G. GALANIS & E. VASSILIOU
429 Sheaves and functions modulo p, L. TAELMAN
430 Recent progress in the theory of the Euler and Navier–Stokes equations, J.C. ROBINSON, J.L. RODRIGO, W. SADOWSKI & A. VIDAL-LÓPEZ (eds)
431 Harmonic and subharmonic function theory on the real hyperbolic ball, M. STOLL
432 Topics in graph automorphisms and reconstruction (2nd Edition), J. LAURI & R. SCAPELLATO
433 Regular and irregular holonomic D-modules, M. KASHIWARA & P. SCHAPIRA
434 Analytic semigroups and semilinear initial boundary value problems (2nd Edition), K. TAIRA
435 Graded rings and graded Grothendieck groups, R. HAZRAT
436 Groups, graphs and random walks, T. CECCHERINI-SILBERSTEIN, M. SALVATORI & E. SAVA-HUSS (eds)
437 Dynamics and analytic number theory, D. BADZIAHIN, A. GORODNIK & N. PEYERIMHOFF (eds)
438 Random walks and heat kernels on graphs, M.T. BARLOW
439 Evolution equations, K. AMMARI & S. GERBI (eds)
440 Surveys in combinatorics 2017, A. CLAESSON *et al* (eds)
441 Polynomials and the mod 2 Steenrod algebra I, G. WALKER & R.M.W. WOOD
442 Polynomials and the mod 2 Steenrod algebra II, G. WALKER & R.M.W. WOOD
443 Asymptotic analysis in general relativity, T. DAUDÉ, D. HÄFNER & J.-P. NICOLAS (eds)
444 Geometric and cohomological group theory, P.H. KROPHOLLER, I.J. LEARY, C. MARTÍNEZ-PÉREZ & B.E.A. NUCINKIS (eds)
445 Introduction to hidden semi-Markov models, J. VAN DER HOEK & R.J. ELLIOTT
446 Advances in two-dimensional homotopy and combinatorial group theory, W. METZLER & S. ROSEBROCK (eds)
447 New directions in locally compact groups, P.-E. CAPRACE & N. MONOD (eds)
448 Synthetic differential topology, M.C. BUNGE, F. GAGO & A.M. SAN LUIS
449 Permutation groups and cartesian decompositions, C.E. PRAEGER & C. SCHNEIDER
450 Partial differential equations arising from physics and geometry, M. BEN AYED *et al* (eds)
451 Topological methods in group theory, N. BROADDUS, M. DAVIS, J.-F. LAFONT & I. ORTIZ (eds)
452 Partial differential equations in fluid mechanics, C.L. FEFFERMAN, J.C. ROBINSON & J.L. RODRIGO (eds)
453 Stochastic stability of differential equations in abstract spaces, K. LIU
454 Beyond hyperbolicity, M. HAGEN, R. WEBB & H. WILTON (eds)
455 Groups St Andrews 2017 in Birmingham, C.M. CAMPBELL *et al* (eds)
456 Surveys in combinatorics 2019, A. LO, R. MYCROFT, G. PERARNAU & A. TREGLOWN (eds)
457 Shimura varieties, T. HAINES & M. HARRIS (eds)
458 Integrable systems and algebraic geometry I, R. DONAGI & T. SHASKA (eds)
459 Integrable systems and algebraic geometry II, R. DONAGI & T. SHASKA (eds)
460 Wigner-type theorems for Hilbert Grassmannians, M. PANKOV
461 Analysis and geometry on graphs and manifolds, M. KELLER, D. LENZ & R.K. WOJCIECHOWSKI
462 Zeta and L-functions of varieties and motives, B. KAHN
463 Differential geometry in the large, O. DEARRICOTT *et al* (eds)
464 Lectures on orthogonal polynomials and special functions, H.S. COHL & M.E.H. ISMAIL (eds)
465 Constrained Willmore surfaces, Á.C. QUINTINO
466 Invariance of modules under automorphisms of their envelopes and covers, A.K. SRIVASTAVA, A. TUGANBAEV & P.A. GUIL ASENSIO
467 The genesis of the Langlands program, J. MUELLER & F. SHAHIDI
468 (Co)end calculus, F. LOREGIAN
469 Computational cryptography, J.W. BOS & M. STAM (eds)
470 Surveys in combinatorics 2021, K.K. DABROWSKI *et al* (eds)
471 Matrix analysis and entrywise positivity preservers, A. KHARE
472 Facets of algebraic geometry I, P. ALUFFI *et al* (eds)
473 Facets of algebraic geometry II, P. ALUFFI *et al* (eds)
474 Equivariant topology and derived algebra, S. BALCHIN, D. BARNES, M. KEDZIOREK & M. SZYMIK (eds)
475 Effective results and methods for Diophantine equations over finitely generated domains, J.-H. EVERTSE & K. GYŐRY
476 An indefinite excursion in operator theory, A. GHEONDEA
477 Elliptic regularity theory by approximation methods, E.A. PIMENTEL
478 Recent developments in algebraic geometry, H. ABBAN, G. BROWN, A. KASPRZYK & S. MORI (eds)
479 Bounded cohomology and simplicial volume, C. CAMPAGNOLO, F. FOURNIER-FACIO, N. HEUER & M. MORASCHINI (eds)
480 Stacks Project Expository Collection (SPEC), P. BELMANS, W. HO & A.J. DE JONG (eds)
481 Surveys in combinatorics 2022, A. NIXON & S. PRENDIVILLE (eds)
482 The logical approach to automatic sequences, J. SHALLIT
483 Rectifiability: a survey, P. MATTILA
484 Discrete quantum walks on graphs and digraphs, C. GODSIL & H. ZHAN
485 The Calabi problem for Fano threefolds, C. ARAUJO *et al*

London Mathematical Society Lecture Note Series: 486

Modern Trends in Algebra and Representation Theory

Edited by

DAVID JORDAN
University of Edinburgh

NADIA MAZZA
Lancaster University

SIBYLLE SCHROLL
University of Cologne

Shaftesbury Road, Cambridge CB2 8EA, United Kingdom

One Liberty Plaza, 20th Floor, New York, NY 10006, USA

477 Williamstown Road, Port Melbourne, VIC 3207, Australia

314–321, 3rd Floor, Plot 3, Splendor Forum, Jasola District Centre, New Delhi – 110025, India

103 Penang Road, #05–06/07, Visioncrest Commercial, Singapore 238467

Cambridge University Press is part of Cambridge University Press & Assessment, a department of the University of Cambridge.

We share the University's mission to contribute to society through the pursuit of education, learning and research at the highest international levels of excellence.

www.cambridge.org
Information on this title: www.cambridge.org/9781009097352

DOI: 10.1017/9781009093750

First published 2023

Printed in the United Kingdom by TJ Books Limited, Padstow Cornwall

A catalogue record for this publication is available from the British Library.

A Cataloging-in-Publication data record for this book is available from the Library of Congress.

ISBN 978-1-009-09735-2 Paperback

Contents

Contributors

Andreas Bode, *University of Calgary*
Ilaria Castellano, *University of Milano-Bicocca*
Nicolas Dupré, *University of Duisburg-Essen*
Sam Gunningham, *Montana State University*
Amit Hazi, *University of York*
David Jordan, *University of Edinburgh*
Rosanna Laking, *University of Verona*
Nadia Mazza, *Lancaster University*
Matthew Pressland, *University of Glasgow*
Sibylle Schroll, *University of Cologne*
Raquel Coelho Simões, *Lancaster University*
Gareth Tracey, *University of Warwick*
Hipolito Treffinger, *Université Paris Cité*
Bart Vlaar, *Tsinghua University, Beijing*
Brian Williams, *Boston University*

Preface

This volume of proceedings follows from the online lectures series held during the LMS Autumn Algebra School 2020 and comprises extended versions of the material presented during the online lectures. The school was supported by the London Mathematical Society Covid Working Group and by ERC grant no. 637618.

The lectures were delivered by early career researchers from three London Mathematical Society Joint Research Groups in the UK, namely ARTIN, BLOC and Functor categories for groups, whose lead organisers are also the Editors of the present volume.

The *Bristol-Leicester-Oxford-City Colloquium (BLOC)* was an LMS Joint Research Group based at the University of Leicester from 1997 up until the closure of the pure mathematics group at the University of Leicester in 2021. The network focused on the representation theory of finite dimensional algebras as well as aspects of modular representation theory of finite groups. *Algebra and Representation Theory in the North (ARTIN)* is an LMS Joint Research Group in the UK founded in 2003 and dedicated to the study of non-commutative algebra, representation theory and quantum groups. *Functor Categories for Groups (FCG)* is an LMS Joint Research Group in the UK founded in 2016 and dedicated to the study of abstract groups and related aspects.

The lectures were hosted online by the International Centre for Mathematical Sciences, and recordings can be found on the ICMS website: www.icms.org.uk/seminars/2020/lms-autumn-algebra-school-2020.

The material written in this book has been prepared by the lecturers and underwent peer-review. The Editors are very grateful to the authors for their engagement throughout the process and the excellent quality of the work produced, and to the mentors and peer-reviewers who helped perfect the chapters.

David Jordan
Nadia Mazza
Sibylle Schroll

Introduction

This volume splits into three main strands in representation theory corresponding to the topics covered by the three LMS networks BLOC, ARTIN and FCG.

The first four chapters contributed by BLOC introduce different aspects of the representation theory of associative algebras focusing on new developments in the subject, in particular, in connection with geometric surface models, cluster theory and model theory.

Raquel Coelho-Simões surveys Auslander–Reiten theory, one of the main tools underlying modern representation theory. Following the classical definitions and, in particular, the introduction of the Auslander–Reiten translate, her summary of geometric surface models describes recent geometric models for the module categories of algebras and cluster-tilted algebras of type A_n and culminates in geometric surface models for the module category of gentle and skew-gentle algebras. This places this survey at the center of recent new trends and developments in the representation theory of finite dimensional algebras.

Hipolito Treffinger gives a comprehensive account of τ-tilting theory that was introduced as such in the early 2010s by Adachi, Iyama and Reiten with τ, the Auslander–Reiten translate of a finite dimensional algebra, playing a central role. The ideas of τ-tilting theory are a generalisation of classical tilting theory and are closely related to categorifications of Cluster algebras and their mutations. Treffinger's contribution explains how τ-tilting theory has been shown to be the right language for many classical concepts in representation theory such as torsion classes, the Brauer–Thrall conjectures and wide subcategories but that it also gives a representation theoretic approach to subjects such as stability conditions and scattering diagrams.

Matthew Pressland in his survey on frieze patterns, cluster algebras and cluster categories gives another representation theoretic interpretation of cluster theory. Namely, he shows in his survey how cluster algebras and cluster categories can be seen as a conceptual explanation of the combinatorics of Coxeter

Conway frieze patterns. More precisely, after an introduction of Coxeter Conway frieze patterns and their relation to cluster algebras, he shows how for a quiver of Dynkin type A_n, the Coxeter Conway frieze patterns of height n are in bijection with the cluster-tilting objects of the corresponding cluster category defined by Buan–Marsh–Reineke–Reiten–Todorov as a quotient of the derived category of the module category of a type A_n quiver.

Rosie Laking gives an introduction to infinite dimensional representations. In contrast to the finite dimensional case, the situation is much more complicated and a different approach and different methods are needed. Through carefully chosen examples, Laking not only gives a good overview of important classes of infinite dimensional representations such as indecomposable pure-injectives, she also explains connections to logic and model theory and why such an abstract approach is useful. She introduces the Ziegler spectrum – a topological space formed by the indecomposable pure-injectives up to isomorphism – and gives, for finite dimensional algebras, an overview of the basic properties of this topological space and how these can be used to characterize the notion of finite representation type of a finite dimensional algebra.

The next four chapters contributed by ARTIN introduce the reader to four important developments of Lie theory in the past two decades. What ties these four strands together is that each one unites a purely algebraic study of representation theory with some central topic in algebraic geometry and/or low-dimensional topology.

The lectures by Sam Gunningham give a modern perspective on Springer theory – the study of the geometry and topology of the flag variety and related homogeneous spaces, and its relation to the Weyl group and Hecke algebras. The chapter proceeds first in the more elementary framework of Borel–Moore homology and convolution algebras, and then retells the story in the more sophisticated framework of perverse sheaves. As presented in these lectures, Springer theory is a gateway to the sprawling field of geometric representation theory, in which one seeks to build representations – of finite groups, Coxeter groups, Lie groups, Lie algebras, quantum groups, W-algebras, Hecke algebras, etc. – as sections of geometrically constructed sheaves on moduli spaces arising in algebraic geometry.

Amit Hazi's lectures introduce the reader to the theory of Soergel bimodules, its algebraic realization via the Bernstein–Gelfand–Gelfand category $\mathcal{O}$ and finally its elementary diagrammatic incarnation. Soergel bimodules and their diagrammatic incarnations define monoidal categorifications of Hecke algebras. The remarkable realization that the intricate algebraic structures of BGG category $\mathcal{O}$ and of Hecke algebras can be encoded efficiently using elementary diagrams on the plane with relations defined locally and diagrammatically points to a deep relation between these categories and low-dimensional topology, one that has yielded major advances in the study of categorified knot invariants, mathematical physics, higher categorical representation theory and many others.

The lectures by Bart Vlaar introduce the reader to the theory of quantum groups and quantum symmetric pairs. Quantum groups were first studied in the context of statistical mechanics, where they described certain generalized symmetries of a statistical lattice model on the plane. Their representation theory was at the heart of the famous Witten–Reshetikhin–Turaev construction of 3-manifold invariants building on Chern–Simons theory, and the full extent of their role in the theory of 3- and 4-dimensional topological field theories is still a very active area of research. These lectures recall some of this history, and focus on recent developments with so-called quantum symmetric pairs. Roughly quantum symmetric pairs describe (quantizations of) involution-fixed-point subgroups, and so they exhibit analogous algebraic structure to real forms of algebraic groups. Their representation theory has proved to be even richer, yet comparable to study, than that of the quantum group itself, and there is an active community of mathematicians fleshing out this theory in recent years.

Brian Williams' lectures recount the algebro–geometric theory of affine Lie algebras and vertex algebras more generally, as factorization algebras on algebraic curves. This approach to vertex algebras in was first developed by Beilinson–Drinfeld and Frenkel–Ben-Zvi, and in these lectures is recast in the modern and more flexible language of holomorphic factorization algebras, after Costello–Gwilliam, Williams, Faonte–Hennion–Kapranov and others. A remarkable consequence of the reformulation as holomorphic factorization algebras, explored in these lectures, is that the definitions naturally extend to higher-dimensional algebraic varieties, which are replete with new classes of examples. An essential new feature of this extension is the deployment of derived algebraic geometry, and the outcome is a new theory of Kac–Moody algebras in higher dimensions.

The final three chapters contributed by FCG introduce the reader to three strikingly different aspects of group theory.

Gareth Tracey introduces the notion of *crowns* in a finite group G, which, loosely, are subquotients of G that are obtained from the non-Frattini chief factors of G. Crowns were introduced by Gaschütz in 1962 in the context of finite soluble groups. Tracey's article is self-contained, including traditional results on the structure of finite groups necessary to understand the theory of crowns. The chapter starts with the review of chief series, chief factors and an equivalence relation between them up to introducing the notion of a crown in a finite group. The chapter concludes with an application of crowns to find the minimum number of generators of a finite group.

Ilaria Castellano's chapter is an introduction to totally disconnected locally compact (TDLC) groups, their properties and applications, with a focus on the compactly generated TDLC groups. A locally compact group is an extension of its connected component containing the identity by a TDLC group. As a consequence of a result by Gleason and Yamabe, connected locally compact groups can be approximated by Lie groups, in a certain sense. Hence it "remains" to study TDLC groups to understand the structure of locally compact groups. TDLC groups are also the automorphism groups of locally finite connected graphs, and they come with a family of Cayley–Abels graphs unique up to quasi-isometry, a geometric invariant of the group. After discussing some intriguing geometric and topological comparisons of the properties of TDLC groups, the final part of Castellano's chapter introduces homological finiteness conditions and presents some recent work of Willis.

Andreas Bode and Nicolas Dupré survey Schneider and Teitelbaum's theory of admissible locally analytic representations of p-adic Lie groups. The study of such representations is motivated by the conjecture of the p-adic Langlands correspondence. Bode and Dupré introduce locally analytic representations, and they describe the construction of the distributions algebra. Then, they explain the notion of a Fréchet–Stein algebra, and they define admissible representations, providing some context on the recent developments.

Clearly, each of these chapters can only scratch the surface of the topics they introduce. However, our hope is that they provide motivated readers – beginning PhD students and subject-adjacent non-experts alike – with a springboard into the vast and growing literature in each area.

1

Auslander–Reiten Theory of Finite-Dimensional Algebras

Raquel Coelho Simões

Introduction

Representation theory is the study of algebraic structures such as groups and rings via their actions on simpler algebraic structures such as vector spaces. In this survey chapter, which served as a basis for a lecture series at the LMS Autumn Algebra School in October 2020, we consider representation theory of finite-dimensional algebras. Quivers provide a useful concrete way to visualise algebras. Given a quiver, i.e. a directed graph, one can associate an algebra, called a *path algebra*, generated by the paths of the quiver. From the point of view of representation theory, the study of finite-dimensional algebras reduces to the study of quotients of path algebras.

A central aim in representation theory of finite-dimensional algebras is to classify all their modules and the morphisms between them. Due to the Krull–Schmidt theorem, the classification of modules can be reduced to the classification of indecomposable modules. That is, in some sense, indecomposable modules are the building blocks of all modules. It is then natural to ask when is a finite-dimensional algebra of *finite representation type*, i.e. when does it have finitely many indecomposable modules up to isomorphism? Gabriel's theorem [22] gives an elegant answer for path algebras of quivers without oriented cycles, also called *hereditary algebras*. This theorem is an example of an ADE classification, i.e. in terms of simply laced Dynkin diagrams. There are many other examples of objects classified by these diagrams, including representation-finite selfinjective algebras [35], irreducible root systems and semisimple Lie algebras (see e.g. [28]) and cluster algebras of finite type [20].

Auslander–Reiten theory gives us a way to visualise the representation theory of a finite-dimensional algebra using a quiver, called the Auslander–Reiten quiver. The vertices of this quiver correspond to the indecomposable modules and the arrows correspond to *irreducible morphisms*, which are the corresponding building blocks for the morphisms.

The aim of this survey chapter is to give a brief introduction to Auslander–Reiten theory and to provide methods for constructing Auslander–Reiten quivers. We present two methods to construct these quivers for some special classes of algebras. The first method is the *knitting algorithm*, which works for instance for hereditary algebras of finite representation type. The second method is a *geometric model* associated to (partial) triangulations of surfaces. This method, which has its origins in cluster-tilting theory [15], encodes the representation theory of an important class of algebras, called *gentle algebras* and more generally *skew-gentle algebras*, which have been the subject of intensive study since the 1980s due to the fact that they remain one of the relatively few classes of algebras for which the representation theory is computationally tractable.

The prerequisites are a basic knowledge of linear algebra and rings and modules. Knowledge of the basic concepts of category theory is beneficial, but not essential. The list of references is not exhaustive, but it includes some of the main references for this subject. We refer the reader to [5, 6, 9, 38] for further study on quiver representations and Auslander–Reiten theory. The language of categories used in these theories is also nicely explained in [6, 38].

Conventions: Throughout this chapter, we consider vector spaces, linear maps and algebras over an algebraically closed field $\mathbf{k}$. Every algebra will be a finite-dimensional associative algebra with unit and every module is considered to be a finite-dimensional right module. For a treatment of infinite-dimensional modules, see Chapter 4.

1.1 Bound Path Algebras

In this section we will associate algebras to quivers, i.e. directed graphs. From a representation-theoretic point of view, we will see that it is enough to study algebras associated to quivers.

Definition 1.1 A *quiver* $Q = (Q_0, Q_1, s, t)$ consists of the following data:

1 a set Q_0 of vertices,

2 a set Q_1 of arrows between vertices,

3 two maps $s, t : Q_1 \to Q_0$, called *source* and *target*, respectively, such that, for each arrow $\alpha : i \to j \in Q_1$, $i = s(\alpha)$ and $j = t(\alpha)$.

A quiver is *finite* if Q_0 and Q_1 are finite sets. Throughout these notes, we will only consider finite and connected quivers.

Definition 1.2 Let Q be a quiver.

1 A *path in Q of length ℓ* is a sequence $p = \alpha_1\alpha_2\cdots\alpha_\ell$, with $\alpha_i \in Q_1$ such that $s(\alpha_i) = t(\alpha_{i-1})$ for each $i = 2,\dots,\ell$. In particular, p has length 1 if and only if $p \in Q_1$.
2 We associate a path ε_i of length 0 to each vertex i of Q, which is called the *stationary path at i*.
3 If $s(\alpha_1) = t(\alpha_\ell)$, then p is said to be an *oriented cycle*. An oriented cycle of length 1 is called a *loop*. An *acyclic* quiver is a quiver with no oriented cycles.

Sometimes we denote a path from i to j by $i \rightsquigarrow j$.
Throughout $\mathbf{k}$ denotes an algebraically closed field.

Definition 1.3 The *path algebra* $\mathbf{k}Q$ of Q is an algebra whose underlying vector space has all the paths of Q as basis and with multiplication defined on two basis elements given by concatenation of paths, i.e. given two paths $p = \alpha_1\cdots\alpha_\ell, p' = \alpha'_1\cdots\alpha'_m$,

$$pp' = \begin{cases} \alpha_1\cdots\alpha_\ell\alpha'_1\cdots\alpha'_m & \text{if } t(\alpha_\ell) = s(\alpha'_1) \\ 0 & \text{otherwise.} \end{cases}$$

Example 1.4 1 Let Q be the quiver:

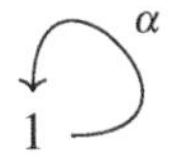

$\mathbf{k}Q$ has basis given by $\{\alpha^t \mid t \geqslant 0\}$, where α^0 denotes the stationary path ε_1. The multiplication is given by $\alpha^s\alpha^t = \alpha^{s+t}$. The algebra $\mathbf{k}Q$ is isomorphic to the algebra $\mathbf{k}[x]$ of polynomials with one indeterminate.

2 Let Q be the quiver

$$1 \xrightarrow{\alpha_1} 2 \xrightarrow{\alpha_2} \cdots \xrightarrow{\alpha_{n-1}} n$$

$\mathbf{k}Q$ is generated by the paths $\varepsilon_i(1 \leqslant i \leqslant n), \alpha_i(1 \leqslant i \leqslant n), \alpha_i\cdots\alpha_j(1 \leqslant i < j \leqslant n)$, and it is isomorphic to the algebra of upper triangular $n \times n$ matrices.

Remark The path algebra $\mathbf{k}Q$ satisfies the following properties:

1 $\mathbf{k}Q$ has an identity $1 = \sum_{i\in Q_0} \varepsilon_i$ if and only if Q_0 is finite.

2 $\mathbf{k}Q$ is an associative algebra.
3 $\mathbf{k}Q$ is finite dimensional if and only if Q is finite and acyclic.

Definition 1.5 Let Q be a finite quiver.

1 The *arrow ideal* R_Q is the two-sided ideal of $\mathbf{k}Q$ generated by all arrows in Q.
2 An *admissible ideal* I is a two-sided ideal of $\mathbf{k}Q$ such that there is $m \geqslant 2$ for which $R_Q^m \subseteq I \subseteq R_Q^2$.
3 Given an admissible ideal I, the quotient algebra $\mathbf{k}Q/I$ is said to be a *bound path algebra*.

The bound path algebra $\mathbf{k}Q/I$ is finite dimensional, since $R_Q^m \subseteq I$ and it is connected (i.e. it is not the direct product of two algebras) because Q is connected and $I \subseteq R_Q^2$.

A *relation* ρ is a linear combination $\rho = \sum_p \lambda_p p$ of paths, all with length at least two, and with same start and same endpoints. It is easy to check that any admissible ideal can be generated by a set of relations.

Example 1.6 Let Q be the quiver

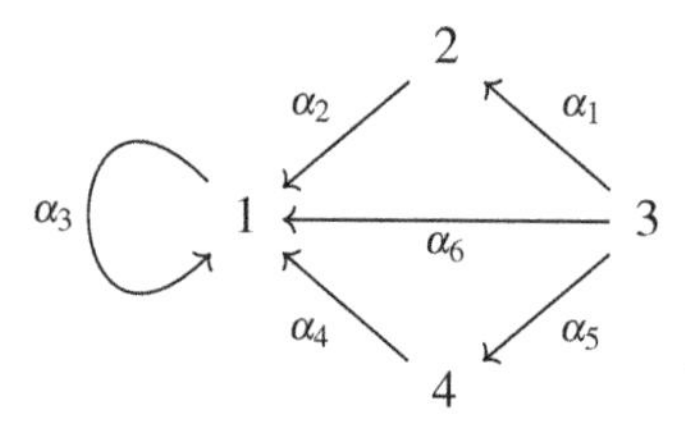

The ideal $I_1 = \langle \alpha_1\alpha_2 - \alpha_5\alpha_4, \alpha_6\alpha_3, \alpha_2\alpha_3, \alpha_3^4\rangle$ is admissible since $R_Q^5 \subseteq I \subseteq R_Q^2$.
The ideal $I_2 = \langle \alpha_1\alpha_2 - \alpha_5\alpha_4, \alpha_6\alpha_3, \alpha_2\alpha_3\rangle$ is not admissible because $\alpha_3^m \notin I_2$ for all $m \geqslant 2$.
The ideal $I_3 = \langle \alpha_1\alpha_2 - \alpha_6\rangle$ is not admissible as $\alpha_1\alpha_2 - \alpha_6 \notin R_Q^2$.

The following theorem is due to [22].

Theorem 1.7 *Any finite-dimensional algebra A is Morita equivalent to a bound path algebra $\mathrm{k}Q/I$, i.e.* $\mathrm{mod}(A) \simeq \mathrm{mod}(\mathrm{k}Q/I)$.

For a proof, see [6, I.6.10, II.3.7].

1.2 Representations of a Bound Path Algebra

In the previous section we saw that quivers provide a nice way to visualise finite-dimensional algebras. Now, we will explain how quivers can be used to visualise also modules and morphisms between modules.

Throughout this section Q denotes a finite, connected quiver and I an admissible ideal. Note that if $I = 0$ is admissible, then Q must be acyclic.

Definition 1.8 A *representation* $M = (M_i, \varphi_\alpha)_{i \in Q_0, \alpha \in Q_1}$ *of* Q is given by:

- $\mathbf{k}$-vector spaces M_i for all $i \in Q_0$, and
- linear maps $\varphi_\alpha : M_{s(\alpha)} \to M_{t(\alpha)}$ for all $\alpha \in Q_1$.

Let $p = \alpha_1 \cdots \alpha_\ell$ be a path in Q and $M = (M_i, \varphi_\alpha)_{i \in Q_0, \alpha \in Q_1}$ be a representation of Q. We denote by φ_p the composition of linear maps $\varphi_p = \varphi_{\alpha_\ell} \cdots \varphi_{\alpha_1}$. Given a relation $\rho = \sum_p \lambda_p p$ in I, we have $\varphi_\rho = \sum_p \lambda_p \varphi_p$.

Definition 1.9 A representation $M = (M_i, \varphi_\alpha)_{i \in Q_0, \alpha \in Q_1}$ of Q is said to be *bound by* I, or to be a *representation of* (Q, I), if $\varphi_\rho = 0$ for all $\rho \in I$.

A representation M is *finite dimensional* if M_i is finite dimensional for all $i \in Q_0$. The *dimension vector* of M is the vector $\underline{\dim} M = (\dim M_i)_{i \in Q_0}$.

Example 1.10 Consider the quiver Q:

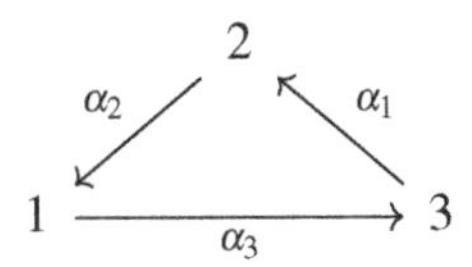

bound by $I = \langle \alpha_1 \alpha_2, \alpha_2 \alpha_3, \alpha_3 \alpha_1 \rangle$. The representation:

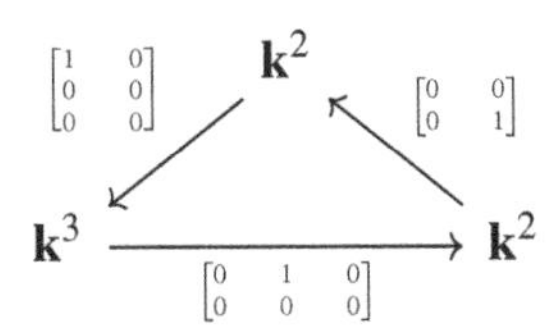

is bound by I. However, the representation given by

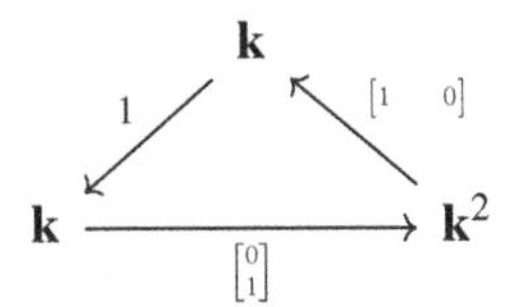

is not bound by I.

Definition 1.11 Let $M = (M_i, \varphi_\alpha), N = (N_i, \psi_\alpha)$ be representations of (Q, I).

1 A *morphism of representations* $f : M \to N$ is a collection $(f_i)_{i \in Q_0}$ of linear maps, $f_i : M_i \to N_i$, such that for each $\alpha : i \to j \in Q_1$, the following diagram commutes:

$$\begin{array}{ccc} M_i & \xrightarrow{\varphi_\alpha} & M_j \\ {\scriptstyle f_i}\downarrow & & \downarrow{\scriptstyle f_j} \\ N_i & \xrightarrow[\psi_\alpha]{} & N_j. \end{array}$$

2 The morphism $f = (f_i)_{i \in Q_0}$ is an *isomorphism* if each f_i is bijective.

Example 1.12 Let Q be the quiver

$$1 \circlearrowleft \alpha$$

The following represents a morphism of representations:

$$\begin{bmatrix} 2 & 0 \\ 0 & 3 \end{bmatrix} \circlearrowright \mathbf{k}^2 \xrightarrow{\begin{bmatrix} 0 & 1 \\ 1 & 1 \end{bmatrix}} \mathbf{k}^2 \circlearrowleft \begin{bmatrix} 3 & 0 \\ 1 & 2 \end{bmatrix}$$

This morphism is bijective with inverse given by

$$\begin{bmatrix} 2 & 0 \\ 0 & 3 \end{bmatrix} \circlearrowright \mathbf{k}^2 \xleftarrow[\begin{bmatrix} -1 & 1 \\ 1 & 0 \end{bmatrix}]{} \mathbf{k}^2 \circlearrowleft \begin{bmatrix} 3 & 0 \\ 1 & 2 \end{bmatrix}$$

We obtain the category $\mathsf{rep}(Q, I)$ of finite-dimensional bound quiver representations of (Q, I), whose objects are finite-dimensional bound quiver representations and maps are given by morphisms of bound quiver representations.

Given a finite-dimensional algebra A, we denote by $\mathsf{mod}(A)$ the category of finite-dimensional right A-modules. Note that we are adopting the same convention as [6] of taking right A-modules and reading paths in a quiver from left to right. Other sources may have the opposite convention.

Theorem 1.13 *There is an equivalence of categories* $\mathrm{mod}(\mathbf{k}Q/I) \simeq \mathsf{rep}(Q, I)$.

Proof Denote the algebra $\mathbf{k}Q/I$ by A and write $e_i = \varepsilon_i + I$. We begin by constructing a functor $F : \mathsf{mod}(A) \to \mathsf{rep}(Q, I)$.

Given $M \in \mathrm{mod}(A)$, we define $F(M)$ to be the representation (M_i, φ_α), where $M_i = Me_i$, and $\varphi_\alpha : M_{s(\alpha)} \to M_{t(\alpha)}$ is the map $me_{s(\alpha)} \mapsto m\bar{\alpha} := m(\alpha + I)$.

Note that each φ_α is a $\mathbf{k}$-linear map since M is an A-module.

In order to check that $F(M) \in \mathrm{rep}(Q,I)$, we need to show that $F(M)$ is bound by I. Given a relation $\rho = \sum_{p:i \rightsquigarrow j} \lambda_p p$ in I, we have

$$\begin{aligned}
\varphi_\rho(me_i) &= \sum_{p:i\rightsquigarrow j} \lambda_p \varphi_p(me_i) \\
&= \sum_{p:i\rightsquigarrow j} \lambda_p m(p+I) \\
&= m \sum_{p:i\rightsquigarrow j} \lambda_p (p+I) \\
&= m(\rho + I) = m0 = 0.
\end{aligned}$$

This defines F on the objects. Now, let $f : M \to N$ be a morphism in $\mathrm{mod}(A)$, and let $F(M) = (M_i, \varphi_\alpha), F(N) = (N_i, \psi_\alpha)$. We define $F(f) = (f_i)_{i \in Q_0}$ by $f_i(me_i) := f(m)e_i$.

We need to check that $f_j \varphi_\alpha = \psi_\alpha f_i$ for each $\alpha : i \to j \in Q_1$. Indeed, given $me_i \in M_i$, we have

$$\begin{aligned}
f_j \varphi_\alpha(me_i) &= f_j(m\bar{\alpha}) = f(m\bar{\alpha})e_j \\
&= f(m)\bar{\alpha}e_j = \psi_\alpha(f(m)e_i) \\
&= \psi_\alpha(f_i(me_i)) = \psi_\alpha f_i(me_i).
\end{aligned}$$

Therefore, $F(f)$ is a morphism of representations.

It is easy to check that $F(f)$ is indeed a (covariant) functor, i.e. that $F(1_M) = 1_{F(M)}$ for any A-module M, and $F(gf) = F(g)F(f)$ for $f : L \to M, g : M \to N \in \mathrm{mod}(A)$.

The next step is to construct a functor $G : \mathrm{rep}(Q,I) \to \mathrm{mod}(A)$. Given $(M_i, \varphi_\alpha) \in \mathrm{rep}(Q,I)$, we define $G(M_i, \varphi_\alpha) = M$ as follows. The underlying vector space of M is $\oplus_{i \in Q_0} M_i$. It is enough to define the right A-action on paths in Q. Let p be a path in Q and $m = (m_i)_{i \in Q_0}$ be an element of M. If $p = \varepsilon_i$ for some i, let $mp := m_i$, and if p has length $\geqslant 1$, we define mp to be the following element in M:

$$(mp)_k := \begin{cases} 0 & \text{if } k \neq t(p) \\ \varphi_p(m_{s(p)}) & \text{if } k = t(p). \end{cases}$$

In order to check that the A-action is well defined, we need to show that if $\rho = \sum_{p:i \rightsquigarrow j} \lambda_p p \in I$, then $m\rho = 0$. Indeed, we have that $m\rho$ is the element in M whose only possible non-zero coordinate is $(m\rho)_j = \sum \lambda_p \varphi_p(m_i)$. But $\sum \lambda_p \varphi_p(m_i) = 0$ since (M_i, φ_α) is bound by I.

The definition of G on morphisms is as follows: given $f = (f_i) : (M_i, \varphi_\alpha) \to (N_i \psi_\alpha)$, we have $G(f) : M \to N$ defined by $G(f)(m) := (f_i(m_i))_{i \in Q_0}$.

Clearly $G(f)$ is linear as each f_i is linear. In order to show that $G(f)$ is a module homomorphism, it is enough to check $G(f)(ma) = G(f)(m)a$ for all $m = m_i \in M_i$ and $a = p + I \in A$, where p is a path from i to j.

On the one hand, we have $(ma)_k = 0$ for $k \neq j$ and $(ma)_j = \varphi_p(m_i)$, and so

$$(G(f)(ma))_k = \begin{cases} 0 & \text{if } k \neq j \\ f_j \varphi_p(m_i) = \psi_p f_i(m_i) & \text{if } k = j. \end{cases}$$

On the other hand, $(G(f)(m))_k = 0$ for $k \neq i$, and $(G(f)(m))_i = f_i(m_i)$, and so according to the definition of A-action,

$$(G(f)(m)a)_k = \begin{cases} 0 & \text{if } k \neq j \\ \psi_p(f_i(m_i)) & \text{if } k = j. \end{cases}$$

It is easy to check that G is indeed a functor and that $FG \simeq 1_{\mathsf{rep}(Q,I)}$ and $GF \simeq 1_{\mathsf{mod}(A)}$, thus giving the required equivalence of categories. □

1.3 Representation Finite Hereditary Algebras

The Krull–Schmidt theorem states that every module over an algebra can be written as a direct sum of indecomposable modules in a unique way (up to isomorphism and changing the order). Therefore, in order to classify all the modules over an algebra, it is sufficient to classify the indecomposable ones.

In this section we discuss representation types of algebras, and discuss the simplest case one can hope for, which is when there are finitely many indecomposable modules.

Definition 1.14

1 Given two representations $M = (M_i, \varphi_\alpha)$, $N = (N_i, \psi_\alpha)$ of Q, we can construct a new representation

$$M \oplus N := \left(M_i \oplus N_i, \begin{bmatrix} \varphi_\alpha & 0 \\ 0 & \psi_\alpha \end{bmatrix} \right),$$

called the *direct sum of M and N*.

2 A representation M is *indecomposable* if $M \neq 0$ and it cannot be written as a direct sum of two non-zero representations.

Example 1.15

1 Let Q be the quiver $1 \longrightarrow 2 \longrightarrow 3$. The representation

$$M = \mathbf{k} \xrightarrow{\begin{bmatrix}1\\0\end{bmatrix}} \mathbf{k}^2 \xrightarrow{\begin{bmatrix}0 & 1\end{bmatrix}} \mathbf{k}$$

is not indecomposable since $M \cong (\ \mathbf{k} \xrightarrow{1} \mathbf{k} \longrightarrow 0\ )$ $\oplus(\ 0 \longrightarrow \mathbf{k} \xrightarrow{1} \mathbf{k}\ )$.

2 Let Q be the quiver $1 \rightrightarrows 2$. We have

$$\mathbf{k} \overset{\begin{bmatrix}1\\2\end{bmatrix}}{\underset{\begin{bmatrix}2\\4\end{bmatrix}}{\rightrightarrows}} \mathbf{k}^2 \cong (\ \mathbf{k} \overset{1}{\underset{2}{\rightrightarrows}} \mathbf{k}\ )\oplus(\ 0 \rightrightarrows \mathbf{k}\ ).$$

Definition 1.16 A connected algebra A is:

1 of *finite representation type* if, up to isomorphism, there are only finitely many indecomposable objects in $\operatorname{mod}(A)$.

2 *hereditary* if $A \cong \mathbf{k}Q$ for some finite, connected and acyclic quiver Q.

Representation finite hereditary algebras have been classified by Gabriel.

Theorem 1.17 (Gabriel's theorem) *An hereditary algebra* $\mathbf{k}Q$ *is of finite representation type if and only if* Q *is an orientation of an ADE diagram, i.e. the underlying graph of* Q *is of one of the following forms:*

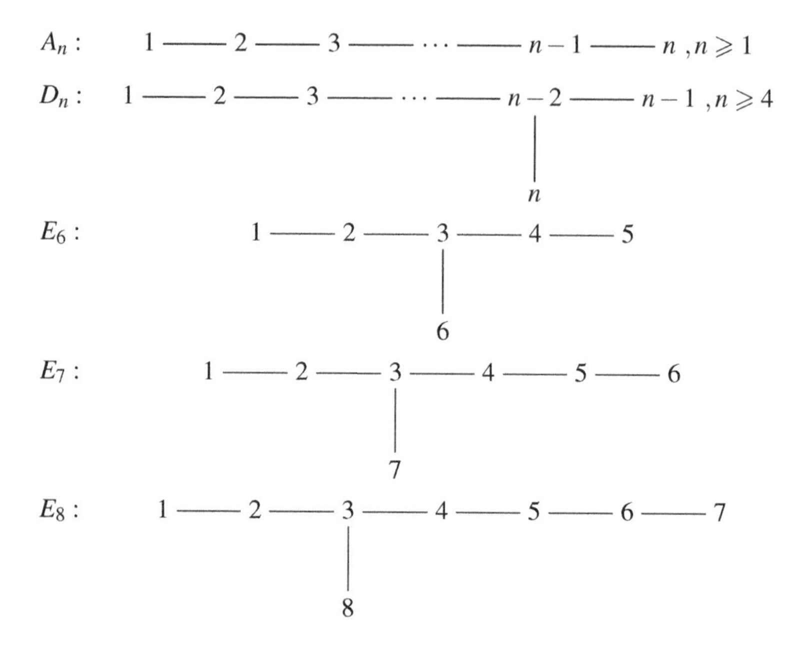

There are two different proofs of this theorem in [6, 38] worth studying. The proof in [6] uses *reflection functors*, which are at the origin of *tilting theory*, where one studies an algebra by comparing its representation theory with that of a simpler algebra. The proof in [38] uses algebraic geometry, namely by studying the space of representations of a quiver with a given dimension vector, which is an algebraic variety.

There are two subtypes of infinite-representation algebras (Drozd's tame-wild dichotomy, 1977):

- **tame type**: infinitely many indecomposable finite-dimensional representations (up to isomorphism), but which are *possible to parametrise*.
- **wild type**: infinitely many indecomposable finite-dimensional representations (up to isomorphism) which *cannot be parametrised.*

Precise definitions of tame and wild algebras can be found for example in [40, Definition XIX.1.3, Definition XIX.3.3]

Hereditary algebras of tame type correspond to acyclic orientations of the Euclidean quivers:

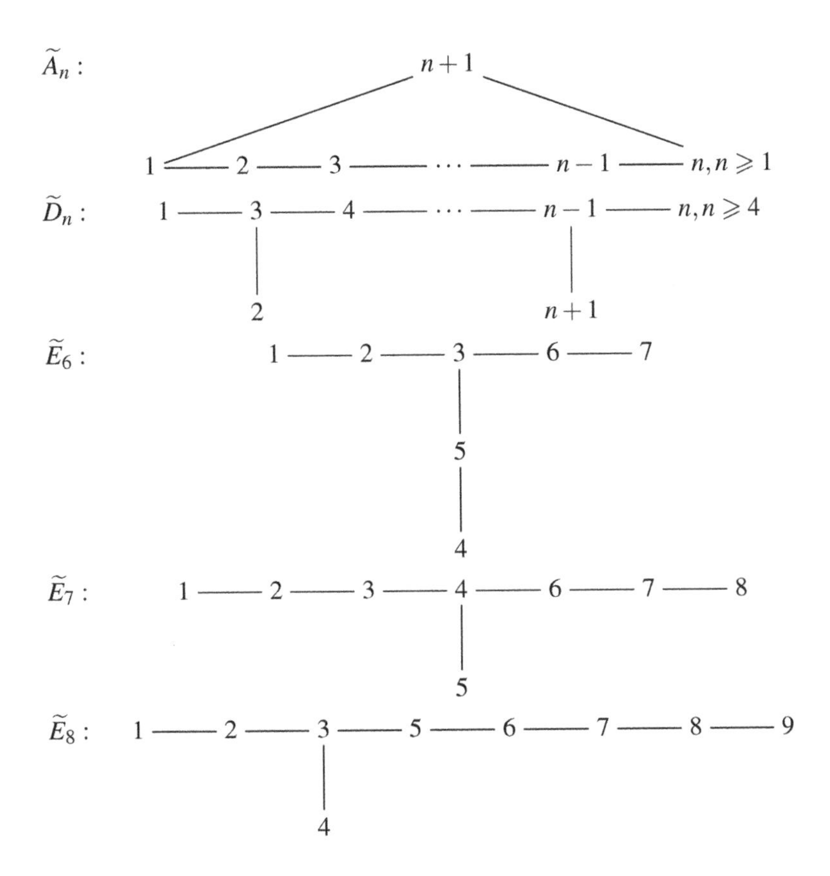

Example 1.18 Let Q be the quiver $1 \rightrightarrows 2$. The indecomposable representations over $\mathbf{k}Q$ are of the following form:

$$\mathbf{k}^n \underset{J_{n,\lambda}}{\overset{1}{\rightrightarrows}} \mathbf{k}^n\ ,\ \mathbf{k}^n \underset{1}{\overset{J_{n,0}}{\rightrightarrows}} \mathbf{k}^n\ ,\ \mathbf{k}^{n+1} \underset{[0\ \ 1]}{\overset{[1\ \ 0]}{\rightrightarrows}} \mathbf{k}^n\ ,\ \mathbf{k}^n \underset{\left[\begin{smallmatrix}0\\1\end{smallmatrix}\right]}{\overset{\left[\begin{smallmatrix}1\\0\end{smallmatrix}\right]}{\rightrightarrows}} \mathbf{k}^{n+1}\ ,$$

where $n > 0$, and $J_{n,\lambda}$ denotes the nilpotent $n \times n$ Jordan block corresponding to the eigenvalue $\lambda \in \mathbf{k}$.

Example 1.19 The path algebra $\mathbf{k}Q$ associated to the quiver:

- $1 \longrightarrow 2$ is of finite type.
- $1 \rightrightarrows 2$ is of tame type.
- $1 \Rrightarrow 2$ is of wild type.

1.4 Auslander–Reiten Theory

In this section we give a brief overview of Auslander–Reiten (AR) theory, giving the basic concepts and main results in order to define the AR-quiver and describe the knitting algorithm, which provides a method to construct the AR-quiver of the representation-finite hereditary algebras.

1.4.1 (Short) Exact Sequences and Extensions

Definition 1.20 Let A be a finite-dimensional algebra.
A sequence of objects and morphisms in $\mathsf{mod}(A)$ of the form

$$\cdots \longrightarrow M_1 \xrightarrow{f_1} M_2 \xrightarrow{f_2} M_3 \xrightarrow{f_3} \cdots$$

is *exact* if $\mathrm{im} f_i = \mathsf{ker} f_{i+1}$ for all i.
A *short exact sequence* (s.e.s. for short) is an exact sequence of the form

$$0 \longrightarrow L \xrightarrow{f} M \xrightarrow{g} N \longrightarrow 0\,.$$

In other words, f is injective, g is surjective and $\mathrm{im} f = \mathsf{ker}\, g$. This is also called an *extension of N by L*.

Note that, in an exact sequence, we have $f_{i+1}f_i = 0$ for all i.

Example 1.21 1 Given a morphism $f : M \to N$ of A-modules, the sequence

$$0 \longrightarrow \mathsf{ker} f \xrightarrow{i} M \xrightarrow{f} N \xrightarrow{p} \mathrm{coker} f \longrightarrow 0\,,$$

where i is the inclusion and p is the projection, is exact, and

$$0 \longrightarrow \mathsf{ker} f \xrightarrow{i} M \xrightarrow{q} M/\mathsf{ker} f \longrightarrow 0$$

is short exact.

2 Let Q be the quiver $1 \longrightarrow 2$, and consider the representations $S(2) := 0 \longrightarrow \mathbf{k}$, $M := \mathbf{k} \xrightarrow{1} \mathbf{k}$ and $S(1) := \mathbf{k} \longrightarrow 0$. Then

$$0 \longrightarrow S(2) \xrightarrow{(0,1)} M \xrightarrow{(1,0)} S(1) \longrightarrow 0 \quad \text{and}$$

$$0 \longrightarrow S(2) \xrightarrow{(0,1)} S(1) \oplus S(2) \xrightarrow{(1,0)} S(1) \longrightarrow 0$$

are short exact sequences, where each component of the pairs $(0,1)$ and $(1,0)$ denotes a linear map between the vector spaces at the corresponding vertex of Q.

The following lemma, known as the splitting lemma, holds for any abelian category (see [6, Definition A.1.5] for the definition of abelian category).

Lemma 1.22 *Given a s.e.s.* $0 \longrightarrow L \xrightarrow{f} M \xrightarrow{g} N \longrightarrow 0$ *in* $\mathrm{mod}(A)$, *the following statements are equivalent:*

1. f *is a split monomorphism (also called a section), i.e. there exists* $h : M \to L$ *such that* $hf = 1_L$.
2. g *is a split epimorphism (also called a retraction), i.e. there exists* $h' : N \to M$ *such that* $gh' = 1_N$.
3. *The sequence is equivalent to the s.e.s.* $0 \longrightarrow L \xrightarrow{i} L \oplus N \xrightarrow{p} N \longrightarrow 0$, *i.e. there is a commutative diagram:*

$$\begin{array}{ccccccccc} 0 & \longrightarrow & L & \xrightarrow{f} & M & \xrightarrow{g} & N & \longrightarrow & 0 \\ & & \| & & \downarrow{\cong} & & \| & & \\ 0 & \longrightarrow & L & \xrightarrow[i]{} & L \oplus N & \xrightarrow[p]{} & N & \longrightarrow & 0. \end{array}$$

In this case, the s.e.s. is said to split.

The set of equivalence classes $\mathrm{Ext}^1(N, L)$ of extensions of N by L, with the equivalence relation defined in Lemma 1.22 (3), is an abelian group, whose zero element is the class of the split extension. For more details, see for instance [37, Section 7.2].

Example 1.23 Let Q be the quiver $1 \rightrightarrows 2$, which is known as the Kronecker quiver. The sequences

$$0 \longrightarrow S(2) \longrightarrow E \longrightarrow S(1) \longrightarrow 0 \quad \text{and}$$

$$0 \longrightarrow S(2) \longrightarrow E' \longrightarrow S(1) \longrightarrow 0 \ ,$$

where $S(1) = \mathbf{k} \rightrightarrows 0$, $S(2) = 0 \rightrightarrows \mathbf{k}$, $E = \mathbf{k} \overset{1}{\underset{0}{\rightrightarrows}} \mathbf{k}$ and $E' = \mathbf{k} \overset{0}{\underset{1}{\rightrightarrows}} \mathbf{k}$ are non-equivalent short exact sequences.

1.4.2 Simple, Projective and Injective Representations

Let $A = \mathbf{k}Q/I$, with Q a finite, connected quiver, and I an admissible ideal.

A *simple A-module* is a non-zero module that has no proper submodules. A *simple representation of* (Q, I) is a representation that corresponds to a simple A-module under the equivalence in Theorem 1.13.

Proposition 1.24 *The simple representations of* (Q,I) *are, up to isomorphism, of the form* $S(i) = (S(i)_j, \varphi_\alpha)$ *for each* $i \in Q_0$*, where* $\varphi_\alpha = 0$ *for all* $\alpha \in Q_1$ *and*

$$S(i)_j = \begin{cases} k & \text{if } i = j \\ 0 & \text{if } i \neq j \end{cases}.$$

An A-module P is *projective* if any s.e.s. ending at P splits, i.e. $\mathrm{Ext}^1(P,-) = 0$. An A-module I is *injective* if any s.e.s. starting at I splits, i.e. $\mathrm{Ext}^1(-,I) = 0$. The reader can find the definition and basic results on Hom and Ext functors in both [6] and [38].

Remark

1 P is projective if and only if for every epimorphism $f : M \to N$ and every morphism $g : P \to N$, there is $g' : P \to M$ such that $g = fg'$. In other words, $\mathrm{Hom}(P,-)$ maps surjective morphisms to surjective morphisms.
2 I is injective if and only if for every monomorphism $u : L \to M$ and every morphism $g : L \to I$, there is $g' : M \to I$ such that $g = g'u$. In other words, $\mathrm{Hom}(-,I)$ maps injective morphisms to surjective morphisms.

Proposition 1.25 *The projective representations of* (Q,I) *are, up to isomorphism, of the form* $P(i) = (P(i)_j, \varphi_\alpha)$ *for each* $i \in Q_0$*, where:*

- $P(i)_j$ *is the vector space generated by* $\{p + I \mid p \text{ path from } i \text{ to } j\}$.
- *Given an arrow* $\alpha : j \to \ell$, $\varphi_\alpha : P(i)_j \to P(i)_\ell$ *is the linear map defined on the basis by composing the paths from* i *to* j *with the arrow* α.

Similarly, the injective representations of (Q,I) *are, up to isomorphism, of the form* $I(i) = (I(i)_j, \varphi_\alpha)$ *for each* $i \in Q_0$*, where:*

- $I(i)_j$ *is the vector space generated by* $\{p + I \mid p \text{ path from } j \text{ to } i\}$.
- *Given an arrow* $\alpha : j \to \ell$, $\varphi_\alpha : P(i)_j \to P(i)_\ell$ *is the linear map defined on the basis by deleting the arrow* α *from the paths from* j *to* i *that start with* α *and sending to zero the remaining paths.*

Example 1.26 Consider the algebra given by the quiver

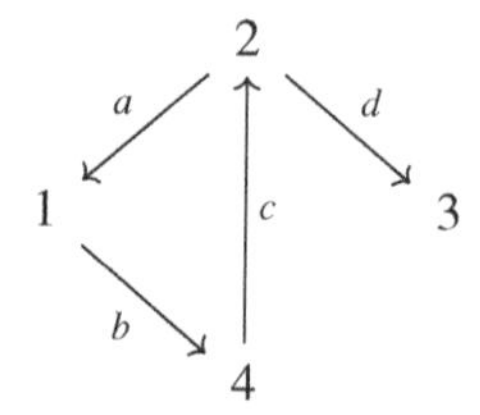

subject to the relations $ca = 0 = ab$.

The projective and injective representations are as follows:

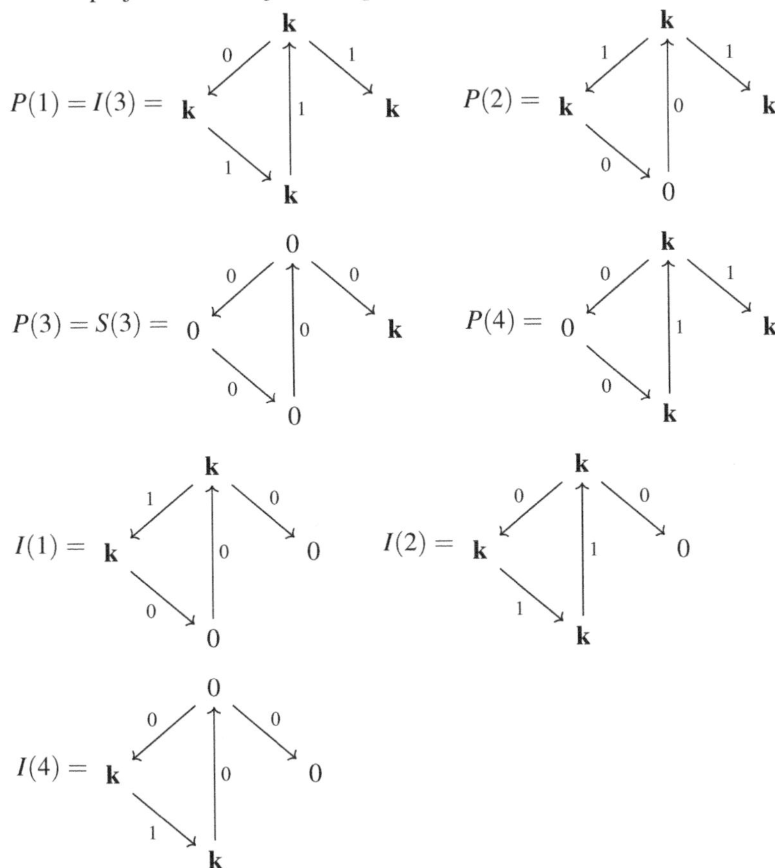

Notation: We will simplify the notation of an indecomposable representation, by encoding their composition series whenever possible. For instance, the projective module $P(1)$ in the example above can be denoted by

$$P(1) = \begin{smallmatrix}1\\4\\2\\3\end{smallmatrix},$$

meaning $P(1)_i = \mathbf{k}$ for all $i \in Q_0$, and there is an identity map from top to bottom, i.e.

$$(P(1))_1 \xrightarrow{1} (P(1))_4 \xrightarrow{1} (P(1))_2 \xrightarrow{1} (P(1))_3 .$$

This module has a unique composition series given by:

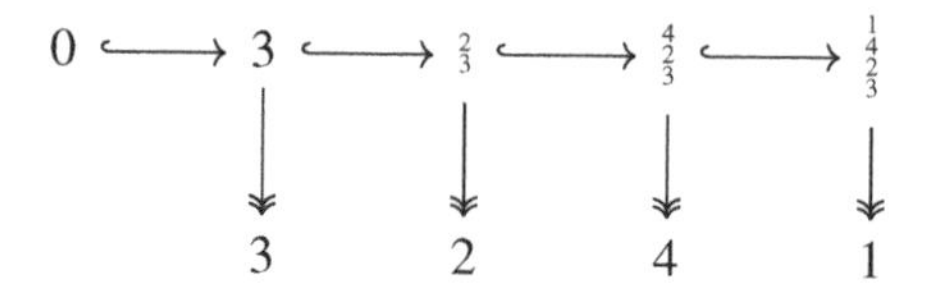

Theorem 1.27 *Given a representation $M = (M_i, \varphi_\alpha)$ in* $\mathsf{rep}(Q,I)$*, we have for all $i \in Q_0$,*

$$\mathrm{Hom}(P(i),M) \simeq M_i \simeq \mathrm{Hom}(M,I(i)).$$

For a proof, see for instance [6, Lemma III.2.11]. Recall the definition of dimension vector in Definition 1.9.

Corollary 1.28 *If* $0 \longrightarrow L \xrightarrow{f} M \xrightarrow{g} N \longrightarrow 0$ *is a s.e.s. in* $\mathsf{rep}(Q,I)$*, then*

$$\underline{\dim} M = \underline{\dim} L + \underline{\dim} N.$$

1.4.3 Irreducible Morphisms and AR-sequences

We now introduce the definition of irreducible morphisms, which are in some sense the building blocks for the morphisms between modules, and we define an important class of short exact sequences, called AR-sequences, which can be defined in terms of irreducible morphisms and indecomposable modules.

Definition 1.29 A morphism $f : M \to N$ is *irreducible* if:

- f is not a split monomorphism,
- f is not a split epimorphism and
- if $f = gh$, then h is a split monomorphism or g is a split epimorphism.

We note that an irreducible morphism is either injective or surjective, but not both. Moreover, the third condition says that an irreducible morphism admits no nontrivial factorisation.

Example 1.30 Let Q be the quiver $1 \longrightarrow 2 \longrightarrow 3$. The map $S(3) \xrightarrow{(0,0,1)} P(2)$ is irreducible. But the map $S(3) \xrightarrow{(0,0,1)} P(1)$ is not irreducible as it factors nontrivially through $P(2)$.

Given two indecomposable A-modules M and N, the set $\mathsf{Irr}(M,N)$ of irreducible morphisms from M to N is a vector space. In fact, $\mathsf{Irr}(M,N)$ is given by the quotient $\mathsf{rad}_A(M,N)/\mathsf{rad}_A^2(M,N)$. For a definition of the (mth power of the) radical of $\mathsf{mod}(A)$, and a proof of this fact, we refer the reader to [6, Section IV.1, Appendix A.3].

Definition 1.31 An s.e.s. $0 \longrightarrow L \xrightarrow{f} M \xrightarrow{g} N \longrightarrow 0$ is an *AR-sequence* if the following conditions hold:

1 L, N are indecomposable;
2 f, g are irreducible morphisms.

Remark An AR-sequence is also known as an *almost-split sequence*, in the sense that any map $u : L \to U$ that is not a split monomorphism (resp. any map $v : V \to N$ that is not split epimorphism) factors through f (resp. g).

Remark

1 An AR-sequence never splits. Therefore, no AR-sequence starts with an injective module or ends with a projective module.
2 An AR-sequence is uniquely determined, up to isomorphism, by each of its end terms.

Theorem 1.32 (Auslander–Reiten theorem) *Let M be an indecomposable A-module.*

1 *If M is non-projective, there is an AR-sequence*
$0 \longrightarrow \tau M \xrightarrow{f} E \xrightarrow{g} M \longrightarrow 0$ *ending at M.*
2 *If M is non-injective, there is an AR-sequence*
$0 \longrightarrow M \xrightarrow{f} E' \xrightarrow{g} \tau^{-1}M \longrightarrow 0$ *starting at M.*

The module τM is called the *AR-translate of M*, and $\tau^{-1}M$ is the *inverse AR-translate of M*. We note that if M is non-projective indecomposable (resp. non-injective indecomposable) then τM (resp. $\tau^{-1}M$) is non-injective indecomposable (resp. non-projective indecomposable).

We recommend [6, Section IV] for a proof of Theorem 1.32. Key tools in this proof are the *AR-formulas*, which describe the relationship between morphisms and extensions. Namely, for any pair of modules $M, N \in \mathsf{mod}(A)$, we have:

$$\mathrm{Ext}^1(M,N) \cong D\underline{\mathrm{Hom}}(\tau^{-1}N, M) \cong D\overline{\mathrm{Hom}}(N, \tau M).$$

Here, D is the standard $\mathbf{k}$-duality $\mathrm{Hom}_{\mathbf{k}}(-,\mathbf{k})$, $\tau^{-1}I = 0$ for all injective module I, $\tau P = 0$ for all projective module P, and the underlining (resp. overlining) means we are considering morphisms that do not factor through projective (resp. injective) modules.

When A is an hereditary algebra, the AR-formulas can be simplified to

$$\mathrm{Ext}^1(M,N) \cong D\mathrm{Hom}(\tau^{-1}N,M) \cong D\mathrm{Hom}(N,\tau M).$$

1.4.4 The AR-quiver and the Knitting Algorithm

Given a finite-dimensional algebra A, we can record the information about $\mathsf{mod}(A)$ in a quiver, called the AR-quiver. In the case when A is of finite representation type, this quiver gives a complete picture of the representation theory of A.

Definition 1.33 The *AR-quiver* $\Gamma(\mathsf{mod}(A))$ of $\mathsf{mod}(A)$ is defined by:

- the vertices of $\Gamma(\mathsf{mod}(A))$ are the isomorphism classes of indecomposable A-modules,
- the arrows correspond to basis elements of the vector space of irreducible morphisms between indecomposable modules.

Note that there are no loops in $\Gamma(\mathsf{mod}(A))$. This follows from the fact that we are dealing with finite-dimensional modules and that any irreducible morphism is either a monomorphism or an epimorphism, but not both. Moreover, in the case when A is representation-finite, the AR-quiver has no multiple arrows, i.e. all the vector spaces of irreducible morphisms between two indecomposable modules have dimension $\leqslant 1$ (cf. [6, Proposition IV.4.9]).

Each AR-sequence $0 \longrightarrow \tau M \longrightarrow L_1 \oplus \cdots \oplus L_r \longrightarrow M \longrightarrow 0$ is represented in the AR-quiver by a *mesh*:

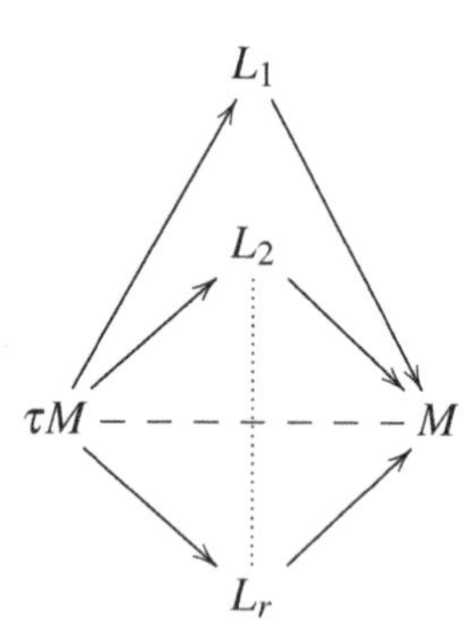

The AR-quiver is a *translation quiver*, i.e. for each arrow $M \to L$, for which $\tau^{-1}M \neq 0$ (resp. $\tau L \neq 0$), there is an arrow $L \to \tau^{-1}M$ (resp. $\tau L \to M$).

Theorem 1.34 (Auslander's Theorem) *[8] If the AR-quiver Γ of a connected finite-dimensional algebra A has a connected component $\mathcal{C}$ such that the lengths* *of the modules in $\mathcal{C}$ are bounded, then A is of finite representation type, and $\Gamma = \mathcal{C}$.*

In particular, the AR-quiver of a representation-finite algebra consists of one finite component. Moreover, in this case the AR-quiver completely describes $\mathsf{mod}(A)$, in the sense that every module is a direct sum of finite-dimensional indecomposable modules and every nonzero non-isomorphism between indecomposable modules is a sum of compositions of irreducible morphisms.

The *knitting algorithm* is an algorithm that allows us to construct, in some special cases, the AR-quiver (or part thereof). One of these special cases is when $A = \mathbf{k}Q$, where the underlying graph of Q is ADE. It owes its name to the fact that it recursively constructs one mesh after the other, from left to right.

What follows is a description of this algorithm. We start by computing all the projective modules and their radicals.

The radical $\mathsf{rad}(M)$ of a module M is the intersection of all maximal submodules of M. The representation $(P(i)'_j, \varphi'_\alpha)$ corresponding to the radical $\mathsf{rad}(P(i))$ of the projective $P(i) = (P(i)_j, \varphi_\alpha)$ at i is such that $P(i)'_j = P(i)_j$ if $i \neq j, P(i)_i$ is the vector space spanned by all nonconstant paths from i to i, and φ'_α is the restriction of φ_α to $P(i)_{s(\alpha)}$.

Proposition 1.35 *Every direct predecessor of $P(i)$ in $\Gamma(\mathsf{mod}(A))$, i.e. every indecomposable module X for which there is an irreducible morphism $X \to P(i)$ is a direct summand of $\mathsf{rad}(P(i))$. In the case when A is hereditary, all predecessors of projective modules are projective modules.*

Base step:

1 Draw a vertex for each simple projective $P(i)$.
2 If $P(i)$ is a summand of $\mathsf{rad}(P)$ for some projective P, then add a vertex corresponding to P and arrows from $P(i)$ to P (the number of arrows equals the multiplicity of $P(i)$ in $\mathsf{rad}(P)$).
3 Add vertices associated to remaining summands R of $\mathsf{rad}(P)$ and arrows $R \to P$ (the number of arrows equals the multiplicity of R in $\mathsf{rad}(P)$).
4 Repeat previous steps for each simple projective.

At this point we get a quiver Δ_0.
Induction Δ_n from Δ_{n-1}:
If $X \in \Delta_{n-1}$ and all its direct predecessors are in Δ_{n-1}, then:

1 if X is a direct summand of $\mathsf{rad}(Q)$ for some projective Q not in Δ_{n-1}, add a vertex associated to Q and arrows $X \to Q$ (the number of arrows equals the multiplicity of X in $\mathsf{rad}(Q)$).

2 if X is not injective, add a vertex corresponding to $\tau^{-1}X$ and for each arrow $X \to Y$, add $Y \to \tau^{-1}X$.

If A is hereditary of finite representation type, it is known that each indecomposable A-module is uniquely determined by its dimension vector. Therefore, in order to calculate $\tau^{-1}X$ in the knitting algorithm, one can simply use the formula $\underline{\dim}\tau^{-1}X = \sum_{X\to Y}\underline{\dim}Y - \underline{\dim}X$, by Corollary 1.28.

Note that there is a dual version of the knitting algorithm where one starts by computing injective modules, and considering the dual of Proposition 1.35 which states that every direct successor of $I(i)$ in $\Gamma(\mathrm{mod}(A))$ is a direct summand of $I(i)/S(i)$, and if A is hereditary then all successors of injective modules are injective modules.

Let A be a hereditary algebra. If A is of finite representation type, then the algorithm terminates when we have reached all the injective modules. If A is not of finite type, then the algorithm does not terminate, and what the algorithm produces is the postprojective component of the AR-quiver (cf. [6, Corollary VIII.2.3]). For the definition of postprojective component, see e.g. [6, Definition VIII.2.2]. Note that some authors refer to postprojective components as preprojective components.

Example 1.36 Let Q be the quiver $1 \longrightarrow 2 \longrightarrow 3 \longleftarrow 4 \longrightarrow 5$ of type A_5. The AR-quiver of $\mathbf{k}Q$ is given by:

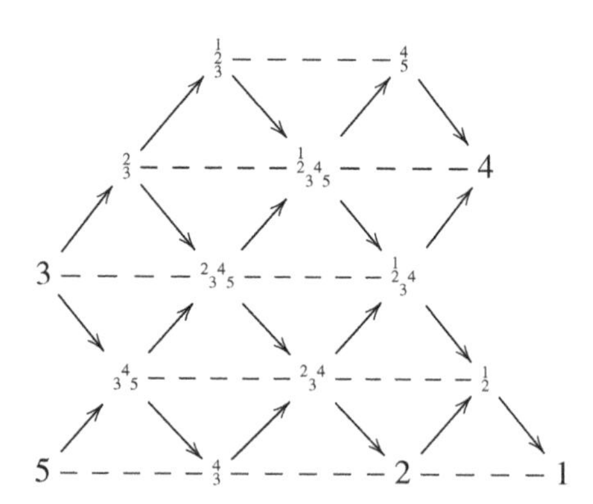

Example 1.37 Let Q be the following quiver of type D_4:

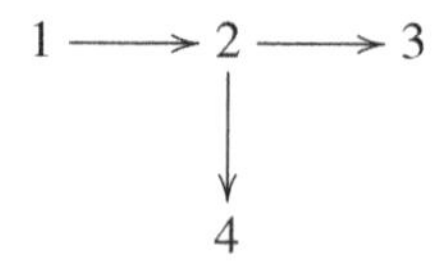

The AR-quiver of $\mathbf{k}Q$ is given by:

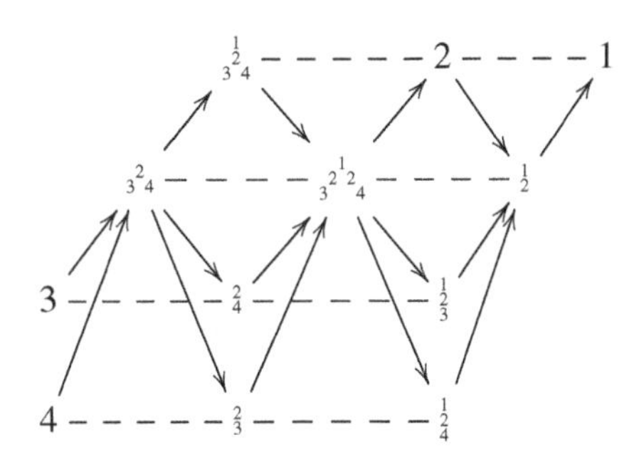

Here we have:

$$\begin{matrix} & 1 & \\ 2 & & 2 \\ 3 & & 4 \end{matrix} \quad = \quad \begin{array}{ccccc} \mathbf{k} & \xrightarrow{\begin{bmatrix}1\\1\end{bmatrix}} & \mathbf{k}^2 & \xrightarrow{\begin{bmatrix}1 & 0\end{bmatrix}} & \mathbf{k} \\ & & \big\downarrow {\scriptstyle \begin{bmatrix}0 & 1\end{bmatrix}} & & \\ & & \mathbf{k} & & \end{array}$$

1.5 Geometric Models

The knitting algorithm might not work when we start with a non-simple projective module.

For instance, consider the quiver Q

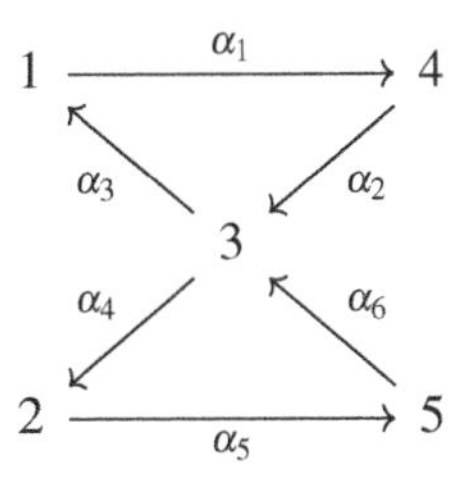

together with the admissible ideal $I = \langle \alpha_1\alpha_2, \alpha_2\alpha_3, \alpha_3\alpha_1, \alpha_4\alpha_5, \alpha_5\alpha_6, \alpha_6\alpha_4 \rangle$, and let $\mathsf{A} = \mathbf{k}Q/I$.

Suppose we start the knitting algorithm with $P(5)$, whose radical is $\mathfrak{m}(P(5)) = \substack{3\\1}$. This module is not the summand of the radical of any other projective module, and so according to the algorithm, we would knit the following mesh:

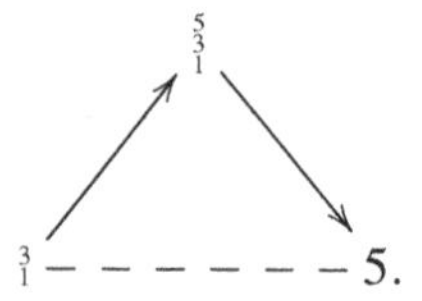

However, this mesh is not correct; the algorithm did not compute the irreducible morphism $\begin{smallmatrix}5\\3\\1\end{smallmatrix} \to 3$.

This section is devoted to a different way of computing the AR-quiver of certain classes of algebras, using the geometry of Riemann surfaces with boundary.

1.5.1 Geometric Model of Type A_n

We start by illustrating how to construct the AR-quiver of a hereditary algebra of type A_n, with the example

$$Q = 1 \longrightarrow 2 \longrightarrow 3 \longleftarrow 4 \longrightarrow 5 .$$

Consider a disc with $8(= n+3)$ marked points on its boundary, together with the triangulation T, i.e. maximal set of non-crossing diagonals, given in Figure 1.1.

Before associating an algebra to these data, we need to introduce some terminology and notation, which follows that of [21]. For further study on the background of combinatorial topology of surfaces we refer the reader to [29].

A *boundary segment* in the marked disc S (or any marked surface) is a segment of a boundary component between two marked points. A *curve* is a continuous map $\gamma\colon [0,1] \to \mathsf{S}$. We always consider curves up to homotopy relative to their endpoints. A curve γ is said to be an *arc* if it satisfies the following properties:

- The endpoints of γ are marked points on the boundary.
- γ intersects the boundary of the surface only in its endpoints.

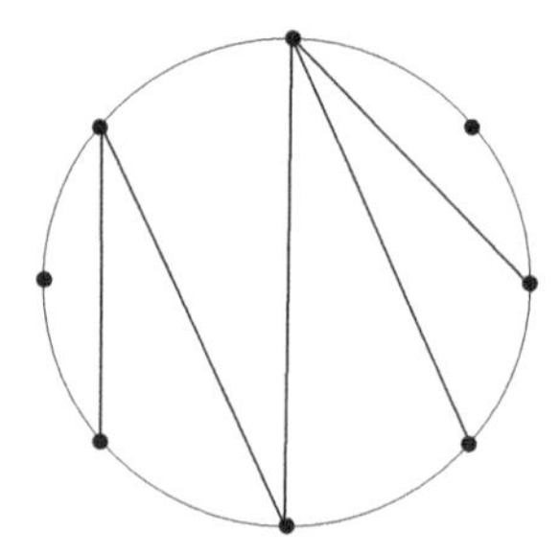

Figure 1.1 A triangulation of an octagon.

- γ is not homotopic to a point or a boundary segment.

Figure 1.2 illustrates all these concepts.

Given a marked point p, let m', m'' be two points in the same boundary component of p such that m', m'' are not marked points and p is the only marked point lying in the boundary segment δ between m' and m''. Draw a curve c homotopic to δ but lying in the interior of the disc except for its endpoints m' and m''. The *complete fan at* p is the sequence of diagonals in T that c crosses in the clockwise order.

We can now associate a quiver Q_{T} to this triangulation, in the following way:

- Vertices of Q_{T} are in one-to-one correspondence with diagonals of T. We will use the same notation for both.
- Given two vertices i and j, there is an arrow $i \to j$ if and only if i and j share a marked point p and j is the immediate successor of i in the complete fan at p.

Note that we can associate a marked point to each arrow of Q_{T}. Namely, using the notation above, the marked point associated to the arrow $i \to j$ is p.

The quiver Q_{T} in Figure 1.3 is indeed Q, and in fact one can obtain any orientation of a Dynkin graph of type A_n from a triangulation of a disc with

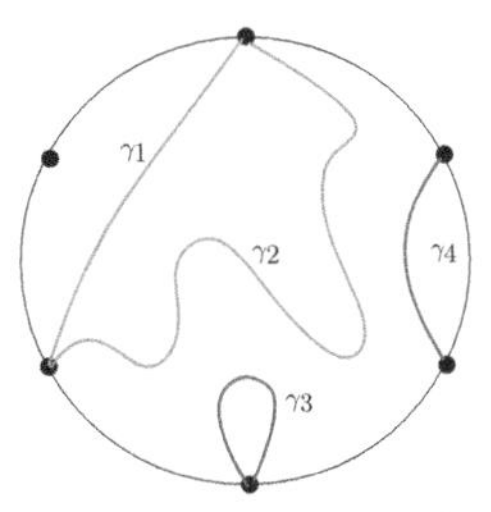

Figure 1.2 The arcs γ_1 and γ_2 are homotopic to each other. The curve γ_3 is homotopic to a point. The curve γ_4 is homotopic to a boundary segment.

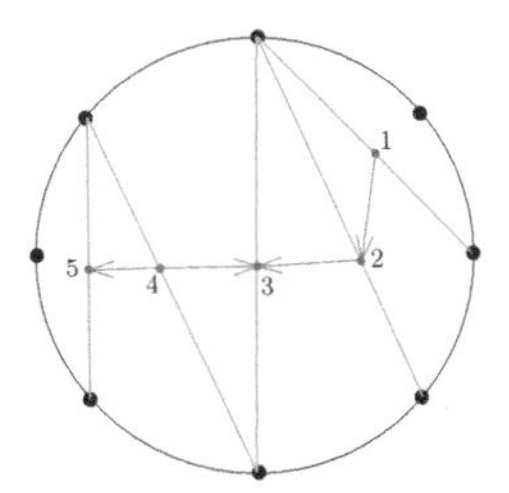

Figure 1.3 The quiver Q_{T} of the triangulation.

$n+3$ marked points on the boundary whose triangles are *outer-triangles*, i.e. triangles with at least one side on the boundary of the disc.

We will now describe how to obtain the AR-quiver of $\mathbf{k}Q$ from this triangulation.

We will always consider arcs up to homotopy relative to their endpoints. Given an arc γ distinct from any diagonal of T, we define a representation $M_\gamma = (M_i, \varphi_\alpha)$ of $\mathbf{k}Q_\mathsf{T}$, as follows:

$$M_i = \begin{cases} \mathbf{k} & \text{if } \gamma \text{ crosses diagonal } i \\ 0 & \text{otherwise,} \end{cases} \qquad \varphi_\alpha = \begin{cases} 1 & \text{if } M_{s(\alpha)} = M_{t(\alpha)} = \mathbf{k} \\ 0 & \text{otherwise.} \end{cases}$$

Irreducible morphisms correspond to pivoting one of the endpoints of an arc to its counterclockwise neighbour (*pivoting elementary move*). Given an arc γ, we define its translate $\tau(\gamma)$ to be the arc obtained from γ by rotating both endpoints to their counterclockwise neighbour. In particular, $M_\gamma = P(i)$ (resp. $M_\gamma = I(j)$) if and only if $\tau\gamma = i$ (resp. $\tau^{-1} = j$).

A presentation of the AR-quiver of $\mathsf{mod}(\mathbf{k}Q_\mathsf{T})$ in terms of these combinatorics is presented in Figure 1.4.

Extensions have a nice description in terms of arcs. Indeed, there is an extension from N to M if and only if the corresponding arcs γ_N and γ_M cross each other as in Figure 1.5.

The summands of the middle term of the extension correspond to the dashed arcs in Figure 1.5.

1.5.2 Geometric Model for Cluster-tilted Algebras of Type A_n

Cluster-tilted algebras arise in the context of cluster-tilting theory. We refer the reader to [4] for a nice survey on this class of algebras.

Cluster-tilted algebras of type A_n are precisely the algebras associated to an arbitrary triangulation of the $(n+3)$-gon.

An arbitrary triangulation T may include *inner triangles*, i.e. triangles whose three boundaries are all diagonals of T. The quiver Q_T is defined as above, but now we include relations $\alpha\beta$, if $s(\alpha), t(\alpha) = s(\beta), t(\beta)$ are the boundaries of an inner triangle.

The algebra A at the start of this section is a cluster-tilted algebra of type A, which can be obtained from the triangulation in Figure 1.6.

Using the same rule for arcs, pivot elementary moves and translates, we are now able to compute the AR-quiver of $\mathsf{mod}(\mathsf{A})$ in terms of the geometric model (see Figure 1.7).

Note that when we have inner triangles, we can get a new type of crossing, see Figure 1.8.

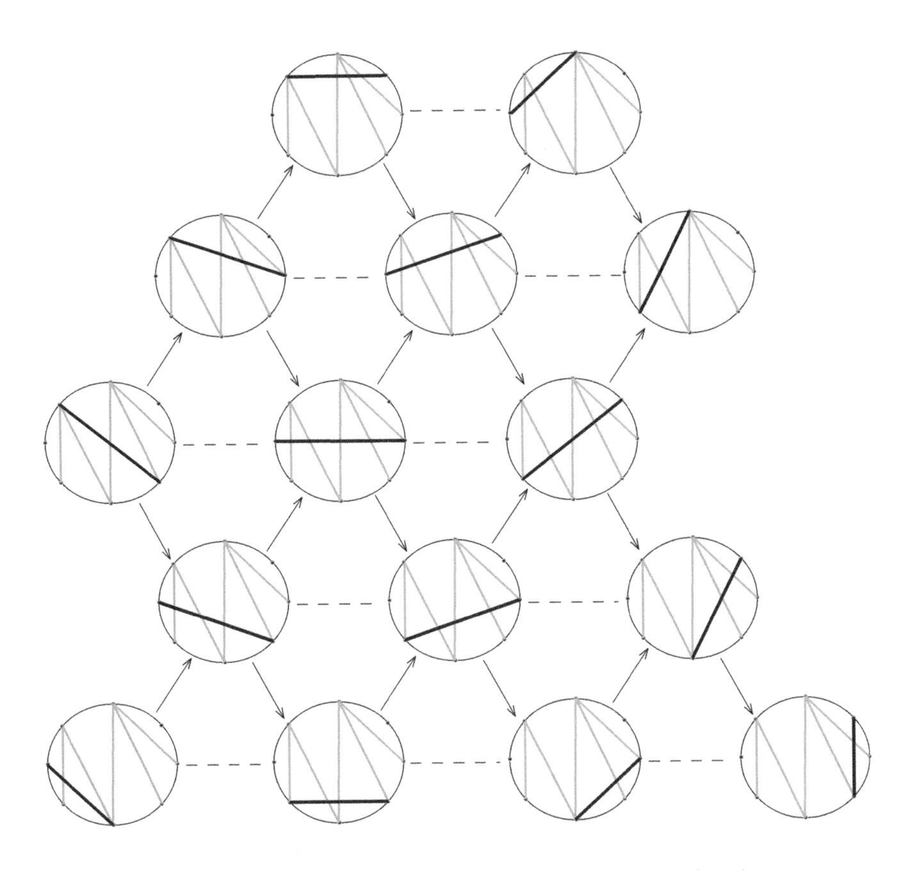

Figure 1.4 The geometric model of the AR-quiver of $\mathsf{mod}(\mathbf{k}Q_\mathrm{T})$.

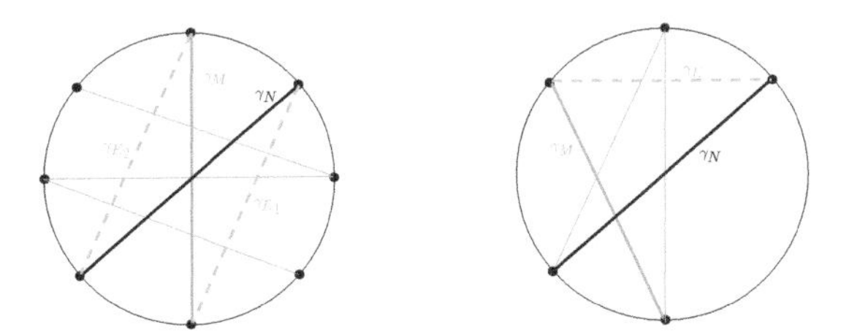

Figure 1.5 Extensions of N by M as crossings of γ_N and γ_M.

However, this type of crossing does not give rise to an extension, and so all extensions are described in the same way as we have seen above. For more details see [16].

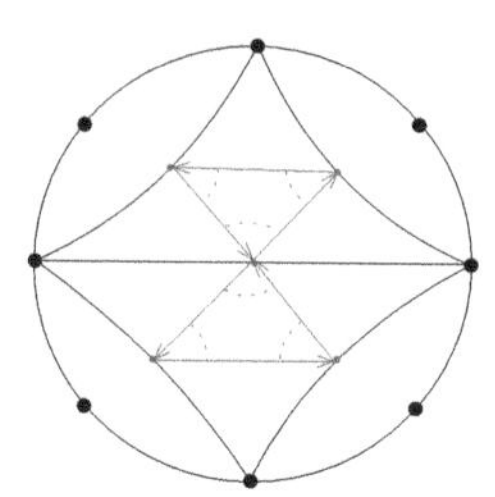

Figure 1.6 The triangulation associated to A.

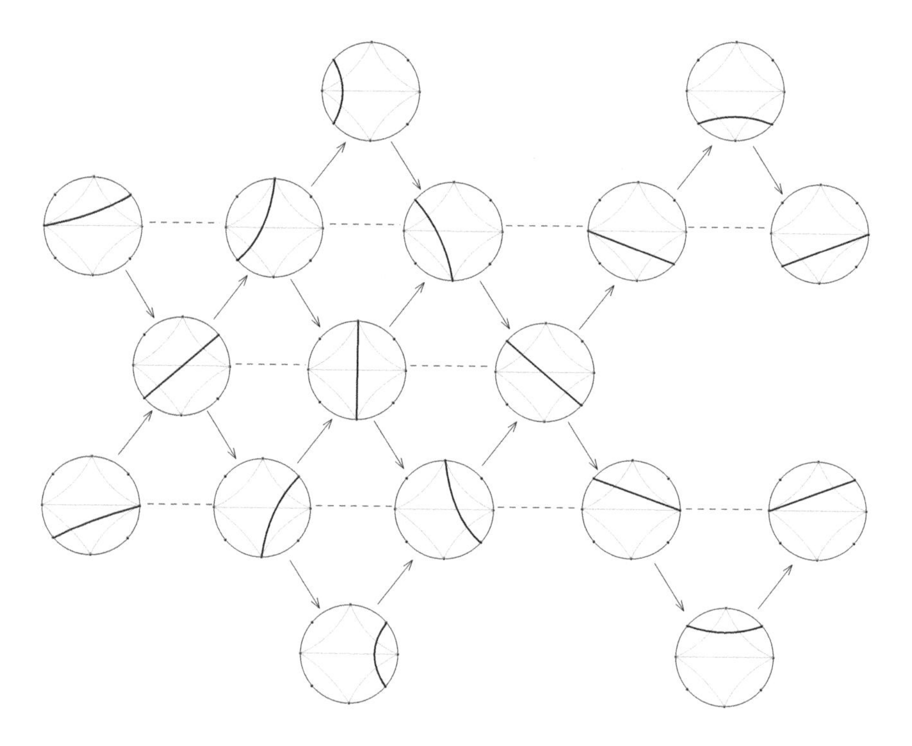

Figure 1.7 The geometric model of the AR-quiver of mod(A).

1.5.3 Geometric Model for Gentle Algebras

We will now consider two possible generalisations of this combinatorial construction: on the one hand we can consider partial triangulations instead (i.e. any set of non-crossing diagonals), and on the other hand we can consider other surfaces.

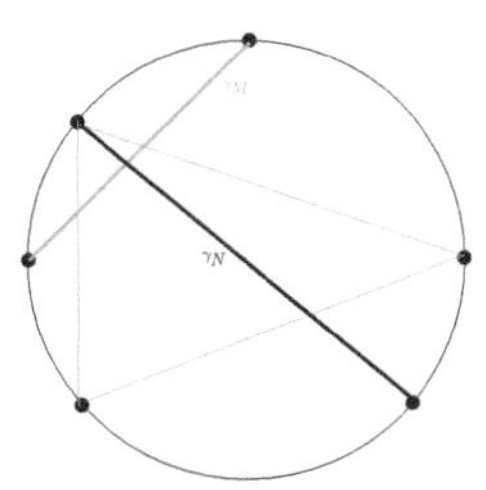

Figure 1.8 Crossing associated to an inner triangle.

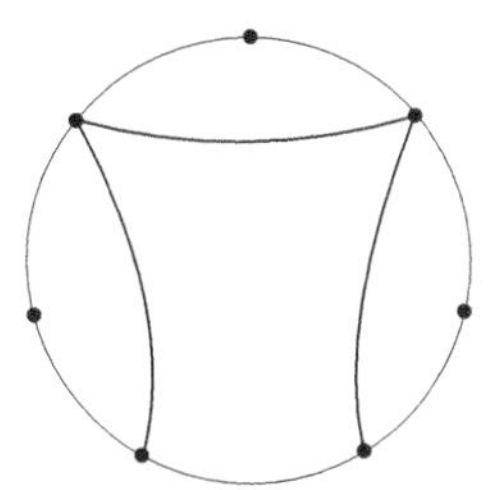

Figure 1.9 The partial triangulation of the algebra C.

Let C be the bound path algebra given by $1 \xrightarrow{\alpha} 2 \xrightarrow{\beta} 3$ bound by $\alpha\beta$. This algebra can be obtained from the partial triangulation of a disc in Figure 1.9.

The quiver is obtained in the same way as before. The relations are given by composition of two arrows in the same region. Note that this rule applied to an arbitrary triangulation of the disc gives rise to the same rule described in the previous subsection.

For partial triangulations, not every arc gives rise to an indecomposable module and two different arcs may give rise to the same indecomposable module. Therefore, we need to define *permissible arcs* and *equivalence of arcs* (this is not the same as homotopy).

An arc is permissible if each consecutive crossing corresponds to an arrow in the quiver. See Figure 1.10 for a counter-example.

Two arcs are isomorphic if they intersect the same diagonals of the partial triangulation (see Figure 1.11).

Indecomposable modules are therefore in bijection with equivalence classes of permissible arcs.

If we perform a pivot elementary move as described in the previous subsections, we may get an isomorphic arc. Hence, an irreducible morphism corresponds to a sequence of pivot elementary moves until one gets a non-equivalent arc.

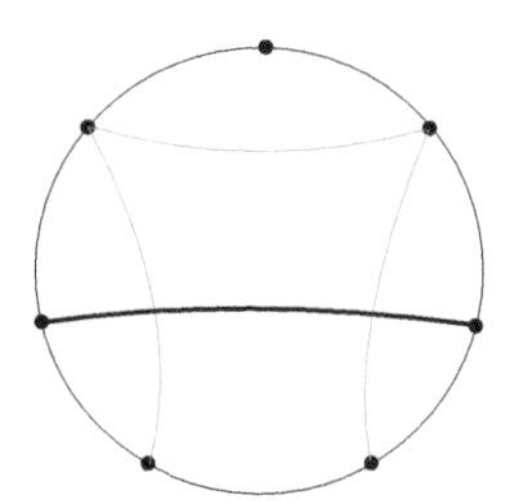

Figure 1.10 An arc that is not permissible.

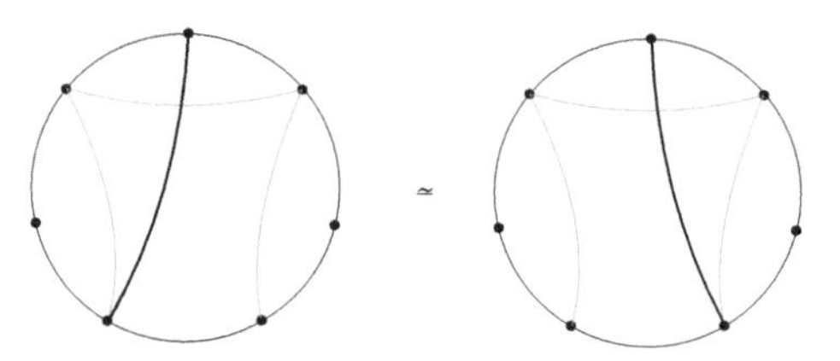

Figure 1.11 Isomorphic permissible arcs.

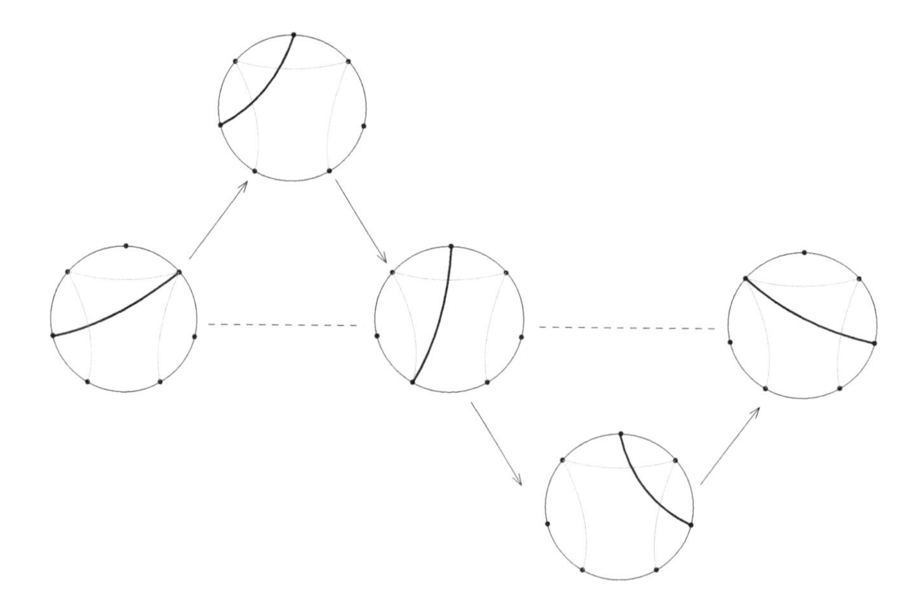

Figure 1.12 The geometric model of the AR-quiver of $\mathsf{mod}(\mathsf{C})$.

The AR-quiver of $\mathsf{mod}(\mathsf{C})$ is given in Figure 1.12.

Now, let us consider an example coming from an annulus (see Figure 1.13). The quiver of the algebra D associated to this partial triangulation is defined

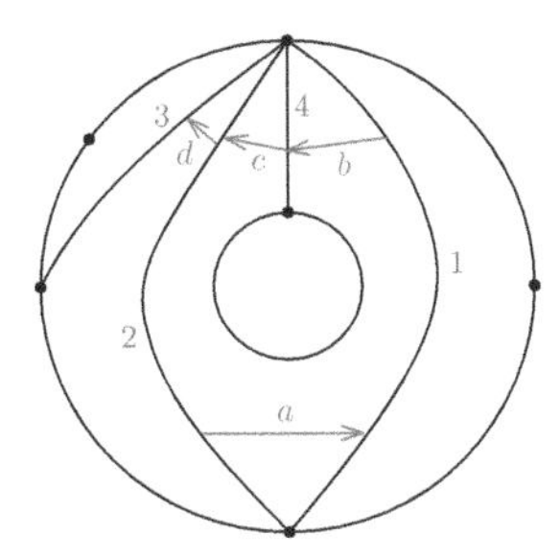

Figure 1.13 A partial triangulation in an annulus and corresponding quiver.

as previously. But we refine the definition of relations as follows: the composition of two arrows with different marked points is zero and if α is a loop, i.e. its start and endpoints correspond to a loop arc of the partial triangulation, then $\alpha^2 = 0$.

The algebra D is then the algebra considered in Example 1.26, given by the quiver

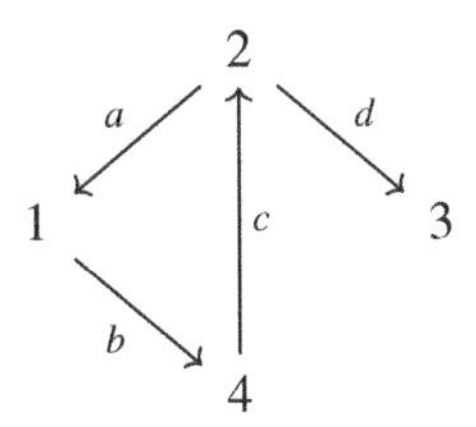

bound by the relations $ca = 0 = ab$. By refining the notions of permissible arcs, equivalence of arcs and pivot elementary moves, we get the AR-quiver of D as in Figure 1.14.

An algebra associated to an unpunctured surface with a finite set of marked points on the boundary is called a *tilting algebra*. It turns out that these algebras are precisely gentle algebras.

Definition 1.38 A finite-dimensional algebra A is *gentle* if it admits a presentation $A = \mathbf{k}Q/I$ satisfying the following conditions:

1 Each vertex of Q is the source of at most two arrows and the target of at most two arrows.
2 For each arrow α in Q, there is at most one arrow β in Q such that $\alpha\beta \notin I$, and there is at most one arrow γ such that $\gamma\alpha \notin I$.
3 For each arrow α in Q, there is at most one arrow δ in Q such that $\alpha\delta \in I$, and there is at most one arrow μ such that $\mu\alpha \in I$.
4 I is generated by paths of length 2.

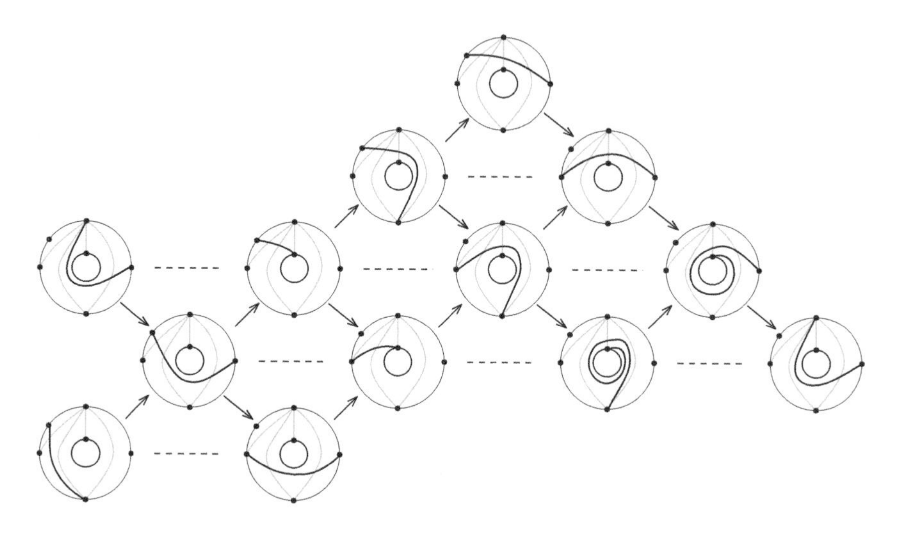

Figure 1.14 The geometric model of the AR-quiver of mod(D).

Gentle algebras first appeared in the context of tilting theory [5] (see also [6, Section IX]), where iterated tilted algebras of types A and $\widetilde{A}$ were observed to satisfy the properties above. Gentle algebras, which are tame, remain one of the relatively few classes of algebras for which the representation theory is computationally tractable. Partly due to this reason, there has been widespread interest in this class of algebras in many different contexts, such as Fukaya categories [25], dimer models [12], enveloping algebras of Lie algebras [27] and cluster theory [7, 23, 30]. The geometric model for the module category of gentle algebras presented above is given in [10]. Derived categories of gentle algebras have also been described geometrically [33] and an important application is a geometric description of derived equivalences of gentle algebras [3]. We refer the reader to [13, 16, 17, 34] for further examples of recent developments in this area.

Ribbon graphs are the bridge between gentle algebras and unpunctured surfaces (cf. [39]). A *ribbon graph* is an undirected finite graph with a cyclic ordering of the half edges at each vertex. Let $A = \mathbf{k}Q/I$ be a gentle algebra, and $\mathcal{M}$ the set of maximal paths in Q avoiding relations together with the stationary paths e_v, for each vertex v of valency 1 or valency 2 such that v is the middle vertex of path of length 2 not in I. Note that each vertex in Q_0 appears twice in the paths in $\mathcal{M}$.

The vertices of the ribbon graph Γ corresponding to A are in one-to-one correspondence with the elements in $\mathcal{M}$. The edges in Γ are in one-to-one correspondence with the vertices of Q. More precisely, given $v \in Q_0$, the corresponding edge connects the two elements in $\mathcal{M}$ containing v. The cyclic ordering of the half edges at each vertex in Γ is determined by the paths in $\mathcal{M}$.

Example 1.39 Let Q be the quiver

$$1 \underset{b}{\overset{a}{\rightrightarrows}} 2 \xrightarrow{c} 3 \circlearrowleft d$$

Let $I = \langle ac, d^2 \rangle$, and $A = \mathbf{k}Q/I$. Then $\mathcal{M} = \{bcd, a\}$ and the corresponding ribbon graph is given in Figure 1.15. Here, the arrows correspond to the path of the corresponding vertex of the ribbon graph.

By replacing edges with oriented strips, vertices with oriented discs and gluing these according to the orientation along the faces of Γ, we get an oriented surface S in such a way that the faces of Γ correspond to the boundary components of S (cf. [32]). The embedding of Γ in the surface defines the partial triangulation. Note that we may have to add marked points to avoid having arcs homotopic to boundary segments. Figure 1.16 shows the surface associated to the gentle algebra in Example 1.39.

1.5.4 Geometric Model for Skew-gentle Algebras

The geometric model given in Subsection 1.5.3 has been recently extended to a wider class of algebras, called skew-gentle algebras, by considering punctured surfaces (see [26]). A skew-gentle algebra can be obtained from a gentle algebra by replacing some relations of the form ε^2, where ε is a loop, by $\varepsilon^2 - \varepsilon$.

Definition 1.40 Let Q be a quiver, I a set of paths in Q, and Sp a subset of Q_0 such that $\mathbf{k}Q^{sp}/I^{sp}$ is a gentle algebra, where Q^{sp} is obtained from Q by adding a loop ε_i at each vertex i in Sp, and $I^{sp} = I \cup \{\varepsilon_i^2 \mid i \in Sp\}$. The vertices in Sp are called *special vertices* and the ε_i are the *special loops*.

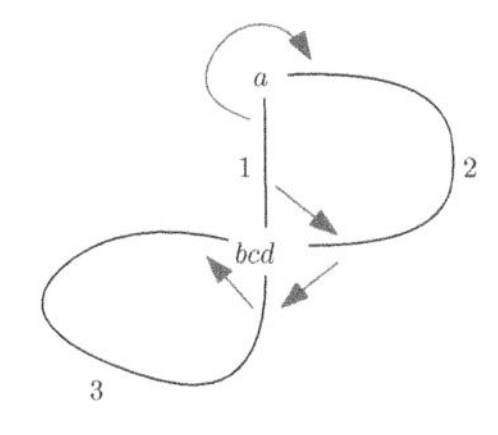

Figure 1.15 The ribbon graph of A.

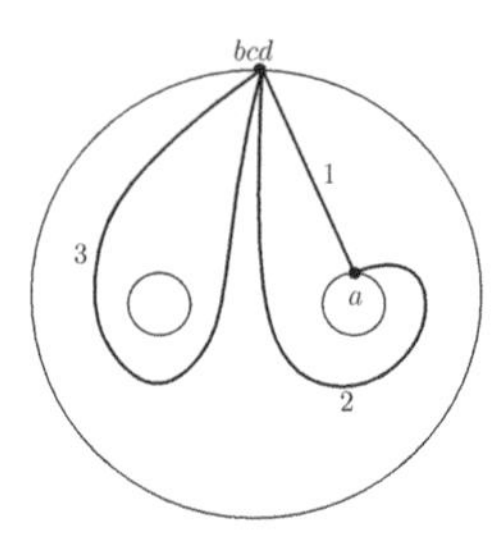

Figure 1.16 The surface and the ribbon graph of A as a partial triangulation.

A finite-dimensional algebra is *skew-gentle* if it admits a presentation of the form $\mathbf{k}Q^{sp}/I^{sg}$, where Q^{sp} comes from a triple (Q, I, Sp) as above and $I^{sg} = I \cup \{\varepsilon_i^2 - \varepsilon \mid i \in Sp\}$.

Note that the presentation I^{sg} is not admissible. However there is an isomorphism $\mathbf{k}Q^{sp}/I^{sg} \simeq \mathbf{k}\hat{Q}/\hat{I}$, with $\hat{I}$ admissible, where Q^{sp} and I^{sg} are defined as follows.

- $\hat{Q}_0$ is obtained from Q_0 by splitting each special vertex i into two vertices i^+ and i^-;
- $\hat{Q}_1$ is obtained from Q_1 by splitting an arrow a for which $s(a) \notin Sp$ and $t(a) \in Sp$ (resp. $s(a) \in Sp$ and $t(a) \notin Sp$) into two arrows a_+ with $t(a_+) = t(a)^+$ and a_- with $t(a_-) = t(a)^-$ (resp. a^+ with $s(a^+) = s(a)^+$ and a^- with $s(a^-) = s(a)^-$);
- given $ab \in I$, if $t(a) \notin Sp$, then all resulting paths in Q_1 of length 2 lie in $\hat{I}$, and if $t(a) \in Sp$, then $a_+b^+ - a_-b^- \in \hat{I}$. All relations in $\hat{I}$ are obtained this way.

Remark 1 Any gentle algebra is skew-gentle; take $Sp = \varnothing$.

2 The linearly oriented quivers of type D and $\widetilde{D}$ are hereditary skew-gentle algebras. Indeed, let $Q = 1 \longrightarrow 2 \longrightarrow \cdots \longrightarrow n$ and $I = \varnothing$. If we set $Sp = \{n\}$ (resp. $Sp = \{1, n\}$), then the corresponding skew-gentle algebra is isomorphic to the hereditary algebra of type D_{n+1} (resp. $\widetilde{D}_{n+1}$) with linear orientation.

Consider the following triangulation of a punctured disc:

We can define a quiver Q associated to this triangulation in the same manner as above, giving us

$$Q = 1 \longrightarrow 2 \longrightarrow 3 \circlearrowright \varepsilon_3 .$$

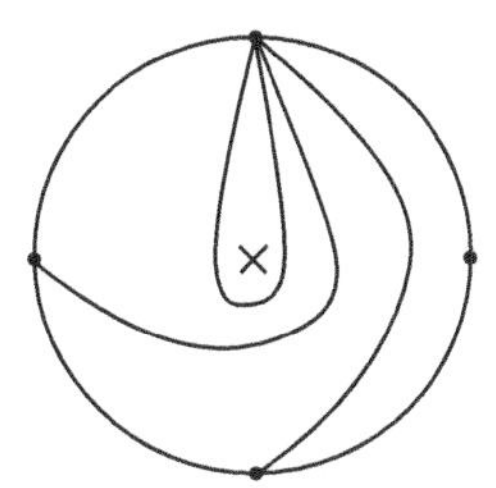

Figure 1.17 A triangulation of a punctured disc.

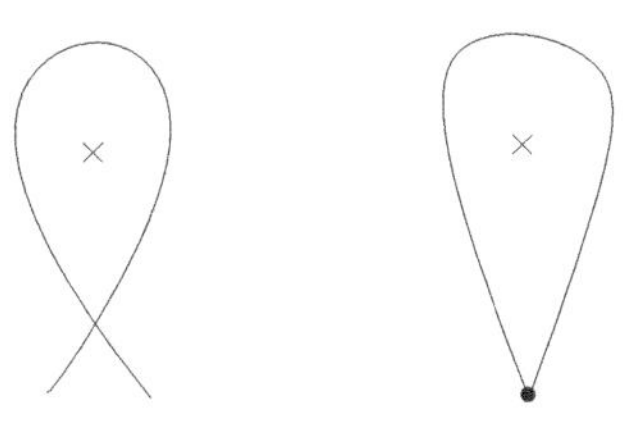

Figure 1.18 Permissible arcs in punctured surfaces do not satisfy these local configurations.

The vertex 3 associated to the loop arc delimiting a monogon with a puncture in its interior is considered to be a special vertex and so the algebra E corresponding to the triangulation in Figure 1.17 is the bound path algebra defined by Q and the relation $\varepsilon_3^2 - \varepsilon_3$. By Remark 1.5.4 (2), this algebra is isomorphic to the hereditary path algebra of the quiver:

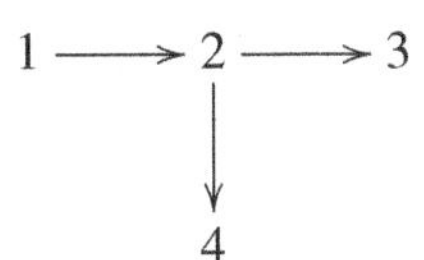

We consider *tagged permissible arcs* in the punctured disc, i.e. pairs (γ, σ), where γ is an arc whose endpoints are marked points in the boundary or the puncture, γ is not an arc in the triangulation and it does not cut out a once-punctured monogon by its self-intersection (see Figure 1.18), and

$$\sigma : \{t \mid \gamma(t) \text{ is a puncture}\} \to \{0, 1\}$$

is a map. If $\sigma(t) = 1$, we put a tag on the arc γ near the puncture.

In what follows, we describe E-modules via representations of the quiver of type D_4. There is a one-to-one correspondence between permissible tagged arcs and the indecomposable E-modules, which is again described via crossings. Given a permissible tagged arc (γ, σ), the corresponding indecomposable E-module $M(\gamma, \sigma)$ is uniquely determined by its support. The support at vertex i

is given by the number of times γ crosses the diagonal indexed by i, if i is not a special vertex. If i is a special vertex, and γ crosses i but is not incident with the puncture, then $(M(\gamma,\sigma))_j = \mathbf{k}$, where $j = 3,4$. If i is a special vertex, γ crosses i and it is incident with the puncture, then

$$(M(\gamma,\sigma))_j = \begin{cases} 0 & \text{if } \sigma = 0, j = 4 \text{ or } \sigma = 1, j = 3 \\ \mathbf{k} & \text{if } \sigma = 0, j = 3 \text{ or } \sigma = 1, j = 4. \end{cases}$$

The pivot elementary moves are described in the same manner as in the gentle case, except for the following cases.

Case 1: If (γ,σ) is an arc incident with the puncture, then the pivot elementary move consists of replacing γ by a loop arc around the puncture and ending at the other endpoint of γ and pivoting one step in the anticlockwise direction the endpoint which does not create a self-crossing.

Case 2: Given a permissible arc γ which is not incident with the puncture, if one obtains a loop arc around the puncture after performing a pivot elementary move, then this move corresponds to two irreducible maps, whose targets correspond to the tagged and untagged arcs incident with the puncture and the endpoint of the loop arc.

The AR-translate is described by clockwise rotation of the endpoints which are marked points in the boundary, and by changing the tag at the puncture.

A geometric model of the AR-quiver of $\mathsf{mod}(\mathsf{E})$ is thus described in Figure 1.19.

One can associate an algebra to a partial triangulation of a punctured surface containing a loop arc around each puncture. These algebras are called *skew-tilting algebras*, and they coincide with the skew-gentle algebras. The definition of permissible arcs and equivalence classes of arcs passes across to the punctured case, permissible arcs incident with punctures can be tagged or untagged and we also allow permissible arcs whose both endpoints are punctures; these will correspond to four non-isomorphic indecomposable modules, determined by their tags at each endpoint. The case where there are tagged permissible arcs whose both endpoints are punctures only show up in the case when the algebra is not of finite representation type.

The following gives an example of a skew-gentle algebra of finite-representation type coming from a partial triangulation of a punctured disc. Consider the path algebra F of the quiver

$$1 \xrightarrow{a} 2 \xrightarrow{b} 3 \quad \text{(with loop } \varepsilon_2 \text{ at } 2)$$

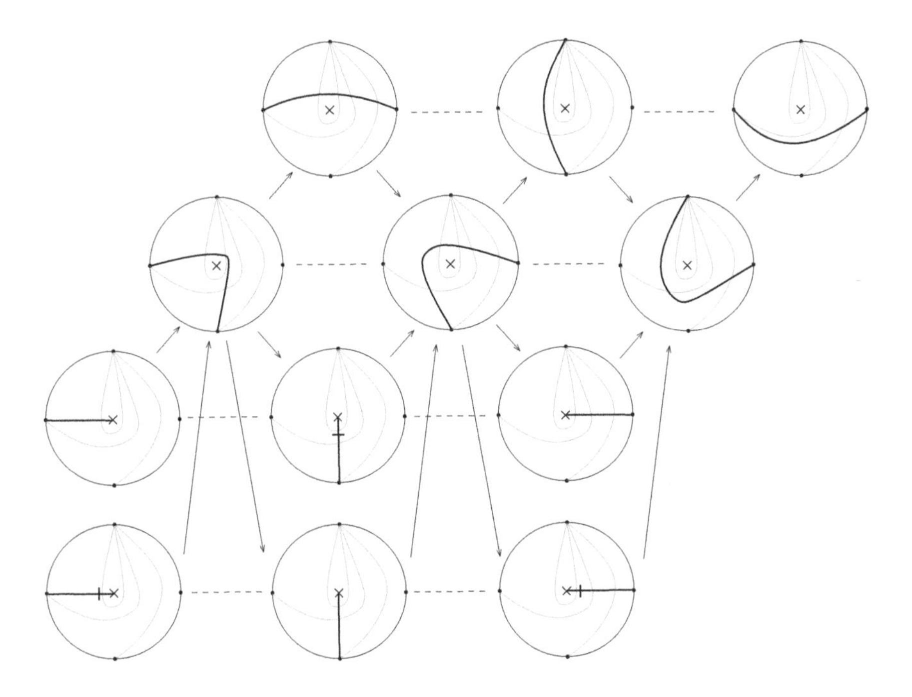

Figure 1.19 The geometric model of the AR quiver of mod(E).

bound by the relations $\varepsilon_2^2 - \varepsilon_2 = 0$ and $ab = 0$. This algebra, which is isomorphic to the bound path algebra of the quiver

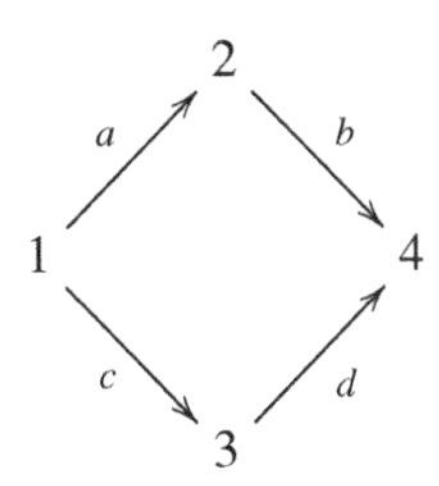

bound by the relation $ab = cd$, is associated to the partial triangulation of the punctured disc in Figure 1.20.

The AR-quiver of F is given in Figure 1.21.

For recent developments of the study of derived categories of skew-gentle algebras via geometric models, see [2, 31].

We note that the geometric models of (skew-)gentle algebras described above are based on the description of the AR theory coming from deep results classifying indecomposable representations of classes of algebras of tame represen-

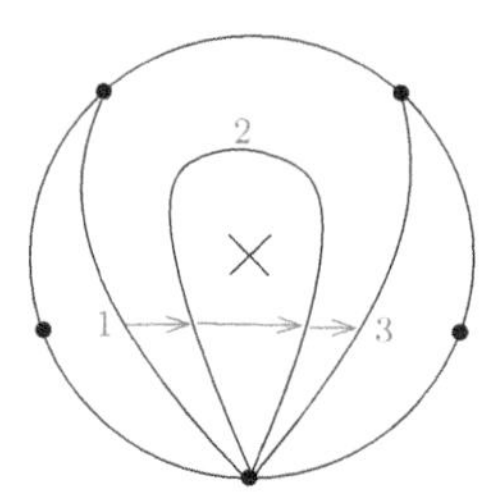

Figure 1.20 The partial triangulation associated to F.

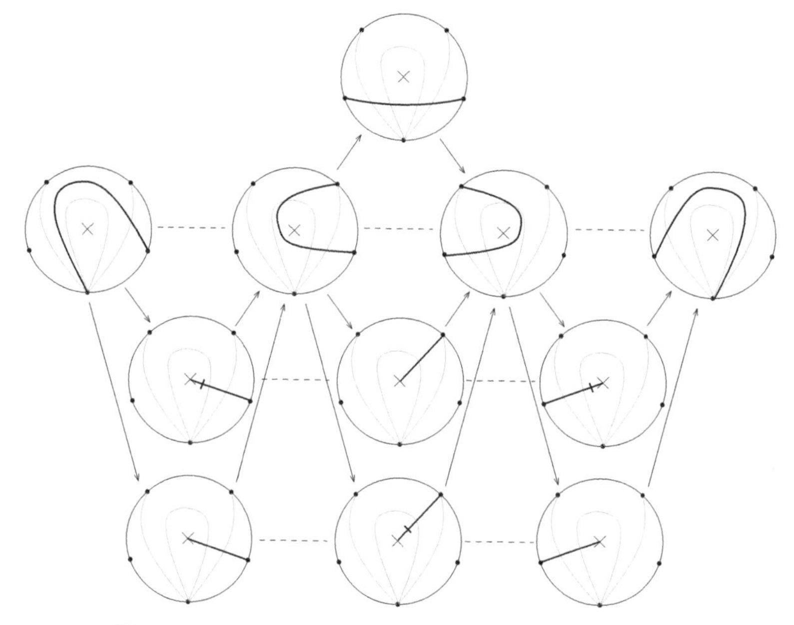

Figure 1.21 The geometric model of the AR-quiver of mod(F).

tation type (cf. [11, 14, 18, 19, 24]). The indecomposable modules are split into two classes: string modules and band modules. All the examples we considered in Section 1.5 are representation-finite, in which case we only have string modules. These correspond to the (tagged) permissible arcs for which at least one endpoint is a marked point. Band modules lying in homogeneous tubes correspond to certain closed curves in the surface and band modules lying in the bottom of tubes of rank 2 correspond to tagged permissible arcs whose both endpoints are punctures. In any case, the set of (tagged) permissible arcs describe in particular τ-rigid modules, which leads us to the final subsection of this chapter.

1.5.5 An Application: τ-tilting Theory

Classical tilting theory compares the representation theory of two algebras, one of which is the endomorphism algebra of a tilting module over the other algebra. In 2012, Adachi, Iyama and Reiten introduced τ-tilting theory, which can be seen as a "mutation closure" of tilting theory [1]. For more on τ-tilting theory, see Chapter 2. In this subsection we use the geometric models described above to give a classification of support τ-tilting modules, the main objects of study in τ-tilting theory. This classification was obtained in [26].

Definition 1.41 Let A be an algebra and M an A-module. Denote by $|A|$ the number of simple A-modules, and by $|M|$ the number of indecomposable summands of M.

1 M is *τ-rigid* if $\mathrm{Hom}_A(M, \tau M) = 0$.
2 M is *τ-tilting* if M is τ-rigid and $|M| = |A|$.
3 M is *support τ-tilting* if there is an idempotent $e \in A$ such that M is a τ-tilting $(A/\langle e \rangle)$-module.

Proposition 1.42 *[1, Proposition 2.3] M is support τ-tilting if and only if M is τ-rigid and there is a projective module P such that* $\mathrm{Hom}_A(P, M) = 0$ *and* $|M| + |P| = |A|$.

Let A be a skew-gentle algebra, S the associated punctured surface and P the associated partial triangulation of S. Given a tagged permissible arc γ, we represent by $[\gamma]$ the arc that is equivalent to γ such that the starting/ending segments have the form represented in Figure 1.22.

A *generalised permissible arc* in S is either $[\gamma]$, where γ is a permissible arc, or an arc whose completion is in P, where the *completion* $\tilde{\gamma}$ of an arc γ is described in Figure 1.23.

Besides the natural definition of crossing of arcs in the interior of the surface, we also need to consider crossings at a puncture. Two tagged generalised

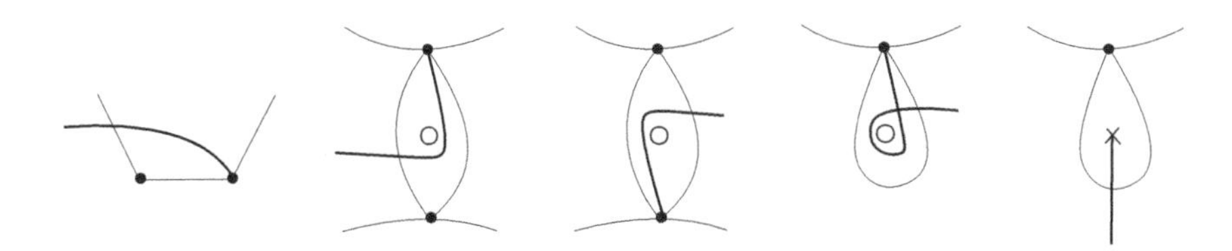

Figure 1.22 Starting/ending segment of $[\gamma]$.

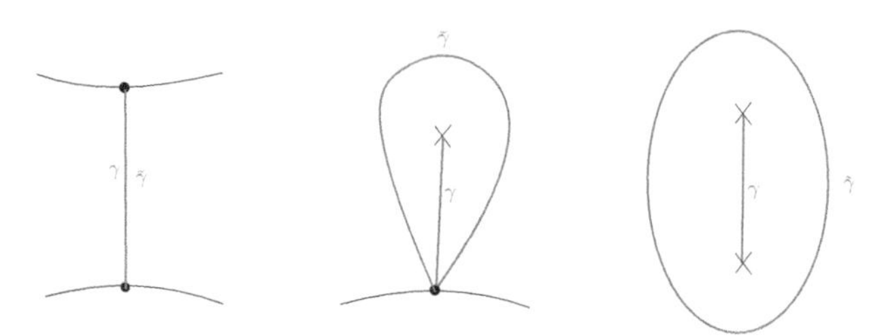

Figure 1.23 Completion of a diagonal in P.

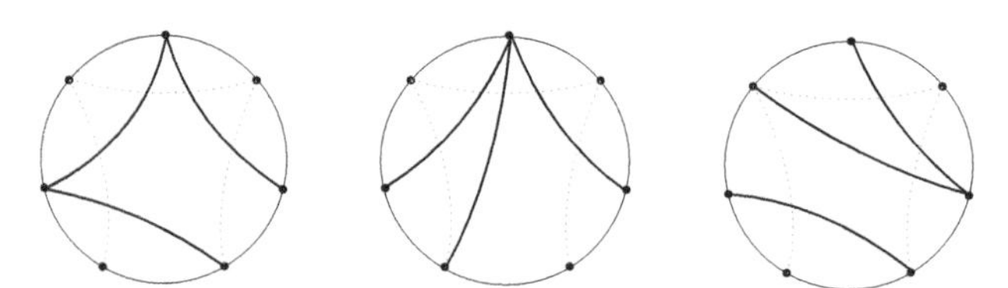

Figure 1.24 τ-tilting C-modules.

permissible arcs cross at a puncture p if p is an endpoint of both arcs, they have different tags at p and if the arcs are homotopic, then the other endpoint of both arcs is also a puncture p' and the tags also differ at p'.

The geometric description of support τ-tilting modules over a skew-gentle algebra is given as follows.

Theorem 1.43 *[26, Corollary 5.9] There is a one-to-one correspondence between the set of maximal collections of noncrossing tagged generalised permissible arcs in S and the set of support τ-tilting A-modules. Moreover, a collection of noncrossing tagged generalised permissible arcs is maximal if and only if its cardinality is $|A|$.*

Example 1.44 Recall the gentle algebra C from the start of Subsection 1.5.3. The τ-tilting modules over C are given by the collection of thick arcs in Figure 1.24. The support τ-tilting C-modules with two summands are given by the collection of thick arcs in Figure 1.25. Here, the thin arc represents the projective module associated to the support τ-tilting C-module. The remaining support τ-tilting C-modules and corresponding projective modules are given by the collection of thick and thin arcs respectively in Figure 1.26.

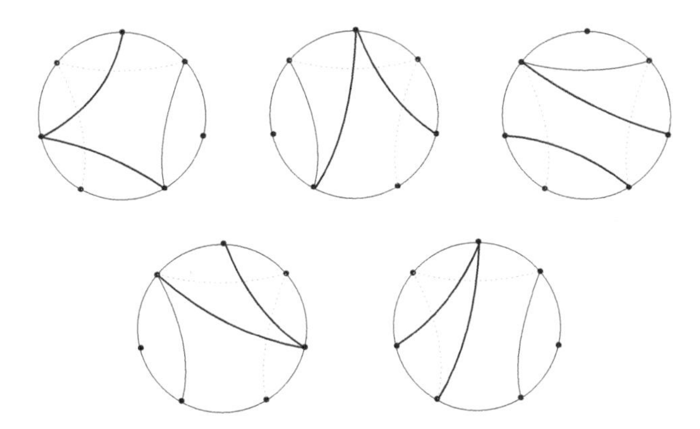

Figure 1.25 Support τ-tilting C-modules with two summands.

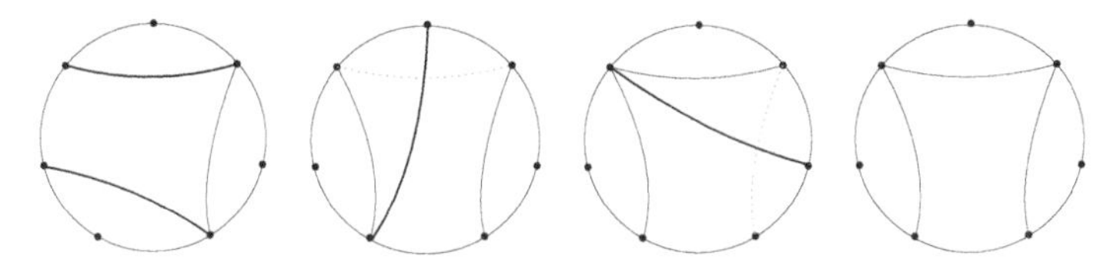

Figure 1.26 The remaining support τ-tilting C-modules.

Acknowledgments

The author would like to thank David Jordan, Nadia Mazza and Sibylle Schroll for the organisation of the LMS Autumn Algebra School 2020. She would also like to thank Jan Grabowski for reading earlier versions of these notes, and the anonymous referee for their useful comments and suggestions. The author is also grateful to the European Union's Horizon 2020 research and innovation programme for financial support through the Marie Skłodowska-Curie Individual Fellowship grant agreement number 838706.

Bibliography

[1] Adachi, T., Iyama, O. and Reiten, I. τ-tilting theory. *Compos. Math.*, 150(3):415–452, 2014.

[2] Amiot, Claire, and Brüstle, Thomas. 2022. Derived equivalences between skew-gentle algebras using orbifolds. *Doc. Math.*, **27**, 933–982.

[3] Amiot, Claire, Plamondon, Pierre-Guy, and Schroll, Sibylle. 2023. A complete derived invariant for gentle algebras via winding numbers and Arf invariants. Selecta Math., **29**(30).

[4] Assem, Ibrahim. 2018. A course on cluster tilted algebras. Pages 127–176 of: *Homological methods, representation theory, and cluster algebras*. CRM Short Courses. Springer, Cham.

[5] Assem, Ibrahim, and Skowroński, Andrzej. 1987. Iterated tilted algebras of type $\tilde{\mathbf{A}}_n$. *Math. Z.*, **195**(2), 269–290.

[6] Assem, I., Simson, D., and Skowroński, A. 2006. *Elements of the Representation Theory of Associative Algebras. Vol. 1.* London Mathematical Society Student Texts, vol. 65. Cambridge University Press, Cambridge. Techniques of representation theory.

[7] Assem, Ibrahim, Brüstle, Thomas, Charbonneau-Jodoin, Gabrielle, and Plamondon, Pierre-Guy. 2010. Gentle algebras arising from surface triangulations. *Algebra Number Theory*, **4**(2), 201–229.

[8] Auslander, Maurice. 1974. Representation theory of Artin algebras. I, II. *Comm. Algebra*, **1**, 177–268; ibid. 1 (1974), 269–310.

[9] Auslander, M., Reiten, I., and Smalø, S. O. 1995. *Representation Theory of Artin Algebras.* Cambridge Studies in Advanced Mathematics, vol. 36. Cambridge University Press, Cambridge.

[10] Baur, Karin, and Coelho Simões, Raquel. 2021. A geometric model for the module category of a gentle algebra. *Int. Math. Res. Not. IMRN*, 11357–11392.

[11] Bondarenko, V. M. 1991. Representations of bundles of semichained sets and their applications. *Algebra i Analiz*, **3**(5), 38–61.

[12] Broomhead, Nathan. 2012. Dimer models and Calabi-Yau algebras. *Mem. Amer. Math. Soc.*, **215**(1011), viii+86.

[13] Brüstle, Thomas, Douville, Guillaume, Mousavand, Kaveh, Thomas, Hugh, and Yıldırım, Emine. 2020. On the combinatorics of gentle algebras. *Canad. J. Math.*, **72**(6), 1551–1580.

[14] Butler, M. C. R., and Ringel, Claus Michael. 1987. Auslander-Reiten sequences with few middle terms and applications to string algebras. *Comm. Algebra*, **15**(1–2), 145–179.

[15] Caldero, P., Chapoton, F. and Schiffler, R. 2006. Quivers with relations arising from clusters (A_n case). *Trans. Amer. Math. Soc.*, **358**(3), 1347–1364.

[16] Çanakçı, İlke, and Schroll, Sibylle. 2017. Extensions in Jacobian algebras and cluster categories of marked surfaces. *Adv. Math.*, **313**, 1–49. With an appendix by Claire Amiot.

[17] Çanakçı, İlke, Pauksztello, David, and Schroll, Sibylle. 2019. Mapping cones in the bounded derived category of a gentle algebra. *J. Algebra*, **530**, 163–194.

[18] Crawley-Boevey, W. W. 1989. Functorial filtrations. II. Clans and the Gelfand problem. *J. London Math. Soc. (2)*, **40**(1), 9–30.

[19] Deng, Bangming. 2000. On a problem of Nazarova and Roiter. *Comment. Math. Helv.*, **75**(3), 368–409.

[20] Fomin, S. and Zelevinsky, A. 2003. Cluster algebras. II. Finite type classification. *Invent. Math.*, **154**(1), 63–121.

[21] Fomin, Sergey, Shapiro, Michael, and Thurston, Dylan. 2008. Cluster algebras and triangulated surfaces. I. Cluster complexes. *Acta Math.*, **201**(1), 83–146.

[22] Gabriel, Peter. 1972. Unzerlegbare Darstellungen. I. *Manuscripta Math.*, **6**, 71–103; correction, ibid. 6 (1972), 309.

[23] Garcia Elsener, A. 2020. Gentle m-Calabi-Yau tilted algebras. *Algebra Discrete Math.*, **30**(1), 44–62.

[24] Geiß, Christof. 1999. Maps between representations of clans. *J. Algebra*, **218**(1), 131–164.

[25] Haiden, F., Katzarkov, L. and Kontsevich, M. 2017. Flat surfaces and stability structures. *Publ. Math. Inst. Hautes Études Sci.*, **126**, 247–318.

[26] He, Ping, Zhou, Yu, and Zhu, Bin. 2020. A geometric model for the module category of a skew-gentle algebra. *2004.11136.*

[27] Huerfano, Ruth Stella, and Khovanov, Mikhail. 2001. A category for the adjoint representation. *J. Algebra*, **246**(2), 514–542.

[28] Humphreys, James E. 1972. *Introduction to Lie algebras and representation theory*. Graduate Texts in Mathematics, Vol. 9. Springer-Verlag, New York-Berlin.

[29] Kinsey, L. C. 1993. *Topology of Surfaces*. Undergraduate Texts in Mathematics. Springer-Verlag, New York.

[30] Labardini-Fragoso, D. 2009. Quivers with potentials associated to triangulated surfaces. *Proc. Lond. Math. Soc. (3)*, **98**(3), 797–839.

[31] Labardini-Fragoso, Daniel, Schroll, Sibylle, and Valdivieso, Yadira. 2022. Derived categories of skew-gentle algebras and orbifolds. *Glasg. Math. J.*, **64**(3), 649–674.

[32] Labourie, François. 2013. *Lectures on representations of surface groups*. Zurich Lectures in Advanced Mathematics. European Mathematical Society (EMS), Zürich.

[33] Opper, Sebastian, Plamondon, Pierre-Guy, and Schroll, Sibylle. 2018. A geometric model for the derived category of gentle algebras. *1801.09659.*

[34] Palu, Yann, Pilaud, Vincent, and Plamondon, Pierre-Guy. 2021. Non-kissing complexes and tau-tilting for gentle algebras. *Mem. Amer. Math. Soc.*, **274**(1343).

[35] Riedtmann, C. 1980. Algebren, Darstellungsköcher, Überlagerungen und zurück. *Comment. Math. Helv.*, **55**(2), 199–224.

[36] Ringel, Claus Michael. 1995. Some algebraically compact modules. I. Pages 419–439 of: *Abelian groups and modules (Padova, 1994)*. Math. Appl., vol. 343. Kluwer Acad. Publ., Dordrecht.

[37] Rotman, Joseph J. 1979. *An introduction to homological algebra*. Pure and Applied Mathematics, vol. 85. Academic Press, Inc. [Harcourt Brace Jovanovich, Publishers], New York-London.

[38] Schiffler, R. 2014. *Quiver representations*. CMS Books in Mathematics/Ouvrages de Mathématiques de la SMC. Springer, Cham.

[39] Schroll, Sibylle. 2015. Trivial extensions of gentle algebras and Brauer graph algebras. *J. Algebra*, **444**, 183–200.

[40] Simson, Daniel, and Skowroński, Andrzej. 2007. *Elements of the representation theory of associative algebras. Vol. 3*. London Mathematical Society Student Texts, vol. 72. Cambridge University Press, Cambridge. Representation-infinite tilted algebras.

2

τ-tilting Theory – an Introduction

Hipolito Treffinger

2.1 Introduction

The term *τ-tilting theory* was coined by Adachi, Iyama and Reiten in [1] at the beginning of the 2010s. In their paper, the authors created a fresh approach to the study of two classical branches of the representation theory of finite dimensional algebras, namely tilting theory and Auslander–Reiten theory. The combination of these two subjects is clearly reflected in the name of this novel theory, where the Greek letter τ represents the Auslander–Reiten translation in the module category of an algebra while the reference to tilting theory is obvious.

Our primary aim with these notes is to give a friendly introduction to τ-tilting theory from a representation theoretic perspective. Here you can find a compilation of many important results on the subject, giving a special emphasis on the close relation between τ-tilting theory and torsion theories. Some background in representation theory and Auslander–Reiten theory is desirable but not necessary to follow this exposition. We note that given the immense amount of work that has been done in the last decade on τ-tilting theory, this is not a complete survey on the topic. For instance, we do not cover the rich connections that τ-tilting theory has with other branches of mathematics, such as combinatorics or algebraic geometry.

This chapter can be divided into five different parts. In the first part, which consists only of Section 2.2, we give a brief historic account of the events leading to the rise of τ-tilting theory in representation theory. We note that this section is not necessary for the understanding of the rest of the chapter and can be skipped. In the second part, which consists of Sections 2.3 and 2.4, we give the basic definitions of the theory. We also describe some of the different forms that τ-tilting can adopt, namely support τ-tilting modules, τ-tilting pairs, functorially finite torsion classes and 2-term silting complexes. The third part, corresponding to Sections 2.5–2.10, is dedicated to compiling general results on the subject, including the so-called τ-tilting reduction and the characterisation of τ-tilting finite algebras. The K-theory of τ-tilting theory is discussed in the fourth part of the chapter, namely Sections 2.12 to 2.14. We finish these notes in Section 2.15, where we show how we can associate to every algebra a geometric object known as its *wall-and-chamber structure* and we explain how this invariant encodes much of its τ-tilting theory.

We warn the reader that we do not include in these notes any proofs; these can be found in the references given by each result. Given the short time that has passed since the introduction of τ-tilting theory, to our knowledge, there is not much material on the subject available other than the original research papers, with the exception of [67]. For background material in representation theory, we recommend the textbooks [9, 17, 95]. For survey materials on more classical tilting theory, the reader is encouraged to see [7, 12].

2.1.1 Notation

In these notes A is always a basic finite dimensional algebra over a field $\mathbb{K}$ that we assume is algebraically closed. For us $\operatorname{mod} A$ is the category of finitely presented right A-modules and τ denotes the Auslander–Reiten translation in $\operatorname{mod} A$.

Given any A-module M, we denote by $|M|$ the number of isomorphism classes of indecomposable direct summands of M. Throughout this document we assume that n is the number of isomorphism classes of simple A-modules. Note that in this case $|A| = n$, since $A = \bigoplus_{i=1}^{n} P(i)$, where $P(i)$ denotes the i-th indecomposable projective module.

Also, unless otherwise specified, every module is assumed to be basic, meaning that the indecomposable direct summands of M are pairwise non-isomorphic.

Acknowledgements

These notes were written to support a series of three lectures entitled *τ-tilting theory for finite-dimensional algebras* framed in the **LMS Autumn Algebra School 2020** organised by David Jordan, Nadia Mazza and Sibylle Schroll and funded by the London Mathematical Society. The author is grateful to the organisers for giving him the opportunity to present these lectures. He also thanks Bethany Marsh and Jenny August and the anonymous referee for their careful reading and their insightful comments in a previous versions of this document. Finally, he acknowledges the financial support of the **Hausdorff Center of Mathematics**. The author is also funded by the Deutsche

Forschungsgemeinschaft (DFG, German Research Foundation) under Germany's Excellence Strategy Grant EXC-2047/1-390685813 and by the European Union's Horizon 2020 research and innovation programme under the Marie Sklodowska-Curie grant agreement No 893654.

2.2 Towards τ-tilting Theory

It can be argued that the modern study of representation theory started with the parallel developments of almost split sequences by Auslander and Reiten [20, 22, 23] (see also [91]) and the theory of quiver representations by Gabriel [55, 56]. Gabriel showed two very important results using quivers. One of these results says that the representation theory of every finite dimensional algebra over an algebraically closed field can be understood using quiver representations. The formal statement is the following.

Theorem 2.1 *[55] Let A be a finite dimensional algebra over an algebraically closed field $\mathbb{K}$. Then A is Morita equivalent to the algebra $\mathbb{K}Q/I$, the path algebra of the quiver Q bounded by an admissible ideal of relations I. Moreover the quiver Q is uniquely determined by A.*

In the literature, people refer to the quiver Q determined by the algebra as the *ordinary quiver* or the *Gabriel quiver* or simply the *quiver* of the algebra. In these notes we take the latter option. The reason to give it such names is that one can associate to each finite dimensional algebra another quiver known as the *Auslander–Reiten quiver* of the algebra, which encodes all the almost split sequences in mod A. For more information about the Auslander–Reiten theory of algebras, the reader is encouraged to see the course on this topic by Raquel Coelho-Simões in this same series. See also [17, IV.5].

The second result of Gabriel we want to mention here is the classification of hereditary algebras of finite representation type by means of Dynkin diagrams as follows.

Theorem 2.2 *[55] Let A be a connected hereditary representation-finite finite dimensional algebra over an algebraically closed field $\mathbb{K}$. Then A is Morita equivalent to $\mathbb{K}\vec{\Delta}$, where $\vec{\Delta}$ is a quiver whose underlying graph is a Dynkin diagram Δ of type $\mathbb{A}, \mathbb{D}$ or $\mathbb{E}$. Moreover there is a one-to-one correspondence between the indecomposable representations of A and the positive roots of the root system associated to Δ.*

When this result appeared, it came as a great surprise since many fundamental properties of the path algebra of a quiver depend on the orientations of the arrows. For instance, if we start with two quivers Q_1, Q_2 that correspond to two different orientations of the same Dynkin diagram Δ then the path algebra $\mathbb{K}Q_1$ is in general not isomorphic to the path algebra $\mathbb{K}Q_2$, not even as vector spaces. Hence, there was no reason to believe that the number of indecomposable representations should be the same.

Example 2.3 For instance, take the algebras A and A' to be the path algebras of the quivers

$$Q_A = 1 \longrightarrow 2 \longrightarrow 3 \qquad Q_{A'} = 1 \longrightarrow 2 \longleftarrow 3$$

of type $\mathbb{A}_3$. A quick calculation shows that $\dim_{\mathbb{K}} A = 6$ while $\dim_{\mathbb{K}} A' = 5$. The Auslander–Reiten quivers of A and A' can be found in Figure 2.1 and Figure 2.2, respectively. Here the arrows correspond to the irreducible morphisms in the module category and the dashed lines correspond to the Auslander–Reiten translation. In these figures we can see that the number of indecomposable representations of A and A' coincide.

As a consequence, explaining this phenomenon became of significant interest. The first explanation was given by Bernstein, Gelfand and Ponomarev in [32] by constructing the so-called reflection functors.

Let Q be a quiver of type Δ and denote by Q_0 the set of vertices of Q. Since every Dynkin diagram is a tree, there is at least one vertex $x \in Q_0$ which is a sink, i.e. a vertex such that all the arrows incident to that vertex are incoming arrows. Now, we construct a quiver $Q_{A'}$ which is identical to Q_A, except for the fact that now the vertex x is a source, which means that every arrow incident to x is an outgoing arrow. One says that Q_A and $Q_{A'}$ are reflections of each other at x. In Example 2.3, Q' is the reflection of Q at the vertex 3. Then, Bernstein, Gelfand and Ponomarev showed the existence of functors, which they called

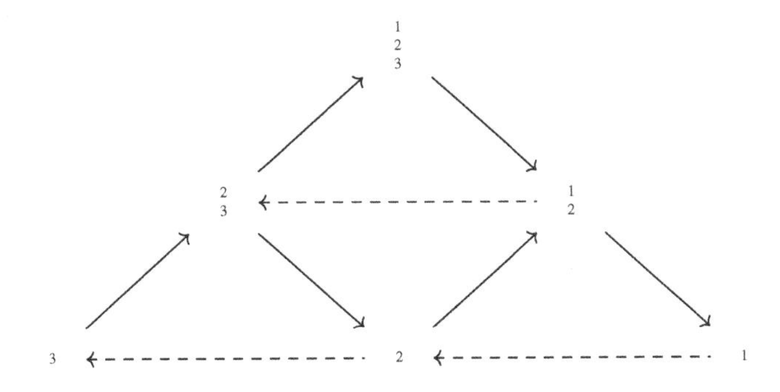

Figure 2.1 The Auslander–Reiten quiver of A.

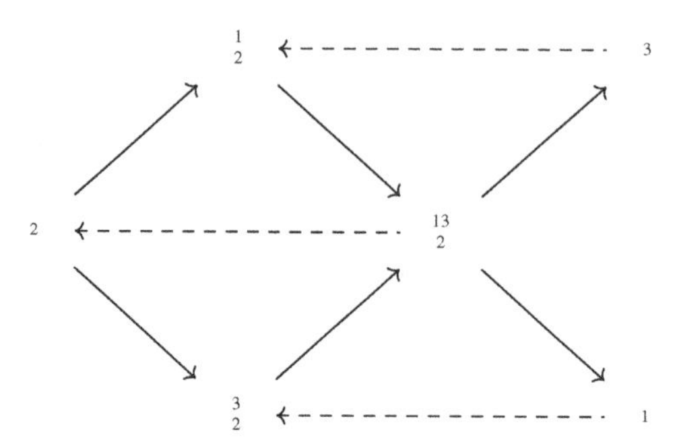

Figure 2.2 The Auslander–Reiten quiver of A'.

reflection functors, between mod $\mathbb{K}Q$ and mod $\mathbb{K}Q'$ that induce a one-to-one correspondence between their indecomposable objects.

Some years after that, Auslander, Platzeck and Reiten [21] realised that these functors were induced by a very specific object in mod $\mathbb{K}Q$. To be more precise, note that the simple module $S(x)$ associated to the vertex $x \in Q_0$ is projective and it is not injective. This implies that the inverse Auslander–Reiten translation $\tau^{-1}S(x)$ of $S(x)$ is a non-zero indecomposable object of mod $\mathbb{K}Q$. Then they showed that the reflection functors described by Bernstein, Gelfand and Ponomarev were equivalent to $\mathrm{Hom}_A(T, -) : \mathrm{mod}\, A \to \mathrm{mod}\, A'$, where T is the module

$$T = \tau^{-1}S(x) \oplus \bigoplus_{x \neq y \in Q_0} P(y). \tag{2.2.1}$$

Thus, T is the direct sum $\tau^{-1}S(x)$ and the direct sum of all of the indecomposable projectives except $S(x)$. Moreover, they showed that $\mathbb{K}Q'$ is isomorphic to $\mathrm{End}_{\mathbb{K}Q}(T)^{op}$. In particular, this approach allowed them to show the existence of reflection-like functors between the module category of any Artin algebra A having a simple projective module and $\mathrm{End}_A(T)$, even when A is not hereditary or even when A is not the quotient of the path algebra of a quiver. Going once again to our running example, the module described by Auslander, Platzeck and Reiten in mod A is $T = 2 \oplus \begin{smallmatrix}1\\2\\3\end{smallmatrix} \oplus \begin{smallmatrix}2\\3\end{smallmatrix}$.

Some years later, Brenner and Butler went further and studied in [34] this phenomenon axiomatically. In this paper they introduce the notion of *tilting modules* as follows.

Definition 2.4 [34] Let A be an algebra and T be an A-module. We say that T is a tilting module if the following holds:

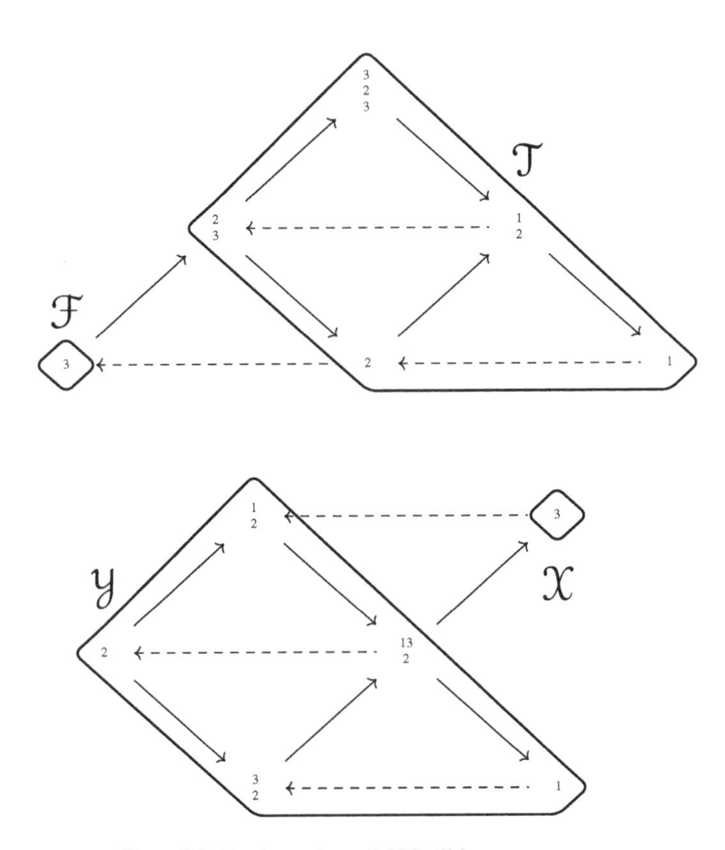

Figure 2.3 Torsion pairs and APR-tilting.

(i) $\mathrm{pd}_A T \leqslant 1$, the projective dimension of T is at most 1.
(ii) T is rigid, that is $\mathrm{Ext}^1_A(T,T) = 0$.
(iii) There exists a short exact sequence of the form

$$0 \to A \to T' \to T'' \to 0,$$

where T', T'' are direct summands of direct sums of T.

In this paper they show that any tilting A-module T acts as a sort of translator between the representation theory of A and $B := \mathrm{End}_A(T)^{op}$, the opposite of the endomorphism algebra of T.

The first thing that they have shown is that a tilting A-module T is also a tilting B-module. Moreover they showed that T induced a torsion pair $(\mathcal{T}, \mathcal{F})$ in $\mathrm{mod}\, A$ and a torsion pair $(\mathcal{X}, \mathcal{Y})$ in $\mathrm{mod}\, B$ such that the functors $\mathrm{Hom}_A(T, -)$:

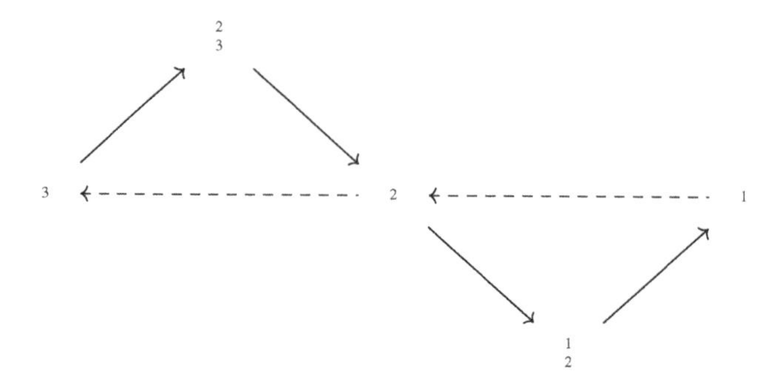

Figure 2.4 The Auslander–Reiten quiver of B.

$\operatorname{mod} A \to \operatorname{mod} B$ and $\operatorname{Ext}^1_A(T,-) : \operatorname{mod} A \to \operatorname{mod} B$ induce equivalences of categories between $\mathcal{T}$ and $\mathcal{Y}$ and between $\mathcal{F}$ and $\mathcal{X}$, respectively. This result of Brenner and Butler can be seen applied to our running example in Figure 2.3. For the precise definition of torsion pair, see Definition 2.17. Also, a more detailed treatment of the tilting theorem will be given in Section 2.5.

Since the module introduced by Auslander, Platzeck and Reiten was their motivating example, one can expect that it satisfies Definition 2.4 (i)–(iii) and indeed this is the case. In fact, nowadays this module is known as the *APR-tilting module*. But, as the reader is already guessing, there are many more examples of tilting modules. Take the module $T = {\scriptstyle 3} \oplus \begin{smallmatrix} 1 \\ 2 \\ 3 \end{smallmatrix} \oplus {\scriptstyle 1}$. One can verify that T is indeed a tilting module. Firstly, the projective dimension of T is less than or equal to one since A is hereditary. Secondly, one can check that T does not admit self-extensions. Finally, the short exact sequence

$$0 \to {\scriptstyle 3} \oplus \begin{smallmatrix} 2 \\ 3 \end{smallmatrix} \oplus \begin{smallmatrix} 1 \\ 2 \\ 3 \end{smallmatrix} \to {\scriptstyle 3} \oplus \begin{smallmatrix} 1 \\ 2 \\ 3 \end{smallmatrix} \oplus \begin{smallmatrix} 1 \\ 2 \\ 3 \end{smallmatrix} \to {\scriptstyle 1} \to 0$$

is such that ${\scriptstyle 3} \oplus \begin{smallmatrix} 1 \\ 2 \\ 3 \end{smallmatrix} \oplus \begin{smallmatrix} 1 \\ 2 \\ 3 \end{smallmatrix}$ and ${\scriptstyle 1}$ are direct summands of direct sums of copies of T.

Now, the algebra $B = \operatorname{End}_A(T)$ is isomorphic to the path algebra of the quiver

$$1 \longrightarrow 2 \longrightarrow 3$$

modulo the ideal generated by the composition of the two arrows. The Auslander–Reiten quiver of B can be seen in Figure 2.4.

As we can see in this example, when we take an arbitrary tilting module A the numbers of indecomposable representations in $\operatorname{mod} A$ and in $\operatorname{mod} B$ are not the same. However, this is not a contradiction of the results of Brenner and

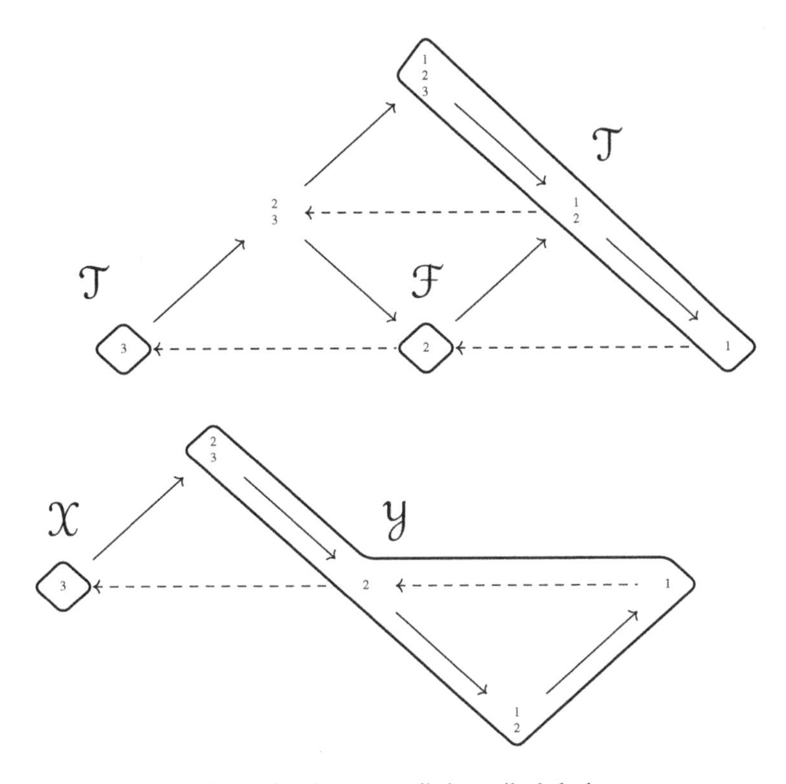

Figure 2.5 The tilting theorem applied to a tilted algebra.

Butler since their result only says what happens inside the torsion pairs induced by T in mod A and mod B. In this particular case, as we can see in Figure 2.5, the indecomposable object $\substack{2\\3}$ does not belong to either of the two subcategories $\mathcal{T}$ and $\mathcal{F}$ induced by the tilting module T.

Although the categories mod A and mod $\mathrm{End}_A(T)$ are not in general equivalent, it was shown by Happel [62] (see also Rickard's generalisation [92]) that these two algebras are *derived equivalent*. Without going into the details, starting from the module category of an algebra, one can construct a triangulated category known as the *derived category* of the algebra that encodes a wealth of homological information of the algebra. Then, the results of Happel and Rickard state that the original algebra and the endomorphism algebra of the tilting module have the same derived category, which implies that they share many homological properties that we will not discuss here. For more information about derived categories and their relationship with cluster algebras, we refer the reader to the notes of Matthew Pressland in this series.

Let us include a small parenthesis here that will be important later. The algebra B is the smallest non-hereditary example of a so-called *tilted algebra.* Tilted algebras were introduced by Happel and Ringel in [63] (see also Bongartz [33]) as the endomorphism algebras of tilting modules over hereditary algebras. The main idea behind their introduction was to use all the information available on hereditary algebras to understand a new class of algebras which had not been studied systematically until that moment.

The study of tilted algebras has sparked a great deal of research which it would be impossible to describe completely here. However, we need to mention two famous developments.

Firstly, note that the tilting theorem of Brenner and Butler does not impose any restriction on the algebra A. So, if we are able to understand some of the representation theory of tilted algebras using the knowledge we have on hereditary algebras, we can repeat the process and understand the representation theory of a new family of algebras using the knowledge we have on tilted algebras via the tilting theorem. These algebras are known as *iterated tilted algebras.* In [18], Assem and Skowroński classified all the iterated tilted algebras of Dynkin type $\tilde{\mathbb{A}}$ in terms of their ordinary quiver and relations, which led them to the definition of the so-called *gentle algebras*. Today, gentle algebras constitute a highly active area of research, deepening our understanding of representation theory of finite dimensional algebras and connecting this topic with various other branches of mathematics such as group theory and Algebraic and differential geometry.

The second is the characterisation of tilted algebras found independently by Liu [74] and Skowroński [100] using the Auslander–Reiten quiver of an algebra. They have shown that an algebra is tilted if and only if there is a structure with specific homological and combinatorial properties in their Auslander–Reiten quiver. Inspired by this characterisation of tilted algebras many families of algebras have been defined and determined by means of their Auslander–Reiten quivers.

Some years later, at the beginning of the twenty-first century, Fomin and Zelevinsky [31, 51, 52, 53] were studying the properties of the canonical bases arising in Lie theory and this study led to the introduction of *cluster algebras*.

These algebras are generated by a set of so-called *cluster variables* that are produced inductively from an *initial seed* via a process called *mutation* that produces new seeds. Even though the process of mutation is iterated an arbitrary (finite) number of times, for some initial seeds there are only finitely many cluster variables that can be constructed. In this case we say that a cluster algebra is of *finite type*. Moreover, for some of these algebras, known as *skew-symmetrisable* cluster algebras, their combinatorial construction can be expressed using quivers. One surprising result shown by Fomin and Zelevinsky in the first of the series of papers where they introduced cluster algebras is the following classification.

Theorem 2.5 *[52] Let $(Q, \{\underline{x}\})$ be the initial seed of a cluster algebra $\mathcal{A}$. Then $\mathcal{A}$ is of finite type if and only if Q is mutation equivalent to a quiver whose underlying graph is a Dynkin diagram.*

The resemblance of this result with Theorem 2.2 is striking and points towards a deep relationship between cluster theory and representations of finite dimensional algebras.

It is very important to remark that for any seed $(Q', \{\underline{x}'\})$ the set of cluster variables $\{\underline{x}\}$ always has the same number of elements, let's call this number n. Then, all the seeds of a cluster algebra can be arranged into an n-regular graph where there is an edge between two seeds if one can be obtained from the other performing a single mutation.

As it turns out, similar phenomena have been described in tilting theory. For instance it was shown by Skowroński in [101] that every basic tilting module has exactly n indecomposable direct summands. Also, Happel and Unger have shown in [64] that every basic partial tilting module having $n-1$ indecomposable direct summands can always be completed into a tilting module and that there are at most two ways in which this can be done.

Hence, one would like to categorify all the cluster phenomena using tilting theory, where the cluster variables are represented by indecomposable partial tilting modules and tilting modules correspond to seeds. However, tilting theory falls short in describing the cluster phenomena for at least two reasons. The first is that there are some examples of almost complete tilting modules that can be completed into a tilting module in exactly one way, which means that we can not reproduce the process of mutation at some indecomposable direct summand of this module.

The second reason, and maybe the most obvious, is that there are fewer indecomposable partial tilting modules than cluster variables. For instance, a hereditary path algebra of type $\mathbb{A}_n$ has exactly $\frac{n(n-1)}{2}$ indecomposable partial tilting modules, while the number of cluster variables in a cluster algebra of type $\mathbb{A}_n$ is $\frac{n(n+1)}{2}$, i.e. there are exactly n more cluster variables than indecomposable partial tilting modules.

Then if one wants to categorify cluster algebras using tilting theory, it is necessary to extend the latter in some way. That is exactly what Buan, Marsh,

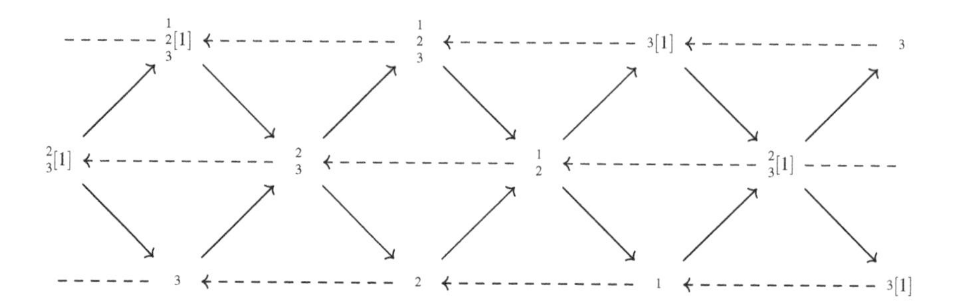

Figure 2.6 The Auslander–Reiten quiver of $\mathcal{C}_A$.

Reineke, Reiten and Todorov did in [39]. In this seminal paper, instead of working with the module category of the algebra, they constructed a slightly larger triangulated category that they called the *cluster category* where everything works perfectly by the definition of the so-called *(partial) cluster-tilting objects*.

See in Figure 2.6 the Auslander–Reiten quiver of the cluster category associated to the algebra A of Example 2.3. The points that are tagged with the same object in the Auslander–Reiten quiver of $\mathcal{C}_A$ should be identified. In particular, we see that the Auslander–Reiten quiver of the cluster category of an algebra of type $\mathbb{A}_3$ is a Möbius strip. In fact the Auslander–Reiten quiver of the cluster category of any algebra of type $\mathbb{A}_n$ is a Möbius strip for every $n \geqslant 2$.

One the one hand, they show that for any orientation of Dynkin quiver there is a one-to-one correspondence between cluster variables and indecomposable partial cluster-tilting objects; that there is a one-to-one correspondence between clusters and cluster-tilting objects; that the mutation is well defined in all the indecomposable direct summands of any cluster-tilting object; and that the mutations of clusters and cluster-tilting objects are compatible. Note that these results were later generalised to the general case [41, 43].

On the other, they showed that there is a natural inclusion of the module category of the path algebra into the cluster category such that every (partial) tilting module in the module category becomes a (partial) cluster-tilting object. Moreover, they show that every possible mutation of tilting modules at the level of the module category becomes a mutation of cluster-tilting modules at the level of the cluster category.

We said before that Happel and Ringel showed that much of the representation theory of tilted algebras can be described from the information we have about the representation theory of the hereditary algebras. Now, the cluster categories associated to hereditary algebras have very nice properties, close to the properties of the hereditary algebras they come from. So Buan, Marsh and Reiten, emulating the construction of tilted algebras, introduced in [40] the *cluster-tilted algebras* as the endomorphism algebras of cluster-tilting objects in a cluster category. In this case, they showed that given a cluster-tilting object T in $\mathcal{C}_A$, the functor $\mathrm{Hom}_{\mathcal{C}_A}(T,-)$ induces an equivalence of categories between mod $(\mathrm{End}_{\mathcal{C}_A}(T))^{op}$ and the quotient of $\mathcal{C}_A$ by the ideal $\mathcal{I}(\tau T)$ of all the morphisms that factor through τT the Auslander–Reiten translation of T.

We have mentioned already that the module category of any hereditary algebra A is naturally immersed in its cluster category $\mathcal{C}_A$. Moreover, if T is a tilting object in mod A, it turns out that T becomes a cluster-tilting object in $\mathcal{C}_A$ when we apply the natural embedding. Then starting from T we can construct a tilted algebra $\mathrm{End}_A(T)$ and a cluster-tilted algebra $\mathrm{End}_{\mathcal{C}_A}(T)$. The relation between $\mathrm{End}_A(T)$ and $\mathrm{End}_{\mathcal{C}_A}(T)$ and their module categories was studied by Assem, Brüstle and Schiffler in a series of papers [13, 14, 15, 16]. Firstly, they showed that one can recover $\mathrm{End}_{\mathcal{C}_A}(T)$ from $\mathrm{End}_A(T)$ via a process that they called *relation extension*, which bypasses the cluster category $\mathcal{C}_A$. Moreover, they have shown that every cluster-tilted algebra is the relation extension of a tilted algebra. They also have characterised all the tilted algebras that have an isomorphic relation extension using particular structures that can be found in the Auslander–Reiten quivers of cluster-tilted algebras which are deeply related to the structures described by Liu [74] and Skowroński [100] for tilted algebras.

In order to start the construction of the cluster category, Buan, Marsh, Reineke, Reiten and Todorov assumed that the quiver in the initial seed of the algebra is acyclic. However, there is no reason why one should start with an acyclic quiver. From a cluster perspective, any quiver is equally valid, so it was expected for a similar cluster category to exist regardless of the quiver we chose at the start. The first problem with the more general quivers arising in cluster theory is that they have cycles, so their path algebras are infinite dimensional. Then in order to use something close to tilting theory, we need to form the quotient of this path algebra by the correct ideal of relations. This problem was solved by Derksen, Weyman and Zelevinsky [48, 49] when they built ideals arising from certain potentials associated to a quiver. They have shown that associated to each quiver there exists a special potential, which they called *non-degenerate*, such that one can categorify the cluster algebra associated to the quiver using their *decorated* representations. Moreover, they went further and showed that there exists a notion of mutation of non-degenerate potentials that is compatible with the cluster mutations of the quivers. We note that the notion of decorated representation gives rise to a rich theory that is closely related to that of τ-tilting theory and it can be considered as a precursor of the latter.

Now that we have the correct algebras associated to the cyclic quivers in cluster theory we would like to have their corresponding cluster categories. To build these categories is not obvious. The main problem being that the construction of Buan, Marsh, Reineke, Reiten and Todorov uses heavily the structure of the derived category of the algebra and some key properties used in their construction fail when the algebra is not hereditary. This problem was overcome by Amiot in [6], where she used the theory of Ginzburg dg-algebras developed by Keller and Yang in [72] to construct a cluster category that is compatible with the other notions of cluster categories existing at that moment.

All the phenomena of cluster algebras and the close parallelism with tilting theory pointed to the existence of another extension of classical tilting theory where we would be always allowed to perform mutations, this time without extending the module category. For hereditary algebras, the construction of this theory was performed by Ingalls and Thomas in [66], where they introduced the so-called *support tilting modules*. To explain the notion of support tilting module, let us come back to the limitations of classical tilting theory.

As we did before consider A to be the path algebra of the linearly oriented $\mathbb{A}_3$ quiver. Then $T = \begin{smallmatrix}1\\2\\3\end{smallmatrix} \oplus {\scriptstyle 1} \oplus {\scriptstyle 3}$ is a tilting module in $\operatorname{mod} A$.

Ideally, given any choice M of an indecomposable direct summand of T, we would like to construct a new tilting module whose indecomposable direct summands other than M are the same as those of T. In other words, we would like to replace each indecomposable direct summand of T by another indecomposable in such a way that the resulting module is again tilting.

The summand ${\scriptstyle 1}$ is replaceable, since we can change it by $\begin{smallmatrix}2\\3\end{smallmatrix}$ to obtain $\begin{smallmatrix}1\\2\\3\end{smallmatrix} \oplus \begin{smallmatrix}2\\3\end{smallmatrix} \oplus {\scriptstyle 3}$ which is tilting.

We can also mutate at the summand ${\scriptstyle 3}$, because it can be replaced by $\begin{smallmatrix}1\\2\end{smallmatrix}$ to obtain the tilting module $\begin{smallmatrix}1\\2\\3\end{smallmatrix} \oplus \begin{smallmatrix}1\\2\end{smallmatrix} \oplus {\scriptstyle 1}$.

However, we cannot replace $\begin{smallmatrix}1\\2\\3\end{smallmatrix}$ by any other indecomposable module to obtain a new tilting module. This is a consequence of a classical result obtained independently by Assem [11] and Smalø [103], which implies that every indecomposable projective-injective object in $\operatorname{mod} A$ is a direct summand of any tilting module in $\operatorname{mod} A$. In particular, $\begin{smallmatrix}1\\2\\3\end{smallmatrix}$ can not be replaced because it is a projective-injective module in $\operatorname{mod} A$.

The solution found by Ingalls and Thomas was to drop $\begin{smallmatrix}1\\2\\3\end{smallmatrix}$ from T altogether to obtain $T' = {\scriptstyle 1} \oplus {\scriptstyle 3}$, which clearly is not tilting. However, it is tilting on its *support algebra*, which is constructed by taking a quotient of A by the ideal generated by the idempotent included in the annihilator $\operatorname{ann} T'$ of T'.

More generally, they showed that for a hereditary algebra the mutation is always possible if we allow our tilting modules not to be supported over every vertex of the algebra.

Now, for more general algebras this construction fails again. For instance, if we take the algebra A to be the path algebra of the quiver

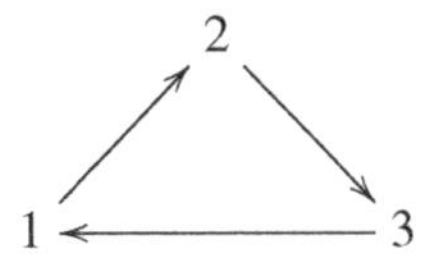

modulo the ideal generated by all paths of length 2, we have that A as a right module over itself is isomorphic to $\begin{smallmatrix}1\\2\end{smallmatrix} \oplus \begin{smallmatrix}3\\1\end{smallmatrix} \oplus \begin{smallmatrix}2\\3\end{smallmatrix}$. Note that in this case every indecomposable projective is also injective. But at the same time we cannot drop any of the direct summands since the sum of the two remaining projective modules is supported on every vertex of the algebra.

Something that we have not said before is that, by construction, in the cluster category we have that

$$\operatorname{Ext}^1_{\mathcal{C}_A}(M,N) \cong \operatorname{Hom}_{\mathcal{C}_A}(N, \tau M).$$

This isomorphism can actually be translated to the module categories of non-hereditary cluster-tilted algebras. So, we can translate the cluster-tilting objects of the cluster category to the module category of a cluster-tilted algebra to get a series of modules that categorify perfectly the corresponding cluster algebra. However, these objects are not in general partial tilting objects because they might be of infinite projective dimension.

Then, Adachi, Iyama and Reiten introduced τ-tilting theory in [1], the object of study of these notes, by dropping the restriction on the projective dimension of the modules into consideration and replacing the classical rigidity with the notion of *τ-rigidity* that we will introduce in the next section. In doing so, as we will see in these notes, Adachi, Iyama and Reiten give a definition which can be easily checked in the module category of every finite dimensional algebra. Moreover, as particular examples of this definition we can find the classical tilting modules and the modules over cluster-tilted algebras that we discussed in the previous paragraph.

Before starting with the material of the lectures notes, we would like to point out that many results of τ-tilting theory were developed independently by Derksen and Fei in [47], where they studied *general presentations* using methods of a more geometric nature.

2.3 τ-tilting Theory: Basic Definitions

In this section we give the basic definitions of τ-tilting theory. We also mention some of the basic relations between τ-tilting theory and classical tilting theory. We start by giving the central definitions of this note: τ-rigid and (support) τ-tilting modules.

Definition 2.6 [1, 26, 27] Let A be an algebra and M be an object in $\operatorname{mod} A$. We say that M is τ-rigid if $\operatorname{Hom}_A(M, \tau M) = 0$.

Definition 2.7 [1] Let A be an algebra. A τ-rigid A-module M is τ-tilting if $|M| = n$. We say that a τ-rigid A-module M is support τ-tilting if there exists an idempotent $e \in A$ such that M is a τ-tilting A/AeA-module, where AeA is the two-sided ideal generated by e in A.

At first glance, τ-tilting and tilting modules have little to do with each other. If we compare Definition 2.4 with Definition 2.7, the only thing that a tilting and a τ-tilting module have in common is that both are A-modules. However, there is a much deeper connection between the two concepts which follows from the so-called *Auslander–Reiten formulas*. Recall that $D(-) := \operatorname{Hom}_k(-, k)$ denotes the classical duality functor, $\mathcal{I}_A(M,N)$ is the vector space of maps from M to N that factor through the injectives A-modules and $\mathcal{P}_A(M,N)$ is the vector space of maps from M to N that factor through the projectives A-modules.

Theorem 2.8 *Let A be an algebra and let M and N be two A-modules. Then there are functorial isomorphisms*

$$\operatorname{Ext}^1_A(M,N) \cong D\left(\frac{\operatorname{Hom}_A(N, \tau M)}{\mathcal{I}_A(N, \tau M)}\right) \cong D\left(\frac{\operatorname{Hom}_A(\tau^- N, M)}{\mathcal{P}_A(\tau^- N, M)}\right).$$

A module M in $\operatorname{mod} A$ is said to be rigid if $\operatorname{Ext}^1_A(M,M) = 0$. An immediate corollary of the Auslander–Reiten formulas is the following.

Corollary 2.9 *Let M be an A-module. If M is τ-rigid then M is rigid.*

In fact this is the first of many other results relating tilting and τ-tilting modules. In the following propositions we compile some properties relating the two notions.

Proposition 2.10 *Let A be an algebra and T be a partial tilting module. Then T is τ-rigid. Moreover, if T is tilting then $|T| = n$.*

Proposition 2.11 *[17] Let M be a τ-rigid module. Then the following hold.*

1 *There are at most n isomorphism classes of indecomposable direct summands of M. In short, $|M| \leqslant n$.*
2 *If the annihilator* $\operatorname{ann}(M)$ *of M is equal to the ideal $\{0\} \subset A$, then M is a partial tilting module.*
3 *If the projective dimension $\operatorname{pd} M$ of M is at most one, then M is a partial tilting module.*
4 *If $|M| = n$ and* $\operatorname{ann}(M) = \{0\}$, *then M is a tilting module.*

Proposition 2.12 *Let A be an algebra. Then an A-module M is tilting if and only if M is τ-tilting and faithful.*

Proposition 2.13 *Let A be an algebra. Then an A-module M is tilting if and only if M is τ-tilting and $\operatorname{pd} M \leqslant 1$.*

The last result can be presented as evidence to the statement that τ-tilting theory is a generalisation of tilting theory, which is independent of the projective dimension of the objects. Following this idea, in the last decade a series of works appeared generalising classical results in tilting theory to τ-tilting theory. Some of these results will be stated in Section 2.5.

However, one needs to be careful when giving such statements, since some results on tilting theory do not hold in the context of τ-tilting theory. Also, before the definition of τ-tilting theory there was at least one other generalisation of tilting theory to higher projective dimensions. We are referring to the generalised tilting modules introduced by Miyashita in [77]. They are defined as follows.

Definition 2.14 [77] Let A be an algebra and T be an A-module and r be a positive integer. We say that T is an r-tilting module if the following holds:

(i) $\operatorname{pd}_A T \leqslant r$, i.e. the projective dimension of T is at most r.
(ii) $\operatorname{Ext}^i_A(T,T) = 0$ for all $1 \leqslant i \leqslant r$.
(iii) There exists a short exact sequence of the form

$$0 \to A \to T^{(1)} \to T^{(2)} \to \cdots \to T^{(r)} \to 0,$$

where $T^{(i)}$ is a direct summand of a direct sum of copies of T for all $1 \leqslant i \leqslant r$.

We now give an example of all the support τ-tilting modules in the module category of an algebra.

Example 2.15 Let A be the path algebra given by the quiver

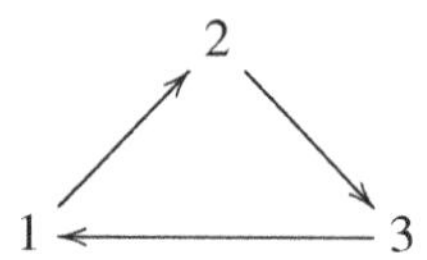

modulo the second power of the ideal generated by all the arrows. The Auslander–Reiten quiver of A can be seen in Figure 2.7.

Note that every module is represented by its Loewy series and both copies of $\substack{3\\1}$ should be identified, so the Auslander–Reiten quiver of A has the shape of a Möbius strip. In the first two columns of Table 2.1 we give a complete list of the support τ-tilting modules in mod A together with their associated idempotents.

Suppose that we are working with the algebra of the previous example and we come across the module $M = \substack{1\\2} \oplus {\scriptstyle 1}$. After a quick calculation we can see that this module is not only τ-rigid, but also support τ-tilting with e_3 as its associated idempotent. But at the same time, M is a direct summand of the support τ-tilting module $\substack{1\\2} \oplus {\scriptstyle 1} \oplus \substack{3\\1}$. So, with this notation we cannot distinguish between the "complete" support τ-tilting module $\substack{1\\2} \oplus {\scriptstyle 1}$ and the "incomplete" τ-rigid module $\substack{1\\2} \oplus {\scriptstyle 1}$.

This can be solved using the notions of τ-rigid and τ-tilting pairs. But before we give their definition, recall that given an idempotent $e \in A$ we have that the right ideal eA is a projective module and that every projective arises this way.

Definition 2.16 Let A be an algebra, M be an A-module and P be a projective module. We say that the pair (M,P) is τ-rigid if M is a τ-rigid module and $\operatorname{Hom}_A(P,M) = 0$. A τ-rigid pair is τ-tilting if $|M| + |P| = n$.

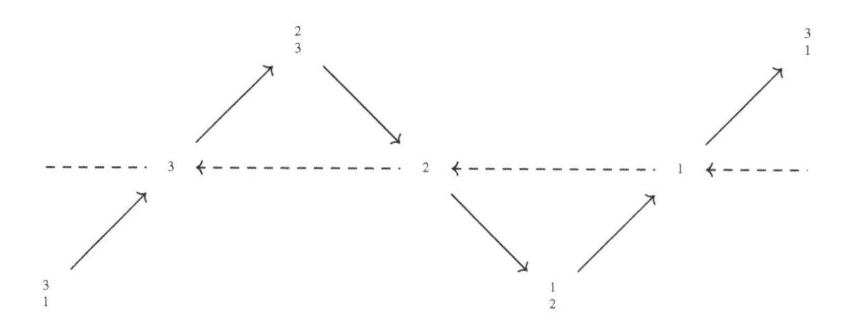

Figure 2.7 The Auslander–Reiten quiver of A.

Table 2.1 *Support τ-tilting modules and τ-tilting pairs in* mod A.

Support τ-tilting module	Idempotent	τ-tilting pair
$\substack{1\\2}\oplus\substack{2\\3}\oplus\substack{3\\1}$	$\varnothing$	$(\substack{1\\2}\oplus\substack{2\\3}\oplus\substack{3\\1},0)$
$\substack{1\\2}\oplus\substack{2\\3}\oplus 2$	$\varnothing$	$(\substack{1\\2}\oplus\substack{2\\3}\oplus 2,0)$
$\substack{1\\2}\oplus\substack{3\\1}\oplus 1$	$\varnothing$	$(\substack{1\\2}\oplus\substack{3\\1}\oplus 1,0)$
$\substack{2\\3}\oplus\substack{3\\1}\oplus 3$	$\varnothing$	$(\substack{2\\3}\oplus\substack{3\\1}\oplus 3,0)$
$\substack{3\\1}\oplus 3$	e_2	$(\substack{3\\1}\oplus 3,\substack{2\\3})$
$\substack{3\\1}\oplus 1$	e_2	$(\substack{3\\1}\oplus 1,\substack{2\\3})$
$\substack{1\\2}\oplus 1$	e_3	$(\substack{1\\2}\oplus 1,\substack{3\\1})$
$\substack{1\\2}\oplus 2$	e_3	$(\substack{2\\3}\oplus 3,\substack{1\\2})$
$\substack{2\\3}\oplus 3$	e_1	$(\substack{2\\3}\oplus 3,\substack{1\\2})$
$\substack{2\\3}\oplus 2$	e_1	$(\substack{2\\3}\oplus 2,\substack{1\\2})$
1	e_2+e_3	$(1,\substack{2\\3}\oplus\substack{3\\1})$
2	e_1+e_3	$(2,\substack{1\\2}\oplus\substack{3\\1})$
3	e_1+e_2	$(3,\substack{1\\2}\oplus\substack{2\\3})$
0	$e_1+e_2+e_3$	$(0,\substack{1\\2}\oplus\substack{2\\3}\oplus\substack{3\\1})$

As you would expect, these two notations are equivalent. Indeed, given a support τ-tiling module M with associated idempotent e we have that (M,eA) is a τ-tilting pair. Conversely, if (M,P) is a τ-tilting pair then we have that $P=eA$ for some idempotent $e\in A$. Then M is a support τ-tilting module with associated idempotent e. The list of all τ-tilting pairs of the algebra in Example 2.15 can be found in the third column of Table 2.1.

2.4 τ-tilting Pairs and Torsion Classes

In this section, after recalling the definition of torsion pairs and their basic properties, we will investigate the deep relation existing between τ-tilting theory and torsion classes.

2.4.1 Torsion Pairs and Torsion Classes

The notion of torsion pairs, also known as torsion theories, started almost with the introduction of abelian categories as a generalisation of a well-known phenomenon in the category of finitely generated abelian groups, one of the most iconic examples of abelian categories.

A classical classification result states that a finitely generated abelian group, up to isomorphism, has a unique torsion subgroup such that the resulting factor group is torsion-free. The extension of this fact to every abelian category was done by Dickson in [50] as follows.

Definition 2.17 [50] Let $\mathcal{A}$ be an abelian category and let $(\mathcal{T},\mathcal{F})$ be a pair of subcategories of $\mathcal{A}$. We say that $(\mathcal{T},\mathcal{F})$ is a torsion pair in $\mathcal{A}$ if the following holds.

1 $\mathrm{Hom}_{\mathcal{A}}(X,Y)=0$ for all $X\in\mathcal{T}$ and $Y\in\mathcal{F}$.
2 For all objects $M\in\mathcal{A}$ there exists, up to isomorphism, a unique short exact sequence

$$0\to tM\to M\to fM\to 0$$

such that tM is an object of $\mathcal{T}$ and fM is an object of $\mathcal{F}$.

If $(\mathcal{T},\mathcal{F})$ is a torsion pair in $\mathcal{A}$ we say that $\mathcal{T}$ is a torsion class and $\mathcal{F}$ is a torsion-free class. Moreover, for each object M of $\mathcal{A}$, we say that

$$0\to tM\to M\to fM\to 0$$

is the canonical short exact sequence of M and that tM is the torsion object of M with respect to the torsion pair $(\mathcal{T},\mathcal{F})$.

The previous definition is valid for an arbitrary abelian category. However, in these notes we are interested in a particular class of abelian categories, namely the categories of finitely generated modules over a finite dimensional algebra. These categories have many extra properties (for example they are length categories) that allow us to describe the torsion pairs more precisely.

Proposition 2.18 *Let A be an algebra. Then the following hold.*

1 *A subcategory $\mathcal{T}$ of $\mathrm{mod}\,A$ is a torsion class if and only if $\mathcal{T}$ is closed under quotients and extensions. Moreover, in this case the torsion-free class associated to $\mathcal{T}$ is*

$$\mathcal{F}=\{Y\in\mathrm{mod}\,A\mid \mathrm{Hom}_A(X,Y)=0\ \textit{for all}\ X\in\mathcal{T}\}.$$

2 *A subcategory $\mathcal{F}$ of $\mathrm{mod}\,A$ is a torsion-free class if and only if $\mathcal{F}$ is closed under subobjects and extensions. Moreover, in this case the torsion-free class associated to $\mathcal{F}$ is*

$$\mathcal{T}:=\{X\in\mathrm{mod}\,A\mid \mathrm{Hom}_A(X,Y)=0\ \textit{for all}\ Y\in\mathcal{F}\}.$$

Suppose that M is an A-module. Then we can ask the following: Is there a minimal torsion class in $\mathrm{mod}\,A$ containing M? The following result answers this question affirmatively.

Proposition 2.19 *Let A be an algebra. Then the intersection of arbitrarily many torsion classes is a torsion class. Likewise, the intersection of arbitrarily many torsion-free classes is a torsion-free class.*

Then, by the previous proposition, the minimal torsion class containing a given object M of $\operatorname{mod} A$ is simply the intersection of all torsion classes containing M. Now, there is a more descriptive answer to this question, but to give that answer we need to introduce some notation.

Let $\mathcal{X}$ be a subcategory of $\operatorname{mod} A$. The category $\operatorname{Filt}(\mathcal{X})$ of objects filtered by $\mathcal{X}$ is defined as the category of all the objects Y in $\operatorname{mod} A$ that admit a filtration

$$0 = Y_0 \subset Y_1 \subset \cdots \subset Y_{r-1} \subset Y_r = Y$$

such that the successive quotients Y_i/Y_{i-1} are objects in $\mathcal{X}$. Note that $\operatorname{Filt}(\mathcal{X})$ is the category of all the objects that can be constructed by making finitely many extensions by objects in $\mathcal{X}$. In other words, $\operatorname{Filt}(\mathcal{X})$ is the extension closure of $\mathcal{X}$.

We define the category $\operatorname{Fac}\mathcal{X}$ as the category of objects Y such that there exists an object X in $\mathcal{X}$ and an epimorphism $p : X \to Y \to 0$. Often in the notes, the category $\mathcal{X}$ we will be the additive category $add\,M$ additively generated by a module M. In this case, by abuse of notation we will write $\operatorname{Fac} M$ instead of $\operatorname{Fac}(add\,M)$. Note that $\operatorname{Fac} M$ can be described as

$$\operatorname{Fac} M = \{Y \in \operatorname{mod} A \mid \exists p : M^r \to Y \to 0 \text{ for some } r \in \mathbb{N}\}.$$

Now we are able to give a better description of the minimal torsion class containing M. As far as we know, the following result was a part of folklore but was first written down formally in [45].

Proposition 2.20 *Let A be an algebra and M be an A-module. Then Filt*$(\operatorname{Fac} M)$ *is the minimal torsion class containing M.*

Remark Note that, in general, $\operatorname{Fac}(\operatorname{Filt} M)$ is **not** a torsion class since it might not be closed under extensions.

2.4.2 Functorially Finite Torsion Pairs and τ-tilting Theory

From the previous subsection we have that to get the minimal torsion class containing M one needs to first calculate $\operatorname{Fac} M$ and then take the extension closure of this category. However, sometimes $\operatorname{Fac} M$ is already closed under extensions, which makes the second step of the construction superfluous.

The following theorem, originally proved by Auslander and Smalø in [26], is arguably the first result on τ-tilting theory, even if this theory was formally introduced thirty years later.

Theorem 2.21 *Let A be an algebra and M be an object in* mod A. *Then* Fac M *is a torsion class if and only if M is τ-rigid. Moreover, in this case*

$$M^{\perp} := \{X \in \operatorname{mod} A \mid \operatorname{Hom}_A(M,X) = 0\}$$

is the torsion-free class such that (Fac $M, M^{\perp}$) *is a torsion pair in* mod A.

As we just said, many years passed between the publication of this result and the start of τ-tilting theory as an independent subject in representation theory. However, this was not the only result that worked with τ-rigid objects. In fact, a well-established technique used in classical tilting theory to determine if an object is tilting was to show that the candidate M was a τ-rigid module such that $pd\,M \leqslant 1$ and $|M| = n$. The interested reader is encouraged to surf the literature to look for such examples.

From a torsion theoretic point of view, the breakthrough made by Adachi, Iyama and Reiten in [1] is that they showed that τ-tilting pairs characterised a particular class of torsion classes, the *functorially finite* torsion classes.

Let $\mathcal{X}$ be a subcategory of an abelian category $\mathcal{A}$ and suppose that X is an object of $\mathcal{X}$ and M is an arbitrary object of $\mathcal{A}$. A morphism $f : X \to M$ is called a *right* $\mathcal{X}$ *-approximation* of M if any map $f' : X' \to M$ with $X' \in \mathcal{X}$ factors through f. Dually, a morphism $g : M \to X$ is called a *left* $\mathcal{X}$*-approximation* of M if any map $g' : M \to X'$ with $X' \in \mathcal{X}$ factors through g. We say that $\mathcal{X}$ is *contravariantly finite* (resp. *covariantly finite*) if any object M in $\mathcal{A}$ admits a right (resp. left) $\mathcal{X}$-approximation. We say that $\mathcal{X}$ is functorially finite if it is both contravariantly finite and covariantly finite.

An important consequence of the uniqueness up to isomorphism of the canonical exact sequence of an object with respect to a torsion pair is the following.

Proposition 2.22 *Let* $(\mathcal{T}, \mathcal{F})$ *be a torsion pair in an abelian category* $\mathcal{A}$ *and let* M *be an object of* $\mathcal{A}$. *If*

$$0 \to tM \to M \to fM \to 0$$

is the short exact sequence of M *with respect to* $(\mathcal{T}, \mathcal{F})$ *then the canonical inclusion* $i : tM \to M$ *is a right* $\mathcal{T}$*-approximation and the canonical projection* $p : M \to fM$ *is a left* $\mathcal{F}$*-approximation. In particular every torsion class in* $\mathcal{A}$ *is contravariantly finite and every torsion-free class* $\mathcal{F}$ *in* $\mathcal{A}$ *is covariantly finite.*

Given a τ-rigid module M, we know by Theorem 2.21 that Fac$M$ is a torsion class. In fact, it turns out that Fac$M$ is functorially finite, as shown by Auslander and Smalø in [26]. Before we give the precise statement of the theorem, we will introduce some notation that will be useful in the rest of the chapter. Given a subcategory $\mathcal{X}$ of mod A, we say that an object M in $\mathcal{X}$ is *Ext-projective* if

$\mathrm{Ext}^1_A(M,X) = 0$ for every object X in $\mathcal{X}$. For every functorially finite torsion class $\mathcal{T}$ of $\mathrm{mod}\, A$ we define $\mathcal{P}(\mathcal{T})$ to be $\mathcal{P}(\mathcal{T}) := T^0_A \oplus T^1_A$, where $f_A : A \to T^0_A$ is the minimal left $\mathcal{T}$-approximation of A as an object of $\mathrm{mod}\, A$ and T^1_A is the cokernel of f_A.

Theorem 2.23 *[26] Let $\mathcal{T}$ be a functorially finite torsion class in* $\mathrm{mod}\, A$. *Then $\mathcal{P}(\mathcal{T})$ is an Ext-projective object in $\mathcal{T}$ such that T is an object in* $\mathrm{add}(\mathcal{P}(\mathcal{T}))$ *for all Ext-projective modules T in $\mathcal{T}$. Moreover $\mathcal{T} = \mathrm{Fac}\, \mathcal{P}(\mathcal{T})$. In particular, $\mathcal{P}(\mathcal{T})$ is a τ-rigid A-module.*

This last result implies that every functorially finite torsion class is generated by a τ-rigid module. This defines a well-defined map $\Phi : \tau\text{-rp-}A \to \text{ftors-}A$ from the set τ-rp-A of all τ-rigid pairs to the set ftors-A of functorially finite torsion classes. The main contribution of [1] to this problem is the proof of the fact that this map is a bijection if we consider restricting Φ to the set $\tau\text{-tp-}A \subset \tau\text{-rp-}A$ of τ-tilting pairs. The precise statement is the following.

Theorem 2.24 *[1] Let A be an algebra. Then the map $\Phi : \tau\text{-tp-}A \to \text{ftors-}A$ defined by*

$$\Phi(M,P) = \mathrm{Fac}\, M$$

is a bijection. Moreover, the inverse $\Phi^{-1} : \text{ftors-}A \to \tau\text{-tp-}A$ is defined as

$$\Phi^{-1}(\mathcal{T}) = (\mathcal{P}(\mathcal{T}), {}^{\perp_P}\mathcal{T}),$$

where ${}^{\perp_P}\mathcal{T}$ is a basic additive generator of the category of projective modules P such that $\mathrm{Hom}_A(P,T) = 0$ *for all $T \in \mathcal{T}$.*

2.5 A (τ-)tilting Theorem

We have mentioned in Section 2.2 that the term *tilting theory* was coined by Brenner and Butler, who showed in [34] what is now known as the tilting theorem. In this section we give a precise statement of the tilting theorem. Afterwards we show how this result can be generalised to τ-tilting theory and mention some of its limits.

2.5.1 A Tilting Theorem

We start by recalling the definition of a tilting module.

Definition 2.25 [34] Let A be an algebra and T be an A-module. We say that T is a tilting module if the following holds:

(i) $\mathrm{pd}_A T \leqslant 1$, the projective dimension of T is at most 1.
(ii) T is rigid, that is $\mathrm{Ext}^1_A(T,T)=0$.
(iii) There exists a short exact sequence of the form

$$0 \to A \to T' \to T'' \to 0,$$

where T', T'' are direct summands of direct sums of T.

It follows from Proposition 2.10 that every tilting module T is τ-tilting. Then we have that T has a torsion pair associated to it, namely $(\mathrm{Fac}\, T, T^{\perp})$. Now, the torsion class $\mathrm{Fac}\, T$ can be characterised homologically as follows.

Proposition 2.26 *Let T be a tilting module. Then*

$$\mathrm{Fac}\, M = \left\{X \in \mathrm{mod}\, A \mid \mathrm{Ext}^1_A(T,X)=0\right\}.$$

Now, for every object M of $\mathrm{mod}\, A$ we have that $\mathrm{End}_A(M)$ is a finite dimensional algebra. In this case, M has a natural structure of a left $\mathrm{End}_A(M)$-module structure. For the rest of the section we denote by $B := \mathrm{End}_A(T)$ the endomorphism algebra of a tilting module T. The following proposition indicates the importance of tilting objects.

Proposition 2.27 *[34] Let T be a tilting A-module. Then T is tilting as a left B-module. Moreover, T induces a torsion pair $(\mathcal{X}(T), \mathcal{Y}(T))$ in the category* $\mathrm{mod}\, B$ *of right B-modules where*

$$\mathcal{X}(T) := \{X \in \mathrm{mod}\, B \mid X \otimes_B T = 0\},$$

$$\mathcal{Y}(T) := \{Y \in \mathrm{mod}\, B \mid \mathrm{Tor}^B_1(Y,T) = 0\}.$$

We are now able to state the tilting theorem of Brenner and Butler.

Theorem 2.28 *[34] Let A be an algebra, T be a tilting A-module and $B =$* $\mathrm{End}_A(T)$. *Then the following hold.*

1 *The algebra* $\mathrm{End}^{op}_B(T)$ *is isomorphic to A.*
2 *The functor* $\mathrm{Hom}_A(T,-) : \mathrm{Fac}\, T \to \mathcal{Y}(T)$ *is an equivalence of categories with quasi-inverse* $-\otimes_B T : \mathcal{Y}(T) \to \mathrm{Fac}\, T$.
3 *The functor* $\mathrm{Ext}^1_A(T,-) : T^{\perp} \to \mathcal{X}(T)$ *is an equivalence of categories with quasi-inverse* $\mathrm{Tor}^B_1(-,T) : \mathcal{X}(T) \to T^{\perp}$.

2.5.2 A τ-tilting Theorem

We have said in various places that τ-tilting theory can be seen as a generalisation of tilting theory. Hence, one would expect the existence of a generalisation of the tilting theorem to τ-tilting theory. This was achieved in [105] building on the results of [68] that we discuss in the next section. Before stating the result, we need to recall some basic facts. The first key observation is that if A and C are algebras such that C is a quotient of A, then $\operatorname{mod} C$ is a full subcategory of $\operatorname{mod} A$. This implies immediately that $\operatorname{End}_A(M) \cong \operatorname{End}_C(M)$ for every C-module M.

In this section, we fix a τ-tilting module T, we denote by $\operatorname{ann} T$ the annihilator of T and by $C := A/\operatorname{ann} T$ the quotient algebra of A by $\operatorname{ann} T$.

Proposition 2.29 *If T is a τ-tilting A-module, then T is a tilting C-module.*

As a consequence of this proposition and Theorem 2.25 we have the existence of a torsion pair $((\operatorname{Fac} T)_C, (T^{\perp})_C)$ in $\operatorname{mod} C$ and a torsion pair $(\mathcal{X}(T), \mathcal{Y}(T))$ in $\operatorname{mod} B$ and equivalences of categories between them. Now, we would like to compare the torsion pair $((\operatorname{Fac} T)_A, (T^{\perp})_A)$ induced by T in $\operatorname{mod} A$ with $(\mathcal{X}(T), \mathcal{Y}(T))$ in $\operatorname{mod} B$. The main ingredient to do that comes from the following two results.

Proposition 2.30 *[1] Let T be a τ-tilting module. Then $T^{\perp} = \operatorname{Sub}(\tau T)$.*

Proposition 2.31 *Let C be a quotient algebra of A and let M be a C-module. Then the Auslander–Reiten translation $\tau_C M$ of M in $\operatorname{mod} C$ is a submodule of the Auslander–Reiten translation $\tau_A M$ of M in $\operatorname{mod} A$.*

Hence, if T is a τ-tilting A-module we have that the torsion pair $(\operatorname{Fac} T, T^{\perp})$ in $\operatorname{mod} A$ coincides with the torsion pair $(\operatorname{Fac} T, T^{\perp})$ in $\operatorname{mod} C$ if and only if $\tau_A M \cong \tau_C M$. Then a τ-tilting version of the tilting theorem of Brenner and Butler reads as follows.

Theorem 2.32 *[105] Let A be an algebra, T be a τ-tilting A-module, $B = \operatorname{End}_A(T)$ and $C = A/\operatorname{ann} T$. Then the following hold.*

1. *The algebra $\operatorname{End}_B^{op}(T)$ is isomorphic to C.*
2. *The functor $\operatorname{Hom}_A(T, -) : \operatorname{Fac} T \to \mathcal{Y}(T)$ is an equivalence of categories with quasi-inverse $- \otimes_B T : \mathcal{Y}(T) \to \operatorname{Fac} T$.*
3. *The functor $\operatorname{Ext}_A^1(T, -) : T^{\perp} \to \mathcal{X}(T)$ is an equivalence of categories with quasi-inverse $\operatorname{Tor}_1^B(-, T) : \mathcal{X}(T) \to T^{\perp}$ if and only if $\tau_A T \cong \tau_C T$.*

Remark We note that there are examples of τ-tilting pairs such that $\tau_A T \cong \tau_C T$, such as the so-called τ-slices introduced in [105].

We also note that there is another generalisation of the tilting theorem of Brenner and Butler for 2-term silting objects proved by Buan and Zhou in [42].

2.6 τ-tilting Reduction

In the last section we explained the deep relationship between τ-tilting pairs and torsion pairs. In this section we will consider the problem of completing a τ-rigid pair, which we consider in two steps. In the first subsection we will show that there are two torsion pairs which are naturally associated to every τ-rigid pair. In the second subsection we give a characterisation of all the completions of a τ-rigid pair developed in [68]. We finish this section by mentioning some bijections between the torsion classes of different categories.

2.6.1 The Bongartz Completion of a τ-rigid Pair

The choice of taking τ-rigid modules to develop τ-tilting theory is arbitrary. In fact, using τ^{-1}-rigid modules, that is, modules N such that $\mathrm{Hom}_A(\tau^{-1}N, N) = 0$, we can develop a completely dual τ^{-1}-tilting theory. See [1]. In this case, the dual of Theorem 2.21 reads as follows.

Theorem 2.33 *[26] Let A be an algebra and N be an object in* mod *A. Then the category*

$$\mathrm{Sub}N := \{Y \in \mathrm{mod}\,A \mid \exists i : 0 \to Y \to N^r \textit{ for some } r \in \mathbb{N}\}$$

is a torsion-free class if and only in N is τ^{-1}-rigid. Moreover, in this case

$$^{\perp}N := \{X \in \mathrm{mod}\,A : \mathrm{Hom}_A(X, N) = 0\}$$

is the torsion class such that $({}^{\perp}N, \mathrm{Sub}N)$ is a torsion pair in mod *A.*

Now, take a non-projective τ-rigid module M. Then it is easy to see that τM is τ^{-1}-rigid. Indeed,

$$\mathrm{Hom}_A(\tau^{-1}\tau M, \tau M) = \mathrm{Hom}_A(M, \tau M) = 0.$$

Hence there are two torsion classes naturally associated to M, namely $\mathrm{Fac}M$ and ${}^{\perp}\tau M$. In the following theorem we give some results regarding the relation between these two torsion classes and M, all of which appeared already in [1].

Theorem 2.34 *[1] Let A be an algebra and M be a τ-rigid A-module. Then the following holds.*

1 $\mathrm{Fac}\,M \subset {}^{\perp}\tau M$.

2 *The torsion classes* Fac M *and* $^{\perp}\tau M$ *coincide if and only if* M *is* τ*-tilting.*
3 *Suppose that* $\mathcal{T}$ *is a functorially finite torsion class. Then* M *is a direct summand of* $\mathcal{P}(\mathcal{T})$ *if and only if* Fac $M \subset \mathcal{T} \subset {}^{\perp}\tau M$.

From the previous theorem we have that $^{\perp}\tau M$ is the maximal torsion class having M as an Ext-projective, which makes the τ-tilting module $\mathcal{P}(^{\perp}\tau M)$ special enough to have a name. We say that $\mathcal{P}(^{\perp}\tau M)$ is the *Bongartz completion* of M. This name was chosen because $\mathcal{P}(^{\perp}\tau M)$ plays an analogous role in τ-tilting theory to that of the Bongartz completion in the classical tilting theory. To be more precise, Bongartz showed in [33] that if T is a partial tilting module, then $\mathcal{P}(^{\perp}\tau T)$ is a tilting module having T as a direct summand. In other words, Bongartz showed that every partial tilting module can be completed to a tilting module.

If we use the language of τ-rigid pairs instead of τ-rigid modules we can be more precise in our statements. From now on, by abuse of notation, we say that a τ-rigid pair (M,P) is a direct summand of (M',P') if M is a direct summand of M' and P is a direct summand of P'.

Theorem 2.35 *[1] Let* A *be an algebra and* (M,P) *be a* τ*-rigid pair in* mod A. *Then the following hold.*

1 $^{\perp}\tau M \cap P^{\perp}$ *is a torsion class and* Fac $M \subset {}^{\perp}\tau M \cap P^{\perp}$.
2 *The torsion classes* $M^{\perp}$ *and* $^{\perp}\tau M \cap P^{\perp}$ *coincide if and only if* (M,P) *is a* τ*-tilting pair.*
3 *Suppose that* $\mathcal{T}$ *is a functorially finite torsion class. Then* (M,P) *is a direct summand of* $\Phi^{-1}(\mathcal{T})$ *if and only if* Fac $M \subset \mathcal{T} \subset {}^{\perp}\tau M \cap P^{\perp}$.

As for τ-rigid modules, we say that $\Phi^{-1}(^{\perp}\tau M \cap P^{\perp})$ is the Bongartz completion of (M,P). But now we can also compute the τ-tilting pair $\Phi^{-1}(\text{Fac}\, M)$, which is the τ-tilting pair generating the smallest torsion class containing M. In this case, we say that $\Phi^{-1}(\text{Fac}\, M)$ is the Bongartz cocompletion of (M,P).

2.6.2 τ-tilting Reduction and Torsion Classes

In this subsection we consider the problem of finding all τ-tilting pairs having a given τ-rigid pair (M,P) as a direct summand. This problem was solved by Jasso in [68] using a procedure that is now known as τ*-tilting reduction*. Here we give a brief summary of that process.

By Theorem 2.35 one knows that (M,P) yields the torsion classes Fac M and $^{\perp}(\tau M) \cap P^{\perp}$. Moreover, Theorem 2.35 states the existence of a τ-tilting pair of the form $(M \oplus M',P)$ such that $\text{Fac}(M \oplus M') = {}^{\perp}(\tau M) \cap P^{\perp}$.

Now define $B_{(M,P)} = \mathrm{End}_A(M \oplus M')$ to be the endomorphism algebra of $M \oplus M'$. In the algebra $B_{(M,P)} = \mathrm{End}_A(M \oplus M')$, there is an idempotent element $e_{(M,P)}$ associated to the $B_{(M,P)}$-projective module $\mathrm{Hom}_A(M \oplus M', M)$. We define the algebra $\tilde{B}_{(M,P)}$ as the quotient of $B_{(M,P)}$ by the ideal generated by $e_{(M,P)}$, that is,

$$\tilde{B}_{(M,P)} := B_{(M,P)}/B_{(M,P)}e_{(M,P)}B_{(M,P)}.$$

Now we are able to state one of the main results of [68].

Theorem 2.36 *[68] Let (M,P) be a τ-rigid pair in* mod A. *Then the functor*

$$\mathrm{Hom}_A(M \oplus M', -) : \mathrm{mod}\,A \to \mathrm{mod}\,B_{(M,P)}$$

induces an equivalence of categories

$$F : M^{\perp} \cap {}^{\perp}\tau M \cap P^{\perp} \to \mathrm{mod}\,\tilde{B}_{(M,P)}$$

between the perpendicular category $M^{\perp} \cap {}^{\perp}\tau M \cap P^{\perp}$ of (M,P) and the module category $\mathrm{mod}\,\tilde{B}_{(M,P)}$.

As a direct consequence of Theorem 2.36 and Theorem 2.35 we obtain the following result.

Theorem 2.37 *[68] Let (M,P) be a τ-rigid pair in* mod A *and $\tilde{B}_{(M,P)}$ as above. Then the functor*

$$\mathrm{Hom}_A(M \oplus M', -) : \mathrm{mod}\,A \to \mathrm{mod}\,B_{(M,P)}$$

induces a bijection between the torsion classes $\mathcal{T}$ in mod A *such that* Fac $M \subset \mathcal{T} \subset {}^{\perp}\tau M \cap P^{\perp}$ *and the torsion classes in* $\mathrm{mod}\,\tilde{B}_{(M,P)}$.

In particular the functor

$$\mathrm{Hom}_A(M \oplus M', -) : \mathrm{mod}\,A \to \mathrm{mod}\,B_{(M,P)}$$

induces a bijection between the τ-tilting pairs in mod A *having (M,P) as a direct summand and the τ-tilting pairs in* $\mathrm{mod}\,\tilde{B}_{(M,P)}$.

Remark Note that Theorem 2.37 does not give a specific number of completions of a given τ-rigid pair (M,P). This is due to the fact that the number of τ-tilting pairs in two algebras might differ hugely.

2.6.3 Bijections of Torsion Classes

Note that the perpendicular category $M^{\perp} \cap {}^{\perp}\tau M \cap P^{\perp}$ defined by Jasso is at the intersection of the torsion class ${}^{\perp}\tau M \cap P^{\perp}$ with the torsion-free class $M^{\perp}$. Moreover, in this case $M^{\perp} \cap {}^{\perp}\tau M \cap P^{\perp}$ is what is called a *wide subcategory* of mod A. A subcategory $\mathcal{X}$ is called wide when it is closed under kernels, cokernels and extensions. In particular, this implies that $\mathcal{X}$ is an abelian category. Then Asai and Pfeiffer found in [10] the following generalisation of Theorem 2.37.

Theorem 2.38 *[10] Let $(\mathcal{T}_1, \mathcal{F}_1)$ and $(\mathcal{T}_2, \mathcal{F}_2)$ be two torsion pairs in* mod A *such that $\mathcal{T}_1 \subset \mathcal{T}_2$. Suppose moreover that $\mathcal{T}_2 \cap \mathcal{F}_1$ is a wide subcategory of* mod A. *Then there is a bijection between the torsion classes $\mathcal{T}$ in* mod A *such that $\mathcal{T}_1 \subset \mathcal{T} \subset \mathcal{T}_2$ and the torsion classes in $\mathcal{T}_2 \cap \mathcal{F}_1$ given by map $\mathcal{T} \mapsto \mathcal{T} \cap \mathcal{T}_1$.*

Note that the intersection of a torsion class with a torsion-free class is not always a wide subcategory. However, it does always have some structure, namely that of a *quasi-abelian* subcategory [104]. The definition of quasi-abelian subcategories is a bit technical and it will be skipped.

However, it is worth mentioning that Tattar showed in [104] that there is a well-defined notion of torsion classes in quasi-abelian subcategories. Moreover he showed that Theorem 2.37 can be generalised to this setting as follows.

Theorem 2.39 *[104] Let $(\mathcal{T}_1, \mathcal{F}_1)$ and $(\mathcal{T}_2, \mathcal{F}_2)$ be two torsion pairs in* mod A *such that $\mathcal{T}_1 \subset \mathcal{T}_2$. Then there is a bijection between the torsion classes $\mathcal{T}$ in* mod A *such that $\mathcal{T}_1 \subset \mathcal{T} \subset \mathcal{T}_2$ and the torsion classes in $\mathcal{T}_2 \cap \mathcal{F}_1$ given by map $\mathcal{T} \mapsto \mathcal{T} \cap \mathcal{T}_1$.*

2.7 Torsion Classes, Wide Subcategories and Semibricks

We have seen already a bijection between τ-tilting pairs and functorially finite torsion classes. In this section we will see the relation of τ-tilting theory with two other notions, namely wide subcategories and semibricks, which were described by Marks and Stovicek [75] and Asai [8], respectively.

2.7.1 Torsion Classes and Wide Subcategories

We start this subsection by recalling the definition of *wide subcategories.*[1]

Definition 2.40 A subcategory $\mathcal{W}$ of mod A is said to be *wide* if it is closed under extensions, kernels and cokernels.

[1] We give here the definition of wide subcategories as usually known in the context of representation theory of algebras. We warn the reader that in algebraic topology the same terminology has a different meaning.

The idea of this section is to describe maps between the set tors-A of torsion classes in $\operatorname{mod} A$ and set wide-A of wide subcategories in $\operatorname{mod} A$. Going from wide-A to tors-A, the map that we take is quite natural. Indeed we define

$$T(-) : \text{wide-}A \to \text{tors-}A,$$

where $T(\mathcal{W})$ is the minimal torsion class in $\operatorname{mod} A$ containing $\mathcal{W}$. We recall from Proposition 2.20 that $T(\mathcal{W}) = \operatorname{Filt}(\operatorname{Fac}\mathcal{W})$. Actually this map has a really nice property.

Proposition 2.41 *[75] The map $T(-) : \text{wide-}A \to \text{tors-}A$ is injective.*

On the other direction we need to build a map $\alpha(-) : \text{tors-}A \to \text{wide-}A$. Given a torsion class $\mathcal{T}$ we define the subcategory $\alpha(\mathcal{T})$ of $\operatorname{mod} A$ as follows.

$$\alpha(\mathcal{T}) := \{X \in \mathcal{T} \mid \ker f \in \mathcal{T} \text{ for all } f \in \operatorname{Hom}_A(Y,X) \text{ with } Y \in \mathcal{T}\}.$$

Proposition 2.42 *[75] Let $\mathcal{T}$ be a torsion class in* $\operatorname{mod} A$. *Then $\alpha(\mathcal{T})$ is a wide subcategory of* $\operatorname{mod} A$.

Inside tors-A there is the subset ftors-A of functorially finite torsion classes, that we have already discussed before. Likewise, inside wide-A there we have the subset of fwide-A of wide subcategories W of $\operatorname{mod} A$ such that they are functorially finite and such that $T(W)$ is a functorially finite torsion class. The following theorem indicates what happens if we restrict the maps $\alpha(-)$ and $T(-)$ to these distinguished subsets.

Theorem 2.43 *[75] The map $T(-) : \text{fwide-}A \to \text{ftors-}A$ is a bijection between the set fwide-A and ftors-A with inverse $\alpha(-) : \text{ftors-}A \to \text{fwide-}A$.*

As an immediate consequence of the previous theorem and Theorem 2.24 we obtain the following corollary.

Corollary 2.44 *For every algebra A there is a one-to-one correspondence between the set τ-tp-A of τ-tilting pairs and the set fwide-A of functorially finite wide subcategories of* $\operatorname{mod} A$.

2.7.2 Semibricks

Let us recall the classical notion of bricks and the more novel notion of semibricks introduced by Asai in [8].

Definition 2.45 We say that an object B in $\operatorname{mod} A$ is a *brick* if its endomorphism algebra $\operatorname{End}_A(B)$ is a division ring. A set $\{B_1, \dots, B_t\}$ of bricks in $\operatorname{mod} A$ is said to be a semibrick if $\operatorname{Hom}_A(B_i, B_j) = 0$ if i is different from j.

The classical example of a semibrick is the set $\{S(1),\dots,S(n)\}$ of non-isomorphic simple A modules. Indeed, the classical Schur's lemma implies that $\{S(1),\dots,S(n)\}$ is a semibrick. In fact, this is a particular case of a more general phenomenon. Recall that an object M in a subcategory $\mathcal{X}$ of mod A is said to be *relatively simple* if the only submodules of M that are in $\mathcal{X}$ are 0 and M.

Let $\mathcal{W}$ be a wide subcategory of mod A and let B be a relative simple object in $\mathcal{W}$. Then it is easy to see that S is necessarily a brick. Indeed, if $f \in \mathrm{End}_A(S)$ then im$f$ is a subobject of B which is in $\mathcal{W}$ because $\mathcal{W}$ is wide. Hence f is either the zero morphism or an isomorphism. A similar argument shows that the set

$$\mathcal{S}(\mathcal{W}) = \{B \mid B \text{ is relative simple in } \mathcal{W}\}$$

is a semibrick. On the other hand, one can see that given a semibrick $\mathcal{S}$ the category Filt($\mathcal{S}$) is wide. In the following result sbrick-A denotes the set of all semibricks in mod A.

Theorem 2.46 *[8] The map*

$$\mathcal{S}(-) : \textit{wide-A} \to \textit{sbrick-A}$$

is a bijection with inverse

$$\textit{Filt}(-) : \textit{sbrick-A} \to \textit{wide-A}.$$

Inside of sbrick-A there are semibricks $\mathcal{S}$ such that the minimal torsion class $T(\mathcal{S})$ containing $\mathcal{S}$ is functorially finite. We denote the set of all such semibricks by fsbrick-A. Combining the last theorem with the results of the previous subsection we obtain the following.

Theorem 2.47 *There are bijections between the sets τ-tp-A, ftors-A, fwide-A and fsbrick-A.*

There are several things that are worth mentioning here. Firstly, the bijections established in Theorem 2.47 are a very small subset of the existing bijections of interesting representation theoretic objects. A very nice paper that covered many of these bijections is the survey article [38] by Brüstle and Yang. We choose to mention these here (and we will mention some other bijections later) because they were shown after the latest update of [38] and are not included there.

We also want to emphasise that in the development of τ-tilting theory a choice was made of working with the Auslander–Reiten translation τ and torsion classes in mod A, instead of working with the inverse Auslander–Reiten translation τ^- and torsion free classes. However, the corresponding dual statements for these results hold. In particular this adds many more bijections to this theory.

Finally, it is worth mentioning that Asai showed in [8] an explicit bijection between fsbricks-A and τ-tp-A. This result was later recovered in [107] using the notions of c-vectors that will be discussed in a later section.

2.8 The Poset Tors-A

The set tors-A of all torsion classes in mod A has a natural poset structure, where the order is given by inclusion. In other words, given a pair of torsion classes $\mathcal{T}$ and $\mathcal{T}'$ we say that $\mathcal{T} \leqslant \mathcal{T}'$ if $\mathcal{T} \subset \mathcal{T}'$. In this section we show some of the basic properties of this poset.

2.8.1 tors-A is a Complete Lattice

The aim of this subsection is to explain its title. In order to do that we need to recall the definitions of a lattice and a complete lattice.

Definition 2.48 A poset $\mathcal{P}$ is a *lattice* if any two elements $x, y \in \mathcal{P}$ admit a greatest common lower bound $x \wedge y$, known as the *meet* of x and y, and a least common upper bound $x \vee y$, known as the *join* of x and y. A lattice $\mathcal{P}$ is said to be *complete* if every subset S of $\mathcal{P}$ admits a lowest common greater bound $\bigvee_{x \in S} x$ (i.e. *a join*) and a greatest lower bound $\bigwedge_{x \in S} x$ (i.e. a *meet*).

There has been a lot of work studying the lattice theoretic properties of the set tors-A for an algebra A. One of the most important results in this direction was obtained by Demonet, Iyama, Reading, Reiten and Thomas in [46].

Theorem 2.49 *[46] The set tors-A is a complete lattice for every algebra A. In this case, given a subset $S \subset$ tors-A the meet and the join of S are defined as follows.*

$$\bigwedge_{\mathcal{T} \in S} \mathcal{T} := \bigcap_{\mathcal{T} \in S} \mathcal{T}$$

$$\bigvee_{\mathcal{T} \in S} \mathcal{T} := T\left(\bigcup_{\mathcal{T} \in S} \mathcal{T}\right)$$

Remark We note that the union of torsion classes is not always a torsion class since in general this union is not closed under extensions or quotients. That is why we need to consider the minimal torsion class containing the union.

Now, suppose that C is a quotient algebra of A. The following theorem tells us how tors-A and tors-C are related.

Theorem 2.50 *If A is an algebra and C is a quotient of A, then there is an epimorphism of lattices p : tors-A $\to$ tors-C. In other words, the lattice tors-C is a quotient of the lattice tors-A.*

One of the main ingredients of the proof of this theorem is the so-called *brick labelling* that we will discuss in the next subsection.

2.8.2 The Hasse Quiver of tors-A

We now shift the focus of our attention to the Hasse diagram of tors-A. In order to do that, let us recall the notion of the Hasse quiver of a poset.

Definition 2.51 Given a poset $\mathcal{P}$, the Hasse quiver $\mathbf{H}(\mathcal{P})$ of $\mathcal{P}$ is an oriented graph whose vertices correspond to the elements of $\mathcal{P}$ and there is an arrow $x \to y$ if $y \leqslant x$ such that $y \leqslant z \leqslant x$ implies that $x = z$ or $y = z$.

In particular the arrows of the Hasse quiver $\mathbf{H}(\text{tors-}A)$ of tors-A correspond to the *maximally included* torsion classes. That is, for two torsion classes $\mathcal{T}$ and $\mathcal{T}'$ we say that $\mathcal{T}$ is maximally included in $\mathcal{T}'$ if $\mathcal{T} \subset \mathcal{T}'$ and $\mathcal{T} \subset \mathcal{T}'' \subset \mathcal{T}'$ implies that $\mathcal{T}'' = \mathcal{T}$ or $\mathcal{T}'' = \mathcal{T}'$. It turns out that maximal inclusions of torsion classes have a very nice characterisation, as shown by Barnard, Carrol and Zhu in [28].

Theorem 2.52 *[28] Let $\mathcal{T}$ and $\mathcal{T}'$ be two torsion classes such that $\mathcal{T}$ is maximally included in $\mathcal{T}'$. Then there exists a brick B in* $\operatorname{mod} A$ *such that $\mathcal{T}' =$* $\operatorname{Filt}(\mathcal{T} \cup \{B\})$. *In this case, we say that B is the minimal extending module of the inclusion $\mathcal{T} \subset \mathcal{T}'$.*

A direct consequence of this result is the following corollary, which is usually known as the *brick labelling* of $\mathbf{H}(\text{tors-}A)$.

Corollary 2.53 *There is a well-defined labelling of the arrows of $\boldsymbol{H}$(tors-A) by bricks in* $\operatorname{mod} A$, *where we label each arrow $\mathcal{T}' \to \mathcal{T}$ of $\boldsymbol{H}$(tors-A) with the minimal extending module corresponding to the inclusion $\mathcal{T} \subset \mathcal{T}'$.*

Remark We note that brick labelling of $\mathbf{H}(\text{tors-}A)$ (or parts of it) can also be deduced as a consequence of several independent works that appeared simultaneously, namely [8, 36, 46, 107].

We have that the set of functorially finite torsion classes ftors-A is a subset of tors-A. So the question is how their Hasse quivers $\mathbf{H}(\text{ftors-}A)$ and $\mathbf{H}(\text{tors-}A)$, respectively, are related. In order to answer that question we need to know what happens when we have a maximal inclusion of torsion classes $\mathcal{T} \subset \mathcal{T}'$ such that either $\mathcal{T}$ or $\mathcal{T}'$ are functorially finite. A complete answer to this question is a consequence of the work of Demonet, Iyama and Jasso [45].

Theorem 2.54 *[45] Let $\mathcal{T} \subset \mathcal{T}'$ be a minimal inclusion of torsion classes in tors-A. Then $\mathcal{T}$ is functorially finite if and only if $\mathcal{T}'$ is functorially finite.*

A direct consequence of the previous theorem is the following.

Corollary 2.55 *The Hasse quiver $\boldsymbol{H}$(ftors-A) of ftors-A and the Hasse quiver $\boldsymbol{H}$(tors-A) of tors-A are locally isomorphic.*

2.9 Mutations of τ-tilting Pairs and Maximal Green Sequences

As we said in in the introduction of these notes, τ-tilting theory was conceived with the goal of completing the classical tilting theory with respect to mutation. In this section we start by discussing the notion of mutation from a torsion theoretic perspective. We finish it by speaking about the notion of maximal green sequences.

2.9.1 Mutations of τ-tilting Pairs

In order to speak about mutation we need to introduce a bit of notation.

Definition 2.56 We say that a τ-rigid pair (M,P) is *almost complete* if $|M| + |P| = n-1$. Given two τ-tilting pairs (M_1,P_1) and (M_2,P_2), we say that (M_1,P_1) is a mutation of (M_2,P_2) if there is an almost complete τ-rigid pair (M,P) which is a direct summand of (M_1,P_1) and (M_2,P_2). By abuse of notation we also say that $\operatorname{Fac} M_1$ is a mutation of $\operatorname{Fac} M_2$ if (M_1,P_1) is a mutation of (M_2,P_2).

The main result about mutation of τ-tilting pairs is the following theorem shown by Adachi, Iyama and Reiten in [1], that can be considered the most important result in τ-tilting theory.

Theorem 2.57 *[1] Let (M,P) be a τ-tilting pair where $M = \bigoplus_{i=1}^{k} M_i$ and $P = \bigoplus_{j=k+1}^{n} P_j$ are basic modules.*

Then for every indecomposable direct summand M_l of M the almost complete τ-rigid pair $(\bigoplus_{i \neq l} M_i, P)$ can be completed to a τ-tilting pair different from (M,P) in a unique way.

Likewise, for every indecomposable direct summand P_l of P the almost complete τ-rigid pair $(M, \bigoplus_{j \neq l} P_j)$ can be completed to a τ-tilting pair different from (M,P) in a unique way.

Remark We note that we can recover Theorem 2.57 as a consequence of Theorem 2.37. Indeed, one can verify that for all almost complete τ-rigid pairs (M,P) the algebra $\tilde{B}_{(M,P)}$ is local, which implies that there is only one isomorphism class of simple modules in $\operatorname{mod}\tilde{B}_{(M,P)}$. As a consequence of this, if S is a simple module in $\operatorname{mod}\tilde{B}_{(M,P)}$ we have that $\operatorname{Hom}_{\tilde{B}_{(M,P)}}(X,S)=0$ implies that X is isomorphic to zero.

Let S be a simple module in $\operatorname{mod}\tilde{B}_{(M,P)}$ and let $\mathcal{T}$ be a torsion class in $\operatorname{mod}\tilde{B}_{(M,P)}$. Then we have two options: either $S\in\mathcal{T}$ or $S\notin\mathcal{T}$. If $S\in\mathcal{T}$, then $X\in\mathcal{T}$ for all $X\in\operatorname{mod}\tilde{B}_{(M,P)}$ since torsion classes are closed under extensions. Otherwise, we have that $\mathcal{T}=\{0\}$ by the argument above.

This shows that every local algebra, no matter how complicated its representation theory, has exactly two torsion classes in its module category, which are the trivial torsion classes. In particular, this implies that there are exactly two τ-tilting pairs in $\operatorname{mod}\tilde{B}_{(M,P)}$ if (M,P) is an almost complete τ-rigid pair. Hence Theorem 2.57 follows from Theorem 2.37.

Another important consequence of the previous argument is the following.

Proposition 2.58 *[1] Let (M_1,P_1) and (M_2,P_2) be τ-tilting pairs that are mutations of each other and let (M,P) be the almost complete τ-rigid pair which is a direct summand of both. Then either* $\operatorname{Fac}M_1\subset\operatorname{Fac}M_2$ *or* $\operatorname{Fac}M_2\subset\operatorname{Fac}M_1$. *In particular,* $\operatorname{Fac}M_1\neq\operatorname{Fac}M_2$. *Moreover, if* $\operatorname{Fac}M_1\subset\operatorname{Fac}M_2$ *then* $\operatorname{Fac}M_1=\operatorname{Fac}M$ *and* $\operatorname{Fac}M_2={}^{\perp}\tau M\cap P^{\perp}$.

Remark From Theorem 2.57 it follows that, given a τ-tilting pair (M,P) and a choice (M_1',P_1') of an indecomposable direct summand of (M,P), there is a unique τ-tilting pair (M_1,P_1) with the same indecomposable direct summands as (M,P) except (M_1',P_1'). We say that (M_1,P_1) is the mutation of (M,P) at (M_1',P_1').

Moreover, if (M_1',P_1') and (M_2',P_2') are distinct indecomposable direct summands of (M,P), then the mutations (M_1,P_1) and (M_2,P_2) of (M,P) at (M_1',P_1') and (M_1',P_1'), respectively, are different. In particular, this implies that for every τ-tilting pair (M,P) there are exactly n τ-tilting pairs which are a mutation of (M,P) (see [1]).

In fact, there are explicit homological formulas to construct (M_1,P_1) from (M_2,P_2) and back but we will not explain them here. The interested reader is encouraged to see [1] for more details on the matter.

We now state a result by Demonet, Iyama and Jasso showing how the mutation of τ-tilting pairs can be seen at the level of torsion theories.

Theorem 2.59 *[45] Let A be an algebra, (M,P) be a τ-tilting pair and $\mathcal{T}$ be a torsion class in* $\operatorname{mod}A$. *Then the following hold.*

1 *If* $\mathcal{T} \subsetneq \operatorname{Fac} M$ *then there exists a mutation* (M', P') *such that* $\mathcal{T} \subset \operatorname{Fac} M' \subsetneq \operatorname{Fac} M$.
2 *If* $\operatorname{Fac} M \subsetneq \mathcal{T}$ *then there exists a mutation* (M'', P'') *such that* $\operatorname{Fac} M \subsetneq \operatorname{Fac} M'' \subset \mathcal{T}$.

Using the notion of mutation of τ-tilting pairs, one can construct a graph **τ-tp-A** where the vertices are the τ-tilting pairs in $\operatorname{mod} A$ and there is an edge between two τ-tilting pairs if and only if they are mutations from each other. Using the results of this section and Theorem 2.24 we can prove the following result.

Proposition 2.60 *The graph* $\tau - tp - \boldsymbol{A}$ *is isomorphic to the undirected graph underlying* $\boldsymbol{H}$*(ftors-A). In particular,* $\boldsymbol{H}$*(ftors-A) is an n-regular quiver.*

2.9.2 Maximal Green Sequences

In the module category of any algebra A there are always at least two torsion classes, sometimes called the trivial torsion classes, which are the whole of $\operatorname{mod} A$ and the torsion class $\{0\}$ containing only the objects that are isomorphic to the zero object. Both of these torsion classes are functorially finite and it is not hard to see that $\Phi(A,0) = \operatorname{mod} A$ and $\Phi(0,A) = \{0\}$.

Clearly $\{0\} \subsetneq \operatorname{mod} A$. So we can apply Theorem 2.59.1 and obtain a τ-tilting pair (M_1, P_1) which is a mutation of $(A,0)$ such that $\{0\} \subset \operatorname{Fac} M_1 \subsetneq \operatorname{mod} A$. If $\operatorname{Fac} M_1$ is not equal to $\{0\}$ we can repeat the process to obtain a mutation (M_2, P_2) of (M_1, P_1) such that $\{0\} \subset \operatorname{Fac} M_2 \subsetneq \operatorname{Fac} M_1 \subsetneq \operatorname{mod} A$. We can repeat this process inductively to obtain a decreasing chain of torsion classes

$$\{0\} \subset \cdots \subsetneq \operatorname{Fac} M_3 \subsetneq \operatorname{Fac} M_2 \subsetneq \operatorname{Fac} M_1 \subsetneq \operatorname{mod} A,$$

which in general can continue forever. However, in some cases this process stops. This leads to the following definition.

Definition 2.61 A *maximal green sequence* of length t in $\operatorname{mod} A$ is a finite set of τ-tilting pairs $\{(M_i, P_i) : 0 \leqslant i \leqslant t\}$ such that $(M_0, P_0) = (A, 0)$, $(M_t, P_t) = (0, A)$ and (M_i, P_i) is a mutation of (M_{i-1}, P_{i-1}) and $\operatorname{Fac} M_i \subset \operatorname{Fac} M_{i-1}$. Equivalently, a maximal green sequence is a finite chain of torsion classes

$$\{0\} = \mathcal{T}_0 \subsetneq \mathcal{T}_1 \subsetneq \cdots \subsetneq \mathcal{T}_{t-1} \subsetneq \mathcal{T}_t = \operatorname{mod} A$$

such that $\mathcal{T}_{i-1}$ is maximally included in $\mathcal{T}_i$ for all $1 \leqslant i \leqslant t$.

Maximal green sequences were originally introduced by Keller in [69] in the context of cluster algebras to give a combinatorial method to calculate certain geometric invariants known as Donaldson–Thomas invariants. The interpretation of maximal green sequences in terms of chains of torsion classes was first used by Nagao in [84]. The definition given here can be considered as a generalisation to the setting of τ-tilting theory of the original definition, since there are many examples of algebras which do not have a cluster counterpart, which first appeared in [36].

We have previously discussed in Section 2.7 that given a maximal inclusion $\mathcal{T} \subset \mathcal{T}'$ of torsion classes $\mathcal{T}, \mathcal{T}'$ there is a brick B in $\operatorname{mod} A$ such that $\mathcal{T}' = \operatorname{Filt}(\mathcal{T} \cup \{B\})$. Then, applying this argument inductively we can see that if we have a maximal green sequence

$$\{0\} = \mathcal{T}_0 \subsetneq \mathcal{T}_1 \subsetneq \cdots \subsetneq \mathcal{T}_{t-1} \subsetneq \mathcal{T}_t = \operatorname{mod} A,$$

we have a set of bricks $\mathcal{B} = \{B_1, B_2, \ldots, B_t\}$ such that $\operatorname{mod} A = \operatorname{Filt}(\mathcal{B})$. As it is, this is not a significant result since a small argument shows that all simple modules belong to $\mathcal{B}$ regardless of our starting maximal green sequence, and we know that simple modules filter every object in $\operatorname{mod} A$. However, one can show that the filtrations given in this way are unique in the following sense.

Theorem 2.62 *Let* $\{0\} = \mathcal{T}_0 \subsetneq \mathcal{T}_1 \subsetneq \cdots \subsetneq \mathcal{T}_{t-1} \subsetneq \mathcal{T}_t = \operatorname{mod} A$ *be a maximal green sequence in* $\operatorname{mod} A$ *and let* $\mathcal{B} = \{B_1, B_2, \ldots, B_t\}$ *be the set of bricks associated to it. Then for every non-zero object* M *of* $\operatorname{mod} A$ *there is a filtration*

$$0 = M_0 \subset M_1 \subset \cdots \subset M_{s-1} \subset M_s = M$$

such that $s \leqslant t$, $M_j/M_{j-1} \in \mathit{Filt}(B_{j_i})$ *and* $j_1 < j_2 < \cdots < j_{s-1} < j_s$. *Moreover, this filtration is unique up-to-isomorphism.*

Remark The previous result is a consequence of two works that appeared independently, namely [70] and [106]. In the appendix of [70], Demonet showed that the set $\mathcal{B}$ is what he calls an *I-chain* and showed that every I-chain induces a unique filtration in every object of $\operatorname{mod} A$.

A more general approach to this problem was given in [106], where it was shown that every chain of torsion classes induces a unique filtration for every object in $\operatorname{mod} A$. In that paper, this filtration was called the *Harder–Narasimhan filtration* induced by the chain of torsion classes since it generalises the Harder–Narasimhan filtrations induced by stability conditions. See also [37].

Remark Note that the word *green* in the name maximal green sequence does **not** make reference to any mathematician of name Green. Instead this word makes reference to the classical colouring in traffic lights. The reason for this is that in cluster algebras there is no evident reason to say that a mutation is going forwards or backwards. However, Keller [69] needed to impose such a direction to mutations in order to get the desired calculation. Then, he came up with a colouring of the vertices of the quiver associated to the cluster algebra in which a vertex is either green or red, which indicates if we are allowed to mutate at the given vertex or not, respectively. In this colouring, every vertex in the quiver of the initial seed is green and we are allowed to mutate at one green vertex at a time. The process finishes if after a finite number of mutations all the vertices in the quiver are red. To learn more about this rich subject, see [70].

2.10 τ-tilting Finite Algebras

To finish this part of the notes we speak about a new class of algebras that originated with the study of τ-tilting theory, the so-called τ-tilting finite algebras. They were introduced by Demonet, Iyama and Jasso as follows.

Definition 2.63 [45] An algebra A is τ-tilting finite if there are only finitely many τ-tilting pairs in $\operatorname{mod} A$.

Even though the class of τ-tilting finite algebras has been recently introduced, they have received a lot of attention. In the following theorem we compile a series of characterisations of τ-tilting finite algebras.

Theorem 2.64 *Let A be an algebra. Then the following are equivalent.*

1. *A is τ-tilting finite.*
2. *There are finitely many indecomposable τ-rigid objects in* $\operatorname{mod} A$.
3. *[45] There are finitely many torsion classes in* $\operatorname{mod} A$.
4. *[46] There are finitely many bricks in* $\operatorname{mod} A$.
5. *[97] The lengths of bricks in* $\operatorname{mod} A$ *are bounded.*
6. *[8] There are finitely many semibricks in* $\operatorname{mod} A$.
7. *[75] There are finitely many wide subcategories in* $\operatorname{mod} A$.
8. *[99] Every brick in* $\operatorname{Mod} A$ *is a finitely presented A-module.*

Remark Note that by Theorem 2.64.3 we have that all torsion classes in the module category of a τ-tilting finite algebra are functorially finite.

As we can see in Theorem 2.64, τ-tilting finite algebras have module categories that are somehow manageable from a torsion theoretic perspective, even if they are wild. As a consequence, there is an ongoing informal programme that aims to classify all the τ-tilting finite algebras. This problem has been attacked by several people in different families of algebras. The following is a list of families of algebras where some progress in understanding the problem has been made. Note that this list is not exhaustive nor efficient, since some families are included in others.

- [55] Hereditary algebras.
- [2] Nakayama algebras.
- [40, 52] Cluster-tilted algebras.
- [1] Radical squared zero algebras.
- [109] Auslander algebras.
- [78] Preprojective algebras.
- [89, 90] Gentle algebras.
- [3] Brauer graph algebras.
- [98] Special biserial algebras.
- [19] Contraction algebras.
- [81] Non-distributive algebras.

2.11 Brauer–Thrall Conjectures and τ-tilting Theory

The systematic study of the representation theory of finite dimensional algebras, one can argue, started around the 1940s. At that time it was already known that every A-module can be written as a direct summand of indecomposable A-modules in essentially one way. As a consequence, people started to classify the algebras in two types, *representation finite* and *representation infinite*, depending on their number of isomorphism classes of indecomposable modules. Much of the motivation in the early days of representation theory of finite dimensional algebras stems from this quest of determining algebras of finite representation type with an important role being played by the first and second Brauer–Thrall Conjectures, which were proved subsequently by Roĭter [93] and Auslander [20] for the first and by Nazarova and Roĭter [87] and Bautista [30] for the second. For a historical survey on Brauer–Thrall conjectures and their influence in representation theory, see [61]. Their statement is the following.

Conjecture 2.1 (First Brauer–Thrall Conjecture) *[20, 93] Let A be an algebra. Then A is of finite representation type if and only if there is a positive integer d such that* $\dim_{\mathbb{K}}(M) \leqslant d$ *for every indecomposable A-module M.*

Conjecture 2.2 (Second Brauer–Thrall Conjecture) *[30, 87] If A is an algebra of infinite representation type, then there is an infinite family of positive integers $\{d_i \mid i \in \mathbb{N}\}$ such that for every d_i there is an infinite family of indecomposable A-modules $\{M_j^{d_i}\}$ where* $\dim_{\mathbb{K}}(M_j^{d_i}) = d_i$ *for all j.*

Then using the characterisation of τ-tilting finite algebras given in Theorem 2.64.4 one can state a τ-tilting analogue of the first Brauer–Thrall Conjecture by restricting the universe of modules to consider from indecomposable to bricks. The statement of the conjecture, which was proven for every finite dimensional algebra over a field in [97], is the following.

Theorem 2.65 (First τ-Brauer–Thrall Conjecture) *[97] Let A be an algebra. Then A is τ-tilting finite if and only if there is a positive integer d such that* $\dim_{\mathbb{K}}(M) \leqslant d$ *for every brick M in* $\operatorname{mod} A$.

The statement of this conjecture appeared independently in [98] and [80]. We note that recently a new proof of the validity of this conjecture was given in [82].

One can see immediately that the τ-tilting version of the first Brauer–Thrall conjecture is a direct translation of the original. However, one can not do the same in the case of the second Brauer–Thrall conjecture, since there are τ-tilting infinite algebras having finitely many infinite families of bricks. The τ-tilting version of the second Brauer–Thrall conjecture is the following.

Conjecture 2.3 (Second τ-Brauer–Thrall Conjecture) *If A is a τ-tilting finite algebra, then there is a positive integer d and an infinite family of bricks $\{M_j^d\}$ in* $\operatorname{mod} A$ *such that* $\dim_{\mathbb{K}}(M_j^d) = d$ *for all j.*

This conjecture is still open in general. However, it has been verified for gentle algebras in [90], special biserial algebras in [80, 98] and for distributive algebras in [81].

Remark We note that if an algebra satisfies the second τ-Brauer–Thrall conjecture then it also satisfies the second Brauer–Thrall conjecture by a result of Smalø [102].

Remark We also note that there are many examples of τ-tilting finite algebras that are of infinite representation type, such as preprojective algebras [78] or contraction algebras [19], to name just a few.

2.12 Dimension Vectors and g-vectors

In the introduction of these notes we said that many developments that occurred in representation theory in the twenty-first century, including τ-tilting theory, were aiming to *categorify* cluster algebras to some extent.

In loose terms, the term *categorification* refers to the process of explaining some combinatorial phenomena by showing the existence of some underlying categorical phenomena. For instance, the bijection between the indecomposable τ-rigid modules in the module category of a hereditary algebra of Dynkin type and the non-initial variables in the cluster algebra of the corresponding Dynkin type is a categorification of cluster variables.

But as there is a process of categorification, there is also a process of *decategorification*, a process where you start with a category and you find some combinatorial or numerical data that reflect the phenomena occurring at the categorical level.

From now on, we shift our focus to a different decategorification of τ-tilting theory of an algebra using integer vectors.

2.12.1 The Grothendieck Group of an Algebra

The most classical decategorification using integer vectors of the representation theory associated to an algebra is the *Grothendieck group* of a category. We start by recalling the definition in the case of arbitrary abelian categories.

Definition 2.66 Let $\mathcal{A}$ be an abelian category. The Grothendieck group $K_0(\mathcal{A})$ of $\mathcal{A}$ is the quotient of the free abelian group generated by the isomorphism classes $[M]$ of all objects $M \in \mathcal{A}$ modulo the ideal generated by the short exact sequences as follows.

$$K_0(\mathcal{A}) = \frac{\langle [M] \mid M \in \mathcal{A} \rangle}{\langle [M] - [L] - [N] \mid 0 \to L \to M \to N \to 0 \rangle}.$$

In these notes we are interested only in the module categories $\operatorname{mod} A$ of finite dimensional algebras A over an algebraically closed field. By abuse of notation, the Grothendieck group $K_0(\operatorname{mod} A)$ of $\operatorname{mod} A$ will be denoted by $K_0(A)$ and we will refer to it as the Grothendieck group of A. As an immediate consequence of the Jordan–Hölder theorem for module categories we have the following result.

Theorem 2.67 *Let A be an algebra. Then $K_0(A)$ is isomorphic to $\mathbb{Z}^n$, where n is the number of isomorphism classes of simple modules in* $\operatorname{mod} A$.

From now on, we fix a complete set $\{[S(1)], \dots, [S(n)]\}$ of isomorphism classes of simple A-modules. Clearly, $\{[S(1)], \dots, [S(n)]\}$ forms a basis of $K_0(A)$. However, this is not the only basis of $K_0(A)$. In these notes, when we speak about the Grothendieck group of A we always assume that the basis chosen to represent our vectors is the basis given by the simple modules with a fixed order.

Theorem 2.68 *Let A be an algebra and $K_0(A)$ be its Grothendieck group having as canonical basis the set $\{[S(1)],\dots,[S(n)]\}$ of isomorphism classes of simple A-modules. Then for every object $M \in \operatorname{mod} A$, we have that*

$$[M] = (\dim_{\mathbb{K}}(\operatorname{Hom}_A(P(1),M)),\dots,\dim_{\mathbb{K}}(\operatorname{Hom}_A(P(n),M))),$$

$$[M] = (\dim_{\mathbb{K}}(\operatorname{Hom}_A(M,I(1)),\dots,\dim_{\mathbb{K}}(\operatorname{Hom}_A(M,I(n))))),$$

where $P(i)$ and $I(i)$ are the projective cover and the injective envelope of the simple $S(i)$, respectively, for all $1 \leqslant i \leqslant n$.

The previous result justifies that the element of the Grothendieck group $[M]$ associated to M is often called the *dimension vector* of M, terminology that we adopt in these notes as well.

Sometimes in the literature one finds the notation $\underline{\dim} M$ for the dimension vector, reserving $[M]$ for the abstract class of M in the Grothendieck group with no preferred basis of $K_0(A)$.

Remark In the previous result we are actually using the hypothesis that A is an algebra over an algebraically closed field. Otherwise, the result is not true in general. We warn the reader that this remark is also valid for several other results in this section.

2.12.2 g-vectors

Another set of integer vectors that can be associated to the category of finitely presented A-modules is the set of g-vectors.

Although the idea of g-vectors has been around for several decades, their systematic study is rather recent since the main motivation behind their study, as the reader might be guessing already, lies in the categorification of cluster algebras.

In fact, the name g-vector itself comes from cluster theory. The g-vectors were introduced by Fomin and Zelevinsky in [53], where they conjectured that cluster variables could be parametrised using g-vectors. Later on, it was shown that g-vectors encoded the projective presentation of τ-rigid A-modules. Their definition is the following.

Definition 2.69 Let M be an A-module. Choose the minimal projective presentation

$$P_1 \longrightarrow P_0 \longrightarrow M \longrightarrow 0$$

of M, where $P_0 = \bigoplus_{i=1}^{n} P(i)^{a_i}$ and $P_1 = \bigoplus_{i=1}^{n} P(i)^{b_i}$. Then the g-vector of M is defined as

$$g^M = (a_1 - b_1, a_2 - b_2, \ldots, a_n - b_n).$$

The g-vector of a τ-rigid pair (M,P) is defined as $g^M - g^P$.

Remark Recently, Nakaoka and Palu introduced in [86] the notion of *extriangulated categories*. These categories are a generalisation of abelian where the notion of *conflations* replaces and generalises the notion of short exact sequences. So, one can define the Grothendieck group of any extriangulated category as we did in Definition 2.66, replacing the short exact sequences by the conflations in this category. In the following section we will see that g-vectors can be thought of as the elements of the Grothendieck group of an extriangulated category denoted by $K^{[-1,0]}(\operatorname{proj} A)$.

2.13 τ-tilting Theory and 2-term Silting Complexes

Associated to any finite dimensional algebra A there is a triangulated category known as the homotopy category of bounded complexes of finitely generated projective A-modules usually denoted by $K^b(\operatorname{proj} A)$. The theory of homotopy categories is very rich, but out of the scope of these notes. The interested reader is encouraged to see [76] for a detailed account on that matter.

In this section we first introduce an abstract category associated to every algebra A that we call $K^{[-1,0]}(\operatorname{proj} A)$ because it can be identified with a full subcategory of $K^b(\operatorname{proj} A)$. Later on, we explain the relation between $K^{[-1,0]}(\operatorname{proj} A)$ and τ-tilting theory.

2.13.1 The Category $K^{[-1,0]}$(proj A)

As one usually does when defining a category, let us start by defining its objects. In this case, an object $\mathbf{P}$ of $K^{[-1,0]}(\operatorname{proj} A)$ is a complex of the form $\mathbf{P} := P_{-1} \xrightarrow{f} P_0$, where P_0 and P_{-1} are two projective A-modules and $f \in \operatorname{Hom}_A(P_{-1}, P_0)$. Now, given two objects $\mathbf{P} := P_{-1} \xrightarrow{f} P_0$ and $\mathbf{Q} := Q_{-1} \xrightarrow{f'} Q_0$ in $K^{[-1,0]}(\operatorname{proj} A)$ we define $g \in \operatorname{Hom}_{K^{[-1,0]}(\operatorname{proj} A)}(\mathbf{P}, \mathbf{Q})$ as a pair $g := (g_{-1}, g_0)$ such that $g_i \in \operatorname{Hom}_A(P_i, Q_i)$ making the following diagram commutative

$$\begin{array}{ccc} P_{-1} & \xrightarrow{f} & P_0 \\ {\scriptstyle g_{-1}}\downarrow & \overset{h}{\swarrow} & \downarrow{\scriptstyle g_0} \\ Q_{-1} & \xrightarrow{f'} & Q_0 \end{array}$$

modulo the equivalence relation given by if $g, g' \in \mathrm{Hom}_{K^{[-1,0]}(\mathrm{proj}\,A)}(\mathbf{P}, \mathbf{Q})$, we impose that g is equal to g' if there exists a map $h : P_0 \to Q_{-1}$ such that $f'h = g_0$ and $hf = g_{-1}$.

We note that in this category one can associate to every pair of objects $\mathbf{P}, \mathbf{Q} \in K^{[-1,0]}(\mathrm{proj}\,A)$ an abelian group $\mathcal{E}_{K^{[-1,0]}(\mathrm{proj}\,A)}(\mathbf{P}, \mathbf{Q})$, known as the group of *conflations* of $\mathbf{P}$ by $\mathbf{Q}$. This group is defined as the group of maps $t \in \mathrm{Hom}_A(P_{-1}, Q_0)$ *up to homotopy*, where $t, t' \in \mathrm{Hom}_A(P_{-1}, Q_0)$ coincide in this category if there are maps $t_i \in \mathrm{Hom}_A(P_i, Q_i)$ such that $t - t' = f'h_1 - h_0 f$.

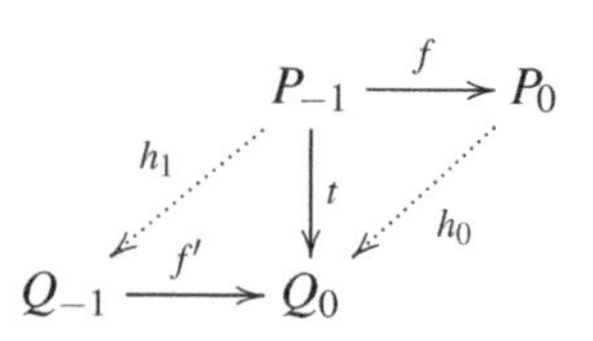

The notion of a conflation in an extriangulated category is a generalisation of that of short exact sequences in abelian categories and distinguished triangles in triangulated categories. As such, a conflation $t \in \mathcal{E}_{K^{[-1,0]}(\mathrm{proj}\,A)}(\mathbf{P}, \mathbf{Q})$ can be realised as

$$\mathbf{Q} \overset{a}{\rightarrowtail} \mathbf{E_t} \overset{b}{\twoheadrightarrow} \mathbf{P},$$

where $a \in \mathrm{Hom}_{K^{[-1,0]}(\mathrm{proj}\,A)}(\mathbf{Q}, \mathbf{E_t})$, $b \in \mathrm{Hom}_{K^{[-1,0]}(\mathrm{proj}\,A)}(\mathbf{E_t}, \mathbf{P})$ and $\mathbf{E_t}$ is in $K^{[-1,0]}(\mathrm{proj}\,A)$. The existence of conflations and their realisation endow the category $K^{[-1,0]}(\mathrm{proj}\,A)$ with the structure of an *extriangulated category*, a notion very recently introduced by Nakaoka and Palu in [86]. This follows from the fact that $K^{[-1,0]}(\mathrm{proj}A)$ is equivalent to an extension-closed subcategory of the triangulated category $K^b(\mathrm{proj}A)$. See [88] for more details.

One can then define the notion of a projective in $K^{[-1,0]}(\mathrm{proj}\,A)$ as an object $\mathbf{P} \in K^{[-1,0]}(\mathrm{proj}\,A)$ such that $\mathcal{E}_{K^{[-1,0]}(\mathrm{proj}\,A)}(\mathbf{P}, \mathbf{Q}) = 0$ for every object $\mathbf{Q} \in K^{[-1,0]}(\mathrm{proj}\,A)$. Likewise, an injective object in $K^{[-1,0]}(\mathrm{proj}\,A)$ is an object $\mathbf{I} \in K^{[-1,0]}(\mathrm{proj}\,A)$ such that $\mathcal{E}_{K^{[-1,0]}(\mathrm{proj}\,A)}(\mathbf{Q}, \mathbf{I}) = 0$ for every object $\mathbf{Q} \in K^{[-1,0]}(\mathrm{proj}\,A)$. As it turns out, the projective and injective objects of $K^{[-1,0]}(\mathrm{proj}\,A)$ can be described explicitly as follows.

$$\mathrm{proj}_{K^{[-1,0]}(\mathrm{proj}\,A)} = \left\{0 \overset{0}{\to} P \mid P \in \mathrm{proj}\,A\right\},$$

$$\mathrm{inj}_{K^{[-1,0]}(\mathrm{proj}\,A)} = \left\{P \overset{0}{\to} 0 \mid P \in \mathrm{proj}\,A\right\}.$$

As we said before, we want to relate the category $K^{[-1,0]}(\mathrm{proj}\,A)$ with the τ-tilting theory of A. To do that, we first need to know how one can go from $\mathrm{mod}\,A$ to $K^{[-1,0]}(\mathrm{proj}\,A)$ and back.

In fact, one can see $\operatorname{mod} A$ inside $K^{[-1,0]}(\operatorname{proj} A)$. This comes from the fact that, by definition, $\operatorname{mod} A$ is the category of finitely presented A-modules. This means that every A-module M admits a minimal projective presentation $P_{-1} \xrightarrow{f} P_0 \to M \to 0$, which induces a map $\mathbf{P}(-) : \operatorname{mod} A \to K^{[-1,0]}(\operatorname{proj} A)$ defined as $\mathbf{P}(M) = P_{-1} \xrightarrow{f} P_0$ on the objects. We note that $\mathbf{P}(-)$ is not a functor in general.

On the other direction, there is functor $H_0(-) : K^{[-1,0]}(\operatorname{proj} A) \to \operatorname{mod} A$, usually known as the *0-th homology*, which in this case consists of $H_0(\mathbf{P}) = \operatorname{coker} f$, where $\mathbf{P} = P_{-1} \xrightarrow{f} P_0$. In fact, one can show that the kernel of the functor H_0 corresponds to the ideal $\langle A \to 0 \rangle$ generated by the injective objects of $K^{[-1,0]}(\operatorname{proj} A)$.

$$\operatorname{mod} A \cong \frac{K^{[-1,0]}(\operatorname{proj} A)}{\langle A \to 0 \rangle}$$

It was noted by Gorsky, Nakaoka and Palu [59] that for every finite dimensional algebra A, the category $K^{[-1,0]}(\operatorname{proj} A)$ is hereditary. In other words, they have shown that for every object $\mathbf{M} \in K^{[-1,0]}(\operatorname{proj} A)$ there is a conflation of the form

$$\mathbf{P}_1 \rightarrowtail \mathbf{P}_0 \twoheadrightarrow \mathbf{M},$$

where $\mathbf{P}_0$ and $\mathbf{P}_1$ are projectives in $K^{[-1,0]}(\operatorname{proj} A)$. This fact is highly surprising, since the global dimension of the module category of an algebra is arbitrary. Moreover, this and the results that we state in the following subsection give a partial explanation of the good behaviour of τ-tilting theory for every algebra.

It follows from the results in this section that the Grothendieck group $K_0(K^{[-1,0]}(\operatorname{proj} A))$ of $K^{[-1,0]}(\operatorname{proj} A)$ is isomorphic to $\mathbb{Z}^n$, where the set $[\mathbf{P}(i)] = [0 \to P(i)]$ form a natural basis to this Grothendieck group. In particular we have that the g-vector g^M of any $M \in \operatorname{mod} A$ is the class $[\mathbf{P}(M)]$ of $\mathbf{P}(M) \in K^{[-1,0]}(\operatorname{proj} A)$ written in the basis $\{[\mathbf{P}(1)], \dots, [\mathbf{P}(n)]\}$.

2.13.2 τ-tilting Theory in $K^{[-1,0]}$(proj A)

As we pointed out before, the objects of $\operatorname{mod} A$ can be seen as objects of $K^{[-1,0]}(\operatorname{proj} A)$. In this subsection we see how the τ-tilting theory of an algebra A can be studied in $K^{[-1,0]}(\operatorname{proj} A)$. For this we start with the following definition.

Definition 2.70 Let A be an algebra and $\mathbf{P}$ be an object in $K^{[-1,0]}(\operatorname{proj} A)$. We say that $\mathbf{P}$ is *presilting* if $\mathcal{E}_{K^{[-1,0]}(\operatorname{proj} A)}(\mathbf{P}, \mathbf{P}) = 0$. Moreover, we say that $\mathbf{P}$ is *silting* if the number of isomorphism classes of indecomposable direct summands of $\mathbf{P}$ is equal to $|A|$.

Remark We said before that $K^{[-1,0]}(\operatorname{proj} A)$ can be identified with a full subcategory of $K^{[b}(\operatorname{proj} A)$. Under this identification, the previous definition coincides with a characterisation of 2-term silting objects in $K^b(\operatorname{proj} A)$ given by Adachi, Iyama and Reiten in [1].

We now state the main result of this section, where we denote by 2-silt(A) the set of silting objects in $K^{[-1,0]}(\operatorname{proj} A)$.

Theorem 2.71 *[1] Let A be an algebra. Then there is a bijection*

$$\boldsymbol{P}(-) : \tau\text{-}tp\text{-}A \to 2\text{-}silt(A)$$

between the set τ-tp-A of τ-tilting pairs in mod A *and the set 2-silt(A) of silting objects in $K^{[-1,0]}(\operatorname{proj} A)$, where the map is defined as*

$$\boldsymbol{P}(M,P) := \boldsymbol{P}(M) \oplus (P \to 0).$$

In particular, if (M,P) is a τ-rigid pair, then $\boldsymbol{P}(M,P)$ is a presilting object in $K^{[-1,0]}(\operatorname{proj} A)$ with the same number of isomorphism classes of indecomposable direct summands.

Remark In Section 2.3 we have discussed the problems that one has to define the basic objects of τ-tilting theory in module categories. However, we can see that τ-tilting pairs adopt a very natural form in a category $K^{[-1,0]}(\operatorname{proj} A)$ since we don't need to distinguish between τ-rigid modules and projective modules in the second entry of the pair. This is one of the arguments supporting the idea that $K^{[-1,0]}(\operatorname{proj} A)$ is the natural environment to study τ-tilting theory. Even if we support this idea, we made the decision of writing this note in terms of τ-tilting pairs to emphasise the relationship of τ-tilting theory and classical tilting theory.

2.14 g-vectors and τ-tilting Theory

In general there are many A-modules having the same projective presentation, which implies that g-vectors are in some sense ambiguous. However this ambiguity disappears when we restrict ourselves to τ-tilting theory.

Theorem 2.72 *Let A be an algebra and let M and M′ be two τ-rigid A-modules. Then $g^M = g^{M'}$ if and only if M is isomorphic to M′.*

Although the spirit of the previous result can be found already in the work of Auslander and Reiten [24], the first appearance of this result stated in these terms was in the work on 2-Calabi–Yau categories of Dehy and Keller [44]. Later, this result was adapted to the context of τ-tilting theory in the works of

Adachi, Iyama and Reiten [1] and later extended by Demonet, Iyama and Jasso in [45] as follows.

Theorem 2.73 *[45] Let A be an algebra and let M and M′ be two τ-rigid A-modules. Suppose that* $(g^M)_i \leqslant (g^{M'})_i$ *for* $1 \leqslant i \leqslant n$. *Then M is a quotient of M′. In particular* $g^M = g^{M'}$ *if and only if M is isomorphic to M′.*

In order to state the next result we need to fix some notation. Given a τ-tilting pair (M,P) we fix a decomposition $M = \bigoplus_{i=1}^{k} M_i$ and $P = \bigoplus_{j=k+1}^{n} P_j$ of M and P, respectively.

Theorem 2.74 *[1] Let* (M,P) *be a τ-tilting pair. Then the set*

$$\{g^{M_1},\dots,g^{M_k},-g^{P_{k+1}},\dots,-g^{P_n}\}$$

forms a basis of $\mathbb{Z}^n$.

2.14.1 g-vectors, Dimension Vectors and the Euler Form

Given a finite dimensional algebra A, one can always associate to it a square matrix known as the Cartan matrix of the algebra as follows.

Definition 2.75 Let A be an algebra and $\{P(1),\dots,P(n)\}$ be a complete set of non-isomorphic indecomposable projective A-modules. The Cartan matrix $\mathbf{C}_A$ of A is the $n \times n$ matrix

$$\mathbf{C}_A := ([P(1)]|[P(2)]|\dots|[P(n)]),$$

where the i-th column is the dimension vector $[P(i)]$ of $P(i)$ for $1 \leqslant i \leqslant n$.

The Euler characteristic of A is a $\mathbb{Z}$-bilinear form

$$\langle -,- \rangle_A : K_0(A) \times K_0(A) \to \mathbb{Z},$$

defined as $\langle [M],[N] \rangle_A = [M]^T \mathbf{C}_A^{-1} [N]$, where $[M]$ and $[N]$ are thought of as column vectors.

An important property of the Euler characteristic of an algebra is that it provides useful homological information, as shown in the following proposition.

Proposition 2.76 *Let A be an algebra of finite global dimension s and let M,N be two A-modules. Then*

$$\langle [M],[N] \rangle_A = \sum_{i=0}^{s} (-1)^i \dim_{\mathbb{K}}(\operatorname{Ext}_A^i(M,N)),$$

where $\mathrm{Ext}^0_A(M,N)$ *stands for* $\mathrm{Hom}_A(M,N)$. *In particular, if A is a hereditary algebra we have that*

$$\langle [M],[N]\rangle_A = \dim_{\mathbb{K}}(\mathrm{Hom}_A(M,N)) - \dim_{\mathbb{K}}(\mathrm{Ext}^1(M,N)).$$

Remark The proof of the previous proposition relies on the fact that $\{[P(1)],\dots,[P(n)]\}$ forms a basis of $\mathbb{Z}^n$ when A is of finite global dimension.

The following theorem was proven at the beginning of the 1980s by Auslander and Reiten in [24] but went unnoticed for several decades. Recently, with the development of τ-tilting theory this result came to light again and it is playing a key role in some of the latest developments of this theory. To state the theorem, we denote by $\langle -,-\rangle : \mathbb{R}^n \times \mathbb{R}^n \to \mathbb{R}$ the classical dot product in $\mathbb{R}$.

Theorem 2.77 *[24] Let M and N be modules over an algebra A. Then*

$$\langle g^M,[N]\rangle = \dim_{\mathbb{K}}(\mathrm{Hom}_A(M,N)) - \dim_{\mathbb{K}}(\mathrm{Hom}(N,\tau M)).$$

As a direct consequence of the previous result and the classical Auslander–Reiten formula we have the following corollary.

Corollary 2.78 *Let A be a hereditary algebra and let M and N be two A-modules. Then*

$$\langle g^M,[N]\rangle = \langle [M],[N]\rangle_A.$$

Based on this last corollary, the author is of the opinion that the pairing between g-vectors and dimension vectors of modules is a τ-tilting version of the Euler form of the algebra.

Remark We note that Theorem 2.77 establishes a natural pairing

$$\langle -,-\rangle : K_0(K^{[-1,0]}(\mathrm{proj}\,A)) \times K_0(\mathrm{mod}\,A) \longrightarrow \mathbb{Z}$$

between $K_0(K^{[-1,0]}(\mathrm{proj}\,A))$ and $K_0(\mathrm{mod}\,A)$. This suggests that the extriangulated category $K^{[-1,0]}(\mathrm{proj}\,A)$ can be thought of as a sort of "dual" category for $\mathrm{mod}\,A$. We will discuss further this duality in Section 2.15.

2.14.2 c-vectors

We have seen in Theorem 2.74 that the set of g-vectors of the indecomposable direct summands of a τ-tilting pair (M,P) forms a basis of $\mathbb{Z}^n$. This fact, together with the so-called *tropical duality* of cluster algebras [49, 53, 85], inspired Fu to introduce in [54] the notion of c-vectors for finite dimensional algebras using τ-tilting theory.

Definition 2.79 [54] Let (M,P) be a τ-tilting pair and let $\{g^{M_1},\dots,g^{M_k}, -g^{P_{k+1}},\dots,-g^{P_n}\}$ be the corresponding basis of g-vectors of $\mathbb{Z}^n$. Define the g-matrix $G_{(M,P)}$ of (M,P) as

$$G_{(M,P)} = \left(g^{M_1},\dots,-g^{P_n}\right) = \begin{pmatrix} (g^{M_1})_1 & \cdots & (-g^{P_n})_1 \\ \vdots & \ddots & \vdots \\ (g^{M_1})_n & \cdots & (-g^{P_n})_n \end{pmatrix}.$$

Then the c-matrix $C_{(M,P)}$ of (M,P) is defined as $C_{(M,P)} = (G^{-1}_{(M,P)})^T$. Each column of $C_{(M,P)}$ is called a *c-vector* of A. Moreover, we say that the i-th column of $C_{(M,P)}$ is the i-th c-vector associated to (M,P).

In the same paper, Fu showed that the c-vectors of certain families of algebras correspond to the dimension vector of bricks. These results were later generalised in [107] to every finite dimensional algebra over an algebraically closed field as follows.

Theorem 2.80 *[107] Let (M,P) be a τ-tilting pair with C-matrix $C_{(M,P)}$. Then there exists a brick B_i in* $\operatorname{mod} A$ *and $\varepsilon_i \in \{0,1\}$ such that $(-1)^{\varepsilon_i}[B_i] = c_i$ is equal to the i-th c-vector associated to (M,P). Moreover, $[B_i] = c_i$ if and only if $B_i \in \operatorname{Fac} M$. Dually, $-[B_i] = c_i$ if and only if $B_i \in M^{\perp}$.*

Remark Note that a direct consequence of this theorem is that every c-vector has either only non-negative coordinates or it has only non-positive entries. This is usually known as *sign-coherence* of c-vectors. If a c-vector c has only non-negative entries we say that c is a *positive* c-vector. Otherwise, we say that c is a *negative* c-vector.

In the previous result we have seen that the sign of the c-vectors associated to a τ-tilting pair is connected with the torsion theory of (M,P). In fact, this connection is very deep, as it can be seen in the following result.

Theorem 2.81 *[8, 107] Let (M,P) be a τ-tilting pair and let $\mathcal{B}_{(M,P)}$ be the set of bricks*

$$\{B_i \mid [B_i] = c_i \textit{ for some } 1 \leqslant i \leqslant n\}.$$

Then $\mathcal{B}_{(M,P)}$ is the unique semibrick in $\operatorname{mod} A$ *such that $T(\mathcal{B}_{(M,P)}) = \operatorname{Fac} M$.*

Remark An interesting consequence of the previous result is that one can label the Hasse quiver $\mathbf{H}(\text{ftors-}A)$ of ftors-A using (positive) c-vectors. This is compatible with the brick labelling discussed in Section 2.9. Indeed, the labelling of c-vectors consists simply of taking the dimension vector of each brick in the brick labelling of ftors-A.

In Section 2.9 we spoke about the fact that we can associate to each maximal green sequence

$$\{0\} = \mathcal{T}_0 \subsetneq \mathcal{T}_1 \subsetneq \cdots \subsetneq \mathcal{T}_{t-1} \subsetneq \mathcal{T}_t = \operatorname{mod} A$$

a set $\mathcal{B} = \{B_1, \dots, B_t\}$ of bricks in $\operatorname{mod} A$. One can show that the dimension vector $[B_i]$ of B_i is a positive c-vector of (M_i, P_i), where (M_i, P_i) is the unique τ-tilting pair such that $\mathcal{T}_i = \operatorname{Fac} M_i$. As a consequence, we can associate to every maximal green sequence a sequence $[\mathcal{B}] = \{[B_1], \dots, [B_t]\}$ of positive c-vectors in $\operatorname{mod} A$. The following result was first shown by Garver, McConville and Serhiyenko for cluster-tilted algebras in [57, 58] and later generalised in [106] to any algebra over an algebraically closed field.

Theorem 2.82 *[57, 58, 106] Every maximal green sequence $\{0\} = \mathcal{T}_0 \subsetneq \mathcal{T}_1 \subsetneq \cdots \subsetneq \mathcal{T}_{t-1} \subsetneq \mathcal{T}_t = \operatorname{mod} A$ is determined by its associated sequence $[\mathcal{B}]$ of c-vectors.*

2.15 The Wall-and-chamber Structure of an Algebra

In this final section of the chapter we introduce a geometric invariant for every algebra A, usually known as the *wall-and-chamber structure* of A, and we explain its relation with the τ-tilting theory of the algebra. In particular, we will see that τ-tilting theory recovers much, if not all, of the stability conditions on the algebra.

2.15.1 Stability Conditions

The study of stability conditions in algebraic geometry started with the introduction of geometric invariant theory by Mumford [83]. In [73], King applied this theory to module categories leading to the following definition of stability conditions. In what follows, we always assume that the Grothendieck group $K_0(A)$ of the algebra A is isomorphic to $\mathbb{Z}^n$ where the canonical basis of $K_0(A)$ is given by the set $\{[S(1)], \dots, [S(n)]\}$. Moreover, by abuse of notation, we identify $[M]$ with the corresponding vector of $\mathbb{Z}^n$ using the previous isomorphism. Recall that we denote by $\langle -, - \rangle : \mathbb{R}^n \times \mathbb{R}^n \to \mathbb{R}$ the classical dot product in $\mathbb{R}$.

Definition 2.83 [73] Let M be an A-module with dimension vector $[M]$ and v be a vector in $\mathbb{R}^n$. We say that M is *v-semistable* if $\langle v, [M] \rangle = 0$ and $\langle v, [L] \rangle \leqslant 0$ for every module L of M different from 0 or M. Similarly, we say that M is *ν-stable* if $\langle v,[M]\rangle = 0$ and $\langle v,[L]\rangle < 0$ for every module L of M different from 0 or M.

Later, Rudakov generalised this definition in [94]. Although Rudakov's notion of stability conditions is outside the scope of this note, we now state some of the algebraic consequences that he deduced in that paper.

Theorem 2.84 *[94] Let v be a vector in $\mathbb{R}^n$. Then the full subcategory* $\operatorname{mod}^v_{ss}A$ *of all the v-semistable modules in* $\operatorname{mod} A$ *is a wide subcategory of* $\operatorname{mod} A$.

It is not difficult to see that the v-stable modules correspond to the relative simple modules in $\operatorname{mod}^v_{ss}A$. The following result shows that every v-semistable module can be built by successive extensions of v-stable modules.

Proposition 2.85 *[94] Fix a vector v in $\mathbb{R}^n$. Then for every v-semistable module M there is a filtration*

$$0 = M_0 \subset M_1 \subset \cdots \subset M_{t-1} \subset M_t = M$$

where M_i/M_{i-1} is a v-stable module. Moreover, all such filtrations have the same length.

Remark If the previous proposition reminds the reader of the classical Jordan–Hölder theorem for the categories of A-modules, this is not a coincidence. We will see later that, for some specific vectors v, these filtrations correspond exactly with the filtrations by simple modules of an object in the module category of a smaller algebra that we can associate to v.

In fact, one can show that stability conditions have two naturally associated torsion pairs. This result, first shown by Baumann, Kamnitzer and Tingley in [29], reads as follows.

Proposition 2.86 *[29] Let v be a vector in $\mathbb{R}^n$. Then there are two torsion pairs $(\mathcal{T}_{\geqslant 0}, \mathcal{F}_{<0})$ and $(\mathcal{T}_{>0}, \mathcal{F}_{\leqslant 0})$ where:*

- $\mathcal{T}_{\geqslant 0}(v) := \{X \in \operatorname{mod} A \mid \langle v,[Y]\rangle \geqslant 0 \,\forall X \to Y \to 0\} \cup \{0\}$,
- $\mathcal{T}_{>0}(v) := \{X \in \operatorname{mod} A \mid \langle v,[Y]\rangle > 0 \,\forall X \to Y \to 0\} \cup \{0\}$,
- $\mathcal{F}_{\leqslant 0}(v) := \{X \in \operatorname{mod} A \mid \langle v,[Z]\rangle \leqslant 0 \,\forall 0 \to Z \to X\} \cup \{0\}$,
- $\mathcal{F}_{<0}(v) := \{X \in \operatorname{mod} A \mid \langle v,[Z]\rangle < 0 \,\forall 0 \to Z \to X\} \cup \{0\}$.

Moreover the category $\operatorname{mod}^v_{ss}A = \mathcal{T}_{\geqslant 0}(v) \cap \mathcal{F}_{\leqslant 0}(v)$.

2.15.2 The wall-and-chamber Structure of an Algebra

As with many other notions in representation theory introduced since the turn of the century, one can find the origins of wall-and-chamber structures in cluster theory.

In this case, the wall-and-chamber structure of an algebra is inspired from the notion of *scattering diagrams* introduced by Gross, Hacking, Keel and Kontsevich in [60] to study cluster algebras from a geometric perspective. The main idea is that one can associate to any cluster algebra $\mathcal{A}$ a geometric object (its scattering diagram) that encodes much of the algebraic properties of $\mathcal{A}$. As we have already mentioned, cluster algebras can be categorified by the representation theory of the Jacobian algebras of quivers with potentials. Based on this connection, Bridgeland showed in [35] that scattering diagrams associated to cluster algebras with an acyclic initial seed can be constructed using the stability conditions of the module category of the corresponding Jacobian algebra. This result was later extended by Mou in [79] for any cluster algebra having a green-to-red sequence.

More precisely, Bridgeland showed in [35] that one can construct a scattering diagram for any finite dimensional algebra and that this scattering diagram is isomorphic to the cluster scattering diagram of [60] if the algebra is hereditary. The complete description of scattering diagrams will be skipped. Instead, we will define now the support of the scattering diagram of an algebra, which is now called the *wall-and-chamber structure* of an algebra. Its definition is as follows.

Definition 2.87 The *stability space* $\mathfrak{D}(M)$ of a module M is the set of vectors

$$\mathfrak{D}(M) := \{v \in \mathbb{R}^n \mid M \text{is} v-\text{semistable}\}.$$

A *chamber* $\mathfrak{C}$ is an open connected component of the set

$$\mathbb{R}^n \setminus \bigcup_{0 \neq M \in \text{mod} A} \mathfrak{D}(M)$$

of all vectors v in $\mathbb{R}^n$ such that there is no non-zero v-semistable module. A *wall* is a stability space $\mathfrak{D}(M)$ of codimension 1, that is, a stability space $\mathfrak{D}(M)$ such that the smallest subspace of $\mathbb{R}^n$ containing it is a hyperplane.

Remark It follows from the previous definition that the wall-and-chamber structure is completely determined by its walls (see [9, 36]). Also, note that every wall is the stability space of a brick in the module category. However, it is not true that every finite brick determines a wall. Also, it follows from the definition that the wall-and-chamber structure of an algebra is always a fan.

Remark We would like to emphasise that the wall-and-chamber structure of a hereditary algebra is equivalent to its *semi-invariant picture* introduced by Igusa, Orr, Todorov and Weymann in [65]. They based their construction on the notion of *semi-invariant of quivers*, a notion of stability condition introduced by Schofield in [96] which is based on the Euler form of the quiver instead of the canonical inner product of $\mathbb{R}^n$.

2.15.3 The g-vector Fan of the Algebra

Recall from Section 2.13 that g-vectors correspond to the elements of the Grothendieck group $K_0(K^{[-1,0]}(\text{proj}\, A))$ of $K^{[-1,0]}(\text{proj}\, A)$ with respect to the basis given by the elements $\{[0 \to P(1)], \dots, [0 \to P(n)]\}$. As such, this group is isomorphic to $\mathbb{Z}^n$. However, in this section we want to relate g-vectors with stability conditions. Hence, we will see every g-vector g^M as a vector in $\mathbb{R}^n$, where $\mathbb{R}^n = K_0(K^{[-1,0]}(\text{proj}\, A)) \otimes \mathbb{R}$ where the canonical basis is given by $\{[0 \to P(1)], \dots, [0 \to P(n)]\}$.

The aim of this subsection is to study the distribution in $\mathbb{R}^n$ of the g-vectors of the indecomposable τ-rigid objects. In order to do that, we start by associating a cone in $\mathbb{R}^n$ to each τ-rigid pair. For that, recall that we are assuming that both M and P are basic objects that can be written as $M = \bigoplus_{i=1}^{k} M_i$ and $P = \bigoplus_{j=k+1}^{t} P_j$.

Definition 2.88 Let A be an algebra and let (M,P) be a τ-rigid pair whose set of g-vectors is

$$\{g^{M_1}, \dots, g^{M_k}, -g^{P_{k+1}}, \dots, -g^{P_t}\}.$$

Then we define the cone $\mathcal{C}_{(M,P)}$ to be the set

$$\mathcal{C}_{(M,P)} = \left\{ \sum_{i=1}^{k} \alpha_i g^{M_i} - \sum_{j=k+1}^{t} \alpha_j g^{P_j} \mid \alpha_i \geqslant 0 \text{ for every } 1 \leqslant i \leqslant t \right\}.$$

Similarly, we denote by $\mathcal{C}^{\circ}_{(M,P)}$ the interior of $\mathcal{C}_{(M,P)}$, that is,

$$\mathcal{C}^{\circ}_{(M,P)} = \left\{ \sum_{i=1}^{k} \alpha_i g^{M_i} - \sum_{j=k+1}^{t} \alpha_j g^{P_j} \mid \alpha_i > 0 \text{ for every } 1 \leqslant i \leqslant t \right\}.$$

Remark We know from Theorem 2.74 that $\{g^{M_1}, \dots, -g^{P_t}\}$ are linearly independent. In particular, this implies that the cone $\mathcal{C}_{(M,P)}$ is of codimension $n-t$.

Given two τ-rigid pairs (M_1,P_1) and (M_2,P_2), we have by definition that $\mathcal{C}_{(M_1,P_1)} \cap \mathcal{C}_{(M_2,P_2)}$ always contains the origin $\mathbf{0} \in \mathbb{R}^n$. Suppose that (M,P) is a

τ-rigid pair which is a direct summand of two different τ-rigid pairs (M_1, P_1) and (M_2, P_2). Then it is clear that $\mathcal{C}_{(M,P)} \subset \mathcal{C}_{(M_1,P_1)}$ and $\mathcal{C}_{(M,P)} \subset \mathcal{C}_{(M_2,P_2)}$. In particular $\mathcal{C}_{(M_1,P_1)} \cap \mathcal{C}_{(M_2,P_2)} \neq \{\mathbf{0}\}$. The following result by Demonet, Iyama and Jasso in [45] states that this is the only way in which this can happen.

Theorem 2.89 *[45] Let A be an algebra and let (M_1, P_1) and (M_2, P_2) be two τ-rigid pairs. Then $\mathcal{C}_{(M_1,P_1)} \cap \mathcal{C}_{(M_2,P_2)} \neq \{0\}$ if and only if there is a τ-rigid pair (M,P) which is a direct summand of both (M_1, P_1) and (M_2, P_2). Moreover, if (M,P) is the maximal common direct summand of (M_1, P_1) and (M_2, P_2) then $\mathcal{C}_{(M_1,P_1)} \cap \mathcal{C}_{(M_2,P_2)} = \mathcal{C}_{(M,P)}$.*

An important consequence of the previous result is that the g-vectors of indecomposable τ-rigid pairs have a geometrical structure known as *polyhedral fan*. Hence, it is common to refer to the set of all g-vectors of indecomposable τ-rigid pairs as the g-vector fan of the algebra.

In particular, the g-vector fan of a τ-tilting finite algebra has finitely many cones associated to its τ-rigid pairs. Moreover, these cones of g-vectors fit together very well, as was shown by Demonet, Iyama and Jasso in [45].

Theorem 2.90 *[45] Let A be an algebra. Then A is τ-tilting finite if and only if*

$$\mathbb{R}^n = \bigcup_{(M,P)\ \tau\text{-rigid}} \mathcal{C}_{(M,P)} = \left(\bigcup_{(M,P)\ \tau\text{-rigid}} \mathcal{C}^{\circ}_{(M,P)} \right) \cup \{\boldsymbol{0}\}.$$

2.15.4 τ-tilting Theory and Stability Conditions

In Section 2.14 we have seen that there is a natural pairing between g-vectors and dimension vectors that provides interesting homological information. For the convenience of the reader, we recall Auslander and Reiten's result here.

Theorem 2.91 *[24] Let M and N be modules over an algebra A. Then*

$$\langle g^M, [N] \rangle = \dim_{\mathbb{K}}(\operatorname{Hom}_A(M,N)) - \dim_{\mathbb{K}}(\operatorname{Hom}(N, \tau M)).$$

Using this pairing as their main tool, it was shown in [36] and, independently, in [108] that we can recover the torsion classes $\operatorname{Fac} M$ and ${}^{\perp}\tau M \cap P^{\perp}$ associated to a τ-rigid pair (M,P) from its g-vectors. The formal statement is the following.

Theorem 2.92 *[36, 108] Let (M,P) be a τ-rigid pair and let $v \in \mathcal{C}^{\circ}_{(M,P)}$. Then $\mathcal{T}_{>0}(v) = \operatorname{Fac} M$ and $\mathcal{T}_{\geqslant 0}(v) = {}^{\perp}\tau M \cap P^{\perp}$. In particular,* $\operatorname{mod}^{v}_{ss} A = M^{\perp} \cap {}^{\perp}\tau M \cap P^{\perp}$.

Combining this result with Theorem 2.36 we obtain the following corollary.

Corollary 2.93 *[36] Let (M,P) be a τ-rigid pair and let $v \in \mathcal{C}^{\circ}_{(M,P)}$. Then the category $\operatorname{mod}^{v}_{ss} A$ of v-semistable modules is equivalent to the module category of the τ-tilting reduction of (M,P). In particular, the number of isomorphism classes of v-stable modules is $n - |M| - |P|$.*

2.15.5 From τ-tilting Theory to the Wall-and-chamber Structure

An important fact that follows from Corollary 2.93 is that the category of v-semistable objects is constant in a given cone $\mathcal{C}^{\circ}_{(M,P)}$. In this subsection we explore some consequences of this result in the wall-and-chamber structure of an algebra.

The first thing to notice is that $\operatorname{mod}^{v}_{ss} A = \{0\}$ for every vector $v \in \mathcal{C}^{\circ}_{(M,P)}$ if and only if (M,P) is a τ-tilting pair. Then, it is easy to see that $\mathcal{C}^{\circ}_{(M,P)}$ is a chamber in the wall-and-chamber structure of A. This fact was first noticed in [36] and it was later shown by Asai in [9] that this is actually a bijection. The formal statement is the following.

Theorem 2.94 *[9, 36] Let (M,P) be a τ-tilting pair. Then $\mathcal{C}^{\circ}_{(M,P)}$ is a chamber in the wall-and-chamber structure of A. Moreover, every chamber arises this way. In particular, there is a one-to-one correspondence between the set of functorially finite torsion classes in* $\operatorname{mod} A$ *and the set of chambers of the wall-and-chamber structure of A.*

One can also see that for every τ-tilting pair (M,P) the set $\mathcal{C}_{(M,P)} \setminus \mathcal{C}^{\circ}_{(M,P)}$ coincides with the union $\bigcup_{1 \leqslant i \leqslant n} \mathcal{C}_{(M_i,P_i)}$ of the cones $\mathcal{C}_{(M_i,P_i)}$ associated to the n almost-complete τ-rigid pairs $\{(M_1,P_1), \ldots, (M_n,P_n)\}$ that are a direct summand of (M,P). Now, it follows from Theorem 2.92 that there exists exactly one brick B_i which is v-stable for every $1 \leqslant i \leqslant n$ and every $v \in \mathcal{C}^{\circ}_{(M_i,P_i)}$. Then we have built a set $\{B_1, \ldots, B_n\}$ of bricks in $\operatorname{mod} A$ associated to (M,P) using stability conditions. In the following result we show that we have encountered this set before.

Theorem 2.95 *[107] Let (M,P) be a τ-tilting pair, $v \in \mathcal{C}^{\circ}_{(M,P)}$ and let $\{B_1, \ldots, B_n\}$ as above. Then the c-matrix $C_{(M,P)}$ of (M,P) is equal to*

$$C_{(M,P)} = (\mathsf{sgn}(\langle v, [B_1] \rangle)[B_1] \quad | \ldots | \quad \mathsf{sgn}(\langle v, [B_n] \rangle)[B_n]).$$

In particular, this implies that we can recover the semibrick associated to (M,P) directly from the wall-and-chamber structure of A, since it is exactly the set $\{B_i : \langle v, [B_i]\rangle > 0\}$ for any $v \in \mathcal{C}^{\circ}_{(M,P)}$.

Given two chambers $\mathfrak{C}_1$ and $\mathfrak{C}_2$, we say that they are *neighbouring* each other if $\overline{\mathfrak{C}_1} \cap \overline{\mathfrak{C}_2}$ is a set of codimension 1 in $\mathbb{R}^n$. In other words, we say that $\mathfrak{C}_1$ and $\mathfrak{C}_2$ are neighbours if they are separated by a wall. Another interesting consequence of Theorem 2.94 is the following.

Proposition 2.96 *[36] Let (M,P) and (M',P') be two τ-tilting pairs. Then (M,P) is a mutation of (M',P') if and only if the chambers $\mathcal{C}^{\circ}_{(M,P)}$ and $\mathcal{C}^{\circ}_{(M',P')}$ associated to (M,P) and (M',P') respectively, are neighbours.*

Note that we can construct a dual graph to the wall-and-chamber structure of an algebra, where the vertices correspond to the chambers and there is an edge between two vertices if and only if the two corresponding chambers are neighbouring each other. Then the previous proposition is equivalent to the following.

Proposition 2.97 *The dual graph of the wall-and-chamber structure of an algebra is isomorphic to the underlying graph* $\mathbf{H}$*(ftors-A), the Hasse quiver of functorially finite torsion classes in* mod A.

In particular, this result implies that crossing a wall in the wall-and-chamber of A corresponds to a mutation of τ-tilting pairs. Hence, one can realise any finite series of mutations of τ-tilting pairs as paths in the wall-and-chamber structure of A. Then, in particular, every maximal green sequence in mod A can be realised as a path in the wall-and-chamber structure of A, which solves the original motivating question of [36].

Given a cone σ in a fan Σ, the star $\mathbf{star}(\sigma)$ of σ is the set of all the cones σ' in Σ having σ as a face, that is, all the cones in Σ such that $\sigma \subset \sigma'$. The following result due to Asai [9] shows that τ-tilting reduction can be also realised at the level of wall-and-chamber structures.

Theorem 2.98 *[9] Let (M,P) be a τ-rigid pair in* mod A. *Then the wall-and-chamber structure of the τ-tilting reduction $\tilde{B}_{(M,P)}$ is isomorphic as a fan to* $\mathbf{star}(\mathcal{C}_{(M,P)})$ *of the cone $\mathcal{C}_{(M,P)}$ associated to (M,P).*

2.15.6 Examples

In this subsection we give a couple of examples illustrating the results of this section.

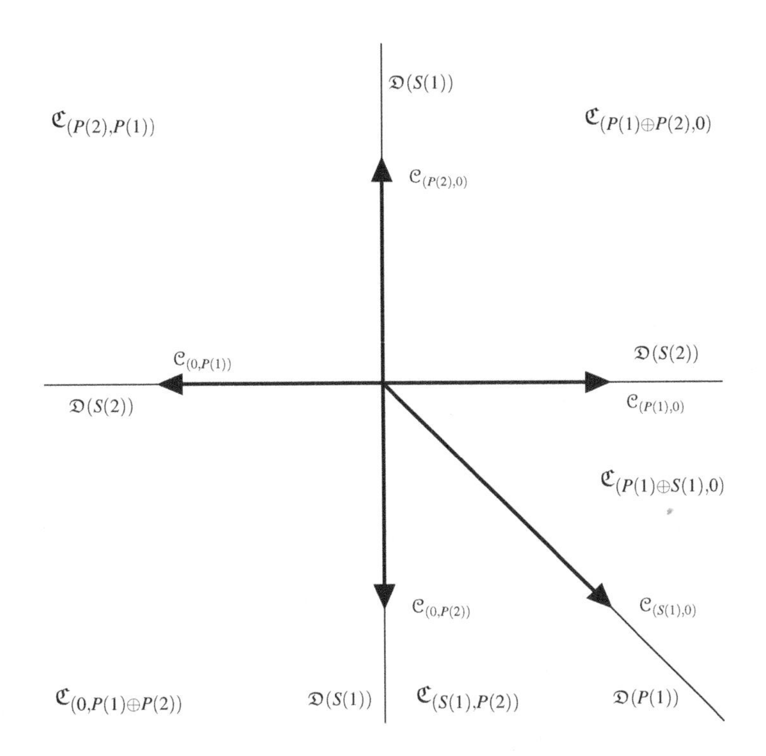

Figure 2.8 Wall and chamber structure for $\mathbb{A}_2$.

Example 2.99 Consider the path algebra $\mathbb{A}_2 = kQ$ of the quiver $Q = \ 1 \longrightarrow 2$. Its Auslander–Reiten quiver is as follows:

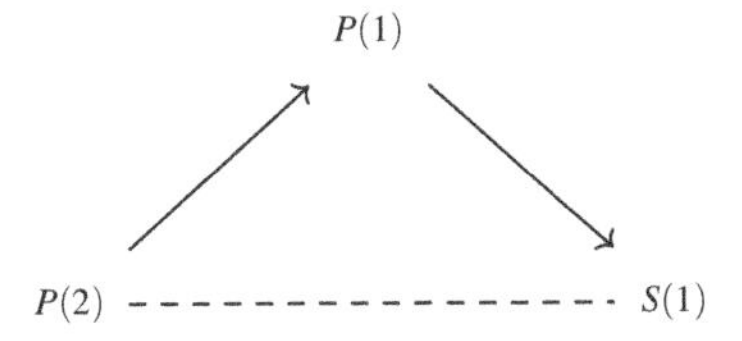

A quick calculation shows that all indecomposable τ-rigid pairs in mod A are

$$\{(P(2),0),(P(1),0),(S(1),0),(0,P(2)),(0,P(1))\}$$

and that the complete list of τ-tilting pairs in mod A is

$$\{(P(1)\oplus P(2),0),(P(1)\oplus S(1),0),(S(1),P(2)),(0,P(1)\oplus P(2)),(P(2),P(1))\}.$$

In Figure 2.8 we have drawn the wall-and-chamber structure of A indicating the cones associated to the indecomposable τ-rigid pairs and the chambers associated to each τ-tilting pair.

Example 2.100 We now take the algebra A that we considered in Example 2.15. That is the path algebra of the quiver

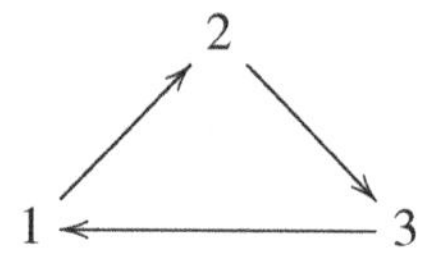

modulo the second power of the ideal generated by all the arrows. The complete list of τ-tilting pairs in mod A can be found in Table 2.1. The wall-and-chamber structure of A is in $\mathbb{R}^3$ since the rank of $K_0(A)$ is equal to 3. For the sake of readability, we choose to draw in Figure 2.9 a stereographic projection of it from a vector in the first orthant.

One can see there that all chambers in the wall-and-chamber structure of A are delimited by exactly three walls. In red we indicate the image of the g-vectors of the indecomposable τ-rigid pairs. One can obtain the τ-tilting pair inducing a particular chamber from the g-vectors situated in the corners of that chamber. For instance, the chamber $\mathfrak{C}_7$ is induced by the τ-tilting pair $(\begin{smallmatrix}2\\3\end{smallmatrix} \oplus 3, \begin{smallmatrix}1\\2\end{smallmatrix})$.

This particular stereographic projection of the wall-and-chamber of A is particularly useful to obtain the c-vectors associated to a given τ-tilting pair, since the convexity of each of the walls indicates the sign of the corresponding c-vector as follows: If the wall is convex when regarded from inside the chamber, then the c-vector corresponding to this wall is a negative c-vector and it is a positive c-vector otherwise. Using again the chamber $\mathfrak{C}_7$, one can see that the c-vectors associated to $(\begin{smallmatrix}2\\3\end{smallmatrix} \oplus 3, \begin{smallmatrix}1\\2\end{smallmatrix})$ are $\{[0,1,0],[0,0,1],[-1,0,-1]\}$. In particular, this allows us to conclude that the semibrick associated to $(\begin{smallmatrix}2\\3\end{smallmatrix} \oplus 3, \begin{smallmatrix}1\\2\end{smallmatrix})$ is $\{2 \oplus 3\}$.

Finally, note that there are two walls incident to the g-vector g^{3}, while there are three which are incident to the g-vector $g^{\begin{smallmatrix}1\\2\end{smallmatrix}}$. This is due to the fact that the τ-tilting reduction of $(3,0)$ is isomorphic to $\mathbb{A}_1 \times \mathbb{A}_1$ while that of $(\begin{smallmatrix}1\\2\end{smallmatrix},0)$ is isomorphic to $\mathbb{A}_2$.

2.15.7 Characterisations of τ-tilting Finite Algebras

We conclude this note by adding to Theorem 2.64 several characterisations of τ-tilting finite algebras that include the notions we have discussed in this last section.

Theorem 2.101 *Let A be an algebra. Then the following are equivalent.*

1. *A is τ-tilting finite.*
2. *There are finitely many indecomposable τ-rigid objects in* mod A.
3. *[45] There are finitely many torsion classes in* mod A.
4. *[46] There are finitely many bricks in* mod A.

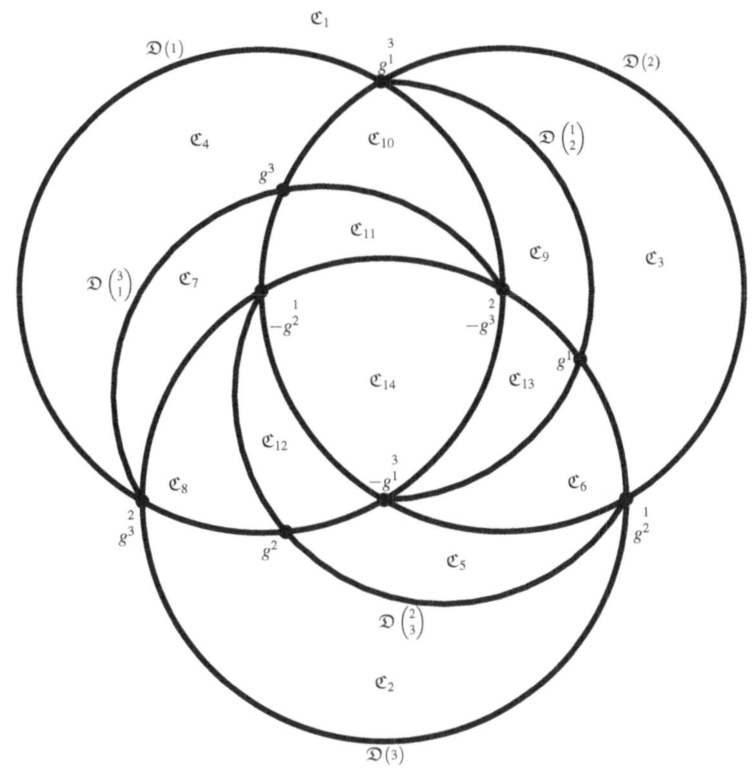

Figure 2.9 The stereographic projection of the wall-and-chamber structure of A.

5. *[97] The lengths of bricks in* mod A *are bounded.*
6. *[8] There are finitely many semibricks in* mod A.
7. *[107] There are finitely many c-vectors in* mod A.
8. *[75] There are finitely many wide subcategories in* mod A.
9. *[99] Every brick in* Mod$A$ *is a finitely presented A-module.*
10. *[45] The g-vector fan of A spans the whole* $\mathbb{R}^n$.
11. *[36] The number of walls in the wall-and-chamber structure of A is finite.*
12. *[36] The number of chambers in the wall-and-chamber structure of A is finite.*
13. *[108] The number of different v-semistable subcategories of* mod A *is finite.*

Bibliography

[1] Adachi, T. 2016. Characterizing τ-tilting finite algebras with radical square zero. *Proc. Am. Math. Soc.*, **144**(11), 4673–4685.

[2] Adachi, T. The classification of τ-tilting modules over Nakayama algebras. *J. Algebra*, 452:227–262, 2016.

[3] Adachi, T., Aihara, T. and Chan, A. Classification of two-term tilting complexes over Brauer graph algebras. *Math. Z.*, 290(1-2):1–36, 2018.

[4] Adachi, T., Iyama, O. and Reiten, I. τ-tilting theory. *Compos. Math.*, 150(3):415–452, 2014.

[5] Aihara, T. and Iyama, O. Silting mutation in triangulated categories. *J. Lond. Math. Soc.* (2) 85 (2012), 633–668.

[6] Amiot, C. Cluster categories for algebras of global dimension 2 and quivers with potential. *Ann. Inst. Fourier*, 59(6):2525–2590, 2009.

[7] Angeleri Hügel, L., Happel, D., and Krause, H., ed. 2007. *Handbook of Tilting Theory*, Volume 332. Cambridge University Press, Cambridge.

[8] Asai, S. Semibricks. *Int. Math. Res. Not.*, 2020(16):4993–5054, 2020.

[9] Asai, S. The wall-chamber structures of the real Grothendieck groups. *Adv. Math.*, 381:45, 2021. Id/No 107615.

[10] Asai, S. and Pfeifer, C. Wide subcategories and lattices of torsion classes. https://arxiv.org/abs/1905.01148.

[11] Assem, I. Torsion theories induced by tilting modules. *Canad. J. Math.*, 36(5):899–913, 1984.

[12] Assem, I. Tilting theory - an introduction. Topics in algebra. Pt. 1: Rings and representations of algebras, Pap. 31st Semester Class. Algebraic Struct., Warsaw/Poland 1988, Banach Cent. Publ. 26, Part 1, 127–180 (1990), 1990.

[13] Assem, I., Brüstle, T. and Schiffler, R. Cluster-tilted algebras and slices. *J. Algebra*, 319(8):3464–3479, 2008.

[14] Assem, I., Brüstle, T. and Schiffler, R. Cluster-tilted algebras as trivial extensions. *Bull. Lond. Math. Soc.*, 40(1):151–162, 2008.

[15] Assem, I., Brüstle, T. and Schiffler, R. On the Galois coverings of a cluster-tilted algebra. *J. Pure Appl. Algebra*, 213(7):1450–1463, 2009.

[16] Assem, I., Brüstle, T. and Schiffler, R. Cluster-tilted algebras without clusters. *J. Algebra*, 324(9):2475–2502, 2010.

[17] Assem, I., Simson, D., and Skowroński, A. 2006. *Elements of the Representation Theory of Associative Algebras. Vol. 1.* London Mathematical Society Student Texts, vol. 65. Cambridge University Press, Cambridge. Techniques of representation theory.

[18] Assem, I. and Skowroński, A. Iterated tilted algebras of type $\tilde{\mathbb{A}}_n$. *Math. Z.* 195, 269–290 (1987; Zbl 0601.16022).

[19] August, J. 2020. The tilting theory of contraction algebras. *Adv. Math.*, **374**, Article 107372.

[20] Auslander, M. Representation theory of Artin algebras. I, II. *Comm. Algebra*, 1:177–310, 1974.

[21] Auslander, M., Platzeck, M. I. and Reiten, I. Coxeter functors without diagrams. *Trans. Amer. Math. Soc.*, 250:1–46, 1979.

[22] Auslander, M. and Reiten, I. Representation theory of artin algebras III almost split sequences. *Communications in Algebra*, 3(3):239–294, 1975.

[23] Auslander, M. and Reiten, I. Representation theory of Artin algebras. IV: Invariants given by almost split sequences. *Commun. Algebra*, 5:443–518, 1977.

[24] Auslander, M. and Reiten, I. Modules determined by their composition factors. *Illinois J. Math.*, 29(2):280–301, 1985.

[25] Auslander, M., Reiten, I., and Smalø, S. O. 1995. *Representation Theory of Artin Algebras*. Cambridge Studies in Advanced Mathematics, vol. 36. Cambridge University Press, Cambridge.

[26] Auslander, M. and Smalø, S. O. Addendum to "Almost split sequences in subcategories", *J. Algebra*, 71:592–594, 1981.

[27] Auslander, M. and Smalø, S. O. Almost split sequences in subcategories. *J. Algebra*, 69:426–454, 1981.

[28] Barnard, E., Carroll, A. and Zhu, S. Minimal inclusions of torsion classes. *Algebr. Comb.*, 2(5):879–901, 2019.

[29] Baumann, P., Kamnitzer, J. and Tingley, P. Affine Mirković-Vilonen polytopes. *Publ. Math. Inst. Hautes Études Sci.*, 120:113–205, 2014.

[30] Bautista, R. On algebras of strongly unbounded representation type. *Comment. Math. Helv.*, 60(3):392–399, 1985.

[31] Berenstein, A., Fomin, S. and Zelevinsky, A. 2005. Cluster algebras. III. Upper bounds and double Bruhat cells. *Duke Math. J.*, **126**(1), 1–52.

[32] Bernšteĭn, I. N., Gelfand, I. M. and Ponomarev, V. A. Coxeter functors, and Gabriel's theorem. *Uspehi Mat. Nauk*, 28(2(170)):19–33, 1973.

[33] Bongartz, K. *Tilted algebras*. Representations of algebras, Proc. 3rd Int. Conf., Puebla/Mex. 1980, Lect. Notes Math. 903, 26–38, 1981.

[34] Brenner, S. and Butler, M. C. R. Generalizations of the Bernstein-Gel′fand-Ponomarev reflection functors. In *Representation theory, II (Proc. Second Internat. Conf., Carleton Univ., Ottawa, Ont., 1979)*, volume 832 of *Lecture Notes in Math.*, pages 103–169. Springer, Berlin-New York, 1980.

[35] Bridgeland, T. 2017. Scattering diagrams, Hall algebras and stability conditions. *Algebr. Geom.*, **4**(5), 523–561.

[36] Brüstle, T., Smith, D. and Treffinger, H. Wall and chamber structure for finite-dimensional algebras. *Adv. Math.*, 354:106746, 31, 2019.

[37] Brüstle, T., Smith, D. and Treffinger, H. Stability Conditions and Maximal Green Sequences in Abelian Categories. https://arxiv.org/abs/1805.04382.

[38] Brüstle, T. and Yang, D. Ordered exchange graphs. In *Advances in representation theory of algebras. Selected papers of the 15th international conference on representations of algebras and workshop (ICRA XV), Bielefeld, Germany, August 8–17, 2012*, pages 135–193. Zürich: European Mathematical Society (EMS), 2014.

[39] Buan, A. B., Marsh, B. R., Reineke, M., Reiten, I. and Todorov, G. 2006. Tilting theory and cluster combinatorics. *Adv. Math.*, **204**(2), 572–618.

[40] Buan, A. B., Marsh, B. R. and Reiten, I. Cluster-tilted algebras. *Trans. Amer. Math. Soc.*, 359(1):323–332 (electronic), 2007.

[41] Buan, A. B., Marsh, B. R., Reiten, I. and Todorov, G. 2007. Clusters and seeds in acyclic cluster algebras. *Proc. Amer. Math. Soc.*, **135**(10), 3049–3060.

[42] Buan, A. B. and Zhou, Y. A silting theorem. *J. Pure Appl. Algebra*, 220(7):2748–2770, 2016.

[43] Caldero, P. and Keller, B. 2006. From triangulated categories to cluster algebras. II. *Ann. Sci. École Norm. Sup. (4)*, **39**(6), 983–1009.

[44] Dehy, R. and Keller, B. On the combinatorics of rigid objects in 2-Calabi-Yau categories. *Int. Math. Res. Not. IMRN*, (11):Art. ID rnn029, 17, 2008.
[45] Demonet, L., Iyama, O. and Jasso, G. τ-tilting finite algebras, bricks, and g-vectors. *Int. Math. Res. Not. IMRN*, (3):852–892, 2019.
[46] Demonet, L., Iyama, O., Reading, N., Reiten, I. and Thomas, H. Lattice theory of torsion classes. *arXiv* (1711.01785), 2017.
[47] Derksen, H. and Fei, J. General presentations of algebras. *Adv. Math.*, 278:210–237, 2015.
[48] Derksen, H., Weyman, J. and Zelevinsky, A. 2008. Quivers with potentials and their representations. I. Mutations. *Selecta Math. (N.S.)*, **14**(1), 59–119.
[49] Derksen, H., Weyman, J. and Zelevinsky, A. Quivers with potentials and their representations II: applications to cluster algebras. *J. Amer. Math. Soc.*, 23(3):749–790, 2010.
[50] Dickson, S. E. A torsion theory for Abelian categories. *Trans. Amer. Math. Soc.*, 121:223–235, 1966.
[51] Fomin, S. and Zelevinsky, A. 2002. Cluster algebras. I. Foundations. *J. Amer. Math. Soc.*, **15**(2), 497–529.
[52] Fomin, S. and Zelevinsky, A. 2003. Cluster algebras. II. Finite type classification. *Invent. Math.*, **154**(1), 63–121.
[53] Fomin, S. and Zelevinsky, A. 2007. Cluster algebras. IV. Coefficients. *Compos. Math.*, **143**(1), 112–164.
[54] Fu, C. c-vectors via τ-tilting theory. *J. Algebra*, 473:194–220, 2017.
[55] Gabriel, P. 1972. Unzerlegbare Darstellungen. I. *Manuscripta Math.*, **6**, 71–103.
[56] Gabriel, P. Indecomposable representations. II. Sympos. math. 11, Algebra commut., Geometria, Convegni 1971/1972, 81–104 (1973).
[57] Garver, A., McConville, T. and Serhiyenko, K. Minimal length maximal green sequences. *Sém. Lothar. Combin.*, 78B:Art. 16, 12, 2017.
[58] Garver, A., McConville, T. and Serhiyenko, K. Minimal length maximal green sequences. *Adv. in Appl. Math.*, 96:76–138, 2018.
[59] Gorsky, M., Nakaoka, H. and Palu, Y. Positive and negative extensions in extriangulated categories https://arxiv.org/abs/2103.12482.
[60] Gross, M., Hacking, P., Keel, S. and Kontsevich, M. 2018. Canonical bases for cluster algebras. *J. Amer. Math. Soc.*, **31**(2), 497–608.
[61] Gustafson, W. H. The history of algebras and their representations. In *Representations of algebras (Puebla, 1980)*, volume 944 of *Lecture Notes in Math.*, pages 1–28. Springer, Berlin-New York, 1982.
[62] Happel, D. *Triangulated categories in the representation theory of finite-dimensional algebras*, volume 119 of *London Mathematical Society Lecture Note Series*. Cambridge University Press, Cambridge, 1988.
[63] Happel, D. and Ringel, C. M. Tilted algebras. *Trans. Amer. Math. Soc.*, 274(2):399–443, 1982.
[64] Happel, D. and Unger, L. Almost complete tilting modules. *Proc. Amer. Math. Soc.*, 107(3):603–610, 1989.
[65] Igusa, K., Orr, K., Todorov, G. and Weyman, J. Cluster complexes via semi-invariants. *Compos. Math.*, 145(4):1001–1034, 2009.
[66] Ingalls, C. and Thomas, H. 2009. Noncrossing partitions and representations of quivers. *Compos. Math.*, **145**(6), 1533–1562.

[67] Iyama, O. and Reiten, I. Introduction to τ-tilting theory. *Proc. Natl. Acad. Sci. USA*, 111(27):9704–9711, 2014.

[68] Jasso, G. Reduction of τ-tilting modules and torsion pairs. *Int. Math. Res. Not. IMRN*, (16):7190–7237, 2015.

[69] Keller, B. On cluster theory and quantum dilogarithm identities. In *Representations of Algebras and Related Topics*, pages 85–116. European Mathematical Society Publishing House, Zurich, Switzerland, 2011.

[70] Keller, B. and Demonet, L. A survey on maximal green sequences, 2020.

[71] Keller, B. and Vossieck, D. Aisles in derived categories. *Bull. Soc. Math. Belg.*, 40 (1988), 239–253.

[72] Keller, B. and Yang, D. Derived equivalences from mutations of quivers with potential. *Adv. Math.*, 226(3):2118–2168, 2011.

[73] King, A. D. Moduli of representations of finite dimensional algebras. *QJ Math*, 45(4):515–530, 1994.

[74] Liu, S. Semi-stable components of an Auslander-Reiten quiver. *J. London Math. Soc. (2)*, 47(3):405–416, 1993.

[75] Marks, F. and J. Šťovíček. Torsion classes, wide subcategories and localisations. *Bull. Lond. Math. Soc.*, 49(3):405–416, 2017.

[76] Miller, H., editor. *Handbook of homotopy theory*. Boca Raton, FL: CRC Press, 2020.

[77] Miyashita, Y. Tilting modules of finite projective dimension. *Math. Z.*, 193:113–146, 1986.

[78] Mizuno, Y. Classifying τ-tilting modules over preprojective algebras of Dynkin type. *Math. Z.*, 277(3-4):665–690, 2014.

[79] Mou, L. Scattering diagrams of quivers with potentials and mutations. https://arxiv.org/abs/1910.13714.

[80] Mousavand, K. τ-tilting finiteness of biserial algebras. https://arxiv.org/abs/1904.11514.

[81] Mousavand, K. τ-tilting finiteness of non-distributive algebras and their module varieties. https://arxiv.org/abs/1910.02251.

[82] Mousavand, K. and Paquette, C. Minimal (τ-)tilting infinite algebras. https://arxiv.org/abs/2103.12700.

[83] Mumford, D. *Geometric invariant theory*. Springer-Verlag, Berlin-New York, ergebnisse edition, 1965.

[84] Nagao, K. Donaldson-Thomas theory and cluster algebras. *Duke Math. J.* 162, No. 7, 1313–1367 (2013).

[85] Nakanishi, T. and Zelevinsky, A. On tropical dualities in cluster algebras. In *Algebraic groups and quantum groups. International conference on representation theory of algebraic groups and quantum groups '10, Graduate School of Mathematics, Nagoya University, Nagoya, Japan, August 2–6, 2010*, pages 217–226. Providence, RI: American Mathematical Society (AMS), 2012.

[86] Nakaoka, H. and Palu, Y. Extriangulated categories, Hovey twin cotorsion pairs and model structures. *Cah. Topol. Géom. Différ. Catég.*, 60(2):117–193, 2019.

[87] Nazarova, L. A. and Roĭter, A. V. Matrix questions and the Brauer-Thrall conjectures on algebras with an infinite number of indecomposable representations. In *Representation theory of finite groups and related topics (Proc. Sympos. Pure Math., Vol. XXI, Univ. Wisconsin, Madison, Wis., 1970)*, pages 111–115, 1971.

[88] Padrol, A., Palu, Y., Pilaud, V. and P.-Plamondon, G. Associahedra for finite type cluster algebras and minimal relations between g-vectors. https://arxiv.org/abs/1906.06861.

[89] Palu, Y., Pilaud, V. and P.-Plamondon, G. Non-kissing and non-crossing complexes for locally gentle algebras. *J. Comb. Algebra*, 3(4):401–438, 2019.

[90] Plamondon, P.-G. τ-tilting finite gentle algebras are representation-finite. *Pac. J. Math.*, 302(2):709–716, 2019.

[91] Reiten, I. The use of almost split sequences in the representation theory of Artin algebras. Representations of algebras, 3rd Int. Conf., Puebla/Mex. 1980, Lect. Notes Math. 944, 29–104 (1982), 1982.

[92] Rickard, J. Morita theory for derived categories. *J. London Math. Soc. (2)*, 39(3):436–456, 1989.

[93] Roĭter, A. V. Unboundedness of the dimensions of the indecomposable representations of an algebra which has infinitely many indecomposable representations. *Izv. Akad. Nauk SSSR Ser. Mat.*, 32:1275–1282, 1968.

[94] Rudakov, A. Stability for an abelian category. *J. Algebra*, 197(1):231–245, 1997.

[95] Schiffler, R. 2014. *Quiver representations*. CMS Books in Mathematics/Ouvrages de Mathématiques de la SMC. Springer, Cham.

[96] Schofield, A. Semi-invariants of quivers. *J. London Math. Soc. (2)*, 43(3):385–395, 1991.

[97] Schroll, S. and Treffinger, H. 2022. A τ-tilting approach to the first Brauer-Thrall conjecture. *Proc. Amer. Math. Soc.*, **150**(11), 4567–4574.

[98] Schroll, S., Treffinger, H. and Valdivieso, Y. 2021. On band modules and τ-tilting finiteness. *Math. Z.*, **299**(3–4), 2405–2417.

[99] Sentieri, F. A brick version of a theorem of Auslander. https://arxiv.org/abs/2011.09253.

[100] Skowroński, A. Generalized standard Auslander-Reiten components without oriented cycles. *Osaka J. Math.*, 30(3):515–527, 1993.

[101] Skowroński, A. Regular Auslander-Reiten components containing directing modules. *Proc. Amer. Math. Soc.*, 120(1):19–26, 1994.

[102] Smalø, S. O. The inductive step of the second Brauer-Thrall conjecture. *Canadian J. Math.*, 32(2):342–349, 1980.

[103] Smalø, S. O. Torsion theories and tilting modules. *Bull. London Math. Soc.*, 16(5):518–522, 1984.

[104] Tattar, A. Torsion pairs and quasi-abelian categories. *Algebr. Represent. Theor.*, 2020.

[105] Treffinger, H. τ-tilting theory and τ-slices. *Journal of Algebra*, 481:362–392, 2017.

[106] Treffinger, H. An algebraic approach to Harder-Narasimhan filtrations. https://arxiv.org/abs/1810.06322.

[107] Treffinger, H. On sign-coherence of c-vectors. *J. Pure Appl. Algebra*, 223(6): 2382–2400, 2019.

[108] Yurikusa, T. Wide subcategories are semistable. *Doc. Math.*, 23:35–47, 2018.

[109] Zhang, X. τ-rigid modules over Auslander algebras. *Taiwanese J. Math.*, 21(4):727–738, 2017.

3

From Frieze Patterns to Cluster Categories

Matthew Pressland

3.1 Introduction

Since their introduction by Fomin and Zelevinsky around the turn of the millennium [41], the theory of cluster algebras has had a significant influence on a wide variety of other areas of mathematics. Originally introduced to study positivity and canonical bases in Lie theory [27, 50, 54, 80], applications have been found in combinatorics [12, 60], Teichmüller theory [34, 37], algebraic geometry and mirror symmetry [5, 16, 53, 86], integrable systems [45, 70], mathematical physics [10, 46] and beyond – the references given here are by no means exhaustive, and more can be found in the introduction to [69]. In these notes we will focus on the influence of cluster algebras on the representation theory of finite-dimensional algebras, leading to the development of cluster categories by Caldero, Chapoton and Schiffler [23, 24] and Buan, Marsh, Reineke, Reiten and Todorov [19] (and later, in more generality than covered here, by Amiot [2]).

We start in Section 3.2 with the combinatorics of frieze patterns, first studied by Coxeter [31] and then Conway–Coxeter [28, 29] several decades before the introduction of cluster algebras. These patterns consist of an infinite (but periodic) strip of positive integers, and exhibit many remarkable combinatorial phenomena. To shed light on these phenomena, we will explain in Section 3.3 how the entries of the frieze can be viewed as specialisations of cluster variables, certain Laurent polynomials which are distinguished elements of a cluster algebra. Continuing this approach, further combinatorial properties of friezes, including the aforementioned periodicity, will be explained in Section 3.4 via categorification, whereby a representation-theoretically defined cluster category is introduced to provide a new perspective on the combinatorics of cluster algebras.

In this way we aim to illustrate how each successively more abstract framework – first passing from friezes to cluster algebras, and then to cluster categories – can lead to clean explanations of phenomena appearing at the previous level. This path can be followed further than we have space to do here, and leads to, for example, the theory of quivers with potential [33] and their more general cluster categories [2], Frobenius cluster categories and their applications to cluster algebras appearing in geometry [18, 49, 63, 77, 78], Adachi–Iyama–Reiten's τ-tilting theory [1] and the theory of stability conditions and scattering diagrams [16, 17, 54].

This chapter is not intended to be a comprehensive exposition of the subjects we cover, but rather a concise introduction which we hope the curious reader will use as a jumping-off point for further study. To that end, we give here a short list of suggestions for further reading covering these topics in more detail, more of which can be found in the references throughout. To the reader looking for a more detailed survey of the representation-theoretic approach to cluster algebras, we recommend Keller [69]. Several books on cluster algebras are available: that by Gekhtman, Shapiro and Vainshtein [51] focusses on connections to Poisson geometry, whereas that by Marsh [72] covers many combinatorial aspects of the theory. A further book, by Fomin, Williams and Zelevinsky, is in preparation, and at the time of writing the first six chapters may be found on arXiv [38, 39, 40]. For up-to-date information concerning the development of the theory of cluster algebras and its many applications, the reader may consult the Cluster Algebras Portal [35].

Schiffler's recent book on representations of quivers [81] contains a section on the properties of cluster-tilted algebras, a class of algebras we will mention briefly in Section 3.4 and which arises from cluster categories. Also recommended are books on representations of quivers, and more generally of finite-dimensional algebras, by Assem–Simson–Skowroński [4] and Auslander–Reiten–Smalø [7], and lecture notes by Angeleri Hügel [3]. All of these sources, particularly [3], have a strong focus on the techniques of Auslander–Reiten theory, which underlie much of the material we cover in Section 3.4.

For a more substantial discussion of the rich combinatorial theory of friezes, we recommend the survey by Morier-Genoud [74] and a paper by Propp [79]. While we restrict to classical friezes in these notes, the recent resurgence of interest in this subject has spawned a number of generalisations. Examples include 2-friezes [73], infinite friezes [8, 11, 55], SL_k-friezes [9, 30], generalised friezes (or $\{0,1\}$-friezes) [15, 59], weak friezes [26], symplectic friezes [75] and friezes with coefficients [32].

3.2 Frieze Patterns

Frieze patterns were first introduced by Coxeter [31] in 1971. Many of their combinatorial properties can be found in a paper by Conway and Coxeter published in two parts, the first [28] consisting of a list of problems, the solutions to which are provided in the second [29].

Definition 3.1 A *frieze pattern* (or simply a frieze) of *height* n consists of $n+2$ rows of positive integers, typically written in a slightly offset fashion as in the following example (with $n=6$):

$$
\begin{array}{ccccccccccccccccccccccc}
\cdots & & \mathbf{1} & & \mathbf{1} & & \mathbf{1} & & \mathbf{1} & & \mathbf{1} & & \mathbf{1} & & \mathbf{1} & & \mathbf{1} & & \mathbf{1} & & \mathbf{1} & & \cdots \\
& 3 & & 1 & & 2 & & 3 & & 2 & & 2 & & 2 & & 1 & & 5 & & 3 & & 1 & \\
\cdots & & 2 & & 1 & & 5 & & 5 & & 3 & & 3 & & 1 & & 4 & & 14 & & 2 & & \cdots \\
& 9 & & 1 & & 2 & & 8 & & 7 & & 4 & & 1 & & 3 & & 11 & & 9 & & 1 & \\
\cdots & & 4 & & 1 & & 3 & & 11 & & 9 & & 1 & & 2 & & 8 & & 7 & & 4 & & \cdots \\
& 3 & & 3 & & 1 & & 4 & & 14 & & 2 & & 1 & & 5 & & 5 & & 3 & & 3 & \\
\cdots & & 2 & & 2 & & 1 & & 5 & & 3 & & 1 & & 2 & & 3 & & 2 & & 2 & & \cdots \\
& \mathbf{1} & & \mathbf{1} & & \mathbf{1} & & \mathbf{1} & & \mathbf{1} & & \mathbf{1} & & \mathbf{1} & & \mathbf{1} & & \mathbf{1} & & \mathbf{1} & & \mathbf{1} &
\end{array}
$$

The defining properties are that

1 every entry in the uppermost or lowermost row is 1, and
2 the entries satisfy the SL_2 *diamond rule*, meaning that every local configuration

$$
\begin{array}{ccc} & b & \\ a & & d \\ & c & \end{array} \tag{3.2.1}
$$

satisfies $ad-bc=1$; in other words, the matrix $\left(\begin{smallmatrix} a & b \\ c & d \end{smallmatrix}\right)$ is an element of $\mathrm{SL}_2(\mathbb{Z})$.

We call the uppermost and lowermost rows of the frieze, consisting only of 1s, *trivial* rows, and will always write their entries in bold to emphasise their fixed value. The height measures the number of non-trivial rows.

Some readers may find Definition 3.1 somewhat informal, preferring to think of a frieze pattern as a function assigning integer values to some set giving the points of the lattice diagram on which these integers are drawn. We will take this viewpoint later in Section 3.4, but for the moment remain circumspect about which set this is.

Because of the SL_2 diamond rule, any three of the frieze entries in a diamond configuration determine the fourth. As a consequence, there are several ways to specify a subset of entries of the frieze in such a way that the rest are determined inductively by this rule.

In this chapter, we will be most interested in specifying a frieze by choosing a *lightning bolt*, consisting of one entry in each row such that the chosen entries in each pair of adjacent rows belong to the same diamond. An example is given by the circled entries in the frieze pattern below.

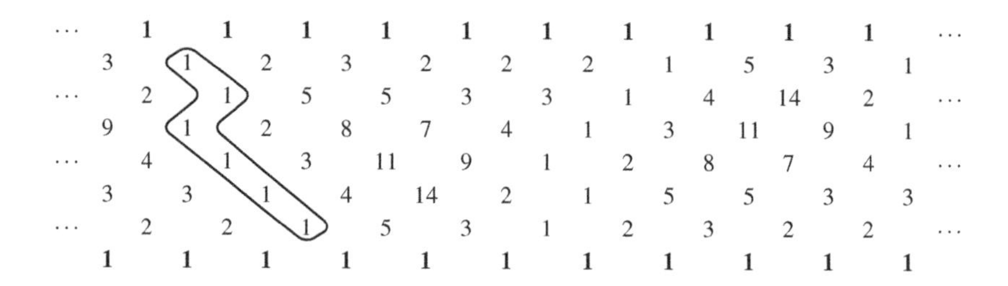

Having chosen positive integers to fill the entries in a lightning bolt, we can compute the other entries of the frieze by repeated use of the diamond rule. (The figures in the proof of Theorem 3.4 below may be helpful in seeing this.) However, this process requires division – for example, if entries a, b and c in the diamond (3.2.1) are known, then

$$d = \frac{1+bc}{a}.$$

Thus, even if a, b and c are all positive integers, as required by the definition of a frieze, there is no guarantee that d will also be an integer (although it will at least be positive). The first phenomenon that we will be concerned with is the following.

Phenomenon 1 *For each lightning bolt, there is a unique frieze pattern such that all entries in the lightning bolt are equal to* 1. *More concretely, starting with this initial set of* 1s *and computing other entries recursively using the diamond rule yields a valid frieze of positive integers.*

Example 3.2 Given Phenomenon 1, one might reasonably ask whether all frieze patterns can be obtained by starting from some lightning bolt with all entries equal to 1. The answer, however, is no; the smallest example exhibiting this is the height 3 frieze shown below.

1 1 1 1 1 1 1 1
… 1 3 1 3 1 3 1 …
2 2 2 2 2 2 2 2
… 3 1 3 1 3 1 3 …
1 1 1 1 1 1 1 1

The alert reader may have already noticed that both of the examples of frieze patterns given so far are highly symmetric; indeed, their rows are periodic with period (dividing) $n+3$. This turns out to be the case for all frieze patterns, and is a consequence of a slightly stronger periodicity.

Phenomenon 2 *Any height n frieze pattern is periodic under a glide reflection, for which a fundamental domain (excluding the trivial rows) consists of a triangle with $n+1$ elements in the base, but missing its peak, as shown in the example below.*

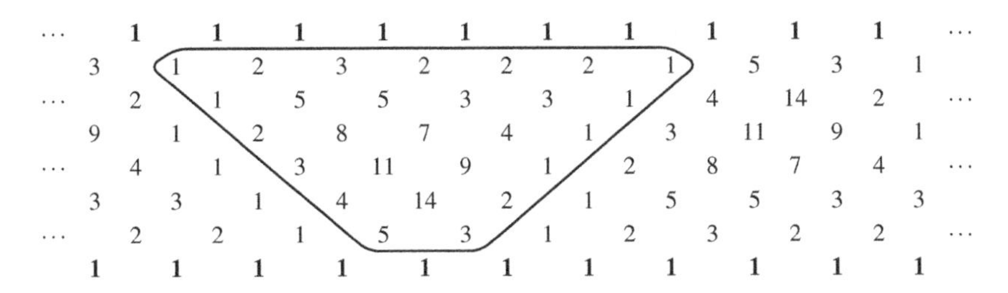

If we do include the trivial rows, a fundamental domain is given simply by a triangle with $n+2$ elements in the base, but the above description will be more recognisable later on in Section 3.4.

We will explain these phenomena, including the appearance of the frieze pattern in Example 3.2, in the coming sections, via the much more general theory of cluster algebras and cluster categories. However, before moving on to these topics, we take a digression to describe another way of generating frieze patterns, and a striking result of Conway and Coxeter.

Definition 3.3 Given a frieze pattern with height n, the (doubly infinite) sequence of integers appearing in the first non-trivial row is called its *quiddity sequence*. Since we have already remarked that this sequence is periodic of period $n+3$, we may equivalently treat it as a finite sequence of length $n+3$, considered up to cyclic reordering, i.e. up to the finest equivalence relation with

$$(a_1, a_2, \ldots, a_{n+3}) \sim (a_{n+3}, a_1, \ldots, a_{n+2}).$$

Just as for the lightning bolt, one can reconstruct a frieze (up to horizontal translation) from its quiddity sequence by repeated use of the diamond rule. Thus, given any potential quiddity sequence $(a_1, a_2, \ldots, a_{n+3})$, we can consider the rows

$$\begin{array}{cccccccccccccccccc} & \mathbf{1} & & \mathbf{1} & & \mathbf{1} & & \mathbf{1} & \cdots & & \mathbf{1} & & \mathbf{1} & & \mathbf{1} & & \mathbf{1} & \\ \cdots & & a_{n+3} & & a_1 & & a_2 & & \cdots & \cdots & & a_{n+3} & & a_1 & & a_2 & & \cdots \end{array}$$

and attempt to complete them to a frieze pattern by computing subsequent rows with the diamond rule. However, now there are many obstructions to success – as well as the problem with the lightning bolt construction, whereby a priori one could obtain non-integer entries, there is no reason why the process of computing further rows of the frieze should not continue indefinitely, rather than terminating with a second trivial row of 1s after the n non-trivial rows we expect. Moreover, we now need to use the diamond rule in the situation in which entries a, b and d of the diamond (3.2.1) are known, and the computation

$$c = \frac{ad - 1}{b}$$

involves subtraction. As a result, we also need to avoid the situation in which $a = d = 1$ (when these are entries of a non-trivial row), because in that case we would compute $c = 0$ in the next row, and thus fail to obtain a valid frieze.

Conway and Coxeter gave a remarkable method for producing all $(n+3)$-periodic sequences which are valid quiddity sequences, involving triangulations of polygons.

Definition 3.4 Consider a convex polygon with $n+3$ sides, drawn in the plane. A *triangulation* of this polygon is a maximal collection of pairwise non-crossing diagonals. In other words, it is a collection of diagonals which cut the polygon into triangles, as in the following example with $n = 6$.

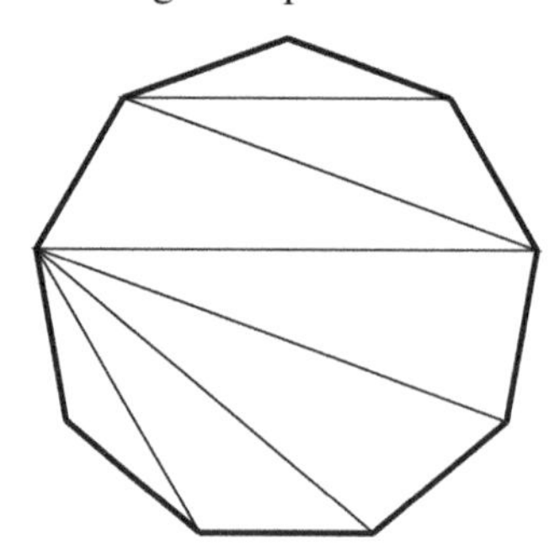

Note that we do not consider the sides of the polygon to be diagonals – if we did, they would be contained in all triangulations.

Each triangulation $\mathcal{T}$ has an associated $(n+3)$-periodic sequence defined as follows. To each vertex i of the polygon, attach the integer a_i one greater than the number of elements of $\mathcal{T}$ incident with i. (Visually, a_i is the number of triangles incident with vertex i.) Since the vertices of the polygon are cyclically ordered (say clockwise), we obtain an $(n+3)$-periodic sequence $(a_1,\ldots,a_{n+3})$. By using these sequences, Conway and Coxeter are able to classify frieze patterns in terms of triangulations.

Theorem 3.1 (Conway–Coxeter [28, 29, (28)–(29)]) *An $(n+3)$-periodic sequence is the quiddity sequence of a height n frieze if and only if it arises from a triangulation of an $(n+3)$-gon as in the preceding paragraph. This gives a bijection between triangulations of polygons (up to rotation) and frieze patterns (up to translation).*

Remark Later on it will be more natural to consider frieze patterns up to the weaker symmetry of the glide reflection appearing in Phenomenon 2. Modifying Theorem 3.1, these friezes are also classified by triangulations of an $(n+3)$-gon, but one in which the vertices are numbered clockwise by $1,\ldots,n+3$ to break the symmetry of this polygon.

For the example of a triangulation in Definition 3.4, we obtain the $(n+3)$-periodic sequence $(1,2,3,2,2,2,1,5,3)$, which is the quiddity sequence of the frieze in Definition 3.1. The frieze pattern in Example 3.2, not arising from a lightning bolt, has quiddity sequence $(1,3,1,3,1,3)$, which corresponds to the triangulation

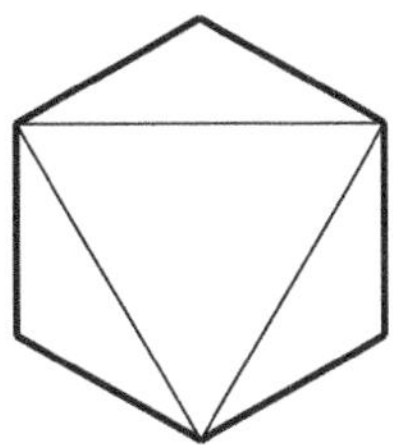

of the hexagon. It is not a coincidence that the frieze not arising from a lightning bolt corresponds to a triangulation featuring an internal triangle, of which no side is an edge of the polygon, and the frieze that does arise from a lightning bolt corresponds to a triangulation with no internal triangles.

We close this section with an experiment. Returning to the lightning bolt method for constructing friezes, rather than setting the entries in our lightning bolt equal to 1, or to other positive integers, we can fill them with formal variables $x_1,\ldots,x_n$. Using the diamond rule, we can then construct a 'frieze' whose entries are rational functions in $x_1,\ldots,x_n$. We do this below for the simplest case of a height two frieze (for which any two lightning bolts are related by translation and reflection).

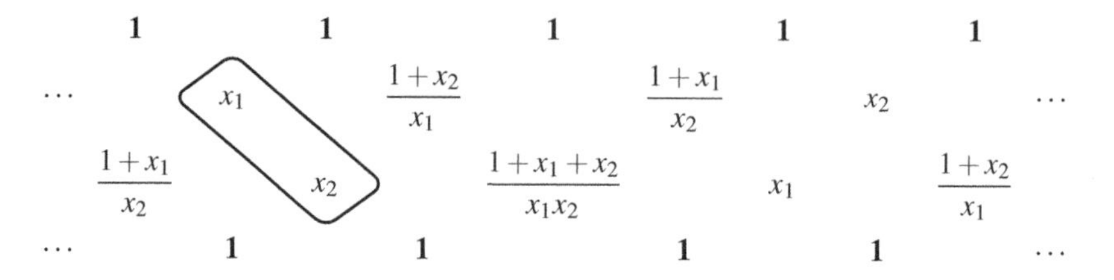

This process exhibits a *Laurent phenomenon*; all of the rational functions appearing are integral Laurent polynomials, meaning they can be expressed as a polynomial with integer coefficients divided by a monomial with coefficient 1. To see that this is not a priori clear, observe that when computing the entry $\frac{1+x_1}{x_2}$ via the SL_2 diamond rule from the three entries to its left, we calculate

$$\frac{1+\frac{1+x_1+x_2}{x_1x_2}}{\frac{1+x_2}{x_1}} = \frac{x_1(1+x_1+x_2+x_1x_2)}{x_1x_2(1+x_2)} = \frac{(1+x_1)(1+x_2)}{x_2(1+x_2)} = \frac{1+x_1}{x_2},$$

and see that the calculation initially produces a non-monomial denominator, but miraculously this has a common factor with the numerator, in such a way that we are left with a monomial again after simplifying. Note that the integrality of Phenomenon 1 follows directly from this Laurent phenomenon – evaluating an integral Laurent polynomial in $x_1,\dots,x_n$ at the point $(1,\dots,1)$ produces a fraction with denominator 1, i.e. an integer.

The five Laurent polynomials appearing in the non-trivial rows of the above 'frieze' turn out to be the five cluster variables of a cluster algebra of type A_2. In the next section, we will explain what cluster algebras and their cluster variables are in some generality, and explain the connection between friezes of height n and cluster algebras of type A_n.

3.3 Cluster Algebras

In this section we introduce cluster algebras, first defined by Fomin–Zelevinsky [41], who (together with Berenstein) developed the theory in a series of seminal papers [14, 42, 43]. For our purposes, it will be sufficient to restrict to the case of cluster algebras arising from quivers, and without coefficients or frozen variables.

A quiver is a directed graph or, more precisely, a tuple $Q = (Q_0, Q_1, h, t)$, where $Q_0 = \{1,\dots,n\}$ is a set of vertices, Q_1 is a set of arrows and

$h, t \colon Q_1 \to Q_0$ are functions specifying the head and tail respectively of each arrow. We represent quivers visually, as in the examples below.

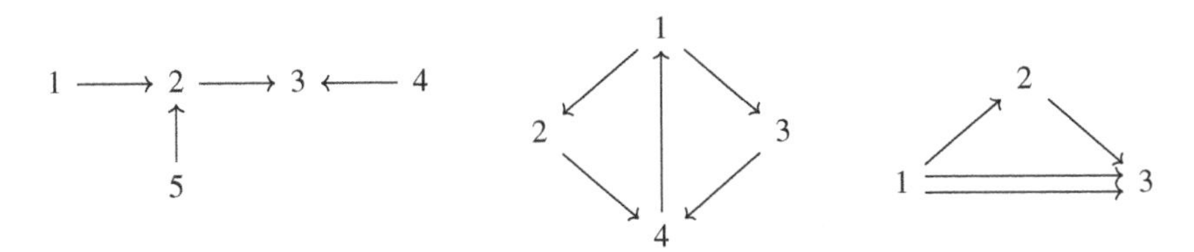

Definition 3.5 A *cluster quiver* is a quiver Q without oriented cycles of length 1 or 2. In other words, no arrow $a \in Q_1$ can have $h(a) = t(a)$ (there are *no loops*) and the configuration $i \rightleftarrows j$ is not permitted (there are *no 2-cycles*).

Remark Given a quiver Q with $Q_0 = \{1, \dots, n\}$, we can consider its *signed adjacency matrix*, the $n \times n$ matrix $M(Q)$ with entries

$$m_{ij} = \#\{\text{arrows } i \to j \text{ in } Q_1\} - \#\{\text{arrows } j \to i \text{ in } Q_1\}. \tag{3.3.1}$$

Note that $M(Q)$ is skew-symmetric, meaning its transpose is equal to its negative. Fomin–Zelevinsky define cluster algebras in terms of such skew-symmetric matrices (or more generally, skew-symmetrisable matrices, defined as those which become skew-symmetric after multiplication with some diagonal matrix), whereas we opt for the more graphical language of quivers. The condition that Q is a cluster quiver means that only one term on the right-hand side of (3.3.1) may be non-zero for a given pair (i, j), and so Q can be reconstructed up to isomorphism from the data of $M(Q)$. Indeed, given a skew-symmetric $n \times n$ matrix M, the quiver $Q(M)$ with vertex set $\{1, \dots, n\}$ and with $\max\{m_{ij}, 0\}$ arrows from i to j is the unique cluster quiver with $M(Q(M)) = M$.

Definition 3.6 Let Q be a cluster quiver and let $k \in Q_0$ be a vertex. The *mutation* of Q at k is the quiver $\mu_k Q$ obtained from Q via the following procedure.

1 For each length 2 path $i \longrightarrow k \longrightarrow j$, add an arrow $i \longrightarrow j$.
2 Reverse the direction of all arrows incident with k.
3 Choose a maximal set of 2-cycles, and remove all arrows appearing in them.

Example 3.7 We mutate the 4-vertex quiver from the earlier list of examples at vertex 1.

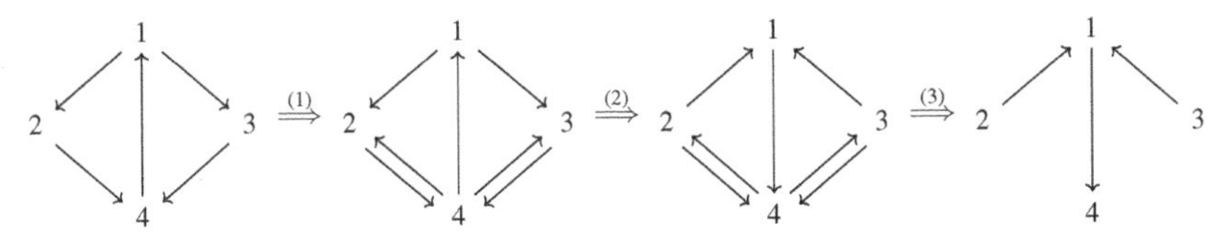

For readers wishing to experiment with further examples, we recommend Keller's Java applet [65].

It is straightforward to check that $\mu_k(\mu_k Q) = Q$, that is, mutating twice at the same vertex recovers the original quiver. Using this mutation operation, we can associate to each cluster quiver a cluster algebra, in the following way.

Definition 3.8 Let $\mathbb{Q}(x_1,\dots,x_n)$ be the field of rational functions, with rational coefficients, in variables x_i for $i \in \{1,\dots,n\}$. A *seed* consists of a cluster quiver Q with vertex set $Q_0 = \{1,\dots,n\}$ and a free generating set $\{f_1,\dots,f_n\} \subseteq \mathbb{Q}(x_1,\dots,x_n)$ indexed by this set of vertices – being a free generating set means that the smallest subfield of $\mathbb{Q}(x_1,\dots,x_n)$ containing $\mathbb{Q}$ and the set $\{f_1,\dots,f_n\}$ is $\mathbb{Q}(x_1,\dots,x_n)$ itself.

We can extend mutation to an operation on seeds. Given a seed $(Q,\{f_i\})$ and a vertex $k \in Q_0$, we define $\mu_k(Q,\{f_i\}) = (\mu_k Q,\{f_i'\})$, where $\mu_k Q$ is the mutated quiver as in Definition 3.6, and

$$f_i' = \begin{cases} f_i, & i \neq k, \\ \dfrac{1}{f_k}\Big(\prod_{k\to j} f_j + \prod_{\ell\to k} f_\ell\Big), & i = k. \end{cases} \tag{3.3.2}$$

In the second case, the products are over arrows with tail, respectively head, k. These products may be over the empty set, in which case they evaluate to 1. Note that mutating twice at the same vertex recovers the original seed.

Given a cluster quiver Q with vertices $Q_0 = \{1,\dots,n\}$, we consider the *initial seed* $s_0 = (Q,\{x_i\})$, whose functions are given by the distinguished generators of $\mathbb{Q}(x_1,\dots,x_n)$. Let $\mathcal{S}_Q$ be the set of all seeds obtained from s_0 by a finite sequence of mutations. The *cluster algebra* A_Q of Q is the $\mathbb{Q}$-subalgebra of $\mathbb{Q}(x_1,\dots,x_n)$ generated by all functions appearing in all seeds in $\mathcal{S}_Q$.

Each function appearing in a seed in $\mathcal{S}_Q$ is called a *cluster variable* of A_Q, and the set $\{f_i\}$ of functions appearing in a single seed $(Q',\{f_i\}) \in \mathcal{S}_Q$ is called a *cluster* of A_Q.

Despite this somewhat esoteric definition, cluster algebras have made surprising appearances in a number of areas of mathematics, as we observed in the introduction. Notably, the coordinate rings of many important algebraic varieties, such as Grassmannians and other varieties of flags, are isomorphic to cluster algebras [14, 49, 82], at least after extending the definition slightly to replace $\mathbb{Q}$ by a field extension (typically $\mathbb{C}$) and declaring a subset of vertices of Q to be frozen, meaning that mutations at these vertices are not permitted when constructing the set $\mathcal{S}_Q$ of seeds – this means in particular that the variable x_i appears in every cluster when i is a frozen vertex.

The main conclusion of this section will be that for certain cluster algebras, the cluster variables give formulae for the entries of a frieze pattern in terms of the entries in a lightning bolt, as the following simple example demonstrates.

Example 3.9 Let Q be the quiver $1 \longrightarrow 2$. Then we can compute that the seeds of A_Q are

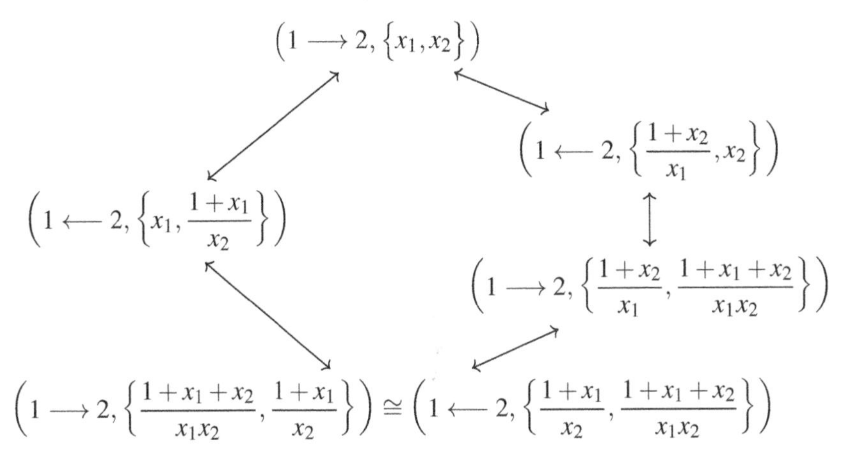

In this diagram, each two-headed arrow represents a mutation. The 'isomorphism' on the final line indicates that the two seeds are related by relabelling the quiver vertices, and correspondingly re-ordering the cluster variables. (If we suppressed this vertex labelling by writing each cluster variable directly on the corresponding quiver vertex, both seeds would consist of an arrow from $\frac{1+x_1+x_2}{x_1x_2}$ to $\frac{1+x_1}{x_2}$.) We see that there are five cluster variables, which are precisely the five Laurent polynomials that appeared in our experiment at the end of Section 3.2.

We now give two important theorems concerning cluster algebras in general, which will shed light on the phenomena concerning frieze patterns from Section 3.2, at least once we have understood how frieze pattern entries are given by specialising cluster variables. The first is the *Laurent phenomenon*, which will completely explain Phenomenon 1.

Theorem 3.2 *Let Q be a cluster quiver. Then every cluster variable in A_Q is an integral Laurent polynomial in the initial variables $\{x_1,\dots,x_n\}$.*

Proof A combinatorial proof is given by Fomin and Zelevinsky [41, Thm. 3.1] in the original paper defining cluster algebras (in more generality than here).

Gross, Hacking and Keel [53, Cor. 3.11] give a deeper geometric argument, which involves realising cluster variables as regular functions on a cluster variety (an algebraic space defined as a union of tori indexed by the set of seeds). □

Since every cluster $\{f_i\}$ in A_Q is a free generating set for $\mathbb{Q}(x_1,\dots,x_n)$, any element of this field has a unique expression as a rational function in the f_i. By change of variables, it follows from Theorem 3.2 that the cluster variables of A_Q are given by integral Laurent polynomials when written as rational functions in any of the clusters, not just the initial cluster $\{x_1,\dots,x_n\}$.

For most quivers Q, the set $\mathcal{S}_Q$ of seeds will be infinite. The second important theorem we mention here concerns when, as in Example 3.9, the number of seeds (and therefore the number of cluster variables) is finite, in which case we say that the cluster algebra has *finite type*.

Theorem 3.3 ([42, Thm. 1.4]) *A cluster algebra A_Q has finite type if and only if Q is related by a sequence of mutations to a quiver obtained by orienting the edges of one of the following graphs.*

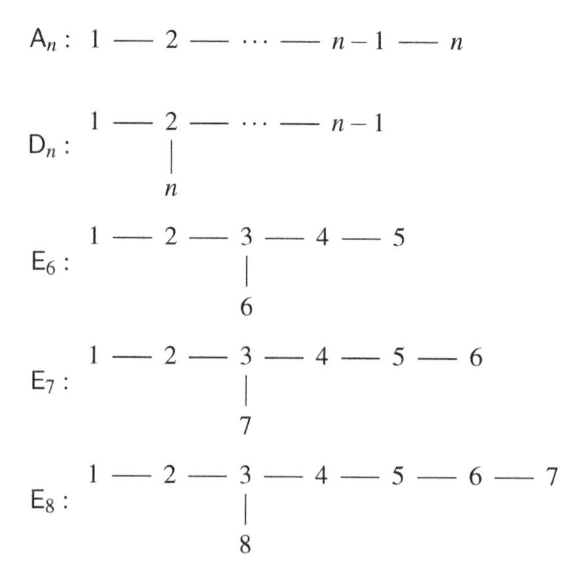

The graphs appearing in Theorem 3.3 are the simply laced Dynkin diagrams, and indeed we call their orientations *Dynkin quivers*. These diagrams (together with their non-simply laced counterparts) often appear as the solution to classification problems, most famously that of semisimple Lie algebras over an algebraically closed field, via the classification of finite root systems. The number of cluster variables in the cluster algebra of a Dynkin quiver turns out to be n greater than the number of positive roots in the corresponding root system, where n is the number of vertices of the quiver [42, Thm. 1.9]. When considering more general cluster algebras from weighted quivers (or skew-symmetrisable matrices, as in [42]), the statement corresponding to Theorem 3.3 also includes the non-simply laced Dynkin diagrams. For an introduction to these Lie-theoretic concepts, we recommend the book by Fulton and Harris [47].

We also remark that any two Dynkin quivers with the same underlying graph are related by a sequence of mutations (indeed, by mutations only at sinks and sources). Thus the cluster algebras associated to two different orientations of the same diagram are related by a change of variables, and so up to isomorphism there is one cluster algebra per Dynkin diagram.

The final observation for us to make in this section is that the formulae expressing general entries of a height n frieze in terms of the entries in a lightning bolt are given by cluster variables in a cluster algebra of type A_n, i.e. the cluster algebra associated to a quiver whose underlying graph is this Dynkin diagram. Then Phenomenon 1 will follow from the Laurent phenomenon in Theorem 3.2. While we will still lack an explanation for the periodicity in Phenomenon 2, this being the topic of the next section, Theorem 3.3 tells us that type A_n cluster algebras have only finitely many cluster variables, and so we will at least see that a frieze pattern can have only finitely many distinct entries.

Theorem 3.4 *Choose a lightning bolt in a frieze of height n. Then every entry of the frieze is obtained by taking a cluster variable of $\mathcal{A}_Q$, where Q is an orientation of the diagram A_n, and specialising the indeterminates $x_1,\dots,x_n$ to the values of the frieze in the lightning bolt.*

Proof We sketch the argument. The first step is to construct a quiver Q from a lightning bolt L. The vertex set is $Q_0 = \{1,\dots,n\}$ as usual, and we associate i to the i-th row of the frieze (treating the upper trivial row as the 0-th row, so that our quiver vertices correspond to the non-trivial rows). There is exactly one arrow a_i between each pair of vertices i and $i+1$, for which $h(a_i) = i$ and $t(a_i) = i+1$ if the entry of L in row $i+1$ is to the left of that in row i, and $h(a) = i+1$, $t(a_i) = i$ otherwise. We then consider the initial seed attached to this quiver. For the lightning bolt given as an example in Section 3.2, we obtain the seed

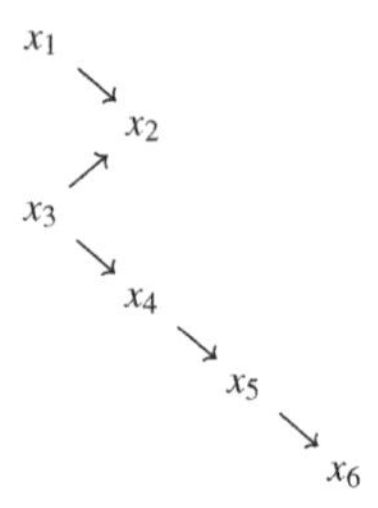

Because Q does not have any oriented cycles, it must have at least one vertex, say k, which is a source. In particular, there are no length 2 paths through k, and so mutating at this vertex only reverses the direction of the incident arrows, making k into a sink. Similarly, when computing the new cluster variable x_k' in the mutated seed via formula (3.3.2), one of the two products is empty, and we obtain

$$x_k' = \frac{1 + x_{k-1}x_{k+1}}{x_k},$$

adopting the convention that $x_0 = x_{n+1} = 1$. In other words, the configuration

$$\begin{matrix} & x_{k-1} & \\ x_k & & x_k' \\ & x_{k+1} & \end{matrix}$$

satisfies the diamond rule of a frieze pattern, and so if we specialise the initial variables x_k to the entries of the frieze in L, the cluster variable x_k' will be specialised to the frieze entry directly to the right of that in the k-th row of L. Moreover, the quiver $\mu_k Q$ of this new seed is exactly the one coming from the lightning bolt $\mu_k L$ obtained from L by replacing the entry in the k-th row by that to its right – the fact that k is a source in Q means that $\mu_k L$ is again a lightning bolt.

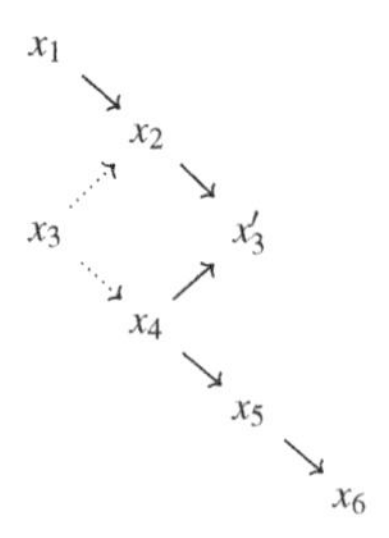

By continuing to mutate at sources in this way, we see that all of the frieze entries to the right of the lightning bolt L are specialisations of cluster variables in A_Q.

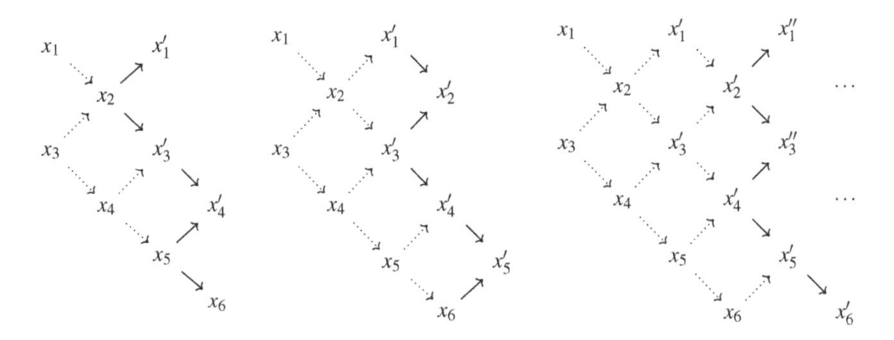

We obtain the same result for the entries to the left of L by instead applying mutations at sinks. □

Note that we did not consider all possible mutations of our initial seed in the proof of Theorem 3.4, and indeed we do not find all seeds of A_Q via mutations only at sources and sinks. On the other hand, it turns out, as we will see in the next section, that we do find all cluster variables of A_Q in this way. Indeed, the cluster algebra of type A_n has $\frac{1}{2}n(n+3)$ cluster variables, which is the number of elements of the fundamental domain appearing in Phenomenon 2.

3.4 Cluster Categories

In this section, we explain how representation theory can be used to study cluster algebras and, by extension, frieze patterns. This will require introducing a number of algebraic concepts relatively quickly: readers wishing to learn more about the representation theory of quivers (and other finite-dimensional algebras) are advised to consult books by Auslander–Reiten–Smalø [7], Assem–Simson–Skowroński [4] or Schiffler [81], the latter also including a section on cluster-tilted algebras, the definition of which is related to the ideas presented in these notes. We rely heavily on techniques from Auslander–Reiten theory, and a concise introduction to this subject may be found in lecture notes by Angeleri Hügel [3], with more extensive discussions in the preceding sources.

We will be particularly brief in our discussion of the derived category, since we will need only to understand a very special example in any detail. A slightly less brief explanation, in a more general context, can be found in Appendix 3.5, and a reader looking for a detailed exposition of this extremely useful and powerful construction is advised to consult Keller's survey [66], or the relevant sections of books by Happel [57] or Gelfand and Manin [52].

A *category* $\mathcal{C}$ consists of a set $\mathrm{Ob}(\mathcal{C})$ of objects, a set $\mathrm{Hom}_{\mathcal{C}}(X,Y)$ of morphisms $X \to Y$ for any pair of objects $X,Y \in \mathrm{Ob}(\mathcal{C})$, and an associative composition law consisting of maps $\circ\colon \mathrm{Hom}_{\mathcal{C}}(Y,Z) \times \mathrm{Hom}_{\mathcal{C}}(X,Y) \to \mathrm{Hom}_{\mathcal{C}}(X,Z)$ for each triple $X,Y,Z \in \mathrm{Ob}(\mathcal{C})$; we write $g \circ f = \circ(g,f)$. Each object X has an identity morphism 1_X, such that $1_X \circ f = f$ and $g \circ 1_X = g$ whenever these compositions are defined. All of our categories will be $\mathbb{K}$-linear, for $\mathbb{K}$ a field, meaning that the sets $\mathrm{Hom}_{\mathcal{C}}(X,Y)$ are $\mathbb{K}$-vector spaces, and the composition maps are $\mathbb{K}$-bilinear. We will, as is common, write $X \in \mathcal{C}$ as shorthand for $X \in \mathrm{Ob}(\mathcal{C})$.

Given categories $\mathcal{C}$ and $\mathcal{D}$, a *functor* $F\colon \mathcal{C} \to \mathcal{D}$ consists of a map $F\colon \mathrm{Ob}(\mathcal{C}) \to \mathrm{Ob}(\mathcal{D})$ and, for each pair of objects $X,Y \in \mathcal{C}$, a ($\mathbb{K}$-linear) map $F\colon \mathrm{Hom}_{\mathcal{C}}(X,Y) \to \mathrm{Hom}_{\mathcal{D}}(FX,FY)$. We require that $F(1_X) = 1_{FX}$ for all $X \in \mathcal{C}$ and $F(g \circ f) = F(g) \circ F(f)$ for any composable morphisms f and g in $\mathcal{C}$. A functor $F\colon \mathcal{C} \to \mathcal{D}$ is an *equivalence* if it admits a weak inverse, a functor $G\colon \mathcal{D} \to \mathcal{C}$ such that both $F \circ G$ and $G \circ F$ are naturally isomorphic [4, §A.2] to the identity functors on $\mathcal{D}$ and $\mathcal{C}$ respectively. This means in particular that $G(F(X)) \cong X$ for any object $X \in \mathcal{C}$, and $F(G(Y)) \cong Y$ for any object $Y \in \mathcal{D}$. An equivalence $F\colon \mathcal{C} \to \mathcal{C}$ is called an *autoequivalence* of $\mathcal{C}$. Note that the notion of an equivalence (or autoequivalence) is weaker than that of an isomorphism (or automorphism), which requires that $F \circ G$ and $G \circ F$ are really equal, not only naturally isomorphic, to the relevant identity functors.

The goal of this section will be to define, for any acyclic quiver Q, a category $\mathcal{C}_Q$, called the cluster category of Q, whose properties reflect those of the cluster algebra A_Q. These categories were originally described by Buan, Marsh, Reineke, Reiten and Todorov [19] (and by Caldero, Chapoton and Schiffler in some special cases [23, 24]). The upshot for us will be that, when Q is an orientation of the graph A_n, we can view a height n frieze pattern as a function from the set $\mathrm{ind}\,\mathcal{C}_Q$, of isomorphism classes of indecomposable objects of $\mathcal{C}_Q$, to the positive integers. This will explain the periodicity in Phenomenon 2, and tell us how to find all height n friezes, not only those arising from lightning bolts.

The construction uses the representation theory of the quiver Q. In the case that Q has underlying graph A_n, relevant to friezes, we will also give a different and more combinatorial explanation, that may be easier to follow for readers not already familiar with quiver representations.

Definition 3.10 A representation (V,f) (often abbreviated to just V) of a quiver Q is an assignment of a $\mathbb{K}$-vector space V_i to each vertex $i \in Q_0$ and a linear map $f_a\colon V_{t(a)} \to V_{h(a)}$ to each arrow $a \in Q_1$. Given representations (V,f) and (W,g) of Q, a morphism $\varphi\colon (V,f) \to (W,g)$ consists of a linear map $\varphi_i\colon V_i \to W_i$ for each i in Q_0, such that the diagram

$$\begin{array}{ccc} V_{t(a)} & \xrightarrow{f_a} & V_{h(a)} \\ \varphi_{t(a)} \downarrow & & \downarrow \varphi_{h(a)} \\ W_{t(a)} & \xrightarrow[g_a]{} & W_{h(a)} \end{array}$$

commutes for any $a \in Q_1$. The morphism φ is an isomorphism if every φ_i is an isomorphism of vector spaces.

We may define the direct sum of two representations pointwise, i.e. by $(V \oplus W)_i = V_i \oplus W_i$, with morphisms

$$\begin{pmatrix} f_a & 0 \\ 0 & g_a \end{pmatrix} : V_{t(a)} \oplus W_{t(a)} \to V_{h(a)} \oplus W_{h(a)}.$$

A representation is *indecomposable* if it is non-zero (i.e. some V_i is not the zero vector space) and not isomorphic to the direct sum of two non-zero representations.

We restrict here to finite-dimensional representations, meaning that the vector spaces attached to quiver vertices are all finite-dimensional. We denote by $\mathsf{rep}Q$ the category whose objects are such representations of Q, and whose morphisms are those described above, with composition $(\psi \circ \varphi)_i = \psi_i \circ \varphi_i$. This category is *abelian*: among other things, this means that it has direct sums as defined above, its morphism spaces are abelian groups (even $\mathbb{K}$-vector spaces), each morphism has a kernel and a cokernel, every injective morphism is a kernel, and every surjective morphism is a cokernel.

Remark Readers familiar with category theory may recognise the commutative diagram in Definition 3.10 as similar to that involved in the definition of a natural transformation of functors [4, §A.2]. Indeed, the quiver Q determines a path category, with objects the quiver vertices, morphisms i to j given by the directed paths from i to j and composition given by concatenating paths. A representation of Q is then nothing but a covariant functor from this path category to the category of $\mathbb{K}$-vector spaces, and a morphism of representations is a natural transformation between two such functors.

The next step towards our desired category $\mathcal{C}_Q$ is to construct the bounded derived category $\mathcal{D}^{\mathrm{b}}(Q)$ of $\mathsf{rep}Q$. This construction, first introduced by Verdier [84], can be made for any abelian category [85] (and even more general categories) but is somewhat complicated. For this reason, we give an *ad hoc* definition in the case of quiver representations, exploiting special properties of the abelian category $\mathsf{rep}Q$ (most importantly that it is hereditary, meaning that every subobject of a projective object [4, Def. I.5.2] is again projective). The more general construction, and its relationship to the definition given now, is briefly described in Appendix 3.5.

Definition 3.11 Given a quiver Q, the *bounded derived category* $\mathcal{D}^{\mathrm{b}}(Q)$ of Q is defined as follows. For each $i \in \mathbb{Z}$ and $V \in \operatorname{rep} Q$, we introduce the formal symbol $\Sigma^i V$, and take the objects of $\mathcal{D}^{\mathrm{b}}(Q)$ to be formal direct sums of these symbols. Each morphism space

$$\operatorname{Hom}_{\mathcal{D}^{\mathrm{b}}(Q)}(\Sigma^i V, \Sigma^j W) := \operatorname{Ext}_Q^{j-i}(V, W) \tag{3.4.1}$$

is given by an extension group [4, §A.4] in $\operatorname{rep} Q$. Since $\operatorname{rep} Q$ is a hereditary abelian category, this vector space may only be non-zero when $j-i=0$ or $j-i=1$; in the first case we have $\operatorname{Ext}_Q^0(V,W) = \operatorname{Hom}_{\operatorname{rep} Q}(V,W)$, and in the second $\operatorname{Ext}_Q^1(V,W)$ is a vector space describing the possible short exact sequences of the form

$$0 \to W \to E \to V \to 0$$

in $\operatorname{rep} Q$ – see [3, §1.7] for a quick introduction. The definition (3.4.1) extends to morphism spaces between formal direct sums via the formulae

$$\operatorname{Hom}_{\mathcal{D}^{\mathrm{b}}(Q)}(X_1 \oplus X_2, Y) = \operatorname{Hom}_{\mathcal{D}^{\mathrm{b}}(Q)}(X_1, Y) \oplus \operatorname{Hom}_{\mathcal{D}^{\mathrm{b}}(Q)}(X_2, Y),$$
$$\operatorname{Hom}_{\mathcal{D}^{\mathrm{b}}(Q)}(X, Y_1 \oplus Y_2) = \operatorname{Hom}_{\mathcal{D}^{\mathrm{b}}(Q)}(X, Y_1) \oplus \operatorname{Hom}_{\mathcal{D}^{\mathrm{b}}(Q)}(X, Y_2),$$

and the composition law is defined using cup product of extensions.

The reason for writing $\Sigma^i V$ to encode the pair (i,V) is that the category $\mathcal{D}^{\mathrm{b}}(Q)$ carries an autoequivalence $\Sigma\colon \mathcal{D}^{\mathrm{b}}(Q) \to \mathcal{D}^{\mathrm{b}}(Q)$, whose action on objects is exactly as suggested by the notation. On morphisms, Σ acts as the identity, noting that

$$\operatorname{Hom}_{\mathcal{D}^{\mathrm{b}}(Q)}(\Sigma^{i+1} V, \Sigma^{j+1} W) = \operatorname{Ext}_Q^{j-i}(V, W) = \operatorname{Hom}_{\mathcal{D}^{\mathrm{b}}(Q)}(\Sigma^i V, \Sigma^j W).$$

One can give $\mathcal{D}^{\mathrm{b}}(\mathcal{A})$ the structure of a triangulated category in which Σ is the suspension functor, part of the data of such a structure. We will not discuss the general definition or theory of triangulated categories, but refer the interested reader to Happel's book [57].

When Q is a Dynkin quiver, $\mathcal{D}^{\mathrm{b}}(Q)$ may be described in very combinatorial terms. First we associate to Q an infinite quiver $\mathbb{Z}Q$, as follows. The vertices of $\mathbb{Z}Q$ are pairs (i,n) with $i \in Q_0$ and $n \in \mathbb{Z}$, and its arrows are $a_n\colon (t(a),n) \to (h(a),n)$ and $a_n^*\colon (h(a),n) \to (t(a),n+1)$ for each $a \in Q_1$ and $n \in \mathbb{Z}$. The quiver $\mathbb{Z}Q$ is highly symmetric – most useful to us is the translation symmetry τ, defined by $\tau(i,n) = (i,n-1)$, $\tau(a_n) = a_{n-1}$ and $\tau(a_n^*) = a_{n-1}^*$.

Each vertex (i,n) of $\mathbb{Z}Q$ gives rise to the *mesh relation*

$$\sum_{a:h(a)=i} a_{n+1}a_n^* - \sum_{b:t(a)=i} b_n^* b_n,$$

a formal linear combination of paths in $\mathbb{Z}Q$ (which we read from right-to-left, like composition of functions). For example, if Q is the quiver

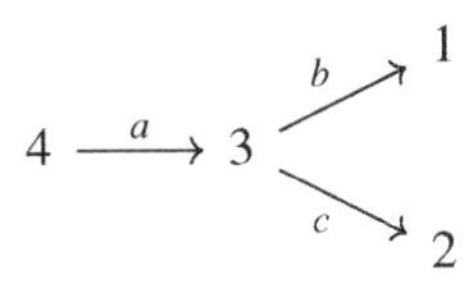

then $\mathbb{Z}Q$ has the local configuration

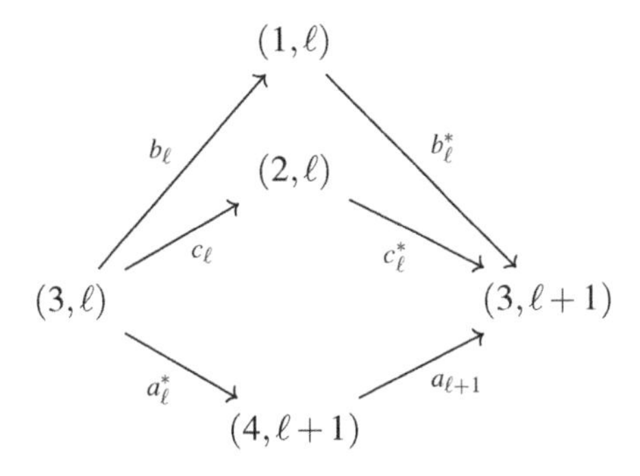

for each $\ell \in \mathbb{Z}$, and the mesh relation is $a_{\ell+1}a_\ell^* - b_\ell^* b_\ell - c_\ell^* c_\ell$.

Now we define a category $\mathcal{D}_Q$ whose objects are formal direct sums of vertices of $\mathbb{Z}Q$. The set $\mathrm{Hom}_{\mathcal{D}_Q}((i,n),(j,m))$ of morphisms from a vertex (i,n) to a vertex (j,m) is the vector space spanned by paths from (i,n) to (j,m), subject to the mesh relations: this means that a linear combination of paths obtained from a mesh relation by postcomposing all terms with a fixed path p and precomposing all terms with a fixed path q is 0. For example, if we see the configuration

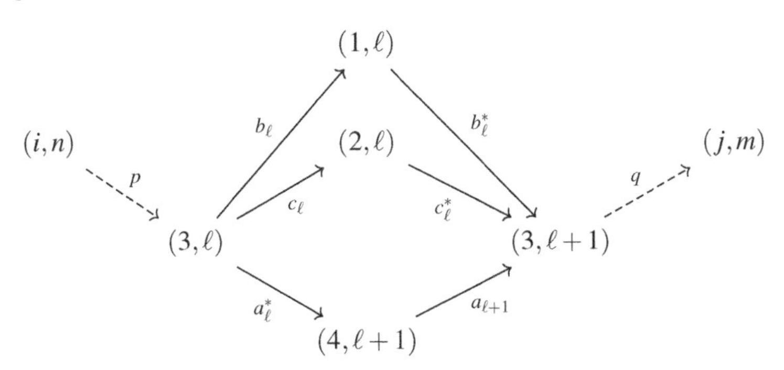

for some paths p and q, then the equation

$$qa_{\ell+1}a_\ell^*p - qb_\ell^*b_\ell p - qc_\ell^*c_\ell p = q(a_{\ell+1}a_\ell - b_\ell^*b_\ell - c_\ell^*c_\ell)p = 0$$

holds in $\mathrm{Hom}_{\mathcal{D}_Q}((i,n),(j,m))$. As above, we extend this definition of morphisms to all objects via the rules

$$\mathrm{Hom}_{\mathcal{D}_Q}(v_1 \oplus v_2, w) = \mathrm{Hom}_{\mathcal{D}_Q}(v_1, w) \oplus \mathrm{Hom}_{\mathcal{D}_Q}(v_2, w),$$
$$\mathrm{Hom}_{\mathcal{D}_Q}(v, w_1 \oplus w_2) = \mathrm{Hom}_{\mathcal{D}_Q}(v, w_1) \oplus \mathrm{Hom}_{\mathcal{D}_Q}(v, w_2).$$

The symmetry τ of $\mathbb{Z}Q$ takes each mesh relation to another mesh relation, and so induces an autoequivalence of $\mathcal{D}_Q$.

Theorem 3.5 ([57, §I.5.6]) *When Q is a Dynkin quiver, there is an equivalence of categories $\mathcal{D}_Q \xrightarrow{\sim} \mathcal{D}^b(Q)$.*

This statement is perhaps a bit misleading for readers already familiar with representations of quivers. Indeed, the most natural equivalence of categories is $\mathcal{D}_{Q^{\mathrm{op}}} \xrightarrow{\sim} \mathcal{D}^{\mathrm{b}}(Q)$, where Q^{op} is the opposite quiver of Q, obtained by reversing the directions of all the arrows, and takes the object $(i,0) \in \mathcal{D}_Q$ to $\Sigma^0 P_i \in \mathcal{D}^{\mathrm{b}}(Q)$, where P_i is the indecomposable projective representation at vertex i. However, by drawing some examples (or looking at the A_6 case shown below) the reader will quickly convince themselves that $\mathcal{D}_Q$ is independent of the orientation of Q, up to equivalence of categories, and so the theorem as stated is also true. In our application, it will be enough to know that we can substitute $\mathcal{D}^{\mathrm{b}}(Q)$ for $\mathcal{D}_Q$ when Q is a quiver of type A_n, and the details of the functor providing the equivalence will not be relevant.

The following picture shows $\mathcal{D}_Q$ in the case that Q is the orientation of A_6 used to illustrate the proof of Theorem 3.4.

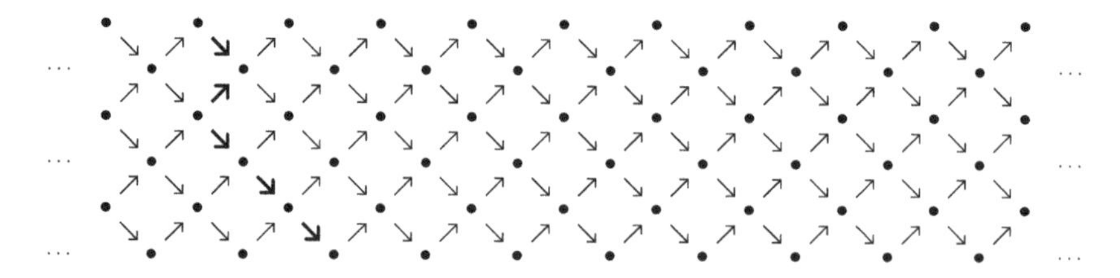

In this picture, each point • represents an indecomposable object in $\mathcal{D}_Q$, i.e. one of the objects (i,n), and each arrow represents an irreducible morphism, a non-isomorphism between indecomposable objects which is not expressible as a product of such morphisms. The arrows between the vertices $(i,0)$ are shown in bold, making the initial quiver Q visible. Comparing to Section 3.2, we see that it is natural to think of a height n frieze pattern (excluding the trivial rows) as a function on the indecomposable objects of $\mathcal{D}_Q$ (or equivalently those of $\mathcal{D}^{\mathrm{b}}(Q)$, by Theorem 3.5) for any orientation Q of the Dynkin diagram A_n.

The mesh relations in $\mathcal{D}_Q$ say that each square

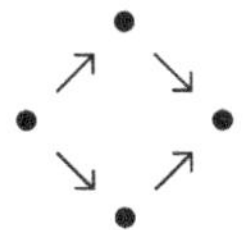

either commutes or anti-commutes, whereas at the bottom of the diagram, a pair of morphisms

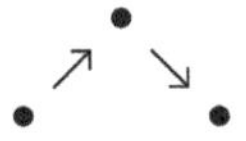

composes to zero (and similarly at the top). In type A_n, it is not difficult to compute the space of morphisms between any pair of objects using these rules, as in the following example.

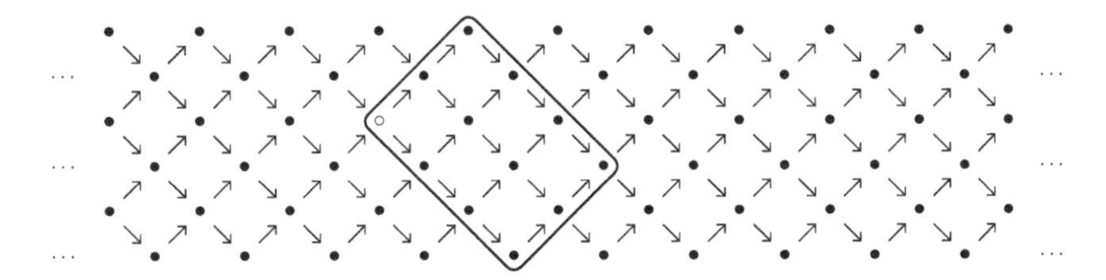

The rectangle shows those indecomposable objects having a non-zero morphism from the fixed object indicated by $\circ$, at the left-hand corner of the rectangle. Moreover, the space of morphisms from $\circ$ to each of these objects is 1-dimensional, spanned by any path from $\circ$ to the given object – any two such paths determine the same morphism (up to sign) because of the mesh relations. In general the morphisms starting at a given indecomposable object $X \in \mathcal{D}_Q$, for Q of type A_n, can be computed by drawing a maximal rectangle with X at its left-hand corner: note that when X is on the upper or lower edge of the figure, this rectangle will degenerate to a line.

The autoequivalence τ acts by translating the picture one step to the left, whereas Σ acts by a glide reflection to the right, with fundamental domain as shown.

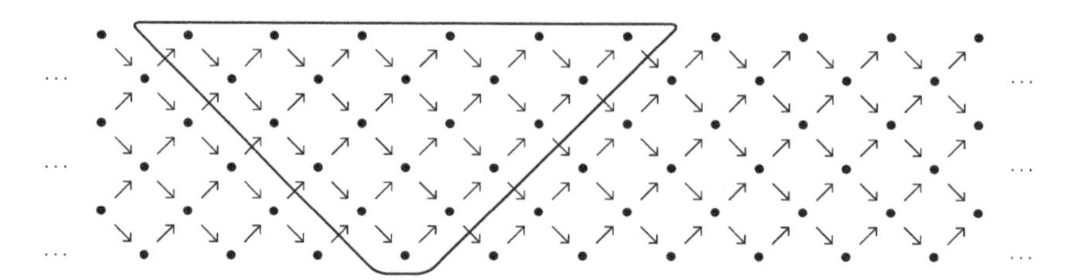

A consequence of Theorem 3.5 is that the autoequivalence τ of $\mathcal{D}_Q$ can be seen as an autoequivalence of $\mathcal{D}^{\mathrm{b}}(Q)$. In fact, this coincides with an autoequivalence of $\mathcal{D}^{\mathrm{b}}(Q)$ defined intrinsically for any acyclic quiver Q, not just the Dynkin quivers – it is closely related to a functor on rep$Q$ called the Auslander–Reiten translation, also typically denoted by τ. (For readers already familiar with this functor on rep$Q$; in $\mathcal{D}^{\mathrm{b}}(Q)$, the equivalence τ takes $\Sigma^n V$ to $\Sigma^n(\tau V)$ when V is indecomposable and not projective, and takes $\Sigma^n P_i$ to $\Sigma^{n-1} I_i$, where P_i, respectively I_i, denotes the indecomposable projective, respectively injective, representation at vertex i.)

The two autoequivalences Σ and τ of $\mathcal{D}^{\mathrm{b}}(Q)$ even commute with each other: $\Sigma \circ \tau = \tau \circ \Sigma$. We will write $\rho = \Sigma^{-1} \circ \tau$, and using this autoequivalence of $\mathcal{D}^{\mathrm{b}}(Q)$ we may finally define the category $\mathcal{C}_Q$. First we give a general construction.

Definition 3.12 Let $\mathcal{C}$ be an additive category, and let $F\colon \mathcal{C} \to \mathcal{C}$ be an automorphism (meaning that there is a functor $F^{-1}\colon \mathcal{C} \to \mathcal{C}$ such that the compositions $F \circ F^{-1}$ and $F^{-1} \circ F$ are both equal to – not only naturally isomorphic to – the identity functor on $\mathcal{C}$). Then the *orbit category* $\mathcal{C}/F$ has the same objects as $\mathcal{C}$, but with morphism spaces

$$\operatorname{Hom}_{\mathcal{C}/F}(X,Y) = \bigoplus_{n\in\mathbb{Z}} \operatorname{Hom}_{\mathcal{C}}(X, F^n Y). \tag{3.4.2}$$

The composition

$$\circ\colon \operatorname{Hom}_{\mathcal{C}/F}(Y,Z) \times \operatorname{Hom}_{\mathcal{C}/F}(X,Y) \to \operatorname{Hom}_{\mathcal{C}/F}(X,Z)$$

has components given by the maps

$$\begin{aligned} \operatorname{Hom}_{\mathcal{C}}(Y,F^nZ) \times \operatorname{Hom}_{\mathcal{C}}(X,F^mY) &\to \operatorname{Hom}_{\mathcal{C}}(X,F^{n+m}Z), \\ (g,f) &\mapsto F^m(g) \circ f, \end{aligned}$$

noting that

$$F^m\colon \operatorname{Hom}_{\mathcal{C}}(Y,F^nZ) \to \operatorname{Hom}_{\mathcal{C}}(F^mY, F^{n+m}Z).$$

If F is only an autoequivalence, rather than an automorphism, then this composition need not be well-defined. On the other hand, in this case one can choose a skeleton $\bar{\mathcal{C}} \subset \mathcal{C}$, consisting of exactly one object in each isomorphism class, with morphisms between such objects defined exactly as in $\mathcal{C}$. Then the inclusion $\bar{\mathcal{C}} \hookrightarrow \mathcal{C}$ is an equivalence, there is an automorphism $\bar{F}\colon \bar{\mathcal{C}} \to \bar{\mathcal{C}}$ related to F in a natural way (for example, $\bar{F}X$ is the unique object of $\bar{\mathcal{C}}$ isomorphic to FX in $\mathcal{C}$), and we may define $\mathcal{C}/F = \bar{\mathcal{C}}/\bar{F}$. More details concerning this subtlety may be found in a paper of Bennett-Tennenhaus and Shah [13] (see also [83, §II.8.1]).

Definition 3.13 ([19, §1]) Given an acyclic quiver Q, its *cluster category* $\mathcal{C}_Q$ is the orbit category

$$\mathcal{C}_Q = \mathcal{D}^{\mathrm{b}}(Q)/\rho$$

for the action of $\rho = \Sigma^{-1} \circ \tau$ on $\mathcal{D}^{\mathrm{b}}(Q)$. For each $X, Y \in \mathcal{D}^{\mathrm{b}}(Q)$, only finitely many of the vector spaces appearing on the right-hand side of (3.4.2) are non-zero, and so the morphism spaces in $\mathcal{C}_Q$ are still finite-dimensional vector spaces. Like $\mathcal{D}^{\mathrm{b}}(Q)$, the cluster category $\mathcal{C}_Q$ is again a triangulated category, by a result of Keller [67].

Remark Since ρ is not an automorphism of $\mathcal{D}^{\mathrm{b}}(Q)$, but only an autoequivalence, one has to take account of the subtlety referred to in Definition 3.12 when defining the cluster category. However, in the special case of Dynkin quivers, the equivalence of categories $\mathcal{D}_Q \xrightarrow{\sim} \mathcal{D}^{\mathrm{b}}(Q)$ from Theorem 3.5 is an isomorphism of $\mathcal{D}_Q$ with a skeleton of $\mathcal{D}^{\mathrm{b}}(Q)$, since isoclasses in $\mathcal{D}_Q$ all consist of single objects. Moreover, on $\mathcal{D}_Q$ the functor ρ (described above as a glide reflection of the quiver $\mathbb{Z}Q$ when Q is of type A_n) is an automorphism, and so one can use the explicit construction from Definition 3.12 to define $\mathcal{C}_Q = \mathcal{D}_Q/\rho$.

While $\mathcal{C}_Q$ may in some sense appear much 'bigger' than $\mathcal{D}_Q$ – it has the same set of objects as this skeleton, and more morphisms between any two – in practical terms it is actually 'smaller', as indecomposable objects which are non-isomorphic in $\mathcal{D}_Q$ can become isomorphic in $\mathcal{C}_Q$. Indeed, any two objects of $\mathcal{D}_Q$ in the same ρ-orbit are isomorphic in $\mathcal{C}_Q$, since for any $n \in \mathbb{Z}$, the morphism space $\operatorname{Hom}_{\mathcal{C}_Q}(X, \rho^n X)$ contains the identity morphism in $\operatorname{Hom}_{\mathcal{D}_Q}(X,X)$, which is an isomorphism $X \to \rho^n X$ in $\mathcal{C}_Q$ (cf. [83, Rem. II.8.3(ii)]). Indeed, these isomorphisms mean that the autoequivalence of $\mathcal{C}_Q$ induced by $\rho = \Sigma^{-1} \circ \tau$ is naturally isomorphic to the identity functor (which is precisely what the orbit category construction is designed to achieve). As a consequence, the autoequivalences of $\mathcal{C}_Q$ induced by τ and Σ are naturally equivalent to each other.

Returning to our A_6 example, we see that a fundamental domain for ρ is as shown (cf. Phenomenon 2).

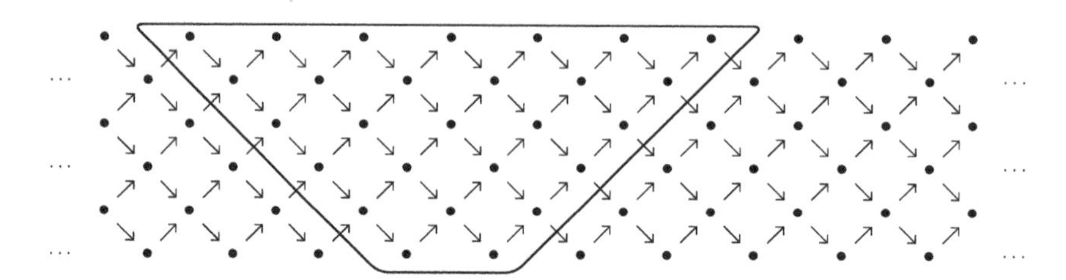

Thus the isomorphism classes of indecomposable objects in $\mathcal{C}_Q$ are in bijection with the points inside this fundamental domain. Note that there are morphisms in $\mathcal{C}_Q$ from objects at the right-hand end of this domain to those at the left-hand end, as the arrows crossing these two ends of the domain are identified in the orbit category – indeed one can think of $\mathcal{C}_Q$ as being drawn on a Möbius band, obtained as a quotient of the strip on which we draw $\mathcal{D}_Q$. If we think of frieze patterns as functions on the indecomposable objects of $\mathcal{D}_Q$, Phenomenon 2 is the observation that these functions descend to $\mathcal{C}_Q$, or in other words that they are constant on ρ-orbits.

As the name suggests, the cluster category $\mathcal{C}_Q$ can be used to study the cluster algebra A_Q: it is a categorification of this cluster algebra. More precisely, it is an additive categorification – cluster algebras may also have monoidal categorifications [58, 64], which have a rather different flavour more in common with categorifications in Lie theory, whereby the cluster algebra is realised as the Grothendieck ring of a monoidal category. The most important feature of $\mathcal{C}_Q$ for us is that certain isomorphism classes of objects in $\mathcal{C}_Q$ are in bijection with the cluster variables of A_Q, whereas others are in bijection with the clusters.

Definition 3.14 We say that two objects $X, Y \in \mathcal{C}_Q$ are *compatible* if $\operatorname{Hom}_{\mathcal{C}_Q}(X, \Sigma Y) = 0$. An object $X \in \mathcal{C}_Q$ is *rigid* if it is compatible with itself.

We write $\operatorname{add} X$ for the set of objects of $\mathcal{C}_Q$ isomorphic to direct sums of direct summands of X, and say that X is *cluster-tilting* if $\operatorname{add} X$ is precisely the set of objects compatible with X. Cluster-tilting objects are sometimes referred to by the longer but more descriptive adjective *maximal* 1*-orthogonal* [61].

Remark For any $X, Y \in \mathcal{C}_Q$, we have

$$\operatorname{Hom}_{\mathcal{C}_Q}(X, \Sigma Y) = \operatorname{Hom}_{\mathcal{C}_Q}(Y, \tau X) = \operatorname{Hom}_{\mathcal{C}_Q}(Y, \Sigma X)^*, \tag{3.4.3}$$

where $(-)^*$ denotes duality for $\mathbb{K}$-vector spaces. Here the first equality holds since Σ and τ are naturally isomorphic autoequivalences of $\mathcal{C}_Q$, and the second is a consequence of the Auslander–Reiten formula for finite-dimensional algebras [6] (see also [3, Thm. 3.4.1]). As a result, the relation of being compatible is symmetric. However, it is not reflexive, since not every object is rigid (although at least every indecomposable object is rigid when Q is a Dynkin quiver), and nor is it transitive. The equality of the outer terms in (3.4.3) means that $\mathcal{C}_Q$ is 2-*Calabi–Yau* as a triangulated category. For more appearances and uses of Calabi–Yau triangulated categories, see Keller's survey [68].

In $\mathcal{C}_Q$, but not in all 2-Calabi–Yau triangulated categories, cluster-tilting objects are precisely the maximal rigid objects, i.e. those rigid objects X for which $X \oplus Y$ is rigid only if $Y \in \operatorname{add} X$.

Theorem 3.6 ([25], [21, Thm. A.1]) *A choice of cluster-tilting object $T = \bigoplus_{i=1}^{n} T_i \in \mathcal{C}_Q$, with a decomposition into indecomposable summands T_i, induces a bijection $X \mapsto \varphi_X^T$ between the rigid indecomposable objects of $\mathcal{C}_Q$ and the cluster variables of A_Q. Under this bijection, compatible pairs of indecomposable rigid objects are sent to cluster variables that appear together in the same cluster. In particular, there is an induced bijection between the cluster-tilting objects of $\mathcal{C}_Q$ and the clusters of A_Q.*

For Q of type A_n, the cluster variables corresponding to indecomposables in a mesh satisfy the SL_2 diamond rule. That is, for each configuration

$$\begin{array}{ccccc} & & B & & \\ & \nearrow & & \searrow & \\ A & & & & D \\ & \searrow & & \nearrow & \\ & & C & & \end{array}$$

the corresponding cluster variables satisfy $\varphi_A^T \varphi_D^T - \varphi_B^T \varphi_C^T = 1$. At the boundary meshes, either B or C will be missing, and we take $\varphi_B^T = 1$ or $\varphi_C^T = 1$ as appropriate in the preceding formula.

The map $X \mapsto \varphi_X^T$ appearing in Theorem 3.6 is obtained by restricting to rigid indecomposable objects the assignment of a Laurent polynomial φ_X^T to any object $X \in \mathcal{C}_Q$ via an explicit formula, known as the Caldero–Chapoton formula [22]; unfortunately, the ingredients in this formula require more representation theory than there is space to discuss here. The interested reader may consult the original paper [22], and a survey by Plamondon [76].

Combining Theorem 3.6 with a classical result in representation theory – Gabriel's theorem [48] – gives an alternative proof of Theorem 3.3. Precisely, Theorem 3.6 shows that A_Q has only finitely many cluster variables if and only if $\mathcal{C}_Q$ has only finitely many indecomposable objects. By construction, this is equivalent to $\operatorname{rep} Q$ having finitely many indecomposable objects (since $\mathcal{C}_Q$ has $n = |Q_0|$ more such objects than $\operatorname{rep} Q$), and Gabriel's theorem states that this happens if and only if Q is an orientation of a simply laced Dynkin diagram.

Theorems 3.4 and 3.6 show that we can consider (the non-trivial rows of) a frieze pattern of height n to be a function on the set $\operatorname{ind}\mathcal{C}_Q$ of indecomposable objects of the cluster category $\mathcal{C}_Q$ for Q an A_n-quiver (all of which are rigid). Indeed, each frieze entry is the specialisation of a cluster variable, related by Theorem 3.6 to an element of $\operatorname{ind}\mathcal{C}_Q$. This connection between cluster categories of type A_n and frieze patterns was first observed by Caldero and Chapoton [22, §5].

Our diagrams in Section 3.2 draw frieze patterns as functions on the set of isoclasses of indecomposable objects of the derived category $\mathcal{D}_Q$, but since indecomposable objects of $\mathcal{D}_Q$ in the same ρ-orbit are isomorphic indecomposable objects in $\mathcal{C}_Q$, and hence correspond to the same cluster variable, a frieze must be constant on each of these orbits. Since ρ acts by a glide reflection, this explains Phenomenon 2.

Our final result is a second classification of height n friezes, in terms of cluster-tilting objects.

Theorem 3.7 *Let Q be any A_n quiver. Then the friezes of height n are in bijection with the (isomorphism classes of) cluster-tilting objects of the cluster category $\mathcal{C}_Q$. Indeed, given a cluster-tilting object T, there is a unique frieze pattern taking the value* 1 *on each indecomposable summand of T, when interpreted as a function on* $\operatorname{ind}\mathcal{C}_Q$.

Proof While this result is well-known, see e.g. [9, Rem. 5.6], we are not able to find a convenient reference or direct proof in the literature. However, the result can be obtained by 'direct calculation', as follows. For a choice of cluster-tilting object T, we change coordinates in A_Q so that $\varphi^T_{T_i}$ is the initial variable x_i for each indecomposable summand T_i of T. We then obtain a frieze by assigning to each indecomposable $X \in \mathcal{C}_Q$ the integer $\varphi^T_X|_{x_i=1}$; this frieze takes the value 1 on indecomposable summands of T by construction. Using the details of the formula φ^T_X, one can check that these are in fact the only entries taking value 1, and so non-isomorphic cluster-tilting objects give different friezes. Then since the number of cluster-tilting objects in $\mathcal{C}_Q$ up to isomorphism (or equivalently the number of clusters in A_Q) is equal to the number of height $n = |Q_0|$ friezes when Q is of type A_n, both being given by the $(n+1)$st Catalan number, it follows that all friezes are obtained in this way. □

Remark Given any Dynkin quiver Q, not necessarily of type A_n, one can extend the definition of frieze pattern to obtain functions $\operatorname{ind}\mathcal{C}_Q \to \mathbb{Z}_{>0}$ satisfying a condition generalising the SL_2 diamond rule. However, in this larger generality, the analogue of Theorem 3.7 is not true – there exist frieze patterns in this sense which do not take the value 1 on all indecomposable summands of any cluster-tilting object. An example, and a discussion of these frieze patterns for Q of type D_n, can be found in a paper of Fontaine and Plamondon [44].

Recall from Remark 3.2 that height n friezes, thought of as functions on $\mathcal{D}_Q$ for Q of type A_n, are classified up to the glide reflection ρ by triangulations of labelled polygons. Comparing to Theorem 3.7, we see that there is a bijection between triangulations of a labelled $(n+3)$-sided polygon and the cluster-tilting objects of $\mathcal{C}_Q$ for Q of type A_n – indeed, this connection was used by Caldero, Chapoton and Schiffler [24] to give a geometric description of $\mathcal{C}_Q$ in these cases. This bijection can be made explicit, and there is a combinatorial way of computing the endomorphism algebra of a cluster-tilting object $T \in \mathcal{C}_Q$ from the corresponding triangulation. These endomorphism algebras are called *cluster-tilted algebras*, and have many interesting properties [20], which are discussed further in Schiffler's book [81].

Example 3.15 The cluster category of type A_6 has a cluster-tilting object whose indecomposable summands are indicated by the white vertices in the following figure.

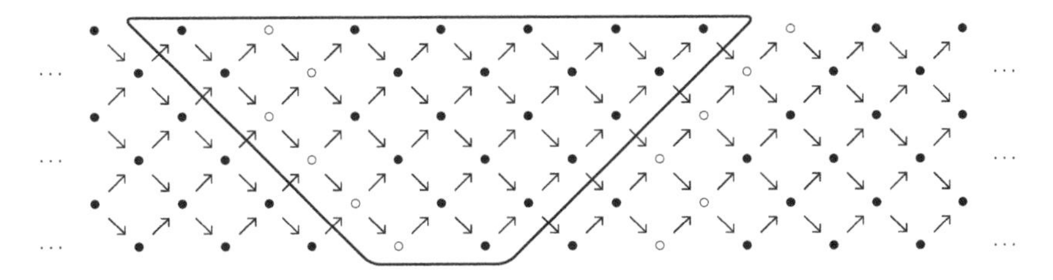

This cluster-tilting object corresponds to our first example of a frieze pattern – observe that it has the same 'shape' as the lightning bolt used to construct that frieze.

The cluster category of type A_3 has a cluster-tilting object indicated by white vertices as shown.

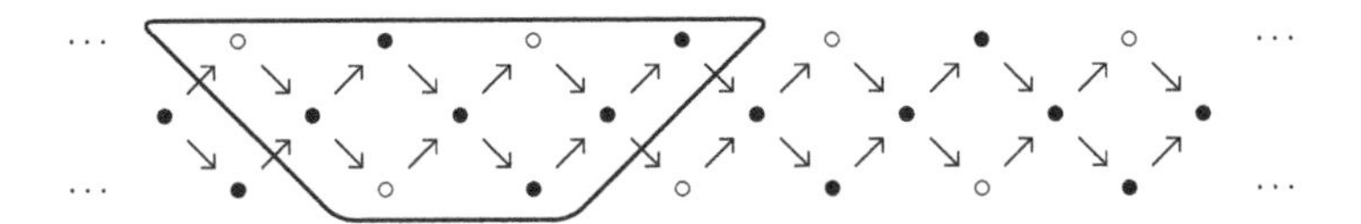

This cluster-tilting object corresponds to the frieze pattern with no lightning bolt observed in Example 3.2. On the level of the cluster algebra, it corresponds to a seed whose quiver has an oriented cycle (and indeed consists entirely of such a cycle, of length 3).

Just as clusters can be mutated, replacing a single cluster variable by a new one, so can cluster-tilting objects, by an operation replacing a single indecomposable summand by a non-isomorphic one. Given what we have already seen, this operation could simply be defined using the bijections of Theorem 3.6. However, mutation of cluster-tilting objects can also be defined intrinsically [19] – indeed Iyama and Yoshino [62] show that this is a general phenomenon of 2-Calabi–Yau triangulated categories – and one can then check that the bijection between clusters and cluster-tilting objects from Theorem 3.6 relates the combinatorial and categorical notions of mutation. Under the bijection with triangulations of the $(n+3)$-gon, when Q has type A_n, mutations correspond to flips: each diagonal separates two triangles whose union is a quadrilateral, and using the other diagonal in this quadrilateral gives a new triangulation.

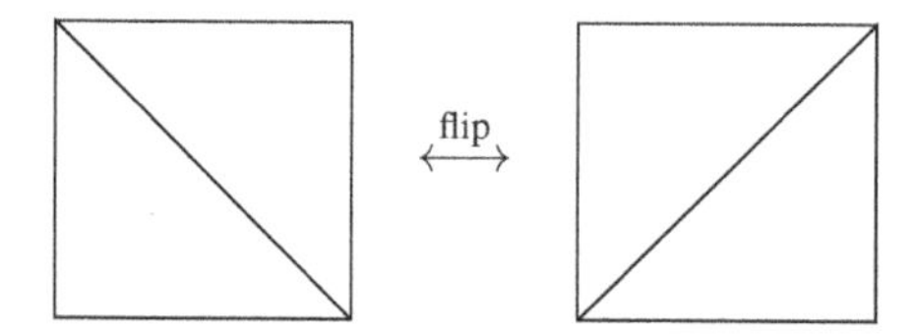

The connection between cluster algebras or categories and triangulations can be greatly generalised [36]. Let S be an oriented surface with boundary, and let $\mathbb{M} \subset \partial S$ be a finite subset of points in the boundary of S, such that each boundary component contains at least one point in $\mathbb{M}$. A triangulation of $(S, \mathbb{M})$ consists of a maximal collection of pairwise non-crossing arcs in S with endpoints in $\mathbb{M}$. Excluding a small number of degenerate cases, any such triangulation determines a cluster-tilting object in a (generalised [2]) cluster category, and mutating this cluster-tilting object corresponds to flipping diagonals in the triangulation. The combinatorial rule for computing endomorphism algebras extends to this generality as well [71].

Acknowledgements

This chapter is adapted from notes written to accompany a series of three lectures given in October 2020 at the LMS Autumn Algebra School, funded by the London Mathematical Society, the European Research Council and the International Centre for Mathematical Sciences in Edinburgh. I would like to thank David Jordan, Nadia Mazza and Sibylle Schroll for organising the lecture series, Karin Baur and Bethany Marsh for their comments on early drafts of the notes and the students attending the lectures for their interest and insightful questions.

3.5 Appendix: The Bounded Derived Category

In this appendix, we will give a more common and more general description of the bounded derived category than that given in Definition 3.11 for the special case of quiver representations.

To start with, we let $\mathcal{A}$ be any abelian category – our main example is the abelian category $\mathsf{rep}Q$, but other examples include categories of representations over more general rings or algebras, the category of abelian groups, or the category of coherent sheaves on an algebraic variety. We can then consider complexes of objects of $\mathcal{A}$, i.e. diagrams

$$V^\bullet\colon \cdots \longrightarrow V^{i-1} \xrightarrow{d^{i-1}} V^i \xrightarrow{d^i} V^{i+1} \longrightarrow \cdots$$

consisting of objects $V^i \in \mathcal{A}$ for each $i \in \mathbb{Z}$ and morphisms $d^i\colon V^i \to V^{i+1}$ between these objects, with the property that $d^i \circ d^{i-1} = 0$ for all i. A morphism of complexes is, similar to a morphism of quiver representations, a commutative diagram

$$\begin{array}{ccccccccc} V^\bullet\colon & \cdots \longrightarrow & V^{i-1} & \xrightarrow{d^{i-1}} & V^i & \xrightarrow{d^i} & V^{i+1} & \longrightarrow \cdots \\ \varphi\downarrow & & \varphi^{i-1}\downarrow & & \varphi^i\downarrow & & \varphi^{i+1}\downarrow & \\ W^\bullet\colon & \cdots \longrightarrow & W^{i-1} & \xrightarrow[\delta^{i-1}]{} & W^i & \xrightarrow[\delta^i]{} & W^{i+1} & \longrightarrow \cdots \end{array}$$

Given a complex $V^\bullet$ of objects of $\mathcal{A}$, we can compute its cohomology groups

$$\mathrm{H}^i(V^\bullet) = \ker(d^i)/\mathrm{im}(d^{i-1}),$$

which are themselves objects of $\mathcal{A}$, and a morphism $\varphi\colon V^\bullet \to W^\bullet$ induces a morphism $\bar{\varphi}^i\colon \mathrm{H}^i(V^\bullet) \to \mathrm{H}^i(W^\bullet)$ in $\mathcal{A}$ for each i. We call φ a *quasi-isomorphism* if all of the maps $\bar{\varphi}^i$ are isomorphisms; note that this does *not* imply the existence of a map $\psi\colon W^\bullet \to V^\bullet$ of complexes such that $\bar{\psi}^i = (\bar{\varphi}^i)^{-1}$.

Definition 3.16 Let $\mathcal{A}$ be an abelian category. The *bounded derived category* $\mathcal{D}^{\mathrm{b}}(\mathcal{A})$ of $\mathcal{A}$ has as objects complexes $V^\bullet$ of objects in $\mathcal{A}$ such that $\mathrm{H}^i(V^\bullet) = 0$ for $i \gg 0$ and $i \ll 0$.[1] The morphism sets in $\mathcal{D}^{\mathrm{b}}(\mathcal{A})$ are obtained from morphisms of complexes by formally adjoining inverses of all quasi-isomorphisms.[2]

[1] Depending on $\mathcal{A}$, sometimes additional restrictions are imposed on these cohomology objects. For example, if $\mathcal{A}$ is the category of all modules over a ring, one sometimes requires that each $\mathrm{H}^i(V^\bullet)$ is finitely generated for $V^\bullet$ to be an object in $\mathcal{D}^{\mathrm{b}}(\mathcal{A})$.

[2] This language is rather sloppy, but will do for our purposes. The reader interested in more details should look up 'localisation of categories', which is a similar construction to localisation of rings. In this language, we define $\mathcal{D}^{\mathrm{b}}(\mathcal{A})$ by taking the category of bounded complexes over $\mathcal{A}$, which has the same objects as $\mathcal{D}^{\mathrm{b}}(\mathcal{A})$ but morphisms given simply by morphisms of complexes, and then localising this category in the set of quasi-isomorphisms.

A little more concretely, this means that a morphism $V^\bullet \to W^\bullet$ in $\mathcal{D}^{\mathrm{b}}(\mathcal{A})$ can be represented (non-uniquely) by a finite sequence

$$\begin{array}{ccccccccccc} & & X_1^\bullet & & & & X_2^\bullet & & \cdots & & X_n^\bullet & & \\ & \overset{\sim}{\swarrow} & & \searrow & & \overset{\sim}{\swarrow} & & \searrow & \overset{\sim}{\swarrow} \quad \searrow & \overset{\sim}{\swarrow} & & \searrow & \\ V^\bullet & & & & Y_1^\bullet & & & & Y_2^\bullet \quad\quad Y_{n-1}^\bullet & & & & W^\bullet \end{array}$$

in which the rightward-pointing arrows are arbitrary maps of complexes, and the leftward-pointing arrows are quasi-isomorphisms.

As mentioned in Section 3.4 for the special case $\mathcal{A} = \mathsf{rep}Q$, we may equip $\mathcal{D}^{\mathrm{b}}(\mathcal{A})$ with the structure of a triangulated category. Part of this structure is the autoequivalence $\Sigma\colon \mathcal{D}^{\mathrm{b}}(\mathcal{A}) \xrightarrow{\sim} \mathcal{D}^{\mathrm{b}}(\mathcal{A})$, which shifts the degrees of the objects and morphisms in a complex in the following way.

$$\begin{array}{rl} V^\bullet\colon & \cdots \longrightarrow V^{i-1} \xrightarrow{d^{i-1}} V^i \xrightarrow{d^i} V^{i+1} \longrightarrow \cdots \\ \Sigma V^\bullet\colon & \cdots \longrightarrow V^{i} \xrightarrow{-d^{i}} V^{i+1} \xrightarrow{-d^{i+1}} V^{i+2} \longrightarrow \cdots \end{array}$$

The category $\mathcal{D}^{\mathrm{b}}(\mathcal{A})$ also has direct sums, given by taking the direct sum of complexes term by term.

For a general abelian category $\mathcal{A}$, understanding $\mathcal{D}^{\mathrm{b}}(\mathcal{A})$ can be rather difficult – indeed, from the description given above it is not clear that it is a category at all, since it is not clear that the morphisms between a pair of objects, which consist of sequences of morphisms of complexes interlaced with formal inverses of quasi-isomorphisms, even form a well-defined set, although it turns out that they do. In some special cases (most notably the case $\mathcal{A} = \mathsf{rep}Q$), simpler descriptions are available. First, we give a general simplification of the description of the morphisms in $\mathcal{D}^{\mathrm{b}}(\mathcal{A})$.

Definition 3.17 We say a morphism $f\colon V^\bullet \to W^\bullet$ of complexes is *null-homotopic* if there are morphisms $h^i\colon V^i \to W^{i-1}$ for each $i \in \mathbb{Z}$, such that $\varphi^i = \delta^{i-1}h^i + h^{i+1}d^i$, as in the following figure.

$$\begin{array}{ccccccccc} V^\bullet\colon \cdots & \longrightarrow & V^{i-1} & \xrightarrow{d^{i-1}} & V^i & \xrightarrow{d^i} & V^{i+1} & \longrightarrow & \cdots \\ \downarrow{\scriptstyle\varphi} & & \downarrow{\scriptstyle\varphi^{i-1}} & \overset{h^i}{\swarrow} & \downarrow{\scriptstyle\varphi^i} & \overset{h^{i+1}}{\swarrow} & \downarrow{\scriptstyle\varphi^{i+1}} & & \\ W^\bullet\colon \cdots & \longrightarrow & W^{i-1} & \xrightarrow[\delta^{i-1}]{} & W^i & \xrightarrow[\delta^i]{} & W^{i+1} & \longrightarrow & \cdots \end{array}$$

The *bounded homotopy category* $\mathcal{K}^{\mathrm{b}}(\mathcal{A})$ has the same objects as $\mathcal{D}^{\mathrm{b}}(\mathcal{A})$, with $\mathrm{Hom}_{\mathcal{K}^{\mathrm{b}}(\mathcal{A})}(V^\bullet, W^\bullet)$ given by the space of maps of complexes $V^\bullet \to W^\bullet$ modulo the subspace of null-homotopic maps.

Proposition 3.8 *The bounded derived category $\mathcal{D}^b(\mathcal{A})$ is equivalent to the category obtained from $\mathcal{K}^b(\mathcal{A})$ by formally inverting (the classes of) quasi-isomorphisms.*

The advantage of describing $\mathcal{D}^{\mathrm{b}}(\mathcal{A})$ as in Proposition 3.8 is that it allows for an easier description of the morphisms, which makes it clear that $\mathrm{Hom}_{\mathcal{D}^{\mathrm{b}}(Q)}(V^\bullet, W^\bullet)$ is indeed a set. Using this description, morphisms $V^\bullet \to W^\bullet$ are represented (still non-uniquely) by *roofs* of the form

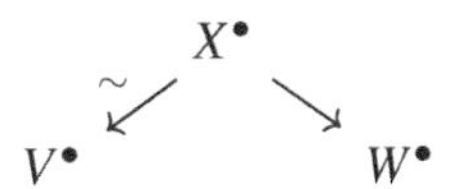

where both arrows represent morphisms in $\mathcal{K}^{\mathrm{b}}(\mathcal{A})$ (that is, homotopy classes of maps of complexes), with the left-hand arrow being a quasi-isomorphism. This advantage derives from the fact that $\mathcal{K}^{\mathrm{b}}(\mathcal{A})$, unlike the category with the same objects but with all morphisms of complexes, is a triangulated category, and the localisation in quasi-isomorphisms can be realised as a Verdier localisation [85] in the triangulated subcategory of acyclic complexes, those $V^\bullet$ for which $\mathrm{H}^i(V^\bullet) = 0$ for all i.

Now let us consider the case that $\mathcal{A} = \operatorname{mod} A$ is the category of finite-dimensional modules over a finite-dimensional algebra A; we abbreviate $\mathcal{D}^{\mathrm{b}}(A) := \mathcal{D}^{\mathrm{b}}(\operatorname{mod} A)$ and $\mathcal{K}^{\mathrm{b}}(A) := \mathcal{K}^{\mathrm{b}}(\operatorname{mod} A)$. (Much weaker, but more technical, assumptions are sufficient for what follows.) Recall that an A-module P is projective if for every surjective A-module map $f\colon X \to Y$ and every A-module map $g\colon P \to Y$, there exists $h\colon P \to X$ with $g = f \circ h$.

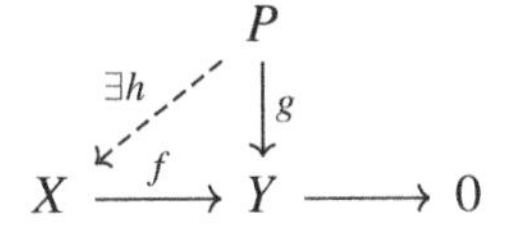

Definition 3.18 Write $\mathcal{K}^{\mathrm{b}}(\operatorname{proj} A)$ for the category whose objects are complexes $P^\bullet$ of A-modules such that P^i is projective for all i and $P^i = 0$ for $i \ll 0$ or $i \gg 0$, and with morphisms given by maps of complexes modulo null-homotopic maps. Using the description of $\mathcal{D}^{\mathrm{b}}(A)$ from Proposition 3.8, we see that there is a natural functor $\mathcal{K}^{\mathrm{b}}(\operatorname{proj} A) \to \mathcal{D}^{\mathrm{b}}(A)$ given by the identity on both objects and morphisms.

Note that the boundedness conditions in Definition 3.18 are different from those in Definition 3.11, since they refer to the terms of the complex, not the cohomology. Indeed, since $\operatorname{proj} A$ is typically not an abelian category, an object

$P^\bullet \in \mathcal{K}^{\mathrm{b}}(\operatorname{proj} A)$ does not have well-defined cohomologies $\mathrm{H}^i(P^\bullet) \in \operatorname{proj} A$ – they are well-defined objects of $\operatorname{mod} A$, but are not typically projective.

Definition 3.19 We say that a finite-dimensional algebra A has *finite global dimension* if there is an integer n such that each $X \in \operatorname{mod} A$ fits into an exact sequence

$$0 \longrightarrow P^n \longrightarrow \cdots \longrightarrow P^0 \xrightarrow{\pi} X \longrightarrow 0 \tag{3.5.1}$$

in which each P^i is a projective A-module. The minimal such n is called the *global dimension* of A, which is instead defined to be ∞ if no such n exists.

Exactness of the complex (3.5.1) is equivalent to the morphism

$$\begin{array}{ccccccccccccc}
\cdots & \longrightarrow & 0 & \longrightarrow & P^n & \longrightarrow & \cdots & \longrightarrow & P^1 & \longrightarrow & P^0 & \longrightarrow & 0 & \longrightarrow & \cdots \\
 & & \downarrow & & \downarrow & & & & \downarrow & & \downarrow{\scriptstyle \pi} & & \downarrow & & \\
\cdots & \longrightarrow & 0 & \longrightarrow & 0 & \longrightarrow & \cdots & \longrightarrow & 0 & \longrightarrow & X & \longrightarrow & 0 & \longrightarrow & \cdots
\end{array}$$

in $\mathcal{K}^{\mathrm{b}}(A)$ being a quasi-isomorphism, so that the lower complex is isomorphic in $\mathcal{D}^{\mathrm{b}}(A)$ to a complex in the image of the natural map $\mathcal{K}^{\mathrm{b}}(\operatorname{proj} A) \to \mathcal{D}^{\mathrm{b}}(A)$. This observation can be upgraded to the following theorem.

Theorem 3.9 ([84, Ch. II, Prop. 1.4]) *If A is a finite-dimensional algebra with finite global dimension, the natural map $\mathcal{K}^b(\operatorname{proj} A) \to \mathcal{D}^b(A)$ is an equivalence.*

Theorem 3.9 is very useful, since the category $\mathcal{K}^{\mathrm{b}}(\operatorname{proj} A)$ has a much simpler description than the (equivalent) category $\mathcal{D}^{\mathrm{b}}(A)$. This theorem covers the case of derived categories of representations of acyclic quivers, which was of interest to us in Section 3.4. Such a quiver Q has a path algebra $\mathbb{K}Q$, which as a vector space is spanned by the paths of Q – this is finite dimensional since Q is acyclic. Multiplication of paths is given by concatenation when this is defined, and 0 when it is not (cf. Remark 3.4). Thus $\mathbb{K}Q$ is a finite-dimensional algebra, and $\operatorname{rep} Q$ is equivalent (even isomorphic) to the category $\operatorname{mod} \mathbb{K}Q$ of finite-dimensional modules over this algebra. The global dimension of $\mathbb{K}Q$ is 1, and so there is an equivalence $\mathcal{K}^{\mathrm{b}}(\operatorname{proj} \mathbb{K}Q) \xrightarrow{\sim} \mathcal{D}^{\mathrm{b}}(\mathbb{K}Q)$ by Theorem 3.9.

An algebra A of global dimension 1 is called *hereditary* (because the property of being a projective A-module is inherited by submodules – that is, any submodule of a projective A-module is again projective). In this case we can simplify our description of $\mathcal{D}^{\mathrm{b}}(A)$ still further, to recover our *ad hoc* description of the objects of $\mathcal{D}^{\mathrm{b}}(Q)$ from Section 3.4.

Proposition 3.10 ([56, Lem. 4.1]) *Let A be a hereditary algebra. Then each object $V^\bullet \in \mathcal{D}^b(A)$ is isomorphic in this category to*

$$\cdots \xrightarrow{0} H^{i-1}(V^\bullet) \xrightarrow{0} H^i(V^\bullet) \xrightarrow{0} H^{i+1}(V^\bullet) \xrightarrow{0} \cdots$$

In other words, if $\Sigma^{-i}M$ denotes the stalk complex *with $M \in \operatorname{mod} A$ in degree i and the zero module in all other degrees, then we have*

$$V^\bullet \cong \bigoplus_{i \in \mathbb{Z}} \Sigma^{-i} H^i(V^\bullet).$$

Our description of morphisms between stalk complexes in $\mathcal{D}^{\mathrm{b}}(Q)$ from Section 3.4 can also be recovered from the results presented in this appendix, but we will not give the details of this. Roughly, the idea is to use Theorem 3.9 to replace each stalk complex by a quasi-isomorphic complex of projectives, and then compare the computation of the morphisms between these complexes up to homotopy to the usual computation of extension groups between the two representations giving the non-zero terms of the stalk complexes.

Bibliography

[1] Adachi, T., Iyama, O. and Reiten, I. τ-tilting theory. *Compos. Math.*, 150(3):415–452, 2014.

[2] Amiot, C. Cluster categories for algebras of global dimension 2 and quivers with potential. *Ann. Inst. Fourier*, 59(6):2525–2590, 2009.

[3] Angeleri Hügel, L. 2006. *An introduction to Auslander–Reiten theory*. http://profs.sci.univr.it/ angeleri/trieste.pdf.

[4] Assem, I., Simson, D., and Skowroński, A. 2006. *Elements of the Representation Theory of Associative Algebras. Vol. 1.* London Mathematical Society Student Texts, vol. 65. Cambridge University Press, Cambridge. Techniques of representation theory.

[5] August, J. 2020. The tilting theory of contraction algebras. *Adv. Math.*, **374**, Article 107372.

[6] Auslander, M. and Reiten, I. 1975. Representation theory of Artin algebras. III. Almost split sequences. *Comm. Algebra*, **3**, 239–294.

[7] Auslander, M., Reiten, I. and Smalø, S. O. 1997. *Representation theory of Artin algebras*. Cambridge Studies in Advanced Mathematics, vol. 36. Cambridge University Press, Cambridge. Corrected reprint of the 1995 original.

[8] Baur, K., Çanakçı, İ., Jacobsen, K. M., Kulkarni, M. C. and Todorov, G. 2020. *Infinite friezes and triangulations of annuli*. Preprint. (arXiv:2007.09411 [math.CO])

[9] Baur, K., Faber, E., Gratz, S., Serhiyenko, K. and Todorov, G. 2021. Friezes satisfying higher SL_k-determinants. *Algebra Number Theory*, **15**(1), 29–68.

[10] Baur, K., King, A. D. and Marsh, B. R. 2016. Dimer models and cluster categories of Grassmannians. *Proc. Lond. Math. Soc. (3)*, **113**(2), 213–260.

[11] Baur, K., Parsons, M. J. and Tschabold, M. 2016. Infinite friezes. *European J. Combin.*, **54**, 220–237.

[12] Bazier-Matte, V., Douville, G., Mousavand, K., Thomas, H. and Yıldırım, E. 2018. *ABHY associahedra and Newton polytopes of F-polynomials for finite type cluster algebras.*

[13] Bennett-Tennenhaus, R. and Shah, A. 2021. Transport of structure in higher homological algebra. *J. Algebra*, **574**, 514–549.

[14] Berenstein, A., Fomin, S. and Zelevinsky, A. 2005. Cluster algebras. III. Upper bounds and double Bruhat cells. *Duke Math. J.*, **126**(1), 1–52.

[15] Bessenrodt, C., Holm, T. and Jørgensen, P. 2014. Generalized frieze pattern determinants and higher angulations of polygons. *J. Combin. Theory Ser. A*, **123**, 30–42.

[16] Bridgeland, T. 2017. Scattering diagrams, Hall algebras and stability conditions. *Algebr. Geom.*, **4**(5), 523–561.

[17] Brüstle, T., Smith, D. and Treffinger, H. 2019. Wall and chamber structure for finite-dimensional algebras. *Adv. Math.*, **354**, Article 106746.

[18] Buan, A. B., Iyama, O., Reiten, I. and Scott, J. 2009. Cluster structures for 2-Calabi–Yau categories and unipotent groups. *Compos. Math.*, **145**(4), 1035–1079.

[19] Buan, A. B., Marsh, B. R., Reineke, M., Reiten, I. and Todorov, G. 2006. Tilting theory and cluster combinatorics. *Adv. Math.*, **204**(2), 572–618.

[20] Buan, A. B., Marsh, B. R. and Reiten, I. 2007. Cluster-tilted algebras. *Trans. Amer. Math. Soc.*, **359**(1), 323–332.

[21] Buan, A. B., Marsh, B. R., Reiten, I. and Todorov, G. 2007. Clusters and seeds in acyclic cluster algebras. *Proc. Amer. Math. Soc.*, **135**(10), 3049–3060.

[22] Caldero, P. and Chapoton, F. 2006. Cluster algebras as Hall algebras of quiver representations. *Comment. Math. Helv.*, **81**(3), 595–616.

[23] Caldero, P., Chapoton, F. and Schiffler, R. 2006. Quivers with relations and cluster tilted algebras. *Algebr. Represent. Theory*, **9**(4), 359–376.

[24] Caldero, P., Chapoton, F. and Schiffler, R. 2006. Quivers with relations arising from clusters (A_n case). *Trans. Amer. Math. Soc.*, **358**(3), 1347–1364.

[25] Caldero, P. and Keller, B. 2006. From triangulated categories to cluster algebras. II. *Ann. Sci. École Norm. Sup. (4)*, **39**(6), 983–1009.

[26] Çanakçı, İ. and Jørgensen, P. 2020. Friezes, weak friezes, and T-paths. *Adv. in Appl. Math.*, **131**, Paper No. 102253.

[27] Cheung, M. W., Gross, M., Muller, G., Musiker, G., Rupel, D., Stella, S. and Williams, H. 2017. The greedy basis equals the theta basis: a rank two haiku. *J. Combin. Theory Ser. A*, **145**, 150–171.

[28] Conway, J. H. and Coxeter, H. S. M. 1973. Triangulated polygons and frieze patterns. *Math. Gaz.*, **57**(400), 87–94.

[29] Conway, J. H. and Coxeter, H. S. M. 1973. Triangulated polygons and frieze patterns. *Math. Gaz.*, **57**(401), 175–183.

[30] Cordes, C. M. and Roselle, D. P. 1972. Generalized frieze patterns. *Duke Math. J.*, **39**, 637–648.

[31] Coxeter, H. S. M. 1971. Frieze patterns. *Acta Arith.*, **18**, 297–310.

[32] Cuntz, M., Holm, T. and Jørgensen, P. 2020. Frieze patterns with coefficients. *Forum Math. Sigma*, **8**, Paper No. e17.

[33] Derksen, H., Weyman, J. and Zelevinsky, A. 2008. Quivers with potentials and their representations. I. Mutations. *Selecta Math. (N.S.)*, **14**(1), 59–119.

[34] Fock, V. V. and Goncharov, A. B. 2009. Cluster ensembles, quantization and the dilogarithm. *Ann. Sci. Éc. Norm. Supér. (4)*, **42**(6), 865–930.

[35] Fomin, S. *Cluster algebras portal.* www.math.lsa.umich.edu/ fomin/cluster.html.

[36] Fomin, S., Shapiro, M. and Thurston, D. 2008. Cluster algebras and triangulated surfaces. I. Cluster complexes. *Acta Math.*, **201**(1), 83–146.

[37] Fomin, S. and Thurston, D. 2018. Cluster algebras and triangulated surfaces Part II: Lambda lengths. *Mem. Amer. Math. Soc.*, **255**(1223).

[38] Fomin, S., Williams, L. and Zelevinsky, A. 2016. *Introduction to Cluster Algebras. Chapters 1–3*. Preprint. (arXiv:1608.05735 [math.CO])

[39] Fomin, S., Williams, L. and Zelevinsky, A. 2017. *Introduction to Cluster Algebras. Chapters 4–5*. Preprint. (arXiv:1707.07190 [math.CO])

[40] Fomin, S., Williams, L. and Zelevinsky, A. 2020. *Introduction to Cluster Algebras. Chapter 6*. Preprint. (arXiv:2008.09189 [math.AC])

[41] Fomin, S. and Zelevinsky, A. 2002. Cluster algebras. I. Foundations. *J. Amer. Math. Soc.*, **15**(2), 497–529.

[42] Fomin, S. and Zelevinsky, A. 2003. Cluster algebras. II. Finite type classification. *Invent. Math.*, **154**(1), 63–121.

[43] Fomin, S. and Zelevinsky, A. 2007. Cluster algebras. IV. Coefficients. *Compos. Math.*, **143**(1), 112–164.

[44] Fontaine, B. and Plamondon, P.-G. 2016. Counting friezes in type D_n. *J. Algebraic Combin.*, **44**(2), 433–445.

[45] Fordy, A. P. and Hone, A. 2014. Discrete integrable systems and Poisson algebras from cluster maps. *Comm. Math. Phys.*, **325**(2), 527–584.

[46] Franco, S. 2012. Bipartite field theories: from D-brane probes to scattering amplitudes. *J. High Energy Phys.*, JHEP11(2012)141.

[47] Fulton, W. and Harris, J. 1991. *Representation Theory: A First Course*. Graduate Texts in Mathematics, vol. 129. Springer-Verlag, New York.

[48] Gabriel, P. 1972. Unzerlegbare Darstellungen. I. *Manuscripta Math.*, **6**, 71–103.

[49] Geiß, C., Leclerc, B. and Schröer, J. 2008. Partial flag varieties and preprojective algebras. *Ann. Inst. Fourier (Grenoble)*, **58**(3), 825–876.

[50] Geiß, C., Leclerc, B. and Schröer, J. 2010. *Cluster algebra structures and semicanonical bases for unipotent groups*. Preprint. (arXiv:math/0703039 [math.RT])

[51] Gekhtman, M., Shapiro, M. and Vainshtein, A. 2010. *Cluster algebras and Poisson geometry*. Mathematical Surveys and Monographs, no. 167. Providence, RI: American Mathematical Society.

[52] Gelfand, S. I. and Manin, Y. I. 2003. *Methods of homological algebra*. Second edition. Springer Monographs in Mathematics. Springer-Verlag, Berlin.

[53] Gross, M., Hacking, P. and Keel, S. 2015. Birational geometry of cluster algebras. *Algebr. Geom.*, **2**(2), 137–175.

[54] Gross, M., Hacking, P., Keel, S. and Kontsevich, M. 2018. Canonical bases for cluster algebras. *J. Amer. Math. Soc.*, **31**(2), 497–608.

[55] Gunawan, E., Musiker, G. and Vogel, H. 2019. Cluster algebraic interpretation of infinite friezes. *European J. Combin.*, **81**, 22–57.

[56] Happel, D. 1987. On the derived category of a finite-dimensional algebra. *Comment. Math. Helv.*, **62**(3), 339–389.

[57] Happel, D. 1988. *Triangulated categories in the representation theory of finite-dimensional algebras*. London Math. Soc. Lecture Note Ser., no. 119. Cambridge: Cambridge University Press.

[58] Hernandez, D. and Leclerc, B. 2010. Cluster algebras and quantum affine algebras. *Duke Math. J.*, **154**(2), 265–341.

[59] Holm, T. and Jørgensen, P. 2015. Generalized friezes and a modified Caldero–Chapoton map depending on a rigid object. *Nagoya Math. J.*, **218**, 101–124.

[60] Ingalls, C. and Thomas, H. 2009. Noncrossing partitions and representations of quivers. *Compos. Math.*, **145**(6), 1533–1562.

[61] Iyama, O. 2007. Higher-dimensional Auslander–Reiten theory on maximal orthogonal subcategories. *Adv. Math.*, **210**(1), 22–50.

[62] Iyama, O. and Yoshino, Y. 2008. Mutation in triangulated categories and rigid Cohen–Macaulay modules. *Invent. Math.*, **172**(1), 117–168.

[63] Jensen, B. T., King, A. D. and Su, X. 2016. A categorification of Grassmannian cluster algebras. *Proc. Lond. Math. Soc. (3)*, **113**(2), 185–212.

[64] Kang, S.-J., Kashiwara, M., Kim, M. and Oh, S.-j. 2018. Monoidal categorification of cluster algebras. *J. Amer. Math. Soc.*, **31**(2), 349–426.

[65] Keller, B. *Quiver mutation in Java*. www.math.jussieu.fr/ keller/quiver-mutation/

[66] Keller, B. 1996. Derived categories and their uses. Pages 671–701 of: *Handbook of algebra, Vol. 1*. Elsevier/North-Holland, Amsterdam.

[67] Keller, B. 2005. On triangulated orbit categories. *Doc. Math.*, **10**, 551–581.

[68] Keller, B. 2008. Calabi–Yau triangulated categories. Pages 467–489 of: *Trends in representation theory of algebras and related topics*. EMS Ser. Congr. Rep. Eur. Math. Soc., Zürich.

[69] Keller, B. 2010. Cluster algebras, quiver representations and triangulated categories. Pages 76–160 of: *Triangulated categories*. London Math. Soc. Lecture Note Ser., no. 375. Cambridge: Cambridge Univ. Press.

[70] Keller, B. 2013. The periodicity conjecture for pairs of Dynkin diagrams. *Ann. of Math. (2)*, **177**(1), 111–170.

[71] Labardini-Fragoso, D. 2009. Quivers with potentials associated to triangulated surfaces. *Proc. Lond. Math. Soc. (3)*, **98**(3), 797–839.

[72] Marsh, B. R. 2013. *Lecture notes on cluster algebras*. Zürich Lectures in Advanced Mathematics. Zürich: European Mathematical Society.

[73] Morier-Genoud, S. 2012. Arithmetics of 2-friezes. *J. Algebraic Combin.*, **36**(4), 515–539.

[74] Morier-Genoud, S. 2015. Coxeter's frieze patterns at the crossroads of algebra, geometry and combinatorics. *Bull. Lond. Math. Soc.*, **47**(6), 895–938.

[75] Morier-Genoud, S. 2019. Symplectic frieze patterns. *SIGMA Symmetry Integrability Geom. Methods Appl.*, **15**, Paper No. 089.

[76] Plamondon, P.-G. 2018. Cluster characters. Pages 101–125 of: *Homological methods, representation theory, and cluster algebras*. CRM Short Courses. Springer, Cham.

[77] Pressland, M. 2017. Internally Calabi–Yau algebras and cluster-tilting objects. *Math. Z.*, **287**(1-2), 555–585.

[78] Pressland, M. 2022. Calabi–Yau properties of Postnikov diagrams. *Forum Math. Sigma*, **10**, Paper No. e56.

[79] Propp, J. 2020. The combinatorics of frieze patterns and Markoff numbers. *Integers*, **20**, Paper No. A12.

[80] Qin, F. 2019. *Bases for upper cluster algebras and tropical points*. Preprint. (arXiv:1902.09507 [math.RT])

[81] Schiffler, R. 2014. *Quiver representations*. CMS Books in Mathematics/Ouvrages de Mathématiques de la SMC. Springer, Cham.

[82] Scott, J. S. 2006. Grassmannians and cluster algebras. *Proc. London Math. Soc. (3)*, **92**(2), 345–380.

[83] Shah, A. 2019. *Partial cluster-tilted algebras via twin cotorsion pairs, quasi-abelian categories and Auslander–Reiten theory*. Ph.D. thesis, University of Leeds.

[84] Verdier, J.-L. 1977. Catégories dérivées: quelques résultats (état 0). Pages 262–311 of: *Cohomologie étale*. Lecture Notes in Math., vol. 569. Springer, Berlin.

[85] Verdier, J.-L. 1996. Des catégories dérivées des catégories abéliennes. *Astérisque*. With a preface by Luc Illusie, edited and with a note by Georges Maltsiniotis.

[86] Wemyss, M. 2018. Flops and clusters in the homological minimal model programme. *Invent. Math.*, **211**(2), 435–521.

4

Infinite-dimensional Representations of Algebras

Rosanna Laking

The aim of these lecture notes is to give an example-driven introduction to a class of modules called *pure-injective* modules, as well as the techniques that allow us to study them systematically. In particular, we will focus on modules over a K-algebra A, where K is a field. Every A-module has an in-built K-vector space structure and this approach will give us access to many interesting examples of modules where this underlying K-vector space is *infinite dimensional*. These are known as infinite-dimensional modules or infinite-dimensional representations.

Studying infinite-dimensional modules in general contains some obvious challenges. It is not easy to make use of the underlying linear algebra – for example, we can prove that any K-vector space has a basis (see Example 4.11) but, since the proof is not constructive, it can be difficult to identify a basis in a given example. It is therefore useful to study K-subspaces of modules, called *finite matrix subgroups* or *pp-definable subgroups*, that are controlled by some finite data. The pure-injective modules are those that behave well with respect to these K-subspaces.

The isomorphism classes of indecomposable pure-injective modules form the underlying set of a topological space, known as the *Ziegler spectrum*. The final part of these notes contains an account of the basic properties of this topological space in the case where A is a finite-dimensional algebra. We will end by making use of these basic properties to characterise finite representation type in terms of the Ziegler spectrum.

We do not assume any prior knowledge of representation theory or category theory. A significant majority of references approaching these topics make use of categorical techniques, often focusing on the connection with functor categories. The idea of these lectures is to demonstrate the usefulness of a more computational viewpoint.

4.1 Algebras and Modules

This section is dedicated to examples of K-algebras and modules over them. This will pave the way to Sections 4.2 and 4.3, which contain some concrete examples of infinite-dimensional modules that will allow us to illustrate the definitions and results covered in the later sections.

4.1.1 K-algebras

Definition 4.1 A **K-algebra** is a K-vector space A with a K-bilinear multiplication $A \times A \to A$, $(a,b) \mapsto ab$ such that there exists an element $1 \in A$ (called the **unit**) such that $1a = a = a1$.

From now on, we will use A to denote an arbitrary K-algebra.

Example 4.2 The one-dimensional vector space K with multiplication given by the field multiplication and the unit given by the multiplicative identity in the field.

Example 4.3 Consider the set

$$K[X] := \{k_0 + k_1 X + k_2 X^2 + \cdots + k_n X^n \mid n \geqslant 0, k_i \in K \text{ for } 0 \leqslant i \leqslant n\}$$

of polynomials with one free variable X and with coefficients in K. This is a K-vector space with a countably infinite basis $\{1, X, X^2, X^3, \dots\}$ and we define multiplication in $K[X]$ to be the usual multiplication of polynomials. The element 1 is the unit.

Example 4.4 Consider the following finite directed graph Q (in this context Q is known as a **quiver**).

$$\alpha \circlearrowleft \bullet^1 \xrightarrow{\beta} \bullet^2 \circlearrowright \delta$$

Let KQ be the vector space with basis given by the paths in Q (including a path of length zero for each vertex denoted by e_i for each vertex i). That is, the elements of KQ are formal K-linear combinations of elements of the set

$$\mathbf{Pa} := \{e_1, e_2, \beta, \alpha^n, \delta^m, \beta\alpha^n, \delta^m\beta, \delta^m\beta\alpha^n \mid n, m \in \mathbb{N}\}.$$

If p, q are paths in **Pa**, then we define their product $p \cdot q$ to be the concatenation of the paths if this is possible and 0 otherwise. Extending this product K-linearly allows us to define a multiplication in KQ. There is a unit element given by $e_1 + e_2$. This algebra is called the **path algebra** of Q.

Example 4.5 Consider the algebra KQ described in Example 4.4 and the ideal I generated by the set $\rho := \{\delta\beta\alpha, \alpha^2, \delta^2\}$. This is an example of an admissible ideal of KQ (see [1, Def. II.2.1]) and a pair (Q,I) consisting of a quiver and an admissible ideal of KQ is called a **bound quiver**.

The quotient algebra KQ/I is called the **path algebra of the bound quiver** (Q,I). Note that the underlying K-vector space of the quotient algebra KQ/I has the following basis:

$$\{e_1, e_2, \beta, \alpha, \beta, \beta\alpha, \delta\beta\}.$$

To learn more about general path algebras of bound quivers see [1].

4.1.2 Modules over a K-algebra

Definition 4.6 Let A be a K-algebra. Then a **(left)** A**-module** is a K-vector space M with an A-action, that is, a K-bilinear map $A \times M \to M$, $(a,m) \mapsto am$ such that, for any $a, b \in A$ and any $m \in M$, we have that $(ab)m = a(bm)$ and $1m = m$.

Unless otherwise specified, the terminology "A-module", will mean "left A-module".

Example 4.7 The definition of a K-module coincides with the definition of a K-vector space.

Example 4.8 Consider the algebra $K[X]$ defined in Example 4.3. By definition, a $K[X]$-module M is a K-vector space together with a $K[X]$-action. Let $p = k_0 + k_1X + k_2X^2 + \cdots + k_nX^n$ be an arbitrary element of $K[X]$. Then, for any element $m \in M$, we have that

$$\begin{aligned} pm &= (k_0 + k_1X + k_2X^2 + \cdots + k_nX^n)m \\ &= k_0m + k_1(Xm) + k_2(X^2m) + \cdots + k_n(X^nm). \end{aligned}$$

It follows that the action of $K[X]$ is determined by the K-vector space structure of M as well as the K-linear endomorphism $\Phi\colon M \to M$ given by $m \mapsto Xm$. Conversely, a K-vector space M together with a K-linear endomorphism Φ uniquely determines a $K[X]$-module. In other words, we can view $K[X]$-modules as representations of the one-loop quiver.

$$M \circlearrowright \Phi$$

Example 4.9 Consider the bound quiver (Q,I) given in Example 4.5. Modules over the path algebra KQ/I are determined by **representations of the bound quiver** (Q,I) (see, for example, [1, Thm. III.1.6]). That is, a pair of vector spaces U_1 and U_2, together with K-linear maps U_α, U_β and U_δ arranged in the following configuration

$$U_\alpha \circlearrowright U_1 \xrightarrow{U_\beta} U_2 \circlearrowleft U_\delta$$

such that $U_\alpha^2 = 0$, $U_\delta^2 = 0$ and $U_\delta U_\beta U_\alpha = 0$.

Given such a representation, we define a KQ/I-module with underlying vector space $U_1 \oplus U_2$. Since every element of KQ/I is a K-linear combination of elements of the set $\{e_1, e_2, \beta, \alpha, \beta, \beta\alpha, \delta\beta\}$, it is enough to specify the action of these basis elements.

- The action of e_1 is given by the matrix $\begin{pmatrix} \mathrm{id}_{U_1} & 0 \\ 0 & 0 \end{pmatrix}$.
- The action of e_2 is given by the matrix $\begin{pmatrix} 0 & 0 \\ 0 & \mathrm{id}_{U_2} \end{pmatrix}$.
- The action of α is given by the matrix $\begin{pmatrix} U_\alpha & 0 \\ 0 & 0 \end{pmatrix}$.
- The action of β is given by the matrix $\begin{pmatrix} 0 & 0 \\ U_\beta & 0 \end{pmatrix}$.
- The action of δ is given by the matrix $\begin{pmatrix} 0 & 0 \\ 0 & U_\delta \end{pmatrix}$.

The action of the remaining two paths in the basis ($\beta\alpha$ and $\delta\beta$) is given by the composition of the relevant matrices

4.1.3 Infinite-dimensional A-modules

Definition 4.10 Let A be a K-algebra. An A-module M is called **finite-dimensional** if the underlying vector space of M is finite-dimensional. An A-module that is not finite-dimensional is called **infinite-dimensional**.

There is a very broad and well-developed body of research devoted to the study of finite-dimensional modules over finite-dimensional algebras. In these lecture notes, however, we will look at certain classes of infinite-dimensional modules. These modules arise naturally in the representation theory of algebras and in future lectures we will begin to explore some of the rich theory surrounding infinite-dimensional pure-injective modules. Before we enter into this framework, we will consider how infinite-dimensional modules over the algebras given in Section 4.1.1 look.

Example 4.11 By definition, any finite-dimensional K-module V of dimension $n \in \mathbb{N}$ has a basis with n elements in it. It follows from this that V is isomorphic to K^n.

Now let us consider an infinite-dimensional K-module W. It is well known that, despite not being finite-dimensional, the vector space W has a basis. Let us sketch an argument to prove this claim. Let $\mathcal{L}$ be the set of linearly independent sets contained in W ordered by inclusion. It is clear that, for any chain $\mathcal{L}_1 \subset \mathcal{L}_2 \subset \mathcal{L}_3 \subset \cdots \subset \mathcal{L}_\alpha \subset \ldots$ in $\mathcal{L}$ indexed by an ordinal β, the union $\bigcup_{\alpha \leqslant \beta} \mathcal{L}_\alpha$ is an upper bound in $\mathcal{L}$. Thus we may apply Zorn's lemma to obtain a maximal linearly independent set $\mathcal{M}$. If $\mathcal{M}$ does not span W, then choose an element $w \in W \setminus \mathrm{Span}(\mathcal{M})$. The set $\mathcal{M} \cup \{w\}$ is linearly independent, contradicting the maximality of $\mathcal{M}$. We have shown that $\mathcal{M}$ spans W and so is a basis. It follows that W is isomorphic to the direct sum $K^{(\mathcal{M})}$ of copies of K indexed by the set $\mathcal{M}$.

From this perspective, the infinite-dimensional K-modules are not much more interesting than the finite-dimensional ones. This kind of behaviour is typical of semi-simple rings (in fact, K is even a simple ring); see [6, Sec. 1.2] for more information about this family of rings.

Example 4.12 In Example 4.8, we saw that a representation (M, Φ) uniquely determines a $K[X]$-module. This is an infinite-dimensional K[X]-module if and only if M is an infinite-dimensional K-vector space.

Example 4.13 In Example 4.9, we saw that KQ/I-modules are determined by representations of the quiver with relations. Such a representation corresponds to an infinite-dimensional KQ/I-module if and only if the K-vector space $U_1 \oplus U_2$ is infinite-dimensional.

4.1.4 Homomorphisms between A-modules

Definition 4.14 Let M and N be A-modules. Then an **A-homomorphism** is a K-linear map $f\colon M \to N$ such that, for any $a \in A$ and $m \in M$, we have that $f(am) = af(m)$.

Example 4.15 The definition of a K-homomorphism coincides with the definition of a K-linear map.

Example 4.16 Let M and N be $K[X]$-modules and suppose $\Phi\colon M \to M$ and $\Psi\colon N \to N$ are the K-linear endomorphisms determined by the actions of X on M and N respectively (see Example 4.8). Then a K-linear map $f\colon M \to N$ is a

$K[X]$-homomorphism if and only if, for any $m \in M$ and $p = k_0 + k_1X + k_2X^2 + \cdots + k_nX^n \in K[X]$, we have that

$$\begin{aligned} k_0 f(m) + \cdots + k_n X^n f(m) &= p f(m) \\ &= f(pm) \\ &= k_0 f(m) + \cdots + k_n f(X^n m) \end{aligned}$$

if and only if we have $Xf(m) = f(Xm)$ for all $m \in M$, i.e. $\Phi \circ f = f \circ \Psi$.

Example 4.17 Consider representations

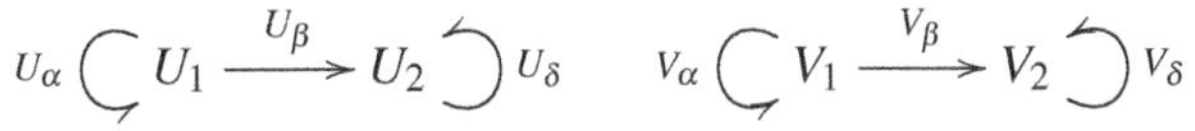

$$U_\alpha \circlearrowright U_1 \xrightarrow{U_\beta} U_2 \circlearrowleft U_\delta \qquad V_\alpha \circlearrowright V_1 \xrightarrow{V_\beta} V_2 \circlearrowleft V_\delta$$

of (Q,I). We saw in Example 4.9 that these representations determine KQ/I-modules with underlying vector spaces $U_1 \oplus U_2$ and $V_1 \oplus V_2$ respectively. It follows from the definition that a K-linear map

$$\begin{pmatrix} a & b \\ c & d \end{pmatrix} : U_1 \oplus U_2 \to V_1 \oplus V_2$$

is a KQ/I-homomorphism if and only if we have that $b = 0 = c$ and the following diagrams commute:

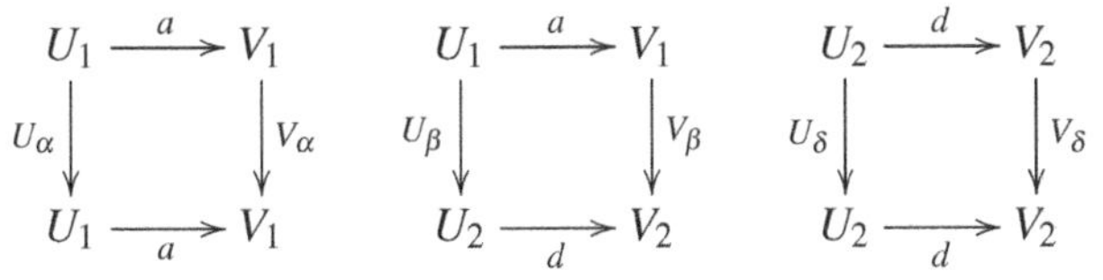

$$\begin{array}{ccc} U_1 & \xrightarrow{a} & V_1 \\ {\scriptstyle U_\alpha}\downarrow & & \downarrow{\scriptstyle V_\alpha} \\ U_1 & \xrightarrow[a]{} & V_1 \end{array} \qquad \begin{array}{ccc} U_1 & \xrightarrow{a} & V_1 \\ {\scriptstyle U_\beta}\downarrow & & \downarrow{\scriptstyle V_\beta} \\ U_2 & \xrightarrow[d]{} & V_2 \end{array} \qquad \begin{array}{ccc} U_2 & \xrightarrow{d} & V_2 \\ {\scriptstyle U_\delta}\downarrow & & \downarrow{\scriptstyle V_\delta} \\ U_2 & \xrightarrow[d]{} & V_2 \end{array}$$

4.2 Direct Limits

In this section we will introduce direct limits as a means to build examples of infinite-dimensional A-modules. The notion of a direct limit allows you to build a new module out of a given family of modules. The word "direct" refers to the fact that the family of modules must form a direct system, i.e. there are A-homomorphisms between the modules that satisfy the next definition.

Definition 4.18 A **directed set** is a nonempty set I with a reflexive and transitive binary relation $\leqslant$ such that, for every $i, j \in I$, there exists $k \in I$ such that $i \leqslant k$ and $j \leqslant k$.

Definition 4.19 Let I be a directed set. A collection of A-modules $\{M_i \mid i \in I\}$ together with a collection of A-homomorphisms

$$\{f_{ij} \colon M_i \to M_j \mid i, j \in I, i \leqslant j\}$$

is called a **direct system** of A-modules if $f_{ii} = \mathrm{id}_{M_i}$ for all $i \in I$ and $f_{jk} f_{ij} = f_{ik}$ for all $i \leqslant j \leqslant k$.

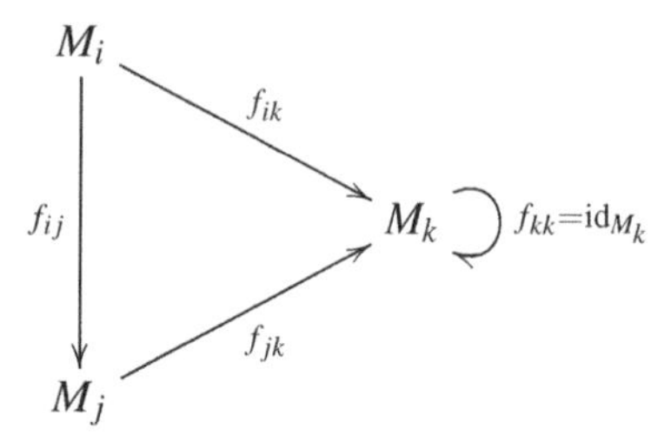

Remark The direct limit (as we define it in Definition 4.20 below) satisfies a universal property that means it is isomorphic to the colimit of the diagram $\mathcal{F} = \{f_{ij} \colon M_i \to M_j \mid i, j \in I, i \leqslant j\}$ in the category of A-modules. See, for example, [11, Sec. IV.8] for more details.

We will usually refer only to the set $\{f_{ij} \colon M_i \to M_j \mid i, j \in I, i \leqslant j\}$ of morphisms as a direct system of A-modules since the existence of the modules $\{M_i \mid i \in I\}$ is implied by this. Define an equivalence relation on the disjoint union $\bigsqcup_{i \in I} M_i$ by declaring that $m_i \sim m_j$ whenever $m_i \in M_i$, $m_j \in M_j$ and there exists $k \geqslant i, j$ such that $f_{ik}(m_i) = f_{jk}(m_j)$.

Definition 4.20 The **direct limit** $\varinjlim_I M_i$ of a direct system $\mathcal{F} = \{f_{ij} \colon M_i \to M_j \mid i, j \in I, i \leqslant j\}$ is the A-module given by the set $\bigsqcup_{i \in I} M_i / \sim$ with the unique A-module operations such that, for every $k \in I$, the canonical map $M_k \to \varinjlim_I M_i$ is an A-homomorphism.

Remark An explicit description of the operations defining the A-module structure of $\varinjlim_I M_i$ can be found in [11, Sec. I.5].

4.2.1 Examples of Direct Limits

Next we will introduce some examples of infinite-dimensional modules that arise as direct limits of finite-dimensional modules.

Remark It is important to observe that a direct limit is not necessarily an infinite-dimensional module. For example, if we take any finite-dimensional module M, then we can define a direct system

$$\{\mathrm{id}_{M_{nm}} \colon M_n \to M_m \mid M_n \cong M, M_m \cong M \text{ for all } n \leqslant m\}$$

where the associated directed set is $I = \mathbb{N}$. Then the direct limit $\varinjlim_I M_i$ is isomorphic to M and hence is finite-dimensional.

Example 4.21 Consider the K-algebra $K[X]$ given in Example 4.3 and let $k \in K$. For each $n \in \mathbb{N}$, consider the vector space K^n and the K-linear endomorphism given by the Jordan block

$$J_{k,n} = \begin{pmatrix} k & 1 & 0 & \cdots & 0 & 0 \\ 0 & k & 1 & \cdots & 0 & 0 \\ \vdots & \vdots & \vdots & \cdots & \vdots & \vdots \\ 0 & 0 & 0 & \cdots & k & 1 \\ 0 & 0 & 0 & \cdots & 0 & k \end{pmatrix}.$$

In Example 4.8, we saw that this defines a $K[X]$-module which we will denote by $M_{k,n}$. The $(n+1 \times n)$-matrix

$$\begin{pmatrix} & I_n & \\ 0 & \cdots & 0 \end{pmatrix}$$

(where I_n is the $(n \times n)$-identity matrix) defines a $K[X]$-homomorphism $f_n \colon M_{k,n} \to M_{k,n+1}$. We consider the directed set $\mathbb{N}$ and the direct system

$$\{g_{nm} := f_{m-1} \circ \cdots \circ f_n \colon M_{k,n} \to M_{k,m} \mid n < m \text{ in } \mathbb{N}\} \cup \{g_{nn} := I_n \mid n \in \mathbb{N}\}.$$

The direct limit $M_{k,\infty} := \varinjlim_{\mathbb{N}} M_{k,n}$ is called the k-**Prüfer module** over $K[X]$.

Proposition 4.22 *The k-Prüfer module $M_{k,\infty}$ is isomorphic to the module with underlying vector space $K^{(\mathbb{N})}$ and with the action of X given by the K-linear endomorphism $J_{k,\infty} \colon K^{(\mathbb{N})} \to K^{(\mathbb{N})}$ defined by $(k_n)_{n\in\mathbb{N}} \mapsto (kk_n + k_{n+1})_{n\in\mathbb{N}}$.*

Proof Let $\Phi \colon \varinjlim_{\mathbb{N}} M_{k,n} \to \varinjlim_{\mathbb{N}} M_{k,n}$ denote the K-linear endomorphism induced by the action of X on $\varinjlim_{\mathbb{N}} M_{k,n}$. Consider the map

$$h \colon \varinjlim_{\mathbb{N}} M_{k,n} \to K^{(\mathbb{N})}$$

that takes an equivalence class $[(k_i)_{i=1}^m]$ with $(k_i)_{i=1}^m \in M_{k,m}$ to the element $(k'_i)_{i\in\mathbb{N}}$ in $K^{(\mathbb{N})}$ with $k'_i := k_i$ for $i \leqslant m$ and $k'_i = 0$ for $i > m$. It is straightforward to check that h is a well-defined $K[X]$-isomorphism. □

Example 4.23 Consider the periodic sequence

$$z = (\ldots \beta^{-1}\delta^{-1}\beta\alpha\beta^{-1}\delta^{-1}\beta\alpha\beta^{-1}\delta^{-1}\beta\alpha)$$

of arrows in (Q,I) and their formal inverses. We may represent this sequence in the following diagram:

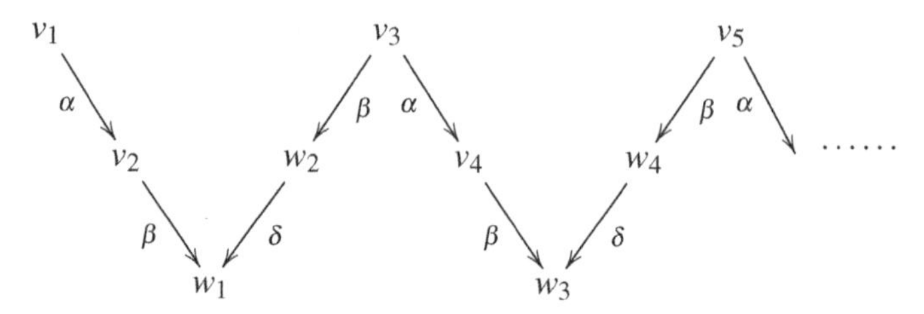

The labels on the starting and ending points of the arrows correspond to basis elements of a representation

$$U_\alpha \circlearrowright U_1 \xrightarrow{U_\beta} U_2 \circlearrowleft U_\delta$$

of (Q,I), which is defined as follows. Define U_1 to be the K-vector space with basis $\{v_i \mid i \in \mathbb{N}\}$ and define U_2 to be the K-vector space with basis $\{w_i \mid i \in \mathbb{N}\}$. Intuitively, we think of the labels on the arrows in the above diagram as corresponding to the K-linear maps that make up this representation. More precisely, define $U_\alpha : U_1 \to U_1$ to be the K-linear map that takes v_i to v_{i+1} when i is odd and takes v_i to zero when i is even. Define $U_\beta : U_1 \to U_2$ to be the K-linear map that take v_1 to zero and v_i to w_{i-1} when $i \geqslant 2$. Define $U_\delta : U_2 \to U_2$ to take w_i to zero when i is odd and takes w_i to w_{i-1} when i is even. The module $M(z)$ determined by this representation is called an **infinite string module** over KQ/I.

The module described in Example 4.23 arises as the direct limit of a direct system of finite-dimensional modules. For each $n \in \mathbb{N}$, consider the submodule $M(z_n)$ of $M(z)$ spanned by the basis elements

$$\{v_i \mid 1 \leqslant i \leqslant 2n\} \cup \{w_i \mid 1 \leqslant i \leqslant 2n-1\}.$$

The module $M(z_n)$ therefore has an underlying vector space that is isomorphic to $K^{2n} \oplus K^{2n-1}$. We will represent a typical element of $M(z_n)$ by $((k_i)_{i=1}^{2n}, (l_i)_{i=1}^{2n-1})$. Consider the direct system of canonical inclusions denoted by $\{\iota_{nm} : M(z_n) \to M(z_m) \mid n \leqslant m\}$. For example, the inclusion ι_{12} is represented by the following diagram

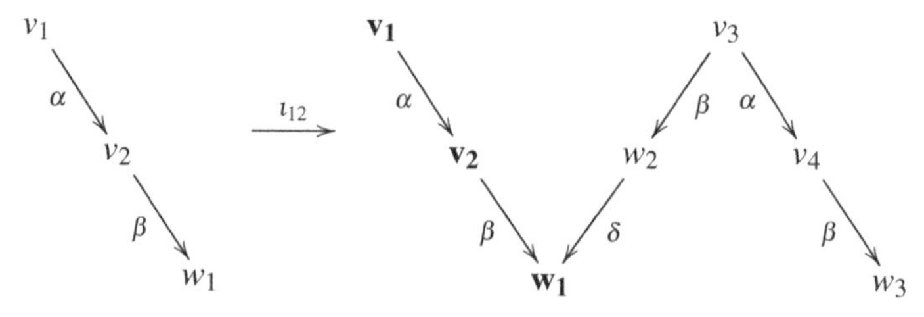

where the bold text indicates the image of ι_{12}.

Proposition 4.24 *The string module $M(z)$ described in Example 4.23 is isomorphic to $\varinjlim_{\mathbb{N}} M(z_n)$.*

Proof Consider the map

$$h\colon \varinjlim_{\mathbb{N}} M(z_n) \to M(z)$$

that takes an equivalence class $[((k_i)_{i=1}^{2n}, (l_i)_{i=1}^{2n-1})]$ to the element $((k_i')_{i\in\mathbb{N}}, (l_i')_{i\in\mathbb{N}})$ where $k_i' = k_i$ for $1 \leqslant i \leqslant 2n$, $k_i' = 0$ for $i > 2n$; $l_i' = l_i$ for $1 \leqslant i \leqslant 2n-1$ and $l_i' = 0$ for $i > 2n-1$. It is straightforward to check that h is a well-defined KQ/I-isomorphism. □

4.3 Duality

In this section we will describe a duality, induced by the usual K-vector space duality, that will allow us to construct new (left) A-modules from right A-modules.

4.3.1 Right A-modules and the Opposite Algebra

The definition of a right A-module is analogous to Definition 4.6 with A acting on the right instead of the left. Another way of viewing right A-modules is as left A^{op}-modules, where A^{op} is the opposite algebra. This perspective will be useful for us when computing examples in this section.

Definition 4.25 Let A be a K-algebra with multiplication $A \times A \to A$ sending (a,b) to ab. Define the **opposite algebra** A^{op} to be K-algebra with the same underlying vector space as A and K-bilinear multiplication $*\colon A^{op} \times A^{op} \to A^{op}$ sending (a,b) to $a * b := ba$.

Suppose M is a left A^{op}-module. Then, by definition, M is a K-vector space with an A^{op}-action $\cdot\colon A^{op} \times M \to M$ such that, for any $a,b \in A^{op}$ and any $m \in M$, we have that $(a*b)\cdot m = a\cdot(b\cdot m)$ and $1\cdot m = m$. We can then define a right A-action $M \times A \to M$ on M to be $ma := a\cdot m$ for all $m \in M$ and $a \in A$. Then we have $m(ab) = m(b*a) = (b*a)\cdot m = b\cdot(a\cdot m) = (a\cdot m)b = (ma)b$ and $m1 = 1\cdot m = m$. We have shown that any left A^{op}-module determines a right A-module. A similar argument yields the converse statement.

Example 4.26 Both K and $K[X]$ are commutative algebras and so they coincide with their opposite algebra. In particular, every right module over K or $K[X]$ is also a left module and vice versa.

Example 4.27 Consider the algebra KQ/I from Example 4.5. The opposite algebra $(KQ/I)^{op}$ is given by the path algebra of the opposite quiver Q^*

$$\alpha^* \circlearrowleft \bullet^{1^*} \xleftarrow{\beta^*} \bullet^{2^*} \circlearrowright \delta^*$$

with relation $\rho^* := \{\alpha^*\beta^*\delta^*, (\alpha^*)^2, (\delta^*)^2\}$. That is, take the K-vector space KQ^* with basis

$$\mathbf{Pa}^* := \{e_1^*, e_2^*, \alpha^*, \beta^*, \delta^*, \alpha^*\beta^*, \beta^*\delta^*, \alpha^*\beta^*\delta^*\}$$

together with the multiplication induced by concatenation of paths. Then $(KQ/I)^{op} = KQ^*/I^*$ where I^* is the ideal of KQ^* generated by ρ^*. By an analogous argument to the one given in Example 4.9, the left KQ^*/I^*-modules (i.e. right KQ/I-modules) are given by representations of (Q^*, I^*).

4.3.2 Dual Modules

We know that any right A-module has an underlying K-vector space structure and so we may consider the dual K-vector space. The following definition yields a canonical way of equipping the dual K-vector space with a left A-module structure.

Definition 4.28 Let A be a K-algebra and let M be a right A-module (equivalently a left A^{op}-module). Then the K**-dual** M^* of M is defined to be the left A-module consisting of the usual K-dual vector space M^* and the A-action $A \times M^* \to M^*$ given by $(a, f) \mapsto af$ where $(af)(m) = f(ma)$ for each $m \in M$.

Example 4.29 Consider a K-vector space V with basis $\mathcal{B}$. We have already observed that $V \cong K^{(\mathcal{B})}$. Then the dual K-module coincides with the dual K-vector space, which is given by the direct product $K^{\mathcal{B}}$ of copies of K indexed by $\mathcal{B}$.

Example 4.30 Let $k \in K$ and consider the k-Prüfer module $M_{k,\infty}$ as a right $K[X]$-module. Then the dual module $M_{k,\infty}^*$ is called the k**-adic module** over $K[X]$ and will be denoted $M_{k,-\infty}$. Moreover, $M_{k,-\infty}$ is isomorphic to the module with underlying vector space given by $K^{\mathbb{N}}$ and with the action of X given by the endomorphism $J_{k,\infty}\colon K^{\mathbb{N}} \to K^{\mathbb{N}}$ defined by $(k_n)_{n\in\mathbb{N}} \mapsto (kk_n + k_{n-1})_{n\in\mathbb{N}}$ where k_0 is defined to be zero.

Example 4.31 Consider the infinite string module over KQ^*/I^* given by periodic sequence

$$y^* = (\ldots(\beta^*)^{-1}(\alpha^*)^{-1}\beta^*\delta^*(\beta^*)^{-1}(\alpha^*)^{-1}\beta^*\delta^*(\beta^*)^{-1}(\alpha^*)^{-1}\beta^*\delta^*),$$

represented by the following diagram:

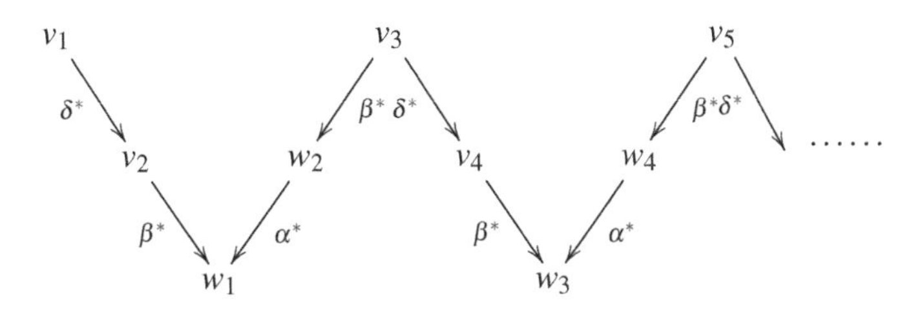

Then the left KQ^*/I^*-module $M(y^*)$ can be considered as a right KQ/I-module. The **dual infinite string module** $M(y^*)^*$ can be described explicitly as follows. Take the periodic sequence of dual arrows $y = (\ldots\beta\alpha\beta^{-1}\delta^{-1}\beta\alpha\beta^{-1}\delta^{-1}\beta\alpha\beta^{-1}\delta^{-1})$ represented by the following diagram:

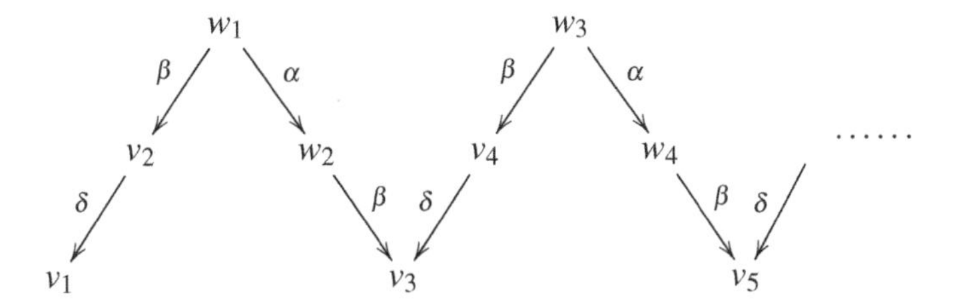

Define a representation

$$V_\alpha \circlearrowleft V_1 \xrightarrow{V_\beta} V_2 \circlearrowright V_\delta$$

of (Q,I) as follows. Define V_1 to be the K-vector space $K^{\mathbb{N}}$ where we label $w_i := (0,\ldots,0,1,0,\ldots)$ with 1 in the ith position. Define V_2 to be the K-vector space $K^{\mathbb{N}}$ and, similarly, we label $v_i := (0,\ldots,0,1,0,\ldots)$ with 1 in the ith position.. The labels on the arrows correspond to the K-linear maps that make up this representation. Define $V_\alpha : V_1 \to V_1$ to be the K-linear map that takes w_i to w_{i+1} when i is odd and take w_i to zero when i is even. Define $V_\beta : V_1 \to V_2$ to be the K-linear map that takes w_i to v_{i+1} for all $i \in \mathbb{N}$. Define $V_\delta : V_2 \to V_2$ to take v_i to zero when i is odd and takes v_i to v_{i-1} when i is even.

4.4 Finite Matrix Subgroups and Pp-definable Subgroups

4.4.1 Finite Matrix Subgroups

In our setting, a matrix subgroup of an A-module M is a K-subspace of the underlying K-vector space of M that can be realised as the trace of an element $x \in L$ in M for some A-module L. In this section we introduce the notion of a finite matrix subgroup, which is a matrix subgroup where the module L is finitely presented.

Definition 4.32 An A-module L is called **finitely presented** if there exist $n, m \in \mathbb{N}$ such that $L \cong A^n/\mathrm{im}(\Phi)$ where $\Phi\colon A^m \to A^n$ and $\mathrm{im}(\Phi)$ denotes the image of Φ. Note that Φ can be represented by an $(m \times n)$-matrix $P = (a_{ji})$ with $a_{ji} \in A$ for $1 \leqslant i \leqslant n$ and $1 \leqslant j \leqslant m$ such that $\Phi\colon \begin{pmatrix} a_1 & \dots & a_m \end{pmatrix} \mapsto \begin{pmatrix} a_1 & \dots & a_m \end{pmatrix} P$.

Example 4.33 If A is a finite-dimensional algebra, then the finitely presented modules coincide with the finite-dimensional modules. Since a finitely presented module is a quotient of the finite-dimensional module A^n, it must be a finite-dimensional module itself. Conversely, if M is an n-dimensional module, then the K-basis is also an A-generating set, so we may define an epimorphism $\Psi\colon A^n \to M$. Then the kernel $\ker(\Psi)$ of Ψ is finite-dimensional and so, by the same argument, there exists an epimorphism $\Omega\colon A^m \to \ker(\Psi)$. Then the composition $\Phi := \iota \circ \Omega$, where $\iota\colon \ker(\Psi) \to A^n$ is the canonical embedding, is the desired presentation of M.

Definition 4.34 Let M be an A-module and let L be a finitely presented A-module. For a fixed $l \in L$, consider the K-subspace $H_{(L,l)}(M) := \{f(l) \mid f\colon L \to M \text{ an } A\text{-homomorphism}\}$. A K-subspace of this form is called a **finite matrix subgroup** of M.

Remark Finite matrix subgroups are usually defined in the context of modules over a ring (that is not necessarily a K-algebra). In that more general setting the set $H_{(L,l)}(M)$ is a subgroup of the underlying abelian group structure of M. This is why $H_{(L,l)}(M)$ is called a finite matrix *subgroup* rather than a finite matrix *subspace*.

Example 4.35 Consider the KQ/I-module $M(z)$ from Example 4.23 and the finite-dimensional submodule $M(z_2)$ with the underlying vector space spanned by $\{v_1, v_2, v_3, v_4, w_1, w_2, w_3\}$. Then the finite matrix subgroup $H_{M(z_2),w_2}(M(z))$ is the K-subspace spanned by

$$\{w_2\} \cup \{w_{2n-1} \mid n \in \mathbb{N}\}.$$

This is witnessed by the fact that there are the following KQ/I-homomorphisms from $M(z_2)$ to $M(z)$:

- The embedding $f_0\colon M(z_2) \to M(z)$ given by $v_i \mapsto v_i$ and $w_j \mapsto w_j$ for $i \in \{1,2,3,4\}$ and $j \in \{1,2,3\}$. Then $f_0(w_2) = w_2$.
- For each $n \in \mathbb{N}$, we have a KQ/I-homomorphism $f_n\colon M(z_2) \to M(z)$ given by $v_3 \mapsto v_{2n}$, $w_2 \mapsto w_{2n-1}$, $v_i \mapsto 0$ and $w_j \mapsto 0$ for $i \in \{1,2,4\}$ and $j \in \{1,3\}$. Then $f_n(w_2) = w_{2n-1}$.

It is an interesting exercise to prove that these KQ/I-homomorphisms $\{f_n \mid n \geqslant 0\}$ form a basis for the K-vector space of KQ/I-homomorphisms from $M(z_2)$ to $M(z)$. Alternatively, we may apply the more general theorem proved in [2, Sec. 1.4].

Remark In general, a finite matrix subgroup $H_{L,l}(M)$ is not an A-submodule of M. Indeed, if we take the KQ/I-module, $Ae_1 = \{a \in A \mid \exists b \in A$ such that $a = be_1\}$. Then $H_{Ae_1,e_1}(M(z))$ coincides with the K-subspace spanned by the set $\{v_n \mid n \text{ odd}\}$. This is not a KQ/I-submodule of $M(z)$ since, for example, $w_2 = \beta v_3$ is not contained in $H_{Ae_1,e_1}(M(z))$.

4.4.2 Pp-definable Subgroups

The notion of a pp-definable subgroup comes from the area of logic called model theory. They are the sets of elements of a module that realise a given positive primitive formula in the language of A-modules. We will not put too much emphasis on the model theoretic perspective in these lectures; however, if you are interested in this subject you can read more in [7].

Definition 4.36 Let M be an A-module and let $\sum_{i=1}^{n} a_{ji}x_i = 0$ where $1 \leqslant j \leqslant m$ is a finite A-linear system. That is, the symbols x_i denote free variables and $a_{ji} \in A$ for each $1 \leqslant i \leqslant n$ and $1 \leqslant j \leqslant m$. Note that the system depends on a $(m \times n)$-matrix $P = (a_{ji})$ with $a_{ji} \in A$ for $1 \leqslant i \leqslant n$ and $1 \leqslant j \leqslant m$. Consider the K-subspace $\varphi_P(M)$ defined by

$$\{u_1 \in M \mid \exists u_2, \ldots, u_n \in M \text{ such that } \sum_{i=1}^{n} a_{ji}u_i = 0 \text{ for all } 1 \leqslant j \leqslant m\}.$$

A K-subspace of this form is called a **pp-definable subgroup** of M.

Remark The symbol φ_P refers to the first order formula

$$\exists x_2 \ldots \exists x_n \left(\sum_{i=1}^{n} a_{1i}x_i = 0 \wedge \cdots \wedge \sum_{i=1}^{n} a_{mi}x_i = 0 \right)$$

that should be read as "there exist x_2 up to x_n such that $\sum_{i=1}^{n} a_{1i}x_i = 0$ and $\sum_{i=1}^{n} a_{2i}x_i = 0\ldots$" and so on up to m. The notation $\varphi_P(M)$ is then used for the solution set of this formula in M. That is, the set of elements u_1 in M such that, when we replace x_1 with u_1, the statement above in quotation marks is true. Note that this coincides with what is written in Definition 4.36.

In parallel to Remark 4.4.1, we observe that, if we made this definition for a module M over a general ring, then $\varphi_P(M)$ would form a subgroup of the underlying abelian group structure of M. In our setting, this is even a K-subspace.

Example 4.37 Consider the following system of KQ/I-linear equations:

$$\begin{aligned} e_1x_1 &= 0, \\ \delta x_1 - x_2 &= 0, \\ \beta x_3 - x_2 &= 0, \\ \alpha x_4 - x_3 &= 0, \\ \beta x_4 &= 0. \end{aligned}$$

We have the corresponding matrix

$$P = \begin{pmatrix} e_1 & 0 & 0 & 0 \\ \delta & -1 & 0 & 0 \\ 0 & -1 & \beta & 0 \\ 0 & 0 & -1 & \alpha \\ 0 & 0 & 0 & \beta \end{pmatrix}.$$

Then the pp-definable subgroup $\Phi_P(M(z))$ of $M(z)$ is the K-subspace spanned by the set

$$\{w_2\} \cup \{w_{2n-1} \mid n \in \mathbb{N}\}.$$

The pp-definable subgroup $\Phi_P(M(y^*)^*)$ of $M(y^*)^*$ is given by the set of (possibly infinite) K-linear combinations of the set $\{v_{2n-1} \mid n \in \mathbb{N}\}$.

If we look at Example 4.35 and Example 4.37, then we find that $H_{M(z_2),w_2}(M(z)) = \Phi_P(M(z))$. In the next proposition we will show that the set of finite matrix subgroups of an A-module M coincides with the set of pp-definable subgroups of M.

Proposition 4.38 *Let U be a K-subspace of an A-module M. The following statements are equivalent.*

1 There exists a finitely presented A-module L and $l \in L$ such that $U = H_{L,l}(M)$.

2 *There exists a finite A-linear system determined by a matrix P such that* $U = \varphi_P(M)$.

Proof Consider an $(m \times n)$-matrix $P = (a_{ji})$ with $a_{ji} \in A$ for $1 \leqslant i \leqslant n$ and $1 \leqslant j \leqslant m$. We have already observed that P determines both an A-linear system $\sum_{i=1}^n a_{ji}x_i = 0$ where $1 \leqslant j \leqslant m$ and a finitely presented module $L := A^n/\mathrm{im}(\Phi)$ where $\Phi \colon A^m \to A^n$ is the A-homomorphism $\begin{pmatrix} a_1 & \dots & a_m \end{pmatrix} \mapsto \begin{pmatrix} a_1 & \dots & a_m \end{pmatrix} P$. We fix the following notation. For each $1 \leqslant j \leqslant m$, let $d_j := (0\dots010\dots0)$ be the element of A^m with 1 in the jth position and zeroes elsewhere. For each $1 \leqslant i \leqslant n$, define $e_i := (0\dots010\dots0)^T$ to be the element of A^n with 1 in the ith position and zeroes elsewhere and let $l_i := \pi(e_i)$ where $\pi \colon A^n \to L$ is the canonical quotient morphism. Observe that $\Phi(d_j) = (a_{j1}\dots a_{jn})^T = \sum_{i=1}^n a_{ji}e_i$.

We will show that $H_{L,l_1}(M) = \varphi_P(M)$ for all A-modules M. First we show that $H_{L,l_1}(M) \subseteq \varphi_P(M)$ so let $u_1 = f(l_1)$ for some A-homomorphism $f \colon L \to M$. Set $u_i := f(l_i)$ for $1 < i \leqslant n$. Then, for each $1 \leqslant j \leqslant m$, we have that

$$\sum_{i=1}^n a_{ji} f(l_i) = f\left(\sum_{i=1}^n a_{ji} l_i\right) = f\pi\left(\sum_{i=1}^n a_{ji} e_i\right) = f\pi\Phi(d_j) = 0$$

since $\pi\Phi = 0$. We therefore have that $u_1 \in \varphi_P(M)$. Next we show the other inclusion $\varphi_P(M) \subseteq H_{L,l_1}(M)$. Let $u_1 \in \varphi_P(M)$ and consider elements $u_2, \dots, u_n \in M$ that satisfy the A-linear equations. The assignment $e_i \mapsto u_i$ extends uniquely to an A-homomorphism $f' \colon A^n \to M$. As

$$f'\Phi(d_j) = f'\left(\sum_{i=1}^n a_{ji} e_i\right) = \sum_{i=1}^n a_{ji} u_i = 0,$$

there exists a unique A-homomorphism $f \colon L \to M$ such that $f\pi = f'$. In particular, we have that $u_1 = f'(e_1) = f(l_1)$ so $u_1 \in H_{L,l_1}(M)$ as desired. □

4.5 Pure Submodules and Pure-injective Modules

The aim of this next section is to define and give examples of the modules that give the points of the Ziegler spectrum. The first definition is that of a pure submodule. These are the submodules that respect the pp-definable (equivalently the finite matrix) subgroups.

Definition 4.39 Let L and M be A-modules such that $L \subseteq M$ is an A-submodule. Then L is a **pure submodule** of M if $\varphi_P(L) = \varphi_P(M) \cap L$ for all $(m \times n)$-matrices P with entries in A. A monomorphism $f \colon L \to M$ such that $\mathrm{im}(f) \subseteq M$ is a pure submodule is called a **pure monomorphism**.

A useful characterisation of a pure monomorphism makes use of the duality defined in Definition 4.28. Notice that, for any A-homomorphism $g\colon M \to N$, the usual K-linear map $g^*\colon N^* \to M^*$ induced by K-vector space duality is an A^{op}-homomorphism. It is well known that a monomorphism $g\colon M \to N$ is pure if and only if there exists an A^{op}-homomorphism $h\colon M^* \to N^*$ such that $g^* \circ h = \mathrm{id}_{M^*}$. To see a proof of this, as well as other characterisations of pure monomorphisms, see [3, Lem. 2.19].

Example 4.40 For any A-module M, the morphism $\delta_M\colon M \to M^{**}$ given by $m \mapsto \mathrm{ev}_m$ where $ev_m(f) = f(m)$ for all $f \in M^*$ is a pure monomorphism. This follows from the discussion preceding this example since $\delta_M^* \circ \delta_{M^*} = \mathrm{id}_{M^*}$.

Definition 4.41 A non-zero A-module N is called **pure-injective** if, for every pure monomorphism $f\colon N \to M$, there exists an A-homomorphism $g\colon M \to N$ such that $gf = \mathrm{id}_N$.

Example 4.42 It follows from Example 4.40 that every pure-injective module is a direct summand of a dual module. It turns out that this, in fact, characterises pure-injective modules (see [3, Thm. 2.27]). In particular, any dual module is pure-injective.

For any finite-dimensional A-module M, we have that $M \cong M^{**}$. It therefore follows that finite-dimensional A-modules are pure-injective.

Example 4.43 The modules defined in Examples 4.21, 4.23, 4.30 and 4.31 are pure-injective modules. The fact that the infinite string module (Example 4.23) and the dual infinite string module (Example 4.31) are pure-injective is proved in [10]. The k-Prüfer module (Example 4.21) is an injective $K[X]$-module and so clearly it is also pure-injective. The k-adic module (Example 4.30) is a dual module and so it is pure-injective by Example 4.42.

4.6 The Ziegler Spectrum

In this section we introduce a topological space called the Ziegler spectrum. The points of the space are the indecomposable pure-injective modules.

Definition 4.44 A non-zero A-module M is called **indecomposable** if, whenever $M \cong N \oplus L$, either $L = 0$ or $N = 0$.

Remark The collection of isomorphism classes of indecomposable pure-injective A-modules has cardinality at most $2^{\kappa+\aleph_0}$ where κ is the cardinality of A. In particular, the isomorphism classes of indecomposable pure-injective modules form a set, which we denote by Zg_A.

The elements of Zg_A are isomorphism classes $[N]$ but we will drop the square brackets and refer instead to the representative N as a **point** of Zg_A. If N is a finite-dimensional module we refer to it as a **finite-dimensional point**. Similarly, if N is infinite-dimensional then we refer to N as an **infinite-dimensional point** of Zg_A.

The Ziegler topology on Zg_A can be defined in many different ways (see, for example, [8, Ch. 5.1]). In these lecture notes we will define the topology in terms of pp pairs.

Definition 4.45 Let (P,Q) be a pair of matrices with entries in A (possibly of different sizes). We will call (P,Q) a **pp pair** if $\varphi_P(M) \subseteq \varphi_Q(M)$ for all A-modules M.

According to Proposition 4.38, a pair (P,Q) of matrices with entries in A determines a pair of pointed finitely presented modules (L,l) and (N,n) such that $\varphi_P(M) = H_{L,l}(M)$ and $\varphi_Q(M) = H_{N,n}(M)$ for all A-modules M. Clearly this means that (P,Q) is a pp pair if and only if $H_{L,l}(M) \subseteq H_{N,n}(M)$ for all A-modules M.

Definition 4.46 Let Zg_A be the set of isomorphism classes of indecomposable pure-injective A-modules. We call a set $\mathcal{U} \subseteq \mathrm{Zg}_A$ **basic open** if there exists a pp pair (P,Q) such that

$$\mathcal{U} = \{M \in \mathrm{Zg}_A \mid \varphi_P(M) \subsetneq \varphi_Q(M)\}.$$

Denote the basic open set corresponding to a pp pair (P,Q) by (φ_P/φ_Q).

Recall that a topological space Z is called **quasi-compact** if, whenever $Z = \bigcup_{i\in I} U_i$ for U_i open sets, we have that there is a finite subset $F \subseteq I$ such that $Z = \bigcup_{i\in F} U_i$. In other words, any open cover of Z has a finite subcover.

Theorem 4.47 (Ziegler, 1984) *The basic open sets form a base of a topology on* Zg_A *and, moreover, the basic open sets are quasi-compact. This topological space is called the* ***Ziegler spectrum*** *of* A.

The proof of the above theorem is originally due to Ziegler and is contained in his landmark paper [12] on the model theory of modules. A more algebraic proof was given later by Herzog using functor categories [4]. See also Krause [5]. Unfortunately, both the model theoretic and more algebraic arguments require material that is beyond the scope of these lectures and so we do not prove the theorem here.

Corollary 4.48 *The Ziegler spectrum of* A *is a quasi-compact topological space.*

Proof By the theorem, it suffices to show that there is a pp pair (P,Q) such that $\mathrm{Zg}_A = (\varphi_P/\varphi_Q)$. If we take P to be the (1×1)-matrix 0 and Q to be the (1×1)-matrix 1, then $\varphi_P(M) = M$ and $\varphi_Q(M) = 0$ for any module M. Thus $\mathrm{Zg}_A = (\varphi_P/\varphi_Q)$ is quasi-compact. □

In Example 4.49 we will describe the points of the Ziegler spectrum of the algebra KQ/I introduced in Example 4.5. In order to do this we describe a way of building a representation of the bound quiver (Q,I) from $K[X]$-modules. Recall from Example 4.8 that each $K[X]$-module is determined by a K-vector space M and a K-linear endomorphism $\Phi\colon M \to M$. Given such a pair (M,Φ), we may define a representation of (Q,I) as follows:

$$W_\alpha \circlearrowleft W_1 \xrightarrow{W_\beta} W_2 \circlearrowright W_\delta$$

where both W_1 and W_2 are isomorphic to $M \oplus M$ and the K-linear maps are given by $W_\alpha := \begin{pmatrix} 0 & 0 \\ \mathrm{id}_M & 0 \end{pmatrix}$, $W_\beta := \begin{pmatrix} \Phi & 0 \\ 0 & \mathrm{id}_M \end{pmatrix}$ and $W_\delta := \begin{pmatrix} 0 & 0 \\ \mathrm{id}_M & 0 \end{pmatrix}$. A KQ/I-module of this kind is known as a **band module** because it can be visualised as follows:

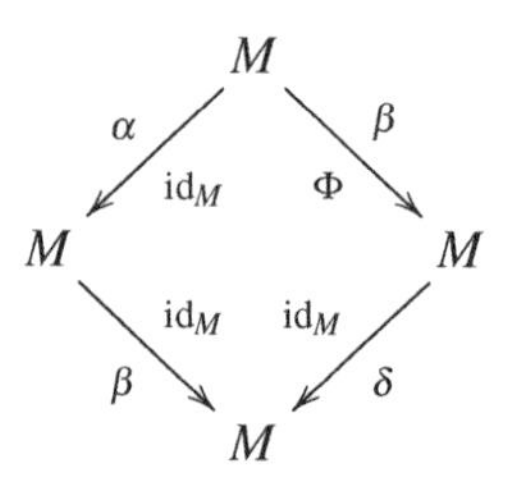

For every $K[X]$-module M, we will denote the corresponding band module over KQ/I by $\mathbf{Ba}(M)$.

Remark The assignment $M \mapsto \mathbf{Ba}(M)$ extends to a functor from the category of $K[X]$-modules to the category of KQ/I-modules.

Example 4.49 Let K be an algebraically closed field. The following is a complete list of the points of the Ziegler spectrum of KQ/I. This classification can be found in [9]:

- The finite-dimensional KQ/I-modules; see Example 4.42.
- The infinite string module $M(z)$ described in Example 4.23 corresponding to the sequence

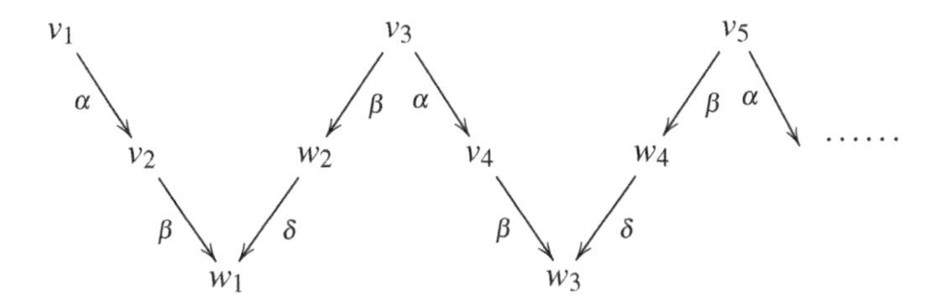

- The submodule of $M(z)$ spanned by $\{w_i \mid i \geqslant 2\} \cup \{v_j \mid j > 2\}$. This module is the infinite string module $M(w)$ associated to the sequence

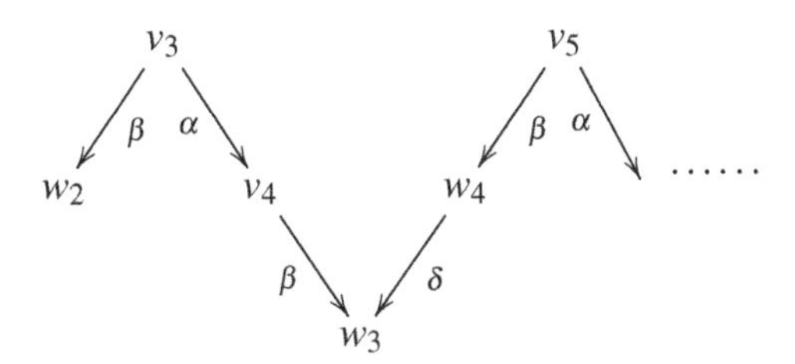

- The dual infinite string module $M(y^*)^*$ described in Example 4.31 corresponding to the dual sequence

$$y^* = (\dots(\beta^*)^{-1}(\alpha^*)^{-1}\beta^*\delta^*(\beta^*)^{-1}(\alpha^*)^{-1}\beta^*\delta^*).$$

The module $M(y^*)^*$ can be visualised as

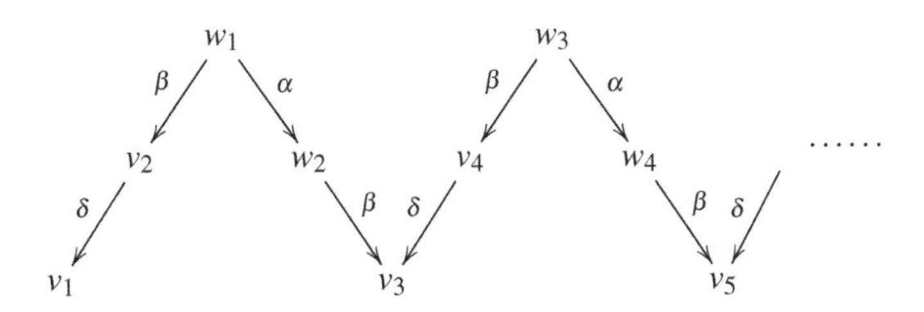

- The submodule of $M(y^*)^*$ consisting of elements of the form

$$\sum_{n\geqslant 2} k_{n+1}v_{n+1} + l_n w_n$$

where $k_n, l_n \in K$. This module is the dual string module $M(x^*)^*$ where $x^* = (\dots(\beta^*)^{-1}(\alpha^*)^{-1}\beta^*\delta^*(\beta^*)^{-1})$ that can be visualised as

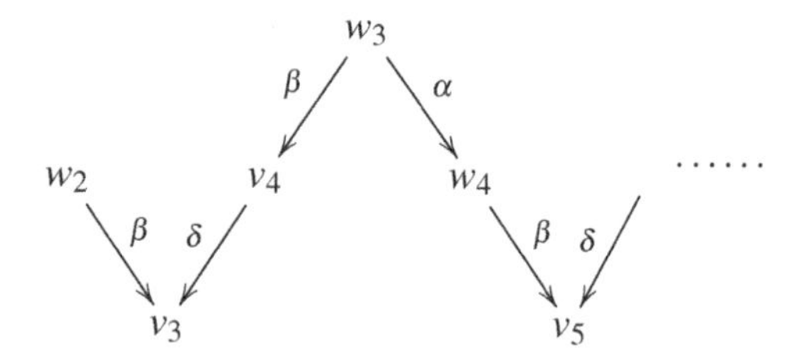

- The following band modules:
  - For each $0 \neq k \in K$, the module $\mathbf{Ba}(M_{k,\infty})$ where $M_{k,\infty}$ is the k-Prüfer module described in Example 4.21.
  - For each $0 \neq k \in K$, the module $\mathbf{Ba}(M_{k,-\infty})$ where $M_{k,-\infty}$ is the k-adic module described in Example 4.30.
  - The module $\mathbf{Ba}(K(X))$ where $K(X)$ denotes the field of rational functions.

4.7 Finite-type and the Ziegler Spectrum

The finite-dimensional A-modules satisfy the following well-known decomposition theorem known as the Krull–Remak–Schmidt Theorem. See, for example, [1, Thm. 4.19].

Theorem 4.50 *Let M be a finite-dimensional A-module. Then $M \cong \bigoplus_{i=1}^{n} M_i$ where M_i is an indecomposable module for each $1 \leqslant i \leqslant n$. Moreover, this decomposition is unique up to isomorphism and reordering of the direct summands.*

This starting point suggests that, if we wish to know about the finite-dimensional A-modules, then we should attempt to understand the indecomposable ones. By Example 4.42, the indecomposable finite-dimensional modules form a subset $\mathfrak{U}_0 \subseteq \mathrm{Zg}_A$ of the Ziegler spectrum of A.

In this section we will consider the case where A is a finite-dimensional K-algebra and consider the topological properties of the subset $\mathfrak{U}_0$. We will make use of the following three important results about finite-dimensional modules over finite-dimensional algebras.

Proposition 4.51 ([8, Cor. 5.3.36, Cor. 5.3.37, Thm. 5.1.12]) *Let A be a finite-dimensional algebra.*

1 *The set $\mathfrak{U}_0$ of finite-dimensional points in Zg_A is* **dense** *in Zg_A. In other words, the closure $\overline{\mathfrak{U}_0}$ of $\mathfrak{U}_0$ is equal to Zg_A.*

2 *The finite-dimensional points in* Zg_A *are* **isolated**. *In other words, the set* $\{M\}$ *is an open set for all* $M \in \mathfrak{U}_0$.
3 *The finite-dimensional points in* Zg_A *are* **closed**. *In other words, the set* $\{M\}$ *is a closed set for all* $M \in \mathfrak{U}_0$.

Using these three facts, together with what we have learned in the previous sections, we can now prove the final theorem of the course, which characterises when the set $\mathfrak{U}_0$ only has finitely many elements. A finite-dimensional algebra A such that the set $\mathfrak{U}_0 \subseteq \mathrm{Zg}_A$ is finite is said to have **finite-representation type**.

Theorem 4.52 *Let A be a finite-dimensional algebra. The following statements are equivalent.*

1 *The algebra A has finite-representation type.*
2 *The Ziegler spectrum* Zg_A *only has finitely many points.*
3 *The Ziegler spectrum* Zg_A *does not contain any infinite-dimensional points.*

Proof First we show that **(1)** and **(2)** are equivalent: The implication **(2)** implies **(1)** is immediate because $\mathfrak{U}_0 \subseteq \mathrm{Zg}_A$. Suppose that **(1)** holds. Then $\mathfrak{U}_0 = \bigcup_{M\in\mathfrak{U}_0}\{M\}$ is a finite union of closed sets by Proposition 4.51(**3**) and therefore $\mathfrak{U}_0$ is a closed set. In particular, we have that $\mathfrak{U}_0 = \overline{\mathfrak{U}_0} = \mathrm{Zg}_A$ by Proposition 4.51(**1**). We have shown that Zg_A is a finite set and so **(2)** holds.

Next we show that **(1)** and **(3)** are equivalent: Note that it was shown in the above paragraph that, if **(1)** holds, then $\mathfrak{U}_0 = \mathrm{Zg}_A$, i.e. **(3)** holds. To show the converse, suppose that **(3)** holds. Then $\mathfrak{U}_0 = \mathrm{Zg}_A$ and so $\mathfrak{U}_0$ is a quasi-compact topological space by Corollary 4.48. We have that $\mathfrak{U}_0 = \bigcup_{M\in\mathfrak{U}_0}\{M\}$ is an open cover by Proposition 4.51(**2**) and clearly this open cover does not have a proper subcover. Since $\mathfrak{U}_0$ is quasi-compact, it follows that $\mathfrak{U}_0$ is a finite set and so **(1)** holds. □

Acknowledgments

The author would like to thank Mike Prest for valuable feedback on the first draft of these lecture notes.

The author was supported by the European Union's Horizon 2020 research and innovation programme under the Marie Skłodowska-Curie grant agreement No 797281.

Bibliography

[1] Assem, I., Simson, D., and Skowroński, A. 2006. *Elements of the Representation Theory of Associative Algebras. Vol. 1.* London Mathematical Society Student

Texts, vol. 65. Cambridge University Press, Cambridge. Techniques of representation theory.

[2] Crawley-Boevey, William. 1998. Infinite-dimensional modules in the representation theory of finite-dimensional algebras. Pages 29–54 of: *Algebras and modules, I (Trondheim, 1996)*. CMS Conf. Proc., vol. 23. Amer. Math. Soc., Providence, RI.

[3] Göbel, Rüdiger, and Trlifaj, Jan. 2012. *Approximations and endomorphism algebras of modules. Volume 1*. extended edn. De Gruyter Expositions in Mathematics, vol. 41. Walter de Gruyter GmbH & Co. KG, Berlin. Approximations.

[4] Herzog, Ivo. 1997. The Ziegler spectrum of a locally coherent Grothendieck category. *Proc. London Math. Soc. (3)*, **74**(3), 503–558.

[5] Krause, Henning. 1997. The spectrum of a locally coherent category. *J. Pure Appl. Algebra*, **114**(3), 259–271.

[6] Lam, T. Y. 1999. *Lectures on modules and rings*. Graduate Texts in Mathematics, vol. 189. Springer-Verlag, New York.

[7] Prest, Mike. 1988. *Model theory and modules*. London Mathematical Society Lecture Note Series, vol. 130. Cambridge University Press, Cambridge.

[8] Prest, Mike. 2009. *Purity, spectra and localisation*. Encyclopedia of Mathematics and its Applications, vol. 121. Cambridge University Press, Cambridge.

[9] Puninski, Gena, and Prest, Mike. 2016. Ringel's conjecture for domestic string algebras. *Math. Z.*, **282**(1-2), 61–77.

[10] Ringel, Claus Michael. 1995. Some algebraically compact modules. I. Pages 419–439 of: *Abelian groups and modules (Padova, 1994)*. Math. Appl., vol. 343. Kluwer Acad. Publ., Dordrecht.

[11] Stenström, Bo. 1975. *Rings of quotients*. Springer-Verlag, New York-Heidelberg. Die Grundlehren der Mathematischen Wissenschaften, Band 217, An introduction to methods of ring theory.

[12] Ziegler, Martin. 1984. Model theory of modules. *Ann. Pure Appl. Logic*, **26**(2), 149–213.

5

The Springer Correspondence

Sam Gunningham

Overview

The goal of these lectures is to introduce the audience to some of the key concepts and tools in the field of *geometric representation theory*, using the *Springer correspondence* as a motivating example.

- In the first lecture, we will go over the background necessary to state the Springer correspondence for an arbitrary semisimple Lie algebra.
- In the second lecture, we will study the notion of convolution in Borel–Moore homology and see how to apply it to the Springer correspondence.
- In the third lecture we will reframe these ideas in the language of perverse sheaves and intersection homology.

These notes are not intended as a detailed reference with complete proofs. Rather, they are designed to give a somewhat informal overview of the subject broadly aimed at new(ish) PhD students.

Acknowledgements I would like to thank Gwyn Bellamy for his generous help with the preparation of these lectures, and an anonymous referee for their useful comments. Any errors are, of course, my own.

Further Reading

Textbooks

- A good place to start is the textbook *Representation Theory and Complex Geometry* by Chriss and Ginzburg [4]. Chapters 2 and 3 form the basis for these lectures.
- There are many good textbooks on algebraic groups and Lie theory, e.g. Springer's book [18] is an appropriate choice.

- For background on derived categories and perverse sheaves, there is the book of Dimca [6], and the (somewhat more technical) classic by Kashiwara and Schapira [12].
- For those looking for some further reading on geometric representation theory, the book *D-modules, Perverse Sheaves, and Representation Theory* [11] by Hotta, Takeuchi and Tanisaki and Takeuchi has good background on D-modules and perverse sheaves and a nice introduction to Kazhdan–Lusztig theory.

Other online resources There are plenty of other lecture notes, theses and the like available online. For example:

- Lecture notes by Zhiwei Yun on Springer theory and orbital integrals (see Lecture I):
  `http://math.mit.edu/~zyun/ZhiweiYunPCMIv2.pdf`
- Senior thesis of Dustin Clausen on the Springer correspondence:
  `www.math.harvard.edu/media/clausen.pdf`
- Survey of Julia Sauter on Springer theory (in a more general sense):
  `https://arxiv.org/abs/1307.0973`
- A great set of notes on perverse sheaves (including representation theoretic applications) by Konni Rietsch:
  `https://arxiv.org/abs/math/0307349`

Original papers Of course, there are also the original papers in which the subject was first developed. We give a partial list here: [17], [19], [13], [3], [16], [10], [14], [7], [15] (the introduction to this last paper of Shoji contains a nice overview of the history of the subject).

5.1 The Statement of the Springer Correspondence

The goal for this lecture We will start by stating the Springer correspondence in type A (i.e. for the symmetric group). Then we will review some of the necessary background from Lie theory to state the Springer correspondence in arbitrary type.

5.1.1 The Springer Correspondence in Type A

Motivation

Let n be a positive integer, and consider the following two sets:

- The set $\mathrm{Irrep}(S_n)$ of isomorphism classes of irreducible (complex) representations of the symmetric group S_n.
- The set Nilp_n of conjugacy classes of $n \times n$ nilpotent matrices.

It is not too difficult to see that both these sets have cardinality equal to the set Part(n) of partitions of n. For example, we know that, in general, the set of irreducible representations of a finite group is in bijection with the set of conjugacy classes, and the conjugacy class of an element in the symmetric group is determined by its cycle type – a partition of n. On the other hand, conjugacy classes of nilpotent matrices are classified by their Jordan type – also a partition of n.

It is natural to ask if we can make this bijection explicit. That is, given a nilpotent conjugacy class can one construct a representation of the symmetric group?

In these lectures, we will discuss a *geometric* approach to this problem, first identified by Tonny Springer in the 1970s [17]. In this theory, the representation of the symmetric group will live in the cohomology of a certain algebraic variety (known as a Springer fibre) associated to a nilpotent matrix. (Recall that a square matrix A is said to be nilpotent if $A^N = 0$ for $N \gg 0$.)

Springer Fibres

To define the Springer fibre, let us recall that a (full) *flag* in the vector space $\mathbb{C}^n$ is defined to be a sequence of linear subspaces

$$0 = V_0 \subsetneq V_1 \subsetneq \cdots \subsetneq V_{n-1} \subsetneq V_n = \mathbb{C}^n$$

with $\dim(V_i) = i$. The set of all flags in $\mathbb{C}^n$ is denoted $\mathcal{F}\ell(n)$. This naturally sits inside a product of Grassmannians as a closed subspace, cut out by polynomial equations. In fact, it has the structure of a smooth projective algebraic variety (and thus a compact Kähler manifold).

Example 5.1 When $n = 2$, we observe that a flag is nothing more than a line in $\mathbb{C}^2$. Thus $\mathcal{F}\ell(2)$ is just the projective line $\mathbb{P}^1$.

The *Springer fibre* $\mathcal{F}\ell(n)^A$ associated to an $n \times n$ matrix A is the subspace of $\mathcal{F}\ell(n)$ consisting of flags $V_\bullet$ such that $A(V_i) \subseteq V_i$ for all $i = 0, \ldots, n$. We will see that the most interesting Springer fibres are those where A is nilpotent.

Example 5.2 If $A = 0$, the Springer fibre $\mathcal{F}\ell(n)^0$ is the entire flag variety $\mathcal{F}\ell(n)$.

Example 5.3 Suppose A is the Jordan normal form $n \times n$ matrix with a single Jordan block. Then $\mathcal{F}\ell(n)^A$ consists of a single point, namely the coordinate flag

$$\langle e_1 \rangle \subsetneq \langle e_1, e_2 \rangle \subsetneq \langle e_1, e_2, e_3 \rangle \subsetneq \cdots \subsetneq \langle e_1, e_2, \ldots, e_n \rangle.$$

Springer fibers are typically singular and have multiple irreducible components, however, they are known to always be equidimensional – that is, every irreducible component has the same dimension $d(A)$. Specifically, if $\lambda = (\lambda_1 \leqslant \dots \lambda_r)$ is the partition corresponding to the lengths of the Jordan blocks of A, consider the dual partition $\mu = (\mu_1 \leqslant \cdots \leqslant \mu_s)$. Then the dimension of the Springer fiber is $\frac{1}{2}\sum_i \mu_i(\mu_i - 1)$ [16] II.5.5.

Example 5.4 Here is a slightly more involved example. Consider the case $n = 3$,

$$A = \begin{pmatrix} 0 & 0 & 1 \\ 0 & 0 & 0 \\ 0 & 0 & 0 \end{pmatrix}.$$

Any flag preserved by A lies in one of the two following families:

$$\begin{array}{ccccccc} \langle e_1 \rangle & \subseteq & \langle e_1, \lambda e_2 + \mu e_3 \rangle & \subseteq & \langle e_1, e_2, e_3 \rangle, & \lambda, \mu \in \mathbb{C}, \\ \langle \lambda e_1 + \mu e_2 \rangle & \subseteq & \langle e_1, e_2 \rangle & \subseteq & \langle e_1, e_2, e_3 \rangle, & \lambda, \mu \in \mathbb{C}. \end{array}$$

Each of these families corresponds to a copy of $\mathbb{P}^1$ in the flag variety, and the two $\mathbb{P}^1$'s intersect at a single point (corresponding to the flag $\langle e_1 \rangle \subseteq \langle e_1, e_2 \rangle \subseteq \langle e_1, e_2, e_3 \rangle$), see Figure 5.1.

The Springer Correspondence

Consider the top non-zero cohomology $H^{2d(A)}(\mathcal{F}\ell(n)^A)$. Cohomology here means singular cohomology of the underlying topological space in the classical topology with rational coefficients. This is a $\mathbb{Q}$-vector space whose dimension is equal to the number of irreducible components in the Springer fibre.

Theorem 5.5 (see e.g. [4], Theorem 3.6.2) *Let n be a positive integer.*

1. *For every nilpotent $n \times n$ matrix A, the vector space $H^{2d(A)}(\mathcal{F}\ell(n)^A)$ carries a natural S_n action, affording an irreducible representation of S_n.*

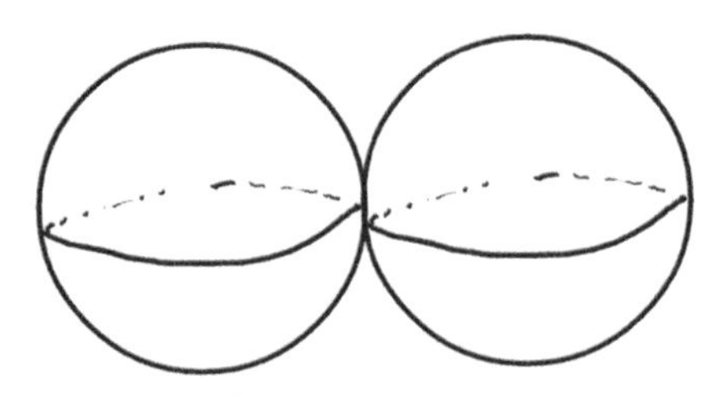

Figure 5.1 A Springer fibre in $\mathcal{F}\ell(3)$.

2 *Each irreducible representation of S_n is isomorphic to $H^{2d(A)}(\mathcal{F}\ell(n)^A)$ for some nilpotent $n \times n$-matrix A. Moreover, the matrix A is uniquely determined up to conjugation.*

In particular, the theorem establishes a bijection between isomorphism classes of irreducible representations, and conjugacy classes of nilpotent matrices as desired.

Remark It is important to note that the action of S_n on the cohomology is not in general induced from an algebraic action on the Springer fibre itself. This is partly what makes the subject so interesting!

Example 5.6 In Example 5.4 we have that $H^2(\mathcal{F}\ell(3)^A)$ carries the unique two-dimensional irreducible representation of S_3. Try to convince yourself that this action cannot arise from automorphisms of $\mathcal{F}\ell(3)^A$.

The Springer representations have been constructed and interpreted in various contexts using convolution algebras, perverse sheaves, D-modules and vanishing cycles. As such, Springer theory provides a fantastic gateway to many of the key concepts and tools in contemporary geometric representation theory. The ideas we will see in these lectures appear all over the subject: in the theory of quiver varieties, cohomological Hall algebras, representations of finite groups of Lie type, Kazhdan-Lusztig theory and Coulomb branches, to name a few such areas.

5.1.2 The Lie Theoretic Set-up

In fact, Springer theory takes place in the wider context of *semi simple Lie algebras* (or algebraic groups) and their associated *Weyl groups*. The Springer correspondence in general exhibits an explicit bijection between the set of irreducible representations of the Weyl group and (a certain refinement of) the set of nilpotent orbits in the Lie algebra. The above example with symmetric group representations and nilpotent matrices corresponds to the special case in which the Lie algebra is $\mathfrak{sl}_n$ and the Weyl group S_n. In what follows we will give a brief outline of this set-up.

Semisimple Lie Algebras

A Lie algebra (over the complex numbers) is simple if it has no proper Lie ideals, and semisimple if it is a direct product of simple Lie algebras. It is quite remarkable that such a short and abstract definition leads to such a deep and intricate theory, as we will now describe.

The classical approach There are a number of ways of approaching the subject of semisimple Lie algebras. In the *classical* approach, we consider symmetries of vector spaces, possibly equipped with bilinear forms. This leads to the following list of examples:

- The *special linear group* SL_n consists of $n \times n$ matrices with determinant 1. Its Lie algebra $\mathfrak{sl}_n$ consists of matrices with trace 0. This is simple for $n \geqslant 2$.
- The *orthogonal group* O_n consists of $n \times n$ orthogonal matrices – those preserving the standard inner product on $\mathbb{C}^n$. Its Lie algebra $\mathfrak{so}_n$ consists of $n \times n$ skew-symmetric matrices. This is simple for $n \geqslant 5$. Moreover, $\mathfrak{so}_4 \cong \mathfrak{sl}_2 \times \mathfrak{sl}_2$ and $\mathfrak{so}_3 \cong \mathfrak{sl}_2$.
- The *symplectic group* Sp_{2n} consists of $2n \times 2n$ symplectic matrices – those preserving the standard symplectic form on $\mathbb{C}^{2n}$:
$$\Omega = \begin{pmatrix} 0 & I_n \\ -I_n & 0 \end{pmatrix}.$$
Its Lie algebra $\mathfrak{sp}_{2n}$ consists of $2n \times 2n$-matrices A such that $\Omega A + A\Omega = 0$. This is simple for all $n \geqslant 1$.

The root-theoretic approach In this route, we start by choosing a Cartan subalgebra $\mathfrak{h}$ of $\mathfrak{g}$ (a maximal abelian subalgebra). The restriction of the adjoint action of $\mathfrak{h}$ on $\mathfrak{g}$ gives a decomposition in to 1-dimensional root spaces $\mathfrak{g}_\alpha$ (together with the fixed space $\mathfrak{h}$ itself):

$$\mathfrak{g} = \mathfrak{h} \oplus \bigoplus_{\alpha \in \Phi} \mathfrak{g}_\alpha.$$

The set $\Phi = \Phi(\mathfrak{g}, \mathfrak{h}) \subseteq \mathfrak{h}^*$ is called the set of roots of $\mathfrak{g}$. The real span E of the roots carries an inner product coming from the Killing form of $\mathfrak{g}$. It turns out that all the information about the semisimple Lie algebra $\mathfrak{g}$ can be encoded in terms of the Euclidean space E together with the set of roots Φ (this data is called the *root system* associated to $\mathfrak{g}$).

If one further specifies a choice of *positive roots* $\Phi_+ \subseteq \Phi$ then we obtain a triangular decomposition:

$$\mathfrak{g} = \mathfrak{n}_- \oplus \mathfrak{h} \oplus \mathfrak{n},$$

where $\mathfrak{n}$ (respectively, $\mathfrak{n}_-$) is spanned by the root spaces of positive (respectively, negative) roots. Given the choice of positive roots, one may define the set $\Delta \subseteq \Phi_+$ of *simple roots* which form a basis of $\mathfrak{h}^*$.

Example 5.7 In the case $\mathfrak{g} = \mathfrak{sl}_n$, we can take $\mathfrak{h}$ to be the diagonal matrices. The set of roots $\alpha_{(i,j)}$ is indexed by pairs $(i, j) \in \{1, \ldots, n\}^2$, $i \neq j$. The root space $\mathfrak{g}_{\alpha_{i,j}}$ consists of matrices whose only possible non-zero entry is in the

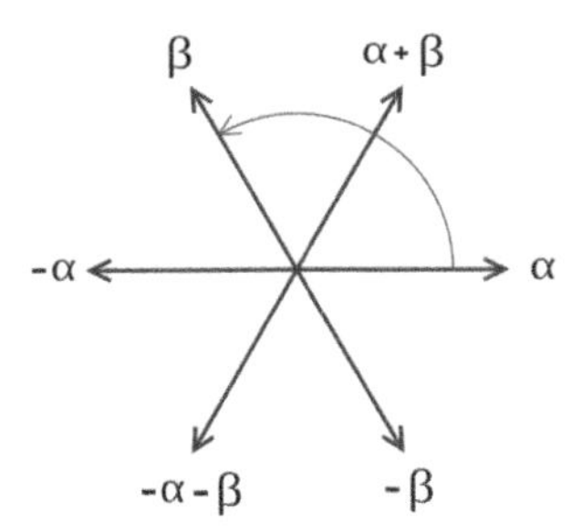

Figure 5.2 The root system of $\mathfrak{sl}_3$ (type A_2). The positive roots with respect to the simple roots $\{\alpha,\beta\}$ are $\{\alpha,\beta,\alpha+\beta\}$. (Image: Wikipedia commons https://commons.wikimedia.org/wiki/)

(i,j)-position. A standard choice for the set of positive roots is to take $\alpha_{(i,j)}$ with $i<j$. The simple roots are $\alpha_{(i,i+1)}$ for $i=1,\ldots,n-1$. With this choice, $\mathfrak{n}$ (respectively $\mathfrak{n}_-$) becomes the set of strictly upper (respectively, strictly lower) triangular matrices.

Borel subalgebras and the canonical Cartan In the above presentation, we needed to pick a Cartan subalgebra $\mathfrak{h}\subseteq\mathfrak{g}$ to get started. Further choosing a subset of positive roots, we obtained a triangular decomposition $\mathfrak{n}_-\oplus\mathfrak{h}\oplus\mathfrak{n}$. The subspace $\mathfrak{b}:=\mathfrak{h}\oplus\mathfrak{n}$ is an example of a *Borel subalgebra*: a maximal solvable Lie subalgebra. In fact, any Borel subalgebra is G-conjugate to this one. There is another approach to this subject, where instead of choosing a Cartan and a set of positive roots, we rather consider all possible Borel subalgebras at once.

More precisely, given a Borel subalgebra $\mathfrak{b}\subseteq\mathfrak{g}$, we consider its nilpotent radical, the ideal $\mathfrak{n}(\mathfrak{b})=[\mathfrak{b},\mathfrak{b}]$ and the corresponding quotient $\mathfrak{H}(\mathfrak{b})=\mathfrak{b}/\mathfrak{n}(\mathfrak{b})$. Thus we get a short exact sequence:

$$0\to\mathfrak{n}(\mathfrak{b})\to\mathfrak{b}\to\mathfrak{H}(\mathfrak{b})\to 0. \tag{5.1.1}$$

This sequence is non-canonically split: choosing a splitting makes $\mathfrak{H}(\mathfrak{b})$ into a Cartan subalgebra of $\mathfrak{g}$ with a choice of positive roots determined by $\mathfrak{b}$.

On the other hand, it turns out that $\mathfrak{H}(\mathfrak{b})$ is actually independent of the choice of $\mathfrak{b}$ in the strongest sense. Namely, suppose $\mathfrak{b}'$ is another Borel subalgebra. Then we can choose $g\in G$ such that $Ad(g)(\mathfrak{b})=\mathfrak{b}'$ (as all Borels are conjugate), which defines an isomorphism:

$$Ad(g):\mathfrak{H}(\mathfrak{b})\cong\mathfrak{H}(\mathfrak{b}').$$

Crucially, this isomorphism is independent of the choice of g (this follows from the fact that B acts trivially on $\mathfrak{H}(\mathfrak{b})$).

We refer to $\mathfrak{H}$ ($= \mathfrak{H}(\mathfrak{b})$ for any Borel $\mathfrak{b}$) as the *canonical Cartan*. Moreover, choosing any splitting of $\mathfrak{H} = \mathfrak{H}(\mathfrak{b})$ into $\mathfrak{b}$ defines a root system in $\mathfrak{H}$ (together with a distinguished choice of positive roots); this root system is independent of choices. Thus we can talk about the Cartan and roots of $\mathfrak{g}$ without having to make any choices.

The Weyl Group

Now suppose $\mathfrak{g}$ is a semisimple Lie algebra. Let us also fix a connected linear algebraic group G with $\mathfrak{g} = \mathrm{Lie}(G)$. Thus G acts on $\mathfrak{g}$ via the adjoint representation (for matrix groups, this is simply the conjugation action).

The *Weyl group* is a certain finite group associated to $\mathfrak{g}$. It plays a central role in our story. There are also a number of different ways to approach its definition.

As a reflection group Recall that the root system on the canonical Cartan $\mathfrak{H}$ associated to $\mathfrak{g}$ determines a Euclidean vector space E together with a distinguished set of root hyperplanes. One definition of the Weyl group $\mathbb{W}$ is as the group generated by the reflections in the root hyperplanes. The reflections corresponding to simple roots give a set of generators for $\mathbb{W}$, giving $\mathbb{W}$ the structure of a Coxeter group. This leads to a presentation of $\mathbb{W}$ as follows:

$$\left\langle s_\alpha, \alpha \in \Delta \mid (s_\alpha s_\beta)^{m(\alpha,\beta)} = 1 \right\rangle,$$

where $m(\alpha,\beta)$ is a certain number in the set $\{1,2,3,4,5\}$ which records the angle $\angle$ formed by α and β according to the following table:

$$m(\alpha,\beta) = \begin{cases} 1 & \text{if } \alpha = \beta, \\ 2 & \text{if } \alpha \perp \beta, \\ 3 & \text{if } \angle(\alpha,\beta) = 120^\circ, \\ 4 & \text{if } \angle(\alpha,\beta) = 135^\circ, \\ 5 & \text{if } \angle(\alpha,\beta) = 150^\circ. \end{cases}$$

It is a remarkable feature of root systems that these are the only possible angles that can occur. This is related to the *crystallographic restriction theorem* (see e.g. [1][Theorem 6.5.12]).

In particular, with these choices, every element $w \in \mathbb{W}$ has a well-defined notion of *length* $\ell(w)$ corresponding to the minimal number of terms appearing in any expression of w as a product of simple reflections (such an expression is called a *reduced word*). There is also a partial order $\leqslant$ on $\mathbb{W}$ characterized by the property that $v \leqslant w$ if and only if there is a reduced word expression for v that sits inside one for w.

Example 5.8 The Weyl group of $\mathfrak{sl}_n$ with respect to the Cartan of diagonal matrices is naturally identified with the symmetric group S_n, acting on $\mathfrak{h}$ by permuting the entries. The root-reflections $s_{\alpha_{(i,j)}}$ correspond to the transpositions $(i \quad j)$. Given the standard choice of positive roots we get the following presentation of S_n:

$$\left\langle s_1,\ldots,s_{n-1} \;\middle|\; \begin{array}{ll} s_is_{i+1}s_i = s_{i+1}s_is_{i+1}, & i=1,\ldots n-2 \\ s_is_j = s_js_i, & i,j=1,\ldots n-1, |i-j| \geqslant 2 \end{array} \right\rangle,$$

where $s_i = s_{\alpha_{i,i+1}}$ corresponds to the transposition $(i \quad i+1)$.

Example 5.9 The Weyl group of $\mathfrak{so}_{2n+1}$ and $\mathfrak{sp}_{2n}$ may both be identified with the *hyperoctahedral group* $(\mathbb{Z}/2\mathbb{Z})^n \rtimes S_n$, realized as the symmetries of an n-dimensional hypercube. The Weyl group of $\mathfrak{so}_{2n}$ is isomorphic to a certain index two subgroup of the hyperoctahedral group, realized as the symmetries of a demihypercube.

Via the normalizer of a Cartan Suppose now we fix a Cartan subalgebra $\mathfrak{h} \subseteq \mathfrak{g}$, and let $H \subseteq G$ denote the centralizer of $\mathfrak{h}$ – this is a maximal torus of G with $\mathrm{Lie}(H) = \mathfrak{h}$. We define $W(\mathfrak{g},\mathfrak{h})$ to be $N_G(\mathfrak{h})/H$. As H acts trivially on $\mathfrak{h}$, the action of $N_G(\mathfrak{h})$ naturally descends to an action of $W(\mathfrak{g},\mathfrak{h})$ on $\mathfrak{h}$. If we further choose a Borel $\mathfrak{b}$ containing $\mathfrak{h}$ (thus giving an identification $\mathfrak{H} \cong \mathfrak{h}$), we obtain an isomorphism $W(\mathfrak{g},\mathfrak{h}) \cong \mathbb{W}$.

The Flag variety Let $\mathcal{F}\ell = \mathcal{F}\ell(\mathfrak{g})$ denote the set of Borel subalgebras $\mathfrak{b} \subseteq \mathfrak{g}$. As any two Borels are G-conjugate, $\mathcal{F}\ell$ is naturally a homogeneous variety for G. The normalizer in G of a given Borel subalgebra $\mathfrak{b}$ is a so-called Borel subgroup $B \subseteq G$ with $\mathrm{Lie}(B) = \mathfrak{b}$. Thus, for any such choice of a basepoint in $\mathcal{F}\ell$, we get an isomorphism:

$$\mathcal{F}\ell \cong G/B.$$

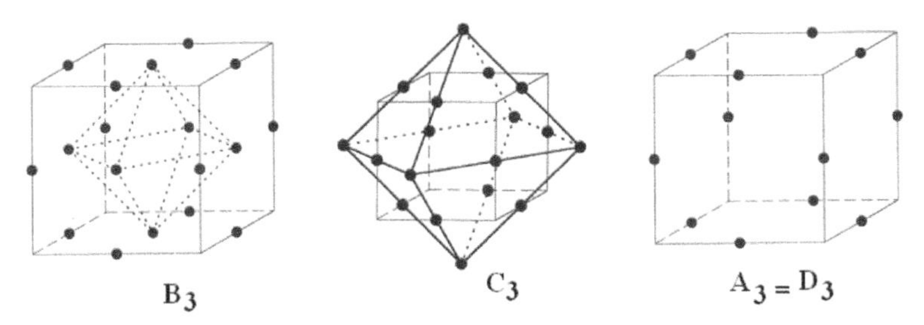

Figure 5.3 Root systems of type B_3, C_3, D_3. (Image: Wikipedia commons https://commons.wikimedia.org/wiki/ File:Root_vectors_b3_c3-d3.png)

In fact, $\mathcal{F}\ell$ carries the structure of a projective complex algebraic variety (in particular, a compact Kähler complex manifold) of (complex) dimension $m = \dim \mathfrak{n}$.

The Bruhat decomposition A fundamental result in this subject is that the orbits of the diagonal G-action on $\mathcal{F}\ell \times \mathcal{F}\ell$ are in bijection with the canonical Weyl group $\mathbb{W}$. To understand why this is, note that if we pick two Borels $\mathfrak{b}_1, \mathfrak{b}_2$, one can choose a Cartan subalgebra in their intersection. Then $\mathfrak{b}_1$ and $\mathfrak{b}_2$ correspond to two choices of a set of positive roots and are thus related by an element of $W(\mathfrak{g},\mathfrak{h})$ (since $W(\mathfrak{g},\mathfrak{h})$ acts simply transitively on the set of such choices). Given a pair of flags $\mathfrak{b}_1, \mathfrak{b}_2$, we say that they are in relative position $w \in \mathbb{W}$ if they lie in the G-orbit corresponding to w.

There are a number of equivalent expressions of this idea, known as the Bruhat decomposition. For example if we fix a Borel subalgebra $\mathfrak{b}$ with normalizer B, then we can identify G-orbits in $\mathcal{F}\ell \times \mathcal{F}\ell$ with B orbits in $\mathcal{F}\ell \cong G/B$ (or equivalently B-double cosets in G). If we further fix a Cartan $\mathfrak{h}$ with corresponding maximal torus H, then we obtain a locally closed decomposition:

$$G = \bigsqcup_{w \in W(\mathfrak{g},\mathfrak{h})} B\dot{w}B,$$

or equivalently

$$G/B = \bigsqcup_{w \in W(\mathfrak{g},\mathfrak{h})} B\dot{w}B/B,$$

where $\dot{w}$ denotes any lift of w to $N_G(H)$. The subsets $B\dot{w}B/B$ are called Bruhat cells – they are affine spaces of dimension $\ell(w)$. These Bruhat cells define a basis for the homology of $\mathcal{F}\ell$.

Example 5.10 Given a pair of flags $U_\bullet, V_\bullet$ in $\mathbb{C}^n$, the numbers:

$$n_{ij} = \dim\left(\frac{U_i \cap V_j}{U_{i-1} \cap V_j + U_i \cap V_{j-1}}\right)$$

define a permutation matrix and thus correspond to an element w of S_n. We say that $U_\bullet, V_\bullet$ are in relative position w.

The Characteristic Polynomial Map

An element $x \in \mathfrak{g}$ is called semisimple if it is contained in some Cartan subalgebra $\mathfrak{h}$. It follows that the set $\mathfrak{c}$ of semisimple conjugacy classes in $\mathfrak{g}$ are in bijection with W-orbits $\mathfrak{c} = \mathfrak{h}/W$ for any given Cartan $\mathfrak{h}$ (or better, in bijection with the canonical $\mathfrak{H}/\mathbb{W}$). It turns out that $\mathfrak{c}$ carries the natural structure of an affine space: it is isomorphic to $\mathbb{C}^r$ where $r := \dim \mathfrak{h}$ is the *rank* of $\mathfrak{g}$. There is a natural G-invariant map:

$$\chi : \mathfrak{g} \to \mathfrak{c},$$

which is defined by taking an element $x \in \mathfrak{g}$ to the unique semisimple conjugacy class in the closure of $G \cdot x$. This is called the *characteristic polynomial map.*

In the language of algebraic geometry, χ is equal to the composite of the quotient map $\mathfrak{g} \to \mathfrak{g}//G := \mathrm{Spec}(\mathbb{C}[\mathfrak{g}]^G)$ with the Chevalley isomorphism $\mathfrak{g}//G \cong \mathfrak{h}/W$, induced by the inclusion of a Cartan $\mathfrak{h} \hookrightarrow \mathfrak{g}$. In other words, the coordinate ring of $\mathfrak{c}$ is identified with the ring of G-invariant polynomial functions on $\mathfrak{g}$ (or alternatively, the ring of $\mathbb{W}$-invariant functions on $\mathfrak{H}$). The fact that $\mathfrak{c}$ is an affine space corresponds to the statement that $\mathbb{C}[\mathfrak{g}]^G \cong \mathbb{C}[\mathfrak{H}]^{\mathbb{W}}$ is a polynomial ring, i.e. is isomorphic to $\mathbb{C}[a_1, \ldots, a_n]$ for some elements $a_1, \ldots, a_n$ (the analogues of the elementary symmetric functions). The degrees (minus 1) of the generators a_i are called the *exponents* of $\mathfrak{g}$.

Example 5.11 The characteristic polynomial map for $\mathfrak{sl}_n$ takes a matrix A to the collection of coefficients of its characteristic polynomial $p_A(t)$ (ignoring the coefficient of t^{n-1}, which is zero by definition) (thought of as an element of the affine space $\mathbb{C}^{n-1}$). Thus, the fibres of χ consist of matrices with a fixed characteristic polynomial (or equivalently, a fixed (multi)set of eigenvalues, counted with multiplicity). In each such fibre, the G-orbits are parameterized by the possible minimal polynomials; if the minimal polynomial has distinct roots, the element is semisimple (i.e. diagonalizable); if the minimal polynomial is equal to the characteristic polynomial, the element is regular (i.e. has maximal size Jordan blocks).

Each fibre of χ is a finite union of G-orbits in $\mathfrak{g}$, and each contains a unique closed orbit (consisting of semisimple elements) and a unique open orbit (consisting of so-called *regular* elements). In particular, the central fibre

$$\mathcal{N} = \chi^{-1}(0)$$

is called the *nilpotent cone* of $\mathfrak{g}$ and its elements are called nilpotent.

Remark The multiplicative group $\mathbb{C}^\times$ naturally acts on $\mathfrak{g}$ (as it does on any vector space). There is a corresponding action on $\mathfrak{c}$ (with certain weights) making χ equivariant and with fixed point $0 \in \mathfrak{c}$. It follows that $\mathcal{N}$ is a *cone*: it carries an action of $\mathbb{C}^\times$ with a unique fixed point $0 \in \mathfrak{g}$.

At the other extreme, the generic fibres of χ consist of a single G-orbit which is both regular and semisimple. Such elements are naturally called regular semisimple. The open subset of regular semisimple elements in $\mathfrak{g}$ is denoted

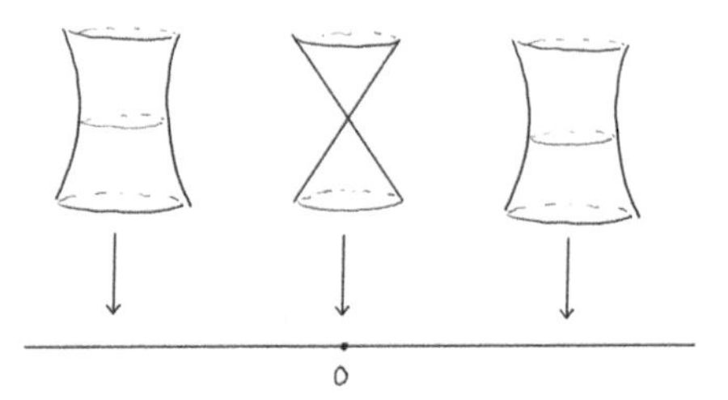

Figure 5.4 A cartoon of the characteristic polynomial map for $\mathfrak{sl}_2$.

$\mathfrak{g}^{rs}$. If $\mathfrak{h}$ is a Cartan subalgebra, the intersection $\mathfrak{g}^{rs} \cap \mathfrak{h}$ is denoted $\mathfrak{h}^{\text{reg}}$; it coincides with the subset of $\mathfrak{h}$ where W acts freely, or equivalently with the complement of the root hyperplanes in $\mathfrak{h}$.

Example 5.12 Regular semisimple elements of $\mathfrak{sl}_n$ are precisely those with distinct eigenvalues. The nilpotent elements are nilpotent matrices in the usual sense (which are characterized by the property that all their eigenvalues are zero, or equivalently, their characteristic polynomial is equal to t^n).

Example 5.13 For $\mathfrak{g} = \mathfrak{sl}_2$ we can be more explicit. The map χ may be identified with

$$\mathfrak{sl}_2 \to \mathbb{C},$$
$$A = \begin{pmatrix} a & b \\ c & -a \end{pmatrix} \mapsto -\det(A) = a^2 - bc.$$

There are two possibilities: if $d = -\det(A) \neq 0$, then the eigenvalues are distinct and A is regular semisimple. In this case $\chi^{-1}(d)$ is a smooth quadric consisting of a single G-orbit. On the other hand $\chi^{-1}(0)$ is a singular conic which is a union of two orbits: the zero orbit $\{0\}$ and the regular nilpotent orbit.

The Killing–Cartan–Dynkin Classification and Exceptional Types

Using the axiomatics of root systems, the simple Lie algebras were classified by Killing and Cartan at the end of the 19th century, and later refined by Dynkin. According to Wikipedia, "the classification is widely considered one of the most elegant results in mathematics" – I would be inclined to agree! The classification consists of four infinite families which correspond to the classical Lie algebras as follows:

- $\mathfrak{sl}_{n+1}$ is type A_n;
- $\mathfrak{so}_{2n+1}$ is type B_n;

Figure 5.5 The finite Dynkin diagrams. (Image: Wikipedia commons https://en.wikipedia.org/wiki/File: Finite_Dynkin_diagrams.svg)

- $\mathfrak{sp}_{2n}$ is type C_n;
- $\mathfrak{so}_{2n}$ is type D_n.

It turns out there are precisely five more "exceptional" Lie algebras which are denoted by E_6, E_7, E_8, F_4, G_2 (the index always refers to the rank). With a bit of work, one can fit the exceptional groups into the classical paradigm using the octonions – see e.g. [2].

We can encode the isomorphism type of $\mathfrak{g}$ in a certain graph called the *Dynkin diagram*. The nodes of the Dynkin diagram correspond to the simple roots, and the number of edges between two nodes is determined by their angle (it is equal to $m(\alpha,\beta)-2$ from the above table). In the case of multiple edges (types B,C,F,G) the two roots have different length, in which case one also draws an arrow going from the long root to the short root.

Example 5.14 The Weyl group of type G_2 is a dihedral group of order 12, acting naturally on the 2-dimensional Cartan $\mathfrak{H}$ as symmetries of a hexagon.

Example 5.15 The Weyl group of type E_8 has order 696729600. It has the unique finite simple group of order 174182400 as a composition factor.

5.1.3 The Springer Correspondence in General Type

The Grothendieck–Springer Simultaneous Resolution

Define the *Grothendieck–Springer space* as follows:

$$\widetilde{\mathfrak{g}} = \{(x,\mathfrak{b}) \in \mathfrak{g} \times \mathcal{F}\ell \mid x \in \mathfrak{b}\}.$$

There are natural maps as indicated below.

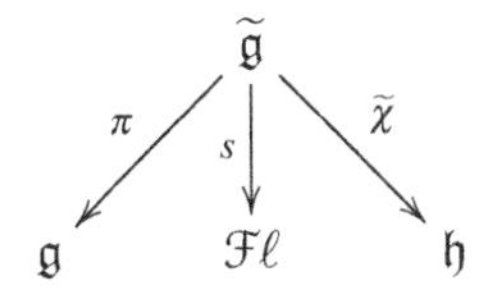

The map s remembers the flag $\mathfrak{b}$ and forgets the element x. This realizes $\widetilde{\mathfrak{g}}$ as a kind of tautological vector bundle over $\mathcal{F}\ell$ (the fibre over the point corresponding a Borel subalgebra $\mathfrak{b}$ is $\mathfrak{b}$ itself).

Remark If we fix a preferred Borel subgroup $B \subseteq G$ with corresponding subalgebra $\mathfrak{b} \subseteq \mathfrak{g}$, then we can write:

$$\widetilde{\mathfrak{g}} = G \times^B \mathfrak{b}.$$

In other words, $\widetilde{\mathfrak{g}}$ is the associated adjoint vector bundle to the B-torsor $G \to G/B \cong \mathcal{F}\ell$.

On the other hand, we can forget the flag and remember x to define the map π. The fibres $\mathcal{F}\ell^x := \pi^{-1}(x)$ are called *Springer fibres*. Explicitly, we have:

$$\mathcal{F}\ell^x = \{\mathfrak{b} \in \mathcal{F}\ell \mid x \in \mathfrak{b}\} \subseteq \mathcal{F}\ell.$$

In other words $\mathcal{F}\ell^x$ is the collection of Borel subalgebras which contain x. More on this later.

Finally, the map $\widetilde{\chi}$ is defined as follows. Given a Borel $\mathfrak{b} \subseteq \mathfrak{g}$, recall that $\mathfrak{b}/[\mathfrak{b}, \mathfrak{b}]$ is identified with the canonical Cartan $\mathfrak{H}$. The map $\widetilde{\chi}$ is defined by taking $(x, \mathfrak{b}) \in \widetilde{\mathfrak{g}}$ to $x \operatorname{mod}[\mathfrak{b}, \mathfrak{b}] \in \mathfrak{H}$.

Remark Fixing a preferred Borel B again, and writing $\mathfrak{n} = [\mathfrak{b}, \mathfrak{b}]$ for the nilpotent radical, we see that there is a short exact sequence of associated vector bundles:

$$G \times^B \mathfrak{n} \to G \times^B \mathfrak{b} \to G \times^B \mathfrak{H}.$$

Note that B acts trivially on $\mathfrak{H}$, so the right-most term is canonically trivial (this is one way to think about the well-definedness of the canonical Cartan):

$$G \times^B \mathfrak{H} \cong G/B \times \mathfrak{H}.$$

The resulting morphism $\widetilde{\mathfrak{g}} \to G/B \times \mathfrak{H}$ is precisely $(s, \widetilde{\chi})$.

We think of $\widetilde{\chi}$ as a lift of the characteristic polynomial map χ from $\mathfrak{c} = \mathfrak{H}/\mathbb{W}$ to $\mathfrak{H}$, as indicated by the following diagram.

$$\begin{array}{ccc} \widetilde{\mathfrak{g}} & \xrightarrow{\widetilde{\chi}} & \mathfrak{H} \\ {\scriptstyle \pi}\downarrow & & \downarrow \\ \mathfrak{g} & \xrightarrow[\chi]{} & \mathfrak{c} \end{array} \tag{5.1.2}$$

Note that $\widetilde{\mathfrak{g}}$ carries an action of G making the maps π and s equivariant, and $\widetilde{\chi}$ invariant.

We have the following key property:

Proposition 5.16 *The map $\widetilde{\chi}$ is a smooth morphism. In particular, the fibres $\widetilde{\chi}^{-1}(t)$ are all smooth.*

The reason this fact is cool is that the original map χ is not smooth – one of the fibres is the nilpotent cone which is generally singular. The diagram (5.1.2) above is referred to as the Grothendieck–Springer simultaneous resolution, because it simultaneously resolves the singularities of (the fibres of) the morphism χ.

Regular semisimple Springer fibres If x is regular semisimple, it is not too hard to show that it is contained in exactly $|\mathbb{W}|$-many Borel subgroups (namely, those Borels containing $\mathfrak{h} = C_{\mathfrak{g}}(x)$). In fact, the collection of such Borels is naturally a torsor for $\mathbb{W}$.

We have the following relative version of this fact:

Proposition 5.17 *There is a free and properly discontinuous action of $\mathbb{W}$ on the locus $\widetilde{\mathfrak{g}}^{rs}$ such that the map*

$$\pi^{rs} : \widetilde{\mathfrak{g}}^{rs} \to \mathfrak{g}^{rs}$$

is identified with the quotient. In other words π^{rs} is a $\mathbb{W}$-Galois covering.

The Springer Resolution

We have seen that the Springer fibres of regular semisimple elements are boring: just discrete sets. At the other end of the spectrum we have the nilpotent cone.

Consider the space

$$\widetilde{\mathcal{N}} := \widetilde{\chi}^{-1}(0) \subseteq \widetilde{\mathfrak{g}}.$$

Note that for an element $(x, \mathfrak{b})$ we have $\widetilde{\chi}(x, \mathfrak{b}) = 0$ if and only if $x \in \mathcal{N}$, or equivalently $x \in \mathfrak{n} := [\mathfrak{b}, \mathfrak{b}]$. In particular, $\widetilde{\mathcal{N}}$ is a vector bundle over $\mathcal{F}\ell$ whose fibre over $\mathfrak{b}$ is $\mathfrak{n}(\mathfrak{b})$.

Restricting π to $\widetilde{\mathcal{N}}$ we get a map

$$\rho : \widetilde{\mathcal{N}} \to \mathcal{N}$$

called the *Springer resolution*. The following proposition establishes the basic properties of the Springer resolution:

Proposition 5.18 *The map ρ is a resolution of singularities. That is:*

1 The variety $\widetilde{\mathcal{N}}$ is smooth (i.e. non-singular).
2 The map ρ is proper (i.e. has compact fibers).
3 There is a dense open subset $U \subseteq \mathcal{N}$ such that $\rho|_{\rho^{-1}(U)}$ is an isomorphism.

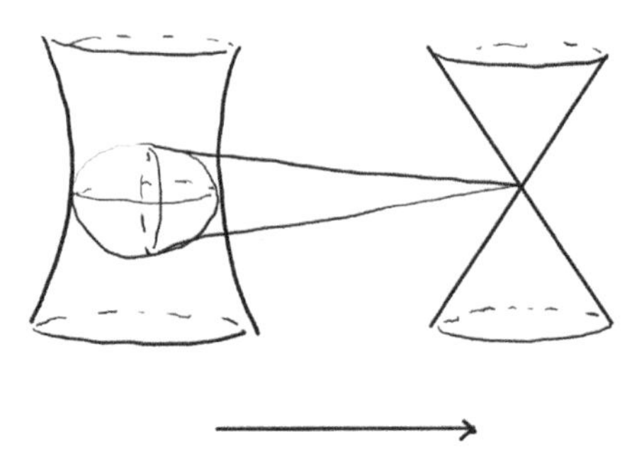

Figure 5.6 A cartoon of the Springer resolution for $\mathfrak{sl}_2$.

The first two statements are clear from what has already been established. For the third, one must show that a regular nilpotent element is always contained in a unique Borel subalgebra.

Remark The Springer resolution has a nice symplectic geometric interpretation. Namely there is a natural identification

$$\widetilde{\mathcal{N}} \cong T^*\mathcal{F}\ell$$

as G-spaces, giving $\widetilde{\mathcal{N}}$ the structure of a Hamiltonian G-space. The map ρ is identified with the moment map

$$T^*\mathcal{F}\ell \to \mathfrak{g}^* \cong \mathfrak{g}.$$

Example 5.19 If $\mathfrak{g} = \mathfrak{sl}_2$, then $\widetilde{\mathcal{N}}$ is the total space of the line bundle $\mathcal{O}(-2)$ over $\mathcal{F}\ell(2) = \mathbb{P}^1$. The map

$$\rho : \widetilde{\mathcal{N}} \to \mathcal{N}$$

just crushes the zero section $\mathbb{P}^1 \subseteq \widetilde{\mathcal{N}}$ to a point.

The following result establishes the key algebro-geometric properties of nilpotent Springer fibres.

Theorem 5.20 (See e.g. [4], Corollary 3.3.24, Remark 3.3.26) *The Springer fibres $\mathcal{F}\ell^x$ for $x \in \mathcal{N}$ are connected and equidimensional (i.e. all the irreducible components have the same dimension). The dimension $d(x)$ is given by the following formula:*

$$d(x) = \frac{1}{2}\dim(C_G(x) - r) = \dim \mathcal{F}\ell - \frac{1}{2}\dim G \cdot x,$$

where $C_G(x)$ denotes the stabilizer and $G \cdot x$ denotes the orbit for the adjoint G-action.

Component Groups

There is one more ingredient required to precisely state the Springer correspondence for arbitrary $\mathfrak{g}$. For each $x \in \mathfrak{g}$ let $A_G(x) = C_G(x)/C_G(x)^\circ$ denote the component group of the centralizer of x in G. It is a finite group.

One shows that the natural action of $C_G(x)$ on $\mathcal{F}\ell^x$ descends to an action of $A_G(x)$ on $H^*(\mathcal{F}\ell^x)$ (preserving the grading). If σ denotes an irreducible representation of $A_G(x)$ and V is any other representation, we denote by

$$V_\sigma := \mathrm{Hom}_{A_G(e)}(\sigma, V)$$

the corresponding multiplicity space.

Remark The group $A_G(x)$ really depends on the choice of group G, not just the Lie algebra $\mathfrak{g}$. In practice, for the purposes of the Springer correspondence, we can take $G = G_{ad}$, the adjoint group, for which the group $A_G(x)$ is smallest.

Remark Recall that any finite dimensional representation of a finite group is a direct sum of irreducible representations. The multiplicity space precisely measures the multiplicity of the given irreducible σ in this decomposition.

The Statement of the Springer Correspondence for Semisimple Lie Algebras

We may now finally state the following:

Theorem 5.21 (See e.g. [4], Theorem 3.5.7) *Let $x \in \mathcal{N}$ be a nilpotent element.*

1 *There is an action of $\mathbb{W}$ on $H^*(\mathcal{F}\ell^x)$, preserving the grading, and commuting with the action of $A_G(x)$.*
2 *For each irreducible representation σ of $A_G(x)$, the multiplicity space*

$$H^{2d}(\mathcal{F}\ell^x)_\sigma$$

is (either zero, or is) an irreducible representation of $\mathbb{W}$. Moreover, up to isomorphism, every irreducible representation appears in this way for a unique pair (x, σ) up to G-conjugation.

Remark Not every irreducible representation of $A_G(x)$ appears in this correspondence. If we allow for G to be the simply connected form then the correspondence is already not one-to-one for $\mathfrak{sl}_2$. There is a beautiful generalization of the Springer correspondence due to Lusztig [14] (called, surprisingly, the generalized Springer correspondence), which accounts for these missing elements in terms of representations of certain other Weyl groups associated to other root systems.

Remark Assuming that all the representations of $A_G(x)$ are defined over $\mathbb{Q}$ (which is the case if we take G to be the adjoint form) then we get that all the representations of $\mathbb{W}$ are also defined over $\mathbb{Q}$. This was not known for all Weyl groups prior to Springer's work.

Example 5.22 (The zero orbit) If $x = 0 \in \mathfrak{g}$ then the Springer fibre $\mathcal{F}\ell^0$ is the entire flag variety $\mathcal{F}\ell$. In this case, the action of $\mathbb{W}$ on $H^*(\mathcal{F}\ell)$ can be described more explicitly as follows. Let H be a maximal torus in G and B a Borel subgroup containing H. There is a map

$$p : G/H \to G/B \cong \mathcal{F}\ell.$$

On the one hand G/H is naturally acted on by $\mathbb{W} \cong W(\mathfrak{g}, \mathfrak{h}) = N_G(H)/H$. On the other hand the map p is a fibration with contractible fibres so induces an isomorphism $H^*(G/H) \cong H^*(G/B)$.

Remark Another way to see this fact is to identify G/B with G_{cpt}/H_{cpt} where G_{cpt} is a maximal compact subgroup of G and $H_{cpt} = H \cap G_{cpt}$ a maximal torus. Then we have $W = N_{G_{cpt}}(H_{cpt})/H_{cpt}$ acts directly on $\mathcal{F}\ell \cong G_{cpt}/H_{cpt}$; the catch is that this action is not holomorphic – it does not respect the complex structure! For example, in the case $\mathfrak{sl}_2$, this action is the antipodal action on $\mathbb{P}^1 \cong S^2$.

It is relatively easy to see that $H^*(G/B)$ has a basis indexed by $w \in \mathbb{W}$, where the degree is given by the length $\ell(w)$. In fact, there is a graded ring isomorphism to the *coinvariant algebra*

$$H^*(G/B) \cong \mathbb{C}[\mathfrak{h}]/\left(\mathbb{C}[\mathfrak{H}]_+^{\mathbb{W}}\right).$$

One verifies that $H^*(G/B)$ is isomorphic to the regular representation of $\mathbb{W}$. The top degree part $H^{2m}(G/B)$ carries the sign character of $\mathbb{W}$.

Example 5.23 (The regular orbit) At the other extreme if $x \in \mathcal{N}$ is regular, then $\mathcal{F}\ell^x \cong pt$. In this case $H^0(\mathcal{F}\ell^x)$ carries the trivial representation.

Example 5.24 (The subregular orbit) One can show that there is a unique G-orbit in $\mathcal{N}$ of dimension $2m - 2$. This is called the subregular orbit. For $x \in \mathcal{N}$ subregular, the Springer fibre $\mathcal{F}\ell^x$ is 1-dimensional, i.e. a (complex) curve. It turns out that it is always a union of $\mathbb{P}^1$'s intersecting according to a certain graph. In the simply laced case (that is $\mathfrak{g}$ is one of the types A, D, or E in the Cartan–Killing–Dynkin classification) this graph is precisely the Dynkin diagram of $\mathfrak{g}$. In the non-simply laced case, one can associate another semisimple Lie algebra $\mathfrak{g}'$ which is simply laced, such that the Dynkin diagram of $\mathfrak{g}$ is obtained from that of $\mathfrak{g}'$ by "folding". Then the graph associated to $\mathcal{F}\ell^x$ is precisely the Dynkin diagram of $\mathfrak{g}'$. Moreover, the diagram automorphism giving rise to the folding is precisely implemented by the action of $A_G(x)$. In general, one can show that the Springer representation $H^2(\mathcal{F}\ell^x)^{A_G(x)}$ associated to subregular x and the trivial representation of $A_G(x)$ is isomorphic to the reflection representation $\mathfrak{h}$.

5.2 Springer Theory via Convolution

The goal for this lecture Last time, we claimed that there is a natural action of the Weyl group on the cohomology of Springer fibres even though the Weyl group does not act on the Springer fibres themselves. So where does this action come from?

In this lecture we will discuss one approach to this problem using convolution in Borel–Moore homology. We will divide the problem into two steps:

1 Construct an algebra A which naturally acts on the cohomology of Springer fibres.
2 Find an algebra isomorphism $\mathbb{Q}[W] \cong A$.

The first part of the lecture will be spent discussing the general properties of Borel–Moore homology. For further details, see [4] Chapter 2.6.

5.2.1 Generalities on Borel–Moore Homology

The Definition

Borel–Moore homology is a certain homology theory for topological spaces. For simplicity, in this section the word *space* shall refer to a suitably nice topological space, say homeomorphic to the complement of a sub CW-complex in a CW complex. Most of the spaces we will consider will be complex algebraic varieties, which all satisfy this condition. If X is a space, $H^*(X)$ (respectively $H_*(X)$) will always denote the singular cohomology (respectively homology) with rational coefficients.

Informally, one can think of a Borel–Moore k-chain on a space X as a possibly non-compact version of an ordinary k-chain. If X is compact then a Borel–Moore chain is just an ordinary chain (and thus Borel–Moore homology agrees with ordinary homology). Borel–Moore homology arises naturally in the study of Poincaré duality in the following form:

Theorem 5.25 (Poincaré duality for Borel–Moore homology) *If X is a smooth oriented manifold of dimension d (not necessarily compact), there is an isomorphism*

$$H_k^{BM}(X) \cong H^{d-k}(X).$$

Equivalently, there is a perfect pairing (called the intersection pairing*)*

$$H_k^{BM}(X) \otimes H_{d-k}(X) \to \mathbb{Q}.$$

There are a few different approaches towards giving a precise definition of Borel–Moore homology. We list some of these below:

1 A singular Borel–Moore chain may be defined as a locally finite sum of singular simplices (i.e. possibly infinite sums which are finite when intersected with any compact subset).
2 If $X \hookrightarrow M$ is an embedding into a closed oriented n-manifold, then
$$H_k^{BM}(X) = H^{n-k}(M, M - X).$$
3 We have
$$H_k^{BM}(X) = H_k(X_+, \{\infty\}),$$
where $X_+ = X \cup \{\infty\}$ is the one-point compactification.
4 $H_*^{BM}(X)$ is the sheaf (hyper)cohomology of the Verdier dualizing complex ω_X (more on this in the next lecture).

Example 5.26 Using any of the above as a definition, we may compute

$$H_k^{BM}(\mathbb{R}^n) = \begin{cases} \mathbb{Q} & \text{if } k = n; \\ 0 & \text{otherwise.} \end{cases}$$

In particular, Borel–Moore homology (like its dual notion, compactly supported cohomology) is not a homotopy invariant (though it is of course a homeomorphism invariant).

Example 5.27 Let us consider the space

$$X = S^1 \times \mathbb{R}.$$

We have:

$*$	$H_*^{BM}(X)$	$H_*(X)$
0	0	$\mathbb{Q}$
1	$\mathbb{Q}$	$\mathbb{Q}$
2	$\mathbb{Q}$	0

One can represent the generators of these groups as (locally finite) cycles. Namely, the generator for H_2^{BM} is the entire space X and the generator for H_1^{BM} is the vertical line $\{*\} \times \mathbb{R}$. Note that this 1-cycle is transverse to the generator for $H_1(X)$. This reflects the perfection of the intersection pairing in the Poincaré duality theorem. See Figure 5.7.

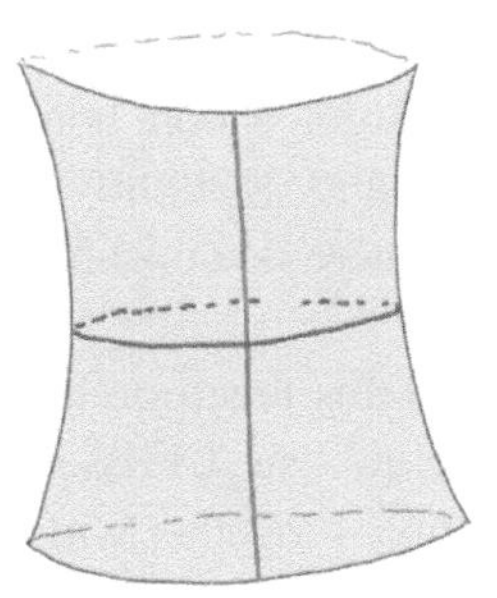

Figure 5.7 The cylinder.

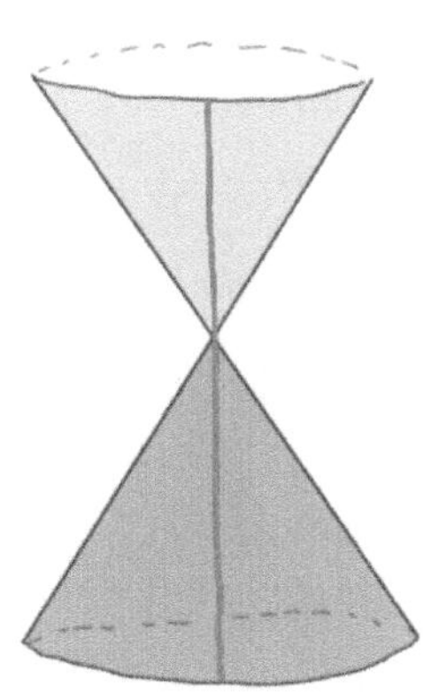

Figure 5.8 The cone.

Example 5.28 Now let $Y = S^1 \times \mathbb{R}/S^1 \times \{0\}$ be a cone. Then we have:

$*$	$H_*^{BM}(Y)$	$H_*(Y)$
0	0	$\mathbb{Q}$
1	$\mathbb{Q}$	0
2	$\mathbb{Q}^2$	0

This time there are two generators in H_2^{BM} represented by the two components of the cone. Note that $H_*^{BM}(Y)$ is quite large even though Y is contractible! See Figure 5.8.

Functoriality

Recall that for a map $f : X \to Y$ of topological spaces, we get an induced push-forward map on homology and an induced pullback map on cohomology. In more categorical terms, homology is *covariantly* functorial and cohomology is *contravariantly* functorial. For Borel–Moore homology (as for compactly supported cohomology) the functoriality is slightly more complicated: sometimes there is a pullback, sometimes there is a pushforward according to the nature of the map f.

Here are the key examples of this functoriality. Most of these can be proved using either definition 2 or 3 below – see Chriss–Ginzburg [4], Chapter 2.6. Alternatively, one can use definition 4 together with the six operations formalism to be discussed later.

Proposition 5.29 *Suppose $f : X \to Y$ is a map of spaces.*

1 *If $f : X \to Y$ is a proper map (i.e. the preimage of a compact set is compact), there is a pushforward map:*

$$f_* : H^{BM}(X) \to H^{BM}(Y).$$

2 *If $f : X \to Y$ is an open embedding there is a restriction map:*

$$f^! : H^{BM}(Y) \to H^{BM}(X).$$

3 *If $f : X \to Y$ is an oriented fibration of relative complex dimension d (that is, a locally trivial fibration, whose fibres are oriented d-manifolds, and the transition maps preserve the orientation), there is a pullback map:*

$$f^! : H_k^{BM}(Y) \to H_{k+d}^{BM}(X).$$

4 *If $f : X \to Y$ is an oriented embedding of a manifold of codimension d (i.e. the normal bundle is oriented), then there is a pullback map:*

$$f^! : H_k^{BM}(Y) \to H_{k-d}^{BM}(X).$$

Remark We will need something a little bit stronger than the last point. Suppose we have a cartesian diagram of spaces (this means that $\widetilde{X}$ is isomorphic to the fibre product $X \times_Y \widetilde{Y}$ such that the maps $\widetilde{f}$ and $\widetilde{g}$ become identified with the projections):

$$\begin{array}{ccc} \widetilde{X} & \xrightarrow{\widetilde{f}} & \widetilde{Y} \\ {\scriptstyle \widetilde{g}}\downarrow & & \downarrow{\scriptstyle g} \\ X & \xrightarrow[f]{} & Y \end{array}$$

Suppose also that f is an oriented embedding of a submanifold of codimension d, so that $f^!$ makes sense. Then there is a pullback map for the base change:

$$\widetilde{f}^! : H_k^{BM}(\widetilde{Y}) \to H_{k-d}^{BM}(\widetilde{X}).$$

Properties and Structures in Borel–Moore Homology

This kind of functoriality may seem strange at first, but it manifests quite naturally in certain situations.

Long exact sequence of an open-closed decomposition For example, suppose $i : Z \hookrightarrow X$ is the inclusion of a closed subset and $j : U = X - Z \hookrightarrow X$ is the inclusion of its open complement. Then i is proper and j is an open embedding. Thus we get a sequence of maps as follows.

$$H_*^{BM}(Z) \xrightarrow{i_*} H_*^{BM}(X) \xrightarrow{j^!} H_*^{BM}(U)$$

In fact, one can further show that these maps come from a short exact sequence of complexes at the chain level. Thus there is an associated long exact sequence at the level of homology:

$$\dots H_{k+1}^{BM}(U) \xrightarrow{\partial} H_k^{BM}(Z) \xrightarrow{i_*} H_k^{BM}(X) \xrightarrow{j^!} H_k^{BM}(U) \xrightarrow{\partial} H_{k+1}^{BM}(Z) \to \dots$$

Remark Suppose we can partition X as a union

$$X = \bigsqcup_{\alpha} X_\alpha$$

of locally closed subsets X_α, with the property that the closure of each X_α is a union of X_β. Suppose also that $H_*^{BM}(X)$ is concentrated in entirely even degrees. Then repeatedly applying the long exact sequence of an open-closed decomposition, and noting that all the boundary maps must vanish, we obtain:

$$H_*^{BM}(X) = \bigoplus_{\alpha} H_*^{BM}(X_\alpha).$$

In particular, this works when X_α is an affine paving, i.e. each X_α is isomorphic to an affine space $\mathbb{C}^k$. This is the case for flag varieties – the affine paving is given by Schubert cells. Less obviously, this is also true for Springer fibres – see [5].

Fundamental classes Suppose U is an oriented d-manifold. Then item 3 of Proposition 5.29 gives us a map:

$$p^! : \mathbb{Q} = H_0^{BM}(pt) \to H_d^{BM}(U).$$

The element $[U] := p^!(\mathbb{Q}) \in H_d^{BM}(U)$ is called the *fundamental class* of the manifold U.

Now suppose that $U \subseteq X$ is embedded as an open dense subset such that the complement $X - U$ has (real) codimension 2 in X (for example U could be the smooth locus of an irreducible complex algebraic variety). Then, by the long exact sequence associated to the open closed decomposition, the restriction map

$$H_d^{BM}(X) \to H_d^{BM}(U)$$

is an isomorphism. It follows that there is a unique element $[X] \in H_d^{BM}(X)$ which maps to $[U]$. We will also refer to this as the fundamental class of X.

Example 5.30 Suppose Z is an algebraic variety of pure complex dimension n, that is, all the irreducible components $Z_1,\ldots,Z_k$ of Z have dimension n. Then the fundamental classes $[Z_1],\ldots,[Z_k]$ form a basis for $H_{2n}^{BM}(Z)$.

Specialization Given a suitable family of spaces X_t, it is possible to specialize Borel–Moore classes from the generic fibre to the special fibre. More precisely, let us fix a manifold S with basepoint s_0. Suppose we have a map of spaces $f : X \to S$ such that it is a locally trivial fibration over $S^* = S - \{s_0\}$. Thus we have a commutative diagram:

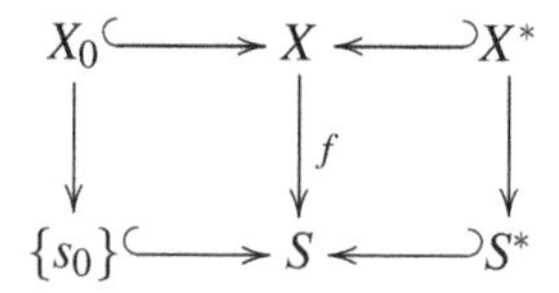

Then there is a natural map:

$$\mathrm{Sp}_{s\to 0} : H_k^{BM}(X^*) \to H_{k-d}^{BM}(X_0).$$

Let us explain how this works in the case $S = [0,\infty)$, $s_0 = 0$ (in fact the general map is constructed by reducing to this case). We assume $X^* \cong X_1 \times (0,\infty)$ is trivialized. In this setting the specialization map is just the boundary map in the long exact sequence associated to the open-closed decomposition $X = X^* \cup X_0$:

$$H_k^{BM}(X^*) \to H_{k-1}^{BM}(X^0).$$

Remark If we assume that $X^* \cong X_1 \times (0,\infty)$ is trivialized then we can interpret specialization as the map

$$H_{k-1}^{BM}(X_1) \to H_k^{BM}(X_1 \times (0,\infty)) \cong H_k^{BM}(X^*) \to H^{BM}(X_0).$$

Thus we are "specializing" a cycle in a generic fibre X_1 to the special fibre X_0.

5.2.2 Convolution Algebras

The set-up Suppose X is a smooth manifold of dimension d with a proper map $f : X \to Y$. We define $Z = X \times_Y X$. Then we have a commutative diagram:

$$\begin{array}{ccccc} Z \times Z & \xleftarrow{\quad s \quad} & Z \times_X Z & \xrightarrow{\quad r \quad} & Z \\ \| & & \| & & \| \\ (X \times_Y X) \times (X \times_Y X) & \xleftarrow[(p_{12}, p_{23})]{} & X \times_Y X \times_Y X & \xrightarrow[p_{13}]{} & X \times_Y X \end{array}$$

Here s is the base change of the diagonal embedding $X \to X \times X$ (of codimension d) and r is proper. Thus the functoriality of Borel–Moore homology defines for us a linear map called *convolution*:

$$* = r_* s^! : H_*^{BM}(Z) \otimes H_*^{BM}(Z) \cong H_*^{BM}(Z \times Z) \to H_{*-d}^{BM}(Z).$$

We denote by $H_*^{BM}(X)[-d]$ the graded vector space where we shift the grading so that $H_d^{BM}(Z)$ lies in degree 0.

The following result can be proved by hand in an elementary fashion, but it also falls out once enough functoriality machinery has been developed (the object Z itself is a monoid object in a suitable category of correspondences, which is a source category for the Borel–Moore homology functor).

Proposition 5.31 *The convolution product $*$ gives $H^{BM}(Z)[-d]$ the structure of a graded associative algebra.*

Semismall Morphisms

Let us assume for the moment that $f : X \to Y$ is a morphism of algebraic varieties. In general, $Z = X \times_Y X$ may be singular and reducible (in the algebro-geometric sense), with components of various dimensions $d = \dim_{\mathbb{R}}(X) \leqslant n \leqslant 2d = \dim_{\mathbb{R}}(X \times X)$. In particular the graded algebra $H_*^{BM}(X)[-d]$ may have graded components of positive and negative degrees.

If it happens that the dimension of Z is equal to the dimension of X (the minimal possible), we say that the map $f : X \to Y$ is *semismall*. In that case, we see that $H_*^{BM}(Z)[-d]$ is supported in positive degrees. Moreover, the degree zero component $H_d^{BM}(Z)$ has a basis given by the fundamental classes of irreducible components of Z. This will be the case in the example of interest to us.

Examples of Convolution

Example 5.32 (The double of a closed oriented manifold) Consider the case when $Y = pt$ and thus X is a compact d-manifold. In this case $Z = X \times X$ and we have

$$H_*^{BM}(Z)[-d] \cong H_*(X) \otimes H_*(X)[-d] \cong H_*(X) \otimes H^*(X) \cong \mathrm{End}(H^*(X)).$$

Here the first isomorphism is by the Künneth theorem and the fact that the Borel–Moore homology agrees with ordinary homology for compact spaces. The second isomorphism is by Poincaré duality. One can check that the convolution structure on $H_*^{BM}(X \times X)[-d]$ corresponds to composition of endomorphisms.

The following example will be useful for our study of the Springer correspondence.

Example 5.33 (Galois covers) Suppose a finite group W acts freely and properly discontinuously on an oriented d-manifold X and let $f: X \to Y = X/W$ be the quotient map. Then there is an identification:

$$Z = X \times_{X/W} X \cong W \times X.$$

In this case the convolution algebra gets identified with the *smash product*:

$$H_*^{BM}(Z)[-d] \cong H^*(X) \sharp \mathbb{Q}[W].$$

Here, the smash product means the algebra whose underlying vector space is $H^*(X) \otimes \mathbb{Q}[W]$ and the multiplication follows the rule as for semidirect products, that when you commute an element of w past a class in $H^*(X)$ you act by w on that class.

The Convolution Action on the Fibre Homology

Recall that $f: X \to Y$ is a proper map of spaces with X an oriented d-manifold. Now let us fix a point $y \in Y$ and consider the fibre $X_y = f^{-1}(y)$, a compact space. Consider the diagram

$$\begin{array}{ccccc}
Z \times X_y & \xleftarrow{\quad s' \quad} & Z \times_X X_y & \xrightarrow{\quad r' \quad} & X_y \\
\| & & \| & & \| \\
(X \times_Y X) \times (X \times_Y \{y\}) & \xleftarrow{(p_{12}, p_{23})} & X \times_Y X \times_Y \{y\} & \xrightarrow{\quad p_{13} \quad} & X \times_Y \{y\}
\end{array}$$

Again we have that r' is proper and s' is a base-change of the diagonal embedding of X. As above, we obtain a map:

$$H_*^{BM}(Z)[-d] \otimes H_*(X_y) \to H_*(X_y).$$

Again, this can be upgraded to the following statement.

Proposition 5.34 *The map defined above equips $H_*(X_y)$ with the structure of a graded $H_*(Z)[-d]$-module.*

5.2.3 The Steinberg Variety

Big vs Small

Recall from Section 5.1.3 the diagram:

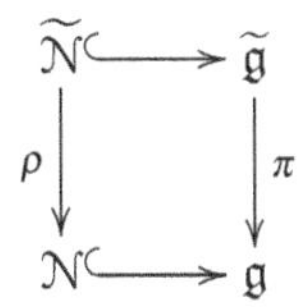

The maps π, ρ are proper, $\widetilde{\mathfrak{g}}$ is smooth of (complex) dimension $d = \dim \mathfrak{g} = 2m + r$ and $\widetilde{\mathcal{N}}$ is smooth of (complex) dimension $2m = \dim \mathcal{N}$.

We define the *big Steinberg variety*

$$\mathrm{St}(\mathfrak{g}) := \widetilde{\mathfrak{g}} \times_{\mathfrak{g}} \widetilde{\mathfrak{g}} = \{(x, \mathfrak{b}_1, \mathfrak{b}_2) \in \mathfrak{g} \times \mathcal{F}\ell \times \mathcal{F}\ell \mid x \in \mathfrak{b}_1 \cap \mathfrak{b}_2\},$$

and the *small (or nilpotent) Steinberg variety*

$$\mathrm{St}(\mathcal{N}) := \widetilde{\mathcal{N}} \times_{\mathcal{N}} \widetilde{\mathcal{N}} = \{(x, \mathfrak{b}_1, \mathfrak{b}_2) \in \mathcal{N} \times \mathcal{F}\ell \times \mathcal{F}\ell \mid x \in \mathfrak{b}_1 \cap \mathfrak{b}_2\}.$$

According to the results of 5.2.2 we have:

Proposition 5.35 *Convolution equips $H_*^{BM}(\mathrm{St})[-2d]$ and $H_*^{BM}(\mathrm{St}(\mathcal{N}))[-4m]$ with a graded algebra structure. Moreover both algebras act canonically on the homology of Springer fibres $H_*(\mathcal{F}\ell^x)$ for $x \in \mathfrak{g}$.*

Let

$$A(\mathfrak{g}) = H_{2d}^{BM}(\mathrm{St}(\mathfrak{g})),$$

and

$$A(\mathcal{N}) = H_{4m}^{BM}(\mathrm{St}(\mathcal{N})).$$

These are algebras with respect to convolution. Each one naturally acts on the homology of Springer fibres. We will see that both of these algebras are in fact isomorphic to $\mathbb{Q}[\mathbb{W}]$, giving the desired action on the homology of Springer fibres.

Remark The way we will present things in this lecture, the isomorphisms are compatible and thus the actions of $\mathbb{W}$ defined using either $A(\mathfrak{g})$ or $A(\mathcal{N})$ are the same. However, we will see in the next lecture that there is another choice for the second isomorphism which causes the two actions to differ by the sign representation of $\mathbb{W}$.

The Components of the Steinberg Variety

Recall that we have a G-equivariant map:

$$s = (s_1, s_2) : \mathrm{St}(\mathfrak{g}) \to \mathcal{F}\ell \times \mathcal{F}\ell,$$

which takes a triple $(x, \mathfrak{b}_1, \mathfrak{b}_2)$ to the pair of flags $(\mathfrak{b}_1, \mathfrak{b}_2)$. Thus $\mathrm{St}(\mathfrak{g})$ is partitioned according to the relative position of $\mathfrak{b}_1$ and $\mathfrak{b}_2$. Accordingly, for each $w \in \mathbb{W}$, we define

$$\mathrm{St}_w(\mathfrak{g}) = s^{-1}((\mathcal{F}\ell \times \mathcal{F}\ell)_w) = \{(x, \mathfrak{b}_1, \mathfrak{b}_2) \mid \mathfrak{b}_1 \text{ and } \mathfrak{b}_2 \text{ are in relative position } w\}$$

and similarly, define $\mathrm{St}_w(\mathcal{N})$.

Stratum by stratum, the Steinberg varieties are relatively easy to understand:

Proposition 5.36 *The projection morphisms*

$$s_w(\mathfrak{g}) : St_w(\mathfrak{g}) \to (\mathcal{F}\ell \times \mathcal{F}\ell)_w$$

and

$$s_w(\mathcal{N}) : St_w(\mathcal{N}) \to (\mathcal{F}\ell \times \mathcal{F}\ell)_w$$

naturally carry the structure of a vector bundle. The fibre of $s_w(\mathfrak{g})$ (respectively, $s_w(\mathcal{N})$) over a pair $(\mathfrak{b}_1, \mathfrak{b}_2)$ is $\mathfrak{b}_1 \cap \mathfrak{b}_2$ (respectively, $\mathfrak{n}(\mathfrak{b}_1) \cap \mathfrak{n}(\mathfrak{b}_2)$).

In particular, $\mathrm{St}_w(\mathfrak{g})$ (respectively, $\mathrm{St}_w(\mathcal{N})$) is a smooth connected variety of dimension $d = \dim(\mathfrak{g})$ (respectively, of dimension $\dim(\mathcal{N}) = 2m$) for each w.

Remark Recall that there is an isomorphism $\widetilde{\mathcal{N}} \cong T^*\mathcal{F}\ell$. Thus $\mathrm{St}(\mathcal{N}) = \widetilde{\mathcal{N}} \times_{\mathcal{N}} \widetilde{\mathcal{N}}$ sits inside $T^*(\mathcal{F}\ell \times \mathcal{F}\ell)$. As such the strata $\mathrm{St}_w(\mathcal{N})$ are identified with the conormal bundles to the orbits $(\mathcal{F}\ell \times \mathcal{F}\ell)_w$.

It follows from the proposition that $\mathrm{St}(\mathfrak{g})$ (respectively, $\mathrm{St}(\mathcal{N})$) itself is equidimensional of dimension d (respectively, $2m$) and the irreducible components are given by the stratum closures. In the terminology introduced in 5.2.2, the Springer resolution is *semismall*.

In particular, the algebras $A(\mathfrak{g})$ and $A(\mathcal{N})$ have bases given by fundamental classes of their components. We denote these bases by

$$\Lambda_w \in A(\mathfrak{g})$$

and

$$T_w \in A(\mathcal{N}),$$

as w ranges over the Weyl group $\mathbb{W}$.

Convolution on the Big Steinberg

Theorem 5.37 *The fundamental classes Λ_w define an isomorphism of algebras $\mathbb{Q}[\mathbb{W}] \cong A(\mathfrak{g})$. In other words, we have*

$$\Lambda_v * \Lambda_w = \Lambda_{vw},$$

for all $v, w \in \mathbb{W}$.

Theorem 5.37 is proved by looking at the open subset

$$\mathrm{St}(\mathfrak{g}^{rs}) := \widetilde{\mathfrak{g}}^{rs} \times_{\mathfrak{g}^{rs}} \widetilde{\mathfrak{g}}^{rs}.$$

Recall from Proposition 5.17 that the morphism

$$\pi^{rs} : \widetilde{\mathfrak{g}}^{rs} \to \mathfrak{g}^{rs}$$

is a $\mathbb{W}$-Galois cover. Following Example 5.33 we see that there is a natural algebra isomorphism:

$$\mathbb{Q}[\mathbb{W}] \cong A(\mathfrak{g}^{rs}) := H_{2d}^{BM}(\mathrm{St}(\mathfrak{g}^{rs})).$$

On the other hand, one observes that the restriction map $A(\mathfrak{g}^{rs}) \to A(\mathfrak{g})$ is an isomorphism of algebras, respecting the fundamental classes of the components, which completes the argument.

Convolution on the Nilpotent Steinberg

While Theorem 5.37 gives an action of $\mathbb{W}$ on the homology of Springer fibres, in order to say something about the nilpotent Springer fibres, we need to understand how this action restricts over the nilpotent cone.

One might first hope then that the linear isomorphism

$$\begin{aligned} \mathbb{Q}[\mathbb{W}] &\to A(\mathcal{N}) \\ w &\mapsto T_w \end{aligned}$$

given by the basis T_w induces an algebra isomorphism. Unfortunately (or perhaps fortunately, as this fact underlies a lot of interesting mathematics!) this map is not an algebra isomorphism. That is,

$$T_v * T_w \neq T_{vw}$$

in general.

To obtain a basis that is compatible with convolution, one must specialize the basis Λ_w from the regular semisimple locus to the nilpotent Steinberg. This procedure is explained in detail in Chriss-Ginzburg [4], Chapter 3.4; we sketch some of the main ideas below.

We let Λ_w^0 denote the elements of $A(\mathcal{N})$ obtained by specializing Λ_w from the big Steinberg. General properties of convolution in Borel–Moore homology can be applied to show that these elements respect the group multiplication in the desired manner. It remains to show that they form a basis. For this, we must compare them to the known basis given by the T_w.

Lemma 5.38 *For each $w, v \in \mathbb{W}$, let n_{vw} be defined by*

$$\Lambda_w^0 = \sum_{v \in \mathbb{W}} n_{vw} T_v.$$

Then

1 $n_{vw} = 0$ if $v \ngeq w$.
2 $n_{ww} = 1$ for all $w \in \mathbb{W}$.

Remark Though it is not obvious from the definitions, the numbers n_{vw} are in fact all non-negative integers.

The first claim in the lemma is easy to check: the specialization construction of Λ_w^0 takes place entirely in the closure $\overline{\mathrm{St}_w(\mathfrak{g})}$. The second claim is less clear, and requires a more careful analysis (see [4], Lemma 3.4.14).

The lemma implies that the matrix (n_{vw}) is upper triangular with 1's along the diagonal. In particular it is invertible. It follows that Λ_w^0 is a basis as required.

This leads to a proof of the following:

Theorem 5.39 *There is an algebra isomorphism:*

$$\mathbb{Q}[\mathbb{W}] \cong A(\mathcal{N}).$$

Once we have this, it is possible to directly prove the Springer correspondence, Theorem 5.20, using a similar kind of geometric analysis – see [4], Section 3.5. Alternatively, we will present another point of view next time using perverse sheaves.

5.3 Springer Theory via Perverse Sheaves

The goal for this lecture We would like to combine the homology of Springer fibres together with their $\mathbb{W}$-action into a single package. This package will be called the *Springer sheaf Spr* and it will live inside a certain category of perverse sheaves on $\mathcal{N}$.

This lecture may require you to take a bit more on faith. If you haven't had much exposure to things like sheaves and the derived category, I recommend you take these things as a black box to begin with (you can enjoy opening up the box and tinkering at a later point).

5.3.1 The Constructible Derived Category

The Constructible Derived Category

In the previous lecture, our main tool was the Borel–Moore homology of a space

$$H_*^{BM}(X).$$

In this lecture our principal player is a certain category

$$D(X)$$

called the *constructible derived category of sheaves.*

We will not have time to discuss the precise definition of this category (see e.g. the books [6],[12] for details). Rather we will attempt to understand this by considering some natural classes of objects and some natural functors out of it.

Remark Very briefly, one can define $D(X)$ as the subcategory of the bounded derived category of sheaves of $\mathbb{Q}$-vector spaces on X whose cohomology sheaves are constructible. Here a sheaf is said to be constructible if there is a stratification $X = \bigsqcup_\alpha X_\alpha$ such that each cohomology sheaf restricted to X_α is a locally constant sheaf of finite rank.

The Case $X = pt$

One can identify the category $D(pt)$ with the category of finite dimensional graded vector spaces. (This is slightly cheating: really $D(pt)$ is the bounded derived category of complexes of vector spaces with finite dimensional cohomology. The identification with graded vector spaces is given by taking a complex to its cohomology.) We think of $D(pt)$ as the home for the (co)homology of a space X. So we have objects $H^*(X) \in D(pt)$ and $H_c^*(X)$ for each X. We can also consider the homology $H_*(X)$ and Borel–Moore homology $H_*^{BM}(X)$ as objects of $D(pt)$ by taking the negative grading (so, e.g. $H_i(X)$ is in degree $-i$).

Measurements: Sections, Stalks, Costalks

Now, given a general space X, there are a bunch of canonical functors to $D(pt)$. These come in two flavours.

Sections We have the functors of *(derived) global sections* and *compactly supported (derived) global sections*:

$$R\Gamma_X, R\Gamma_{X,c} : D(X) \to D(pt).$$

More generally, if U is an open subset of X, there is a canonical restriction functor

$$(-)|_U : D(X) \to D(U),$$

and we can compose with the sections on U to get functors:

$$R\Gamma_U, R\Gamma_{U,c} : D(X) \to D(pt).$$

Stalks On the other hand, for any point $x \in X$, we have two functors called the *stalk* and *costalk*, respectively:

$$i_x^*, i_x^! : D(X) \to D(pt).$$

Given any object $\mathcal{F} \in D(X)$ we can attempt to understand it by analyzing its global sections, stalks and costalks.

Objects

Constant and dualizing sheaves Given a space X we have two basic objects:

- the *constant sheaf* $\mathbb{Q}_X$ and
- the *dualizing complex* ω_X.

Both are preserved under restriction to an open subset $U \subseteq X$. One can think that $\mathbb{Q}_X$ is representing local cochains on X and ω_X is representing local Borel–Moore chains. More precisely, we have the following isomorphisms (in fact, the left-hand side could be taken as a definition of the right-hand side).

$$\begin{aligned} R\Gamma_U(\mathbb{Q}_X) &\cong H^*(U), \\ R\Gamma_U(\omega_X) &\cong H_*^{BM}(U), \\ R\Gamma_{U,c}(\mathbb{Q}_X) &\cong H_c^*(U), \\ R\Gamma_{U,c}(\omega_X) &\cong H_*(U). \end{aligned}$$

The constant sheaf (respectively, dualizing complex) has the property that its stalks $i_x^*(\mathbb{Q}_X)$ (respectively, costalks $i_x^!(\omega_X)$) are isomorphic to the 1-dimensional vector space $\mathbb{Q} \in D(pt)$.

Local systems A local system $\mathcal{L}$ on X (also known as a locally constant sheaf) is an object of $D(X)$ which is a twisted form of the constant sheaf. The sections $R\Gamma_U$ and $R\Gamma_{U,c}$ measure the cohomology and compactly supported cohomology with local coefficients in $\mathcal{L}$.

For each $x \in X$, the stalk $i_x^*(\mathcal{L})$ is a single vector space L_x in degree 0, and it carries an action of the fundamental group $\pi_1(X,x)$. In fact, the category of local systems on a connected space X is equivalent to the category of representations of the fundamental group.

The objects of geometric origin Now, given a map of spaces

$$f : X \to Y,$$

we have certain objects $f_!(\mathbb{Q}_X)$ and $f_*(\omega_X)$ of $D(Y)$. These objects are designed to measure the various (co)homology theories on the fibres X_y of f. More precisely, for each $y \in Y$ we have the following isomorphisms.

$$i_y^* f_!(\mathbb{Q}_X) \cong H_c^*(X_y),$$
$$i_y^! f_*(\omega_X) \cong H_*^{BM}(X_y).$$

Example 5.40 Consider the case where X is a cylinder $S^1 \times \mathbb{R}$, and $f : X \to Y$ is obtained by pinching the subspace $S^1 \times \{0\}$ to a point $y_0 \in Y$, making a cone. Then $\mathcal{F} := f_!(\mathbb{Q}_X)$ can be thought of in the following way. Over the open subset $Y - \{y_0\}$ we get a copy of the constant sheaf $\mathbb{Q}_Y$. But over the special point y_0, the stalk of $\mathcal{F}$ is equivalent to $H^*(S^1)$.

The Formalism of the Six Operations

A neat way to package this stuff is via the so-called *six operations*. In general, if $f : X \to Y$ is a map of spaces, we have the following four functors (one should add to these the functors of internal Hom and tensor product to make six).

$$f_*, f_! : D(X) \leftrightarrows D(Y) : f^!, f^*.$$

We gather the fundamental properties of these functors here:

Proposition 5.41 *Suppose $f : X \to Y$ is a map of spaces.*

1 *The functor f^* is left adjoint to f_* and $f^!$ is right adjoint to $f_!$.*
2 *If f is proper, then we have a natural isomorphism $f_* \cong f_!$.*

3 Suppose we have a Cartesian diagram:

$$\begin{array}{ccc} \widetilde{X} & \xrightarrow{\widetilde{f}} & \widetilde{Y} \\ {\scriptstyle \widetilde{g}}\downarrow & & \downarrow{\scriptstyle g} \\ X & \xrightarrow[f]{} & Y \end{array}$$

Then there are natural isomorphisms:

$$g^! f_* \cong \widetilde{f}_* \widetilde{g}^!,$$
$$g^* f_! \cong \widetilde{f}_! \widetilde{g}^*.$$

Remark These functors subsume all the objects and functors already defined. For example, if $p : X \to pt$ is the unique map, we have

$$\mathbb{Q}_X = p^*(\mathbb{Q}), \quad \omega_X = p^!(\mathbb{Q}),$$

and

$$R\Gamma_X = p_*, \quad R\Gamma_{X,c} = p_!.$$

If $j : U \hookrightarrow X$ is the inclusion of an open subset, we have $j^* = j^! = (-)|_U$.

Grading shift For each integer n we get an autoequivalence of $D(X)$,

$$\mathcal{F} \mapsto \mathcal{F}[n]$$

called *shifting degree by n*. All the functors we will consider will intertwine the operations of shifting degree.

In the case of $X = pt$, the functor $[n]$ has the effect of shifting the grading degree so that if $V = \bigoplus V_k$ is a graded vector space, the degree k part of $V[n]$ is V_{k+n}.

Verdier and Poincaré Duality

The six operation formalism offers a nice way of packaging the idea of Poincaré duality. Namely we have the following:

Theorem 5.42 (Poincaré duality) *Suppose X is a smooth oriented d-manifold. Then there is a canonical isomorphism $\omega_X \cong \mathbb{Q}_X[d]$.*

The more traditional forms of Poincaré duality can essentially be recovered from this fact, together with the properties of the six operations.

One advantage of this setting is that we can formulate a relative version of the above statement:

Theorem 5.43 (Relative Poincaré duality) *Suppose $f : X \to Y$ is a smooth oriented fibration of relative dimension d. Then there is a canonical isomorphism*

$$f^! \cong f^*[d].$$

The reader may have noticed a certain symmetry in this subject, between constant and dualizing, or ! and $*$. This symmetry is realized by a contravariant duality functor called the *Verdier duality functor*

$$\mathbb{D}_X : D(X) \to D(X)^{op},$$

such that $\mathbb{D}^2 \cong Id$. The basic property of this functor is that it exchanges the constant sheaf $\mathbb{Q}_X$ with the dualizing sheaf ω_X. More generally, given a map $f : X \to Y$ we have

$$\mathbb{D}_Y f_* \cong f_! \mathbb{D}_X,$$
$$\mathbb{D}_X f^* \cong f^! \mathbb{D}_Y.$$

The functor $\mathbb{D}_{pt}$ is just the usual duality for graded vector spaces.

5.3.2 Perverse Sheaves and Intersection Homology

Motivation

Now suppose X is a smooth algebraic variety of (pure, complex) dimension d (thus it is a smooth $2d$-manifold). Then we have seen that there is a Poincaré duality isomorphism

$$\omega_X \simeq \mathbb{Q}_X[2d].$$

To put it more symmetrically, we have:

$$\omega_X[-d] \simeq \mathbb{Q}_X[d].$$

Note that this grading shift is only possible on an even real dimensional manifold (e.g. a complex manifold).

Yet another way to express this fact is to say that for a smooth d-dimensional variety X the object $\mathbb{Q}_X[d]$ is canonically Verdier self-dual.

More generally, if $\mathcal{L}$ is a local system on a smooth variety X of dimension d, then we have

$$\mathbb{D}(\mathcal{L}[d]) \cong \mathcal{L}^\vee[d],$$

where $\mathcal{L}^\vee$ is the dual local system (corresponding to the dual representation of the fundamental group).

If X is singular, then this of course fails in general. However, one can still ask the following:

Question 1 Is there an object $IC_X \in D(X)$ such that:

1 IC_X is Verdier self-dual, i.e. $\mathbb{D}(IC_X) \simeq IC_X$?
2 If $U \subseteq X$ is a smooth, open, dense subvariety of X (e.g. the entire smooth locus), then $IC_X|_U \simeq \mathbb{Q}_U[d]$?

Moreover, can one make this construction suitably canonical and functorial?

It turns out the answer is: yes, there is a such an object. It is called the *intersection complex* (more precisely, we are using the so-called *middle perversity*, and shifting the grading so that all our complexes are perverse sheaves). We will present a characterization below.

Remark Historically, the complex $R\Gamma(IC_X)$, called the intersection complex, was defined by Goresky and MacPherson [8] using the concept of *perversity* and *allowable cycles*, where the manner in which the cycles intersect with the singularities is restricted in a particular fashion.

Characterization of Intersection Cohomology

Suppose $U \subseteq X$ is an open subvariety of X which is smooth of pure dimension d. Suppose we fix a local system $\mathcal{L}$ on U.

Theorem 5.44 *([9]) There is an object $IC_X(\mathcal{L})$ together with an isomorphism:*

$$IC_X(\mathcal{L})|_U \cong \mathcal{L}[d],$$

and such that:

1

$$\dim\{x \in X - U \mid H^k(i_x^* IC_X) \neq 0\} < -k,$$

2

$$\dim\{x \in X - U \mid H^k(i_x^! IC_X) \neq 0\} < k.$$

Moreover, the object $IC_X(\mathcal{L})$ is uniquely characterized by these properties (up to unique isomorphism in $D(X)$).

The object $IC_X(\mathcal{L})$ defined by the above theorem satisfies the desired Verdier duality property.

Theorem 5.45 ([9]) *Given $X, U, \mathcal{L}$ as above, we have*

$$\mathbb{D}(IC_X(\mathcal{L})) \cong IC_X(\mathcal{L}^\vee).$$

We define

$$IH_*(X;\mathcal{L}) = R\Gamma_{X,c} IC_X(\mathcal{L})[d]$$

and

$$IH_*^{BM}(X;\mathcal{L}) = R\Gamma_X IC_X(\mathcal{L})[d].$$

Note that here we have shifted the grading back to lie in the traditional (rather than perverse) degrees.

Corollary 5.46 (Poincaré duality for intersection homology) *There is a perfect pairing:*

$$IH_k^{BM}(X;\mathcal{L}) \otimes IH_{2d-k}(X;\mathcal{L}^\vee) \to \mathbb{Q}.$$

In particular, if X is proper and we take $\mathcal{L}$ to be trivial we get a perfect pairing:

$$IH_k(X) \otimes IH_{2d-k}(X) \to \mathbb{Q}.$$

Example 5.47 (The cone revisited) Let $Y = S^1 \times \mathbb{R}/S^1 \times \{0\}$, the cone. We have:

$*$	$H_*^{BM}(Y)$	$H_*(Y)$	$IH_*^{BM}(Y)$	$IH_*(Y)$
0	0	$\mathbb{Q}$	0	$\mathbb{Q}^2$
1	$\mathbb{Q}$	0	0	0
2	$\mathbb{Q}^2$	0	$\mathbb{Q}^2$	0

The chain generating H_1^{BM} is no longer "allowable" in intersection homology. Thus we are left with either the two fundamental classes in IH_2^{BM} or the classes of two points in IH_2 – see Figure 5.9.

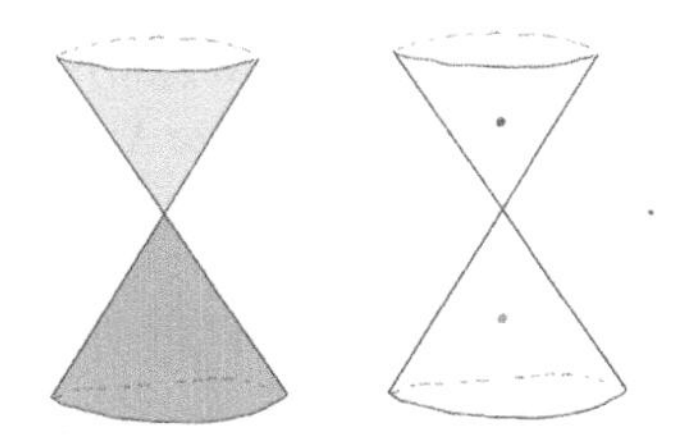

Figure 5.9 IH_*^{BM} and IH_* for the cone.

Definition of Perverse Sheaf

If one relaxes slightly the dimension bounds in the definition of the IC complex, one obtains the definition of a perverse sheaf:

Definition 5.48 An object $\mathcal{F}$ in $D(X)$ is called a *perverse sheaf* if:

1 $\dim\{x \in X \mid H^k(i_x^* \mathcal{F}) \neq 0\} \leqslant -k$ and
2 $\dim\{x \in X \mid H^k(i_x^! \mathcal{F}) \neq 0\} \leqslant k$.

We denote by $\mathrm{Perv}(X)$ the full subcategory of $D(X)$ whose objects are perverse sheaves.

Examples of perverse sheaves on X include the objects $IC_Z(\mathcal{L})$ for any closed subvariety Z of X and local system $\mathcal{L}$ on an open dense subset of Z.

Theorem 5.49 *The category* $\mathrm{Perv}(X)$ *is abelian and every object has finite length. The simple objects are given by* $IC_Z(\mathcal{L})$*, where* Z *is a closed subvariety, and* $\mathcal{L}$ *is an irreducible local system defined on a dense open subset of* Z*.*

Small and Semismall Maps

Recall from last time that we said a morphism $f : X \to Y$ of algebraic varieties was said to be *semismall* if the dimension of $X \times_Y X$ was equal to the dimension of X. This can be reformulated as follows:

Definition 5.50 Let X be a smooth variety of dimension d. A morphism $f : X \to Y$ of algebraic varieties is said to be *semismall* if

$$\dim\{y \in Y \mid \dim f^{-1}(y) \geqslant k\} \leqslant d - k,$$

for all $k \geqslant 0$, and *small* if

$$\dim\{y \in Y \mid \dim f^{-1}(y) \geqslant k\} \leqslant d - k,$$

for $k > 0$.

Notice the similarities between the definition of small (respectively, semismall) and IC complexes (respectively, perverse sheaves). This observation leads to the following key result.

Proposition 5.51 (See e.g. [11], Proposition 8.2.30) *Suppose* $f : X \to Y$ *is a proper morphism of algebraic varieties and that* X *is smooth of dimension* d*.*

1 *If* f *is small, then*

$$f_* \mathbb{Q}_X[d] \cong IC_Y(\mathcal{L}),$$

where $\mathcal{L} = (f|_U)_* \mathbb{Q}_U[d]$*, and* $U \subseteq X$ *is an open dense subset such that* $f|_U$ *is a covering map.*

2 *If f is semismall then there is an isomorphism:*

$$f_*\mathbb{Q}_X[d] \cong \bigoplus_{\alpha=1}^{n} IC_{Y_\alpha}(\mathcal{L}_\alpha),$$

where $Y_\alpha \subseteq Y$ are closed subvarieties together with irreducible local systems $\mathcal{L}_\alpha$ on the smooth locus Y_α^{sm} for each $\alpha = 1, \dots, n$.

5.3.3 The Springer Sheaf

Let us return again to the Lie theoretic setting of Section 5.1.2.

Big vs Small

We define the *big Springer sheaf*

$$\mathcal{S}_\mathfrak{g} := \pi_*\mathbb{Q}_{\tilde{\mathfrak{g}}}[d]$$

and the *small* or *nilpotent Springer sheaf*

$$\mathcal{S}_\mathcal{N} := \rho_*\mathbb{Q}_{\tilde{\mathcal{N}}}[2m].$$

The stalks of $\mathcal{S}_\mathfrak{g}$ and of $\mathcal{S}_\mathcal{N}$ at nilpotent elements $x \in \mathcal{N}$ both record the homology of Springer fibres. In particular, by restriction, any endomorphism of $\mathcal{S}_\mathfrak{g}$ or $\mathcal{S}_\mathcal{N}$ defines an endomorphism of the homology of Springer fibres.

The basic properties of the six operations allow us to relate the endomorphism algebras of these objects to the convolution algebras considered in the previous lecture.

Proposition 5.52 *There are isomorphisms of graded algebras:*

$$H_*^{BM}(\mathrm{St}(\mathfrak{g}))[-2d] \cong R\mathrm{Hom}_{D(\mathfrak{g})}(\mathcal{S}_\mathfrak{g}, \mathcal{S}_\mathfrak{g}),$$
$$H_*^{BM}(\mathrm{St}(\mathcal{N}))[-4m] \cong R\mathrm{Hom}_{D(\mathcal{N})}(\mathcal{S}_\mathcal{N}, \mathcal{S}_\mathcal{N}).$$

In particular we get isomorphisms of algebras:

$$A(\mathfrak{g}) \cong \mathrm{Hom}_{D(\mathfrak{g})}(\mathcal{S}_\mathfrak{g}, \mathcal{S}_\mathfrak{g}),$$
$$A(\mathcal{N}) \cong \mathrm{Hom}_{D(\mathcal{N})}(\mathcal{S}_\mathcal{N}, \mathcal{S}_\mathcal{N}).$$

Thus we can rephrase Theorem 5.37 and Theorem 5.39 as statements about the endomorphism algebra of Springer sheaves.

The (Semi) smallness of the Springer Maps

The dimension formula for Springer fibres directly implies the following crucial result:

Proposition 5.53 *1 The morphism*

$$\pi : \widetilde{\mathfrak{g}} \to \mathfrak{g}$$

is small and its restriction π^{rs} to the regular semisimple locus is a $\mathbb{W}$-Galois covering.

2 The morphism

$$\rho : \widetilde{\mathcal{N}} \to \mathcal{N}$$

is semismall and birational (i.e. an isomorphism over an open set).

In particular, it follows that both $\mathcal{S}_{\mathfrak{g}}$ and $\mathcal{S}_{\mathcal{N}}$ are perverse sheaves.

The Structure of the Big Springer Sheaf

Let

$$\mathcal{K} = \pi_*^{rs} \mathbb{Q}_{\widetilde{\mathfrak{g}}^{rs}} \cong \mathcal{S}_{\mathfrak{g}}|_{\mathfrak{g}^{rs}}.$$

As π^{rs} is a $\mathbb{W}$-Galois cover, it follows that $\mathcal{K}$ is a local system on $\mathfrak{g}^{rs}$ of rank $|\mathbb{W}|$, and the endomorphisms of $\mathcal{K}$ are precisely the group algebra $\mathbb{Q}[\mathbb{W}]$. More or less equivalently, we have a decomposition

$$\mathcal{K} \cong \bigoplus_{L \in \mathrm{Irrep}(\mathbb{W})} L \otimes \mathcal{K}_L,$$

where $\mathcal{K}_L$ is an irreducible local system on $\mathfrak{g}^{rs}$ (of rank $\dim L$). The smallness of the map π then implies that endomorphisms extend uniquely from the regular semisimple locus, giving the following sheaf-theoretic version of Theorem 5.37:

Theorem 5.54 *We have a canonical isomorphism*

$$\mathrm{End}_{\mathrm{Perv}(\mathfrak{g})}(\mathcal{S}_{\mathfrak{g}}) \cong \mathbb{Q}[\mathbb{W}]$$

and the perverse sheaf $\mathcal{S}_{\mathfrak{g}}$ decomposes as a direct sum:

$$\mathcal{S}_{\mathfrak{g}} \cong \bigoplus_{L \in \mathrm{Irrep}(\mathbb{W})} L \otimes IC_{\mathfrak{g}}(\mathcal{K}_L).$$

The Structure of the Small Springer Sheaf

The semismallness of the map ρ implies that there is some decomposition

$$\mathcal{S}_{\mathcal{N}} = \bigoplus_i M_i \otimes IC_{Z_i}(\mathcal{E}_i)$$

as a direct sum of IC sheaves $IC_{Z_i}(\mathcal{E}_i)$, where the Z_i are closed subsets of $\mathcal{N}$, M_i are multiplicity vector spaces and $\mathcal{E}_i$ local systems on some open dense subset of Z_i. Moreover (by absorbing repeating factors into M_i), we may assume that the $IC_{Z_i}(\mathcal{E}_i)$ are pairwise non-isomorphic.

As ρ is G-equivariant, the closed subset $Z = \overline{\mathcal{O}}$ must be the closure of some G-orbit $\mathcal{O}$ and $\mathcal{E}$ must be a G-equivariant local system on $\mathcal{O}$. Note that G-equivariant local systems on an orbit $G \cdot x$ are precisely given by representations σ of $A_G(x)$. Thus, each of the factors $IC_{Z_i}(\mathcal{E}_i)$ above are of the form $IC_{\overline{G \cdot x}}(\sigma)$ for some pair (x, σ).

Reinterpreting the Springer Correspondence

Now suppose, for a moment, we assume Theorem 5.39, i.e. that there is an isomorphism

$$\mathbb{Q}[\mathbb{W}] \cong \mathrm{End}_{\mathrm{Perv}(\mathcal{N})}(\mathcal{S}_{\mathcal{N}}). \tag{5.3.1}$$

This then gives a precise enumeration of the decomposition of $\mathcal{S}_{\mathcal{N}}$ into simple objects: namely, there are pairwise non-isomorphic simple summands for each irreducible representation L, and the multiplicity space of each such summand is again given by L. In other words, there is an injective map of sets

$$L \mapsto (\mathcal{O}_L, \sigma_L)$$

from $\mathrm{Irrep}(\mathbb{W})$ to the set of pairs $(\mathcal{O}, \sigma)$ of a nilpotent orbit $\mathcal{O} = G \cdot x$ and an irreducible representation σ of $A_G(x)$. This is the Springer correspondence, reinterpreted in the language of perverse sheaves!

From here, it is not too hard to show (using the dimension formula) the traditional statement of the Springer correspondence, that L matches up with the σ_L-multiplicity space of $H^{2d(x)}(\mathcal{F}\ell^x)$ where $G \cdot x = \mathcal{O}_L$ (see e.g. Section 4.1 in Clausen's notes).

Two Parameterizations of the Springer Correspondence

Let $i_{\mathcal{N}} : \mathcal{N} \hookrightarrow \mathfrak{g}$ denote the inclusion. We have an isomorphism:

$$\mathcal{S}_{\mathcal{N}} \cong i^!_{\mathcal{N}}[r]\mathcal{S}_{\mathfrak{g}},$$

where r is the rank of the Lie algebra $\mathfrak{g}$ (i.e. the dimension of a Cartan subalgebra). Thus the functor $i^!_{\mathcal{N}}[r]$ induces an algebra homomorphism:

$$\mathrm{End}_{\mathrm{Perv}(\mathfrak{g})}(\mathcal{S}_{\mathfrak{g}}) \to \mathrm{End}_{\mathrm{Perv}(\mathcal{N})}(\mathcal{S}_{\mathcal{N}}). \tag{5.3.2}$$

It is possible to prove directly that this is an isomorphism; it is equivalent to proving that the simple objects $IC(\mathfrak{g}, \mathcal{K}_L)$ appearing in Theorem 5.54 restrict to pairwise non-isomorphic simple objects in $\mathrm{Perv}(\mathcal{N})$ via $i^!_{\mathcal{N}}[r]$ (namely, the objects $IC_{\overline{\mathcal{O}_L}}(\sigma_L)$ appearing above). This leads to the same isomorphism

$A(\mathfrak{g}) \cong A(\mathcal{N})$ as in Theorem 5.39, and thus the same Springer correspondence as in the previous lecture.

However, there is also another approach. The *Fourier transform* (or Fourier–Deligne transform) is a certain involutive endofunctor (the superscript mon denotes that we only consider the full subcategory of objects which are equivariant for the action of the scaling torus $\mathbb{C}^\times$):

$$\mathbb{F} : \mathrm{Perv}^{\mathrm{mon}}(\mathfrak{g}) \to \mathrm{Perv}^{\mathrm{mon}}(\mathfrak{g}).$$

It turns out that we have $\mathbb{F}(\mathcal{S}_\mathfrak{g}) \cong (\mathcal{S}_\mathcal{N})$ (where the latter is considered a perverse sheaf on $\mathfrak{g}$ via pushforward under the closed embedding). This leads to another isomorphism as in (5.3.2) and thus another identification $\mathbb{Q}[\mathbb{W}] \cong \mathrm{End}_{\mathrm{Perv}(\mathcal{N})}(\mathcal{S}_\mathcal{N})$ and finally to another Springer correspondence! It is possible to show that the two parameterizations of the Springer correspondence differ by the sign character of $\mathbb{W}$.

Bibliography

[1] Artin, Michael. 2011. *Algebra.* Boston, MA: Pearson Education.

[2] Baez, John C. 2001. The octonions. *Bulletin of the American Mathematical Society*, **39**(02), 145–206.

[3] Borho, Walter, and MacPherson, Robert. 1983. Partial resolutions of nilpotent varieties. Pages 23–74 of: *Analysis and topology on singular spaces, II, III (Luminy, 1981).* Astérisque, vol. 101. Paris: Soc. Math. France.

[4] Chriss, Neil, and Ginzburg, Victor. 1997. *Representation theory and complex geometry.* Boston, MA: Birkhäuser Boston Inc.

[5] Concini, De, C., Lusztig, G. and Procesi, C. 1988. Homology of the Zero-Set of a Nilpotent Vector Field on a Flag Manifold. *Journal of the American Mathematical Society*, **1**(1), 15.

[6] Dimca, Alexandru. 2004. *Sheaves in Topology.*

[7] Ginsburg, V. 1987. "Lagrangian" construction for representations of Hecke algebras. *Advances in Mathematics*, **63**(1), 100–112.

[8] Goresky, Mark, and MacPherson, Robert. 1980. Intersection homology theory. *Topology*, **19**(2), 135–162.

[9] Goresky, Mark, and MacPherson, Robert. 1983. Intersection homology. II. *Invent. Math.*, **72**(1), 77–129.

[10] Hotta, Ryoshi, and Kashiwara, Masaki. 1984. The invariant holonomic system on a semisimple Lie algebra. *Inventiones Mathematicae*, **75**(2), 327–358.

[11] Hotta, Ryoshi, Takeuchi, Kiyoshi, and Tanisaki, Toshiyuki. 2008. *D-modules, perverse sheaves, and representation theory.* Progress in Mathematics, vol. 236. Boston, MA: Birkhäuser Boston Inc. Translated from the 1995 Japanese edition by Takeuchi.

[12] Kashiwara, Masaki, and Schapira, Pierre. 1990. *Sheaves on Manifolds.* Berlin, Heidelberg: Springer Berlin Heidelberg Imprint Springer.

[13] Kazhdan, David, and Lusztig, George. 1980. A topological approach to Springer's representations. *Advances in Mathematics*, **38**(2), 222–228.

[14] Lusztig, G. 1984. Intersection cohomology complexes on a reductive group. *Inventiones Mathematicae*, **75**(2), 205–272.

[15] Shoji, Toshiaki. 1988. Geometry of orbits and Springer correspondence. Pages 61–140 of: *Orbites unipotentes et représentations - I. Groupes finis et Algèbres de Hecke*. Astérisque, no. 168. Société mathématique de France.

[16] Spaltenstein, N. 1982. *Classes Unipotentes et Sous-groupes de Borel*. Springer-Verlag GmbH.

[17] Springer, T. A. 1976. Trigonometric sums, Green functions of finite groups and representations of Weyl groups. *Inventiones Mathematicae*, **36**(1), 173–207.

[18] Springer, T. A. 2008. *Linear Algebraic Groups*. Birkhauser Boston Inc.

[19] Springer, Tonny A. 1978. A construction of representations of Weyl groups. *Inventiones Mathematicae*, **44**(3), 279–293.

6

An Introduction to Diagrammatic Soergel Bimodules

Amit Hazi[1]

6.1 Motivation

Let $\mathfrak{g}$ be a semisimple Lie algebra over $\mathbb{C}$, with a Cartan subalgebra $\mathfrak{h}$ and Borel subalgebra $\mathfrak{b}$. The Cartan subalgebra $\mathfrak{h}$ gives rise to a root system $\Phi \subset \mathfrak{h}^*$, and the choice of Borel subalgebra corresponds to a selection of simple roots Σ and positive roots Φ^+ inside Φ. The root system Φ induces a Weyl group W generated by the set S of reflections in the simple roots Σ. (We will later generalize this situation in Definition 6.3.) Inside $\mathfrak{h}^*$ we also have

$$\Lambda = \{\lambda \in \mathfrak{h}^* : \langle \alpha^\vee, \lambda \rangle \in \mathbb{Z} \text{ for all } \alpha \in \Sigma\} \tag{6.1.1}$$

$$\cup$$

$$\Lambda^+ = \{\lambda \in \mathfrak{h}^* : \langle \alpha^\vee, \lambda \rangle \in \mathbb{Z}_{\geqslant 0} \text{ for all } \alpha \in \Sigma\}. \tag{6.1.2}$$

Finally let $U\mathfrak{g}$ denote the universal enveloping algebra of $\mathfrak{g}$. We will consider $\mathfrak{g}$-modules and $U\mathfrak{g}$-modules interchangeably. A standard reference for all the facts about $\mathfrak{g}$-modules in this section is [11].

For each $\lambda \in \Lambda$, define the *Verma module* $M(\lambda) = U\mathfrak{g} \otimes_{U\mathfrak{b}} \mathbb{C}_\lambda$. (Here $U\mathfrak{b}$ denotes the universal enveloping algebra of $\mathfrak{b}$, while $\mathbb{C}_\lambda$ denotes the 1-dimensional $\mathfrak{b}$-module given by $\mathfrak{b} \to \mathfrak{h} \xrightarrow{\lambda} \mathbb{C}$.) Each Verma module $M(\lambda)$ has a unique simple quotient $L(\lambda)$, which is the unique simple weight module of highest weight λ. The simple module $L(\lambda)$ is finite dimensional if and only if $\lambda \in \Lambda^+$.

The category $U\mathfrak{g}$-mod of *all* $U\mathfrak{g}$-modules is too large to be useful. Instead we restrict our attention to a smaller category which contains Verma modules and highest weight simple modules.

Definition 6.1 Let $\lambda \in \Lambda^+$, and write $U\mathfrak{g}\text{-mod}_{U\mathfrak{h}\text{-ss}}$ for the category of $\mathfrak{g}$-modules which are semisimple as $\mathfrak{h}$-modules. (In other words, $U\mathfrak{g}\text{-mod}_{U\mathfrak{h}\text{-ss}}$

[1] Supported by the Royal Commission for the Exhibition of 1851.

is the category of weight modules.) We define $\mathcal{O}_\lambda$ to be the minimal full subcategory of $U\mathfrak{g}\text{-mod}_{U\mathfrak{h}\text{-ss}}$ that contains $M(\lambda)$ and is closed under submodules, quotients and extensions.

It is obvious that $\mathcal{O}_\lambda$ is an abelian category. It is somewhat less obvious that $\mathcal{O}_\lambda$ is in fact a *finite* abelian category, with finite length objects, finitely many isomorphism classes of simple objects and finite-dimensional Hom-spaces.

Remark The above definition of $\mathcal{O}_\lambda$ is non-standard. Most treatments (e.g. [11]) first define the *BGG category* $\mathcal{O}$ which contains all Verma modules and all highest weight simple modules. Then $\mathcal{O}_\lambda$ is defined for arbitrary $\lambda \in \mathfrak{h}^*$ as a subcategory of $\mathcal{O}$ with a certain prescribed action of the centre $Z\mathfrak{g}$ of $U\mathfrak{g}$. In general $\mathcal{O}_\lambda$ is a union of blocks of $\mathcal{O}$, and when $\lambda \in \Lambda^+$, one can show that $\mathcal{O}_\lambda$ is the block containing $L(\lambda)$.

Example 6.2 Suppose $\mathfrak{g} = \mathfrak{sl}_2$. The corresponding root system Φ is of Dynkin type A_1, with Weyl group $W = \{1, s\}$. Within $\mathfrak{h}^*$ there are obvious identifications $\Lambda \cong \mathbb{Z}$ and $\Lambda^+ \cong \mathbb{Z}_{\geqslant 0}$. Let $n \in \mathbb{Z}_{\geqslant 0}$. The indecomposable objects in $\mathcal{O}_n$ are $L(n)$, $L(-n-2) = M(-n-2)$, $M(n) = P(n)$ and $P(-n-2)$. The structures of the last two modules are given by the exact sequences

$$0 \longrightarrow L(-n-2) \longrightarrow M(n) \longrightarrow L(n) \longrightarrow 0,$$

$$0 \longrightarrow M(n) \longrightarrow P(-n-2) \longrightarrow M(-n-2) \longrightarrow 0.$$

Let $\rho = \frac{1}{2}\sum_{\alpha \in \Phi^+} \alpha$ be the half-sum of the positive roots. For $w \in W$ and $\lambda \in \mathfrak{h}^*$, we define the following shift

$$w \cdot \lambda = w(\lambda + \rho) - \rho \tag{6.1.3}$$

of the usual Weyl group action, called the *dot action*. The dot action parametrizes several sets of modules in $\mathcal{O}_\lambda$.

Theorem 6.1 *There are bijections*

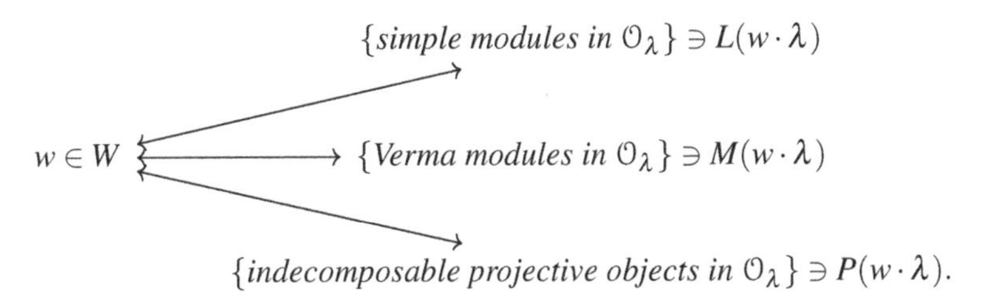

Here $P(w \cdot \lambda)$ denotes the projective cover of $L(w \cdot \lambda)$ in $\mathcal{O}_\lambda$.

We can say a little more about the structure of the indecomposable projective objects.

Proposition 6.2 *For $w \in W$ there is a sequence of submodules*

$$0 = P_0 < P_1 < \cdots < P_n = P(w \cdot \lambda)$$

such that $P_n/P_{n-1} \cong M(w \cdot \lambda)$, and for each $1 \leqslant i < n$, there is some $w_i \in W$ with $\ell(w_i) < \ell(w)$ such that $P_i/P_{i-1} \cong M(w_i \cdot \lambda)$.

In particular, from the case $w = 1$ we conclude that $P(\lambda) = M(\lambda)$.

Since $\mathcal{O}_\lambda$ is a finite abelian category, it is equivalent to the category of finite-dimensional right modules over some finite-dimensional algebra. In fact, it can be shown that this algebra is not dependent on λ!

Theorem 6.3 *There is a finite-dimensional algebra A such that for any $\lambda \in \Lambda^+$, $\mathcal{O}_\lambda \simeq \mathrm{mod}_{\mathrm{fd}} - \mathrm{A}$.*

It is evident that the algebra A is only well defined up to Morita equivalence. A natural problem is to find a concrete presentation of A. Since A is Morita equivalent to

$$\mathrm{End}_{\mathcal{O}_\lambda}\left(\bigoplus_{w \in W} P(w \cdot \lambda)\right),$$

this problem is equivalent (in some sense) to understanding projective objects and morphisms between them. Counter-intuitively, it is more effective to investigate *functors* acting on the category of projective objects and morphisms (i.e. natural transformations) between them.

Proposition 6.4 *For each $s \in S$ there is an exact self-adjoint functor $\theta_s : \mathcal{O}_\lambda \longrightarrow \mathcal{O}_\lambda$ with the following properties:*

1. *θ_s preserves projective objects;*
2. *if $w \in W$ with $\ell(ws) > \ell(w)$ there is an exact sequence*

$$0 \to M(w \cdot \lambda) \to \theta_s(M(w \cdot \lambda)) \cong \theta_s(M(ws \cdot \lambda)) \to M(ws \cdot \lambda) \to 0;$$

3. *if $st \cdots u$ is a reduced expression for some $w \in W$ in terms of simple reflections in S, then $P(w \cdot \lambda)$ is a direct summand of $\theta_s \theta_t \cdots \theta_u(M(\lambda))$.*

Since $M(\lambda)$ is itself projective, every natural transformation

$$\theta_s \theta_t \cdots \theta_u \longrightarrow \theta_{s'} \theta_{t'} \cdots \theta_{u'}$$

for reduced expressions $st \cdots u$ and $s't' \cdots u'$ induces a homomorphism

$$\theta_s \theta_t \cdots \theta_u(M(\lambda)) \longrightarrow \theta_{s'} \theta_{t'} \cdots \theta_{u'}(M(\lambda))$$

of projective objects. In fact, it can be shown that every homomorphism between such projective objects is induced in this way [11, Theorem 10.7]. So to find a concrete presentation of the algebra A, it is enough to describe

$$\mathrm{Hom}(\theta_s\theta_t\cdots\theta_u, \theta_{s'}, \theta_{t'}\cdots\theta_{u'}),$$

the space of all natural transformations between the functors $\theta_s\theta_t\cdots\theta_u$ and $\theta_{s'}\theta_{t'}\cdots\theta_{u'}$.

Theorem 6.5 ([15, 17]) *There are $\mathbb{C}[\mathfrak{h}]$-$\mathbb{C}[\mathfrak{h}]$ bimodules $\{B_s\}_{s\in S}$ such that for any reduced expressions $st\cdots u$ and $s't'\cdots u'$, there is an isomorphism*

$$\mathrm{Hom}(\theta_s\theta_t\cdots\theta_u, \theta_{s'}, \theta_{t'}\cdots\theta_{u'}) \cong \\ C\otimes\mathrm{Hom}(B_s\otimes B_t\otimes\cdots\otimes B_u, B_{s'}\otimes B_{t'}\otimes\cdots\otimes B_{u'})\otimes\mathbb{C},$$

where all tensor products are over $\mathbb{C}[\mathfrak{h}]$ and C denotes the coinvariant algebra $\mathbb{C}[\mathfrak{h}]/\mathbb{C}[\mathfrak{h}]\mathbb{C}[\mathfrak{h}]^W_+$, i.e. the quotient of $\mathbb{C}[\mathfrak{h}]$ by the ideal generated by positive degree W-invariants. (Here we are using the fact that the space of bimodule homomorphisms between two bimodules is itself a bimodule.)

The bimodules $\{B_s\}_{s\in S}$ (and more generally any direct summand of a tensor product of such bimodules) are today called *(classical) Soergel bimodules*, and can be used to give a presentation of A as follows. Fix a reduced expression for each $w\in W$. Then A is Morita equivalent to

$$\bigoplus_{\substack{w,w'\in W\\ w=st\cdots u\\ w'=s't'\cdots u'}} C\otimes\mathrm{Hom}(B_s\otimes B_t\otimes\cdots\otimes B_u, B_{s'}\otimes B_{t'}\otimes\cdots\otimes B_{u'})\otimes\mathbb{C},$$

where $st\cdots u$ and $s't'\cdots u'$ are the fixed reduced expressions for w and w'.

6.2 The Diagrammatic Category $\mathcal{D}$ of Soergel Bimodules

It is an amazing fact that Soergel bimodules make sense for *arbitrary* Coxeter groups, not just Weyl groups. This suggests that we should define "category $\mathcal{O}_\lambda$" for arbitrary Coxeter groups in terms of Soergel bimodules.

Theorem 6.6 ([13], [5, 6, 9]) *The monoidal category of Soergel bimodules has an explicit diagrammatic presentation.*

Equivalently, the finite-dimensional algebra A above has a presentation as a *diagram algebra*. In this context, a *diagrammatic presentation* means a presentation of a (strict) monoidal category using string diagrams. The essence of this approach is summarized in Table 6.1. In short, a morphism in a monoidal category corresponds to a diagram or a linear combination of diagrams. The sequence of colours of the edges which meet the bottom and top of the diagram give the domain and codomain of the corresponding morphism respectively. Vertical concatenation of diagrams corresponds to composition of morphisms, while horizontal concatenation corresponds to the tensor product of morphisms.

There are several advantages of the diagrammatic approach to Soergel bimodules over classical Soergel bimodules. In general, presenting a monoidal category diagrammatically makes bifunctoriality of the tensor product visually obvious through *rectilinear isotopy* of diagrams. Informally, we say that two diagrams are equivalent up to rectilinear isotopy if we can deform one diagram into the other by continuously moving vertices and stretching or shrinking edges, without moving edges or vertices past other edges and without introducing "caps" or "cups" in any edges. (See the left-hand sides of (6.2.1) for pictures of cap/cup diagrams. For a more formal description of rectilinear isotopy, see [10, (7.5)–(7.8)].) In the specific case of Soergel bimodules, there are several other "visually intuitive" relations which we will see later. More importantly, classical Soergel bimodules sometimes behave poorly over fields of positive characteristic, while diagrammatic Soergel bimodules remain well behaved. For applications to modular representation theory it is therefore best to work in the diagrammatic category.

From now on, we generalize from the setting of semisimple Lie algebras and assume that (W,S) is an arbitrary Coxeter system. In other words, W is a group with a presentation

$$W = \langle S \mid \forall s,t \in S,\ (st)^{m_{st}} = 1 \rangle$$

for certain positive integers m_{st}, with $m_{st} = m_{ts}$ and $m_{ss} = 1$ for all $s,t \in S$. The natural replacement for the Cartan subalgebra in this setting is called a *realization*.

Definition 6.3 Let $\Bbbk$ be an integral domain. A *realization* of (W,S) over $\Bbbk$ consists of a free, finite rank $\Bbbk$-module $\mathfrak{h}$ along with subsets $\{\alpha_s^\vee : s \in S\} \subset \mathfrak{h}$ and $\{\alpha_s : s \in S\} \subset \mathfrak{h}^* = \mathrm{Hom}(\mathfrak{h}, \Bbbk)$ such that

(i) $\langle \alpha_s^\vee, \alpha_s \rangle = 2$ for all $s \in S$;
(ii) the assignment

$$s(\lambda) = \lambda - \langle \alpha_s^\vee, \lambda \rangle \alpha_s$$

for all $s \in S$ and $\lambda \in \mathfrak{h}^*$ defines a representation of W on $\mathfrak{h}^*$;
(iii) the technical condition [9, (3.3)] is satisfied.

Table 6.1 *A comparison of some of the monoidal categories seen in §6.1. We note that in the diagrammatic approach, each diagram represents a morphism, so it is customary to use the diagram for the identity morphism on an object to represent an object diagrammatically.*

Category of functors on $\mathcal{O}_\lambda$	Category of (classical) Soergel bimodules	Diagrammatic category of Soergel bimodules	
functor θ_s	bimodule B_s	vertical edge	$[= \mathrm{id}_{B_s}]$
functor composition $\theta_s\theta_t$	tensor product $B_s \otimes B_t$	horizontal concatenation	 $[= \mathrm{id}_{B_s \otimes B_t}]$
natural transformation $\theta_s\theta_t \xrightarrow{\alpha} \theta_t\theta_s$	bimodule homomorphism $B_s \otimes B_t \xrightarrow{f} B_t \otimes B_s$	vertex joining edges	f 
(vertical) composition of natural transformations $\theta_s\theta_s \xrightarrow{\beta} \theta_s \xrightarrow{\gamma} \theta_s\theta_s$	composition of homomorphisms $B_s \otimes B_s \xrightarrow{g} B_s \xrightarrow{h} B_s \otimes B_s$	vertical concatenation	h g 
(horizontal) composition of natural transformations $\theta_s\theta_t\theta_s\theta_s \xrightarrow{\alpha * \beta} \theta_t\theta_s\theta_s$	tensor product of homomorphisms $B_s \otimes B_t \otimes B_s \otimes B_s \xrightarrow{f \otimes g} B_t \otimes B_s \otimes B_s$	horizontal concatenation	h g 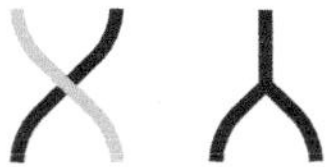
interchange law	bifunctoriality of $\otimes$	rectilinear isotopy	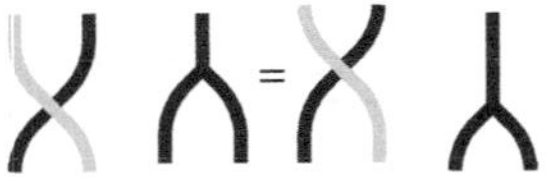 =

Example 6.4

1 Let $\mathfrak{g}$ be a complex semisimple Lie algebra, and let $\mathfrak{b}$ be a choice of Borel subalgebra. The Cartan subalgebra $\mathfrak{h}$ with the usual simple roots and coroots is a $\mathbb{C}$-realization of the Weyl group W.
2 Let $\Bbbk$ be an algebraically closed field of characteristic $p > 0$, and let G be a semisimple algebraic group over $\Bbbk$ with maximal torus T and cocharacter group $X(T) = \mathrm{Hom}(\mathbb{G}_m, T)$. The space $\mathfrak{h} = \Bbbk \otimes_{\mathbb{Z}} X(T)$, with the images of the usual roots and coroots, is a $\Bbbk$-realization of the Weyl group W.

We will use the data of a realization to construct the category $\widetilde{\mathcal{D}}_{\mathrm{BS}}$ below, the first step towards our goal of defining the diagrammatic category $\mathcal{D}$ of Soergel bimodules. As the construction of $\widetilde{\mathcal{D}}_{\mathrm{BS}}$ is entirely diagrammatic, it will be useful to identify the set S of simple generators with a set of colours for the purposes of drawing string diagrams. In the diagrams below, we will colour the generator s black and the generator t grey.

Definition 6.5 ($\widetilde{\mathcal{D}}_{\mathrm{BS}}$: generators) Let $\mathfrak{h}$ be a $\Bbbk$-realization of (W,S). Set $R = \mathrm{Sym}(\mathfrak{h}^*)$, the symmetric algebra of $\mathfrak{h}^*$, with $\deg \mathfrak{h}^* = 2$. The category $\widetilde{\mathcal{D}}_{\mathrm{BS}}$ is the $\Bbbk$-linear graded strict monoidal category defined as follows.

- The objects of $\widetilde{\mathcal{D}}_{\mathrm{BS}}$ are the formal (tensor) products of form $B_s \otimes B_t \otimes \cdots \otimes B_u$ for $s, t, \ldots, u \in S$.
- The morphisms in $\widetilde{\mathcal{D}}_{\mathrm{BS}}$ are generated (under $\Bbbk$-linear combinations, compositions and tensor products) by the following elementary morphisms.
  - For each homogeneous $f \in R$, there is a morphism

$$f : \mathbf{1} \longrightarrow \mathbf{1}$$

$$f$$

    of degree $\deg(f)$.
  - For each $s \in S$ there are morphisms

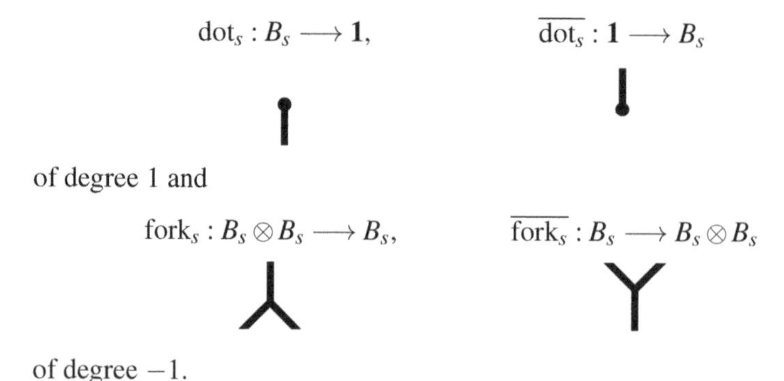

$$\mathrm{dot}_s : B_s \longrightarrow \mathbf{1}, \qquad \overline{\mathrm{dot}_s} : \mathbf{1} \longrightarrow B_s$$

    of degree 1 and

$$\mathrm{fork}_s : B_s \otimes B_s \longrightarrow B_s, \qquad \overline{\mathrm{fork}_s} : B_s \longrightarrow B_s \otimes B_s$$

    of degree -1.

– For each pair $(s,t) \in S \times S$ with $s \neq t$ and $m_{st} < \infty$, there is a morphism

$$\mathrm{braid}_{st} : \underbrace{B_s \otimes B_t \otimes B_s \otimes \cdots \otimes B_s}_{m_{st}} \longrightarrow \underbrace{B_t \otimes B_s \otimes B_t \otimes \cdots \otimes B_t}_{m_{st}}$$

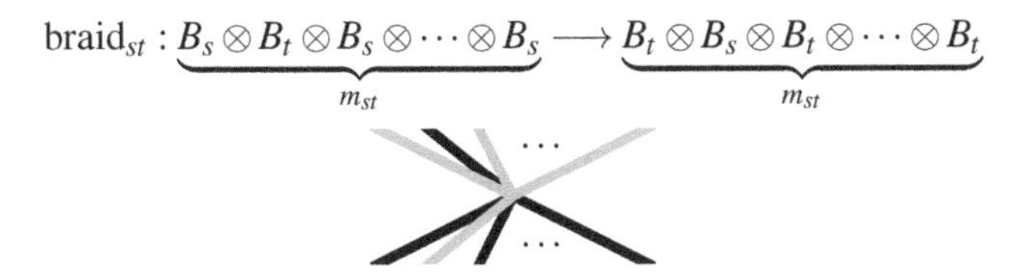

when m_{st} is odd, or

$$\mathrm{braid}_{st} : \underbrace{B_s \otimes B_t \otimes B_s \otimes \cdots \otimes B_t}_{m_{st}} \longrightarrow \underbrace{B_t \otimes B_s \otimes B_t \otimes \cdots \otimes B_s}_{m_{st}}$$

when m_{st} is even, of degree 0.

These morphisms are subject to a number of relations, which can be found in [1, §2.2], or (in a slightly different form) [9, (5.1)–(5.12)].

For convenience we will also use the following shorthand

$$\mathrm{cap}_s = \mathrm{dot}_s \circ \mathrm{fork}_s : B_s \otimes B_s \to \mathbf{1}, \quad \mathrm{cup}_s = \overline{\mathrm{fork}_s} \circ \overline{\mathrm{dot}_s} : \mathbf{1} \to B_s \otimes B_s. \tag{6.2.1}$$

In an entirely standard way, we change our point of view slightly so that we allow grade shifts of objects in $\widetilde{\mathcal{D}}_{\mathrm{BS}}$ but only consider homogeneous (i.e. degree 0) morphisms.

Definition 6.6 The *diagrammatic category of Bott–Samelson bimodules* is the $\Bbbk$-linear monoidal category $\mathcal{D}_{\mathrm{BS}}$ defined as follows.

- The objects of $\mathcal{D}_{\mathrm{BS}}$ are the formal symbols $B(m)$, for $B \in \mathrm{Obj}\,\widetilde{\mathcal{D}}_{\mathrm{BS}}$ and $m \in \mathbb{Z}$, with tensor product $B(m) \otimes B'(n) = (B \otimes B')(m+n)$.
- The morphisms in $\mathcal{D}_{\mathrm{BS}}$ are given by

$$\mathrm{Hom}_{\mathcal{D}_{\mathrm{BS}}}(B(m), B'(n)) = \mathrm{Hom}^{n-m}_{\widetilde{\mathcal{D}}_{\mathrm{BS}}}(B, B'),$$

with composition and tensor product defined via $\widetilde{\mathcal{D}}_{\mathrm{BS}}$.

Objects in $\mathcal{D}_{\mathrm{BS}}$ are called *(diagrammatic) Bott–Samelson bimodules*. As we will see below, Bott–Samelson bimodules are the prototypical Soergel bimodules, from which all others are constructed.

Definition 6.7 The *diagrammatic category $\mathcal{D}$ of Soergel bimodules* is the Karoubi envelope of $\mathcal{D}_{\mathrm{BS}}$. In other words $\mathcal{D}$ is the closure of $\mathcal{D}_{\mathrm{BS}}$ with respect to all finite direct sums and all direct summands of objects and morphisms in $\mathcal{D}_{\mathrm{BS}}$.

Objects in $\mathcal{D}$ are called (diagrammatic) Soergel bimodules. It can be shown that under some mild conditions on the realization $\mathfrak{h}$, $\mathcal{D}$ is a Krull–Schmidt category, i.e. every Soergel bimodule decomposes uniquely into a direct sum of indecomposable Soergel bimodules [9, Lemma 6.25]. The indecomposable Soergel bimodules then play the same role in $\mathcal{D}$ as the indecomposable projective objects in $\mathcal{O}_\lambda$. As one might expect these objects are highly dependent on characteristic, since idempotent decompositions of the identity in the endomorphism algebra of a Bott–Samelson bimodule are usually characteristic-dependent.

6.3 Some Diagrammatic Relations

In this section we will investigate a subset of the relations which define $\widetilde{\mathcal{D}}_{\mathrm{BS}}$.

Polynomial Relations

Regions labelled by polynomials add and multiply in the usual way, i.e. for any $f, g \in R$ we have

$$\left(f\right) + \left(g\right) = \left(f+g\right), \qquad \left(f\right) \otimes \left(g\right) = \left(fg\right), \qquad \left(f\right) \circ \left(g\right) = \left(fg\right). \tag{6.3.1}$$

(Here we use dashed circles for emphasis around a single diagram without strings, e.g. the left-hand side of the first equation consists of a sum of two diagrams, while the right-hand side is a single diagram.)

For each $s \in S$ we also have

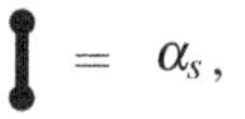

(6.3.2)

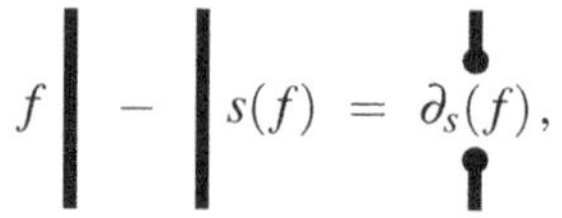

(6.3.3)

where $\partial_s(f) = \alpha_s^{-1}(f - s(f))$.

One-colour Relations

For each $s \in S$ we have

$$= \quad = \quad , \qquad = \quad = \quad , \tag{6.3.4}$$

$$= \quad = \quad , \tag{6.3.5}$$

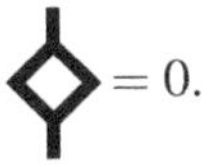

$$= 0. \tag{6.3.6}$$

These relations give *all* the relations defining $\widetilde{\mathcal{D}}_{\mathrm{BS}}$ in a few special cases.

Definition 6.8 ($\widetilde{\mathcal{D}}_{\mathrm{BS}}$: relations (no finite dihedral parabolics)) Suppose (W,S) is a Coxeter system with no finite dihedral parabolic subgroups (i.e. $m_{st} = \infty$ whenever $s \neq t$). Then (6.3.1)–(6.3.6) is a full list of relations defining $\widetilde{\mathcal{D}}_{\mathrm{BS}}$.

Thus we have defined enough relations to understand Soergel bimodules for the smallest Lie algebra $\mathfrak{sl}_2$ $\big(W = \{1, s\}\big)$.

Other Diagrammatic Relations

In general, the definition of $\widetilde{\mathcal{D}}_{\mathrm{BS}}$ requires more diagrammatic relations than (6.3.1)–(6.3.6). Perhaps unsurprisingly, the remaining relations all involve the morphism braid_{st}, which only exists when $m_{st} < \infty$. They come in two flavours, depending on how many colours of strings appear in the diagrams.

The *two-colour relations* are defined for all distinct $s, t \in S$ such that $m_{st} < \infty$, i.e. whenever braid_{st} exists. The most important of these, the *Jones–Wenzl relation*, is closely related to the Temperley–Lieb algebra.

The *three-colour relations* are defined for all distinct $s, t, u \in S$ which generate a finite parabolic subgroup. These relations involve three different kinds of braids, but no other generating morphisms. The form of the relation also only depends on the Coxeter type of the resulting parabolic subgroup. The most complicated forms (in types A_3, B_3 and H_3) are sometimes called the *Zamolodchikov relations*.

6.4 Some Consequences and Applications

Proposition 6.7 *Any two diagrams which are isotopic correspond to equal morphisms in* $\widetilde{\mathcal{D}}_{BS}$. *In other words, we may freely deform the edges of any diagram without changing the morphism in* $\widetilde{\mathcal{D}}_{BS}$.

Proof (Sketch) We first must show that the zig-zag relations hold, i.e.

(6.4.1)

This ultimately follows by first applying (6.3.5) and then applying (6.3.4) twice:

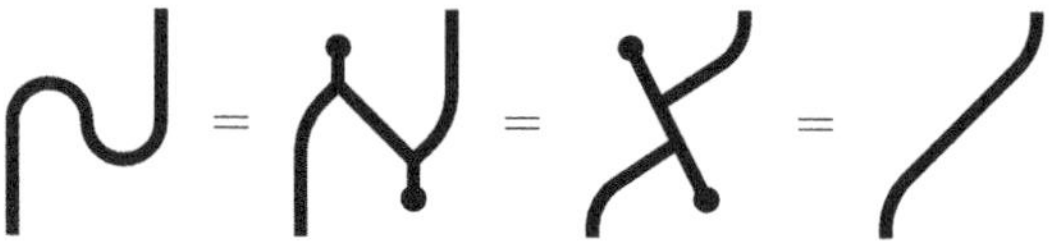

and similarly for the second equality.

Next we must show that dot_s and fork_s twist to their barred counterparts, i.e.

(6.4.2)

and

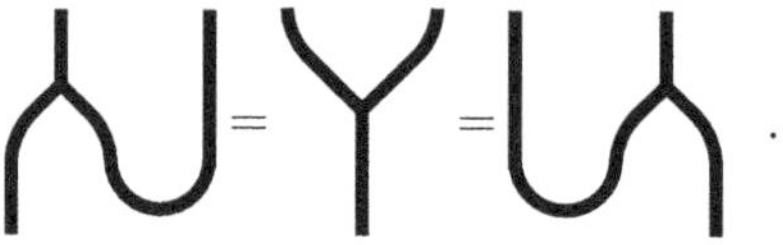

(6.4.3)

Proving (6.4.2) involves only one application of (6.3.4):

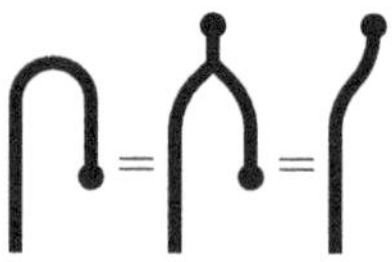

and similarly for the second equality. The proof of (6.4.3) is almost identical to that of (6.4.1):

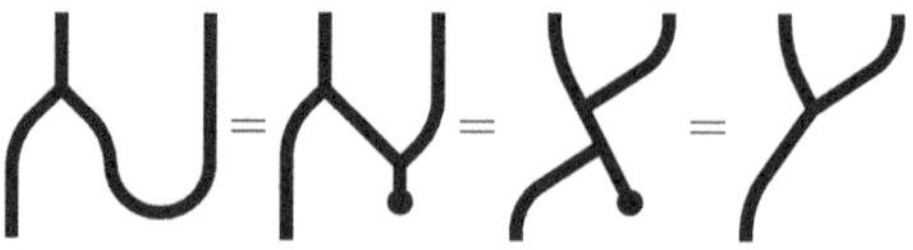

and similarly for the second equality.

Since we already have the zig-zag relations, this also means that the barred counterparts of these two relations also hold, i.e. the vertical flips of equations (6.4.2) and (6.4.3) also hold. For $s,t \in S$ with $m_{st} < \infty$, the corresponding equation for the braid morphism braid_{st} is already a relation in $\widetilde{\mathcal{D}}_{\mathrm{BS}}$ [1, §2.2(7)]. These twisting relations are enough to ensure that *any* isotopy of string diagrams is a relation in $\widetilde{\mathcal{D}}_{\mathrm{BS}}$. For more discussion on this, see [6, Proposition 3.2]. □

Lemma 6.8 *For $s \in S$ we have an idempotent decomposition*

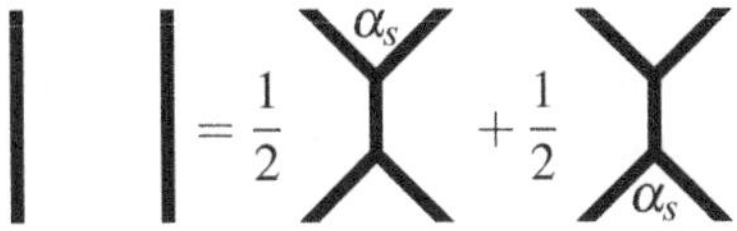

Proof First, we show that each of the terms on the right-hand side are idempotents:

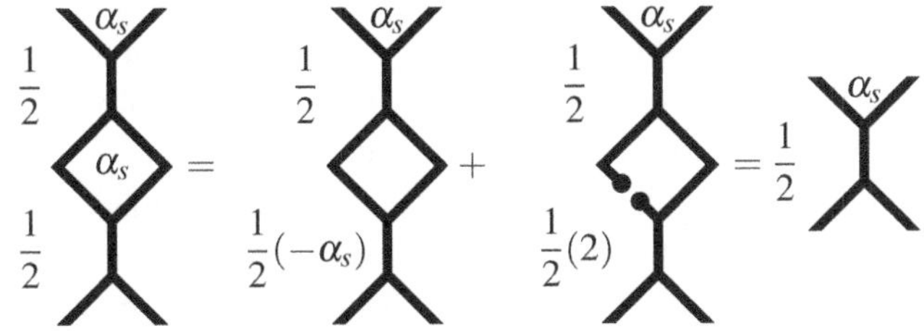

Next, we verify the decomposition by applying relations (6.3.3)–(6.3.5):

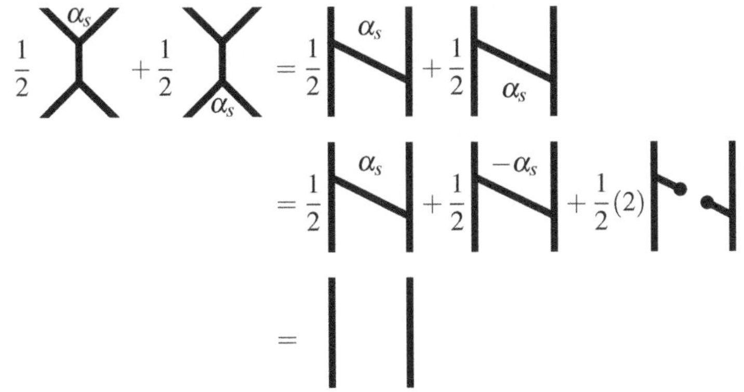

□

From this lemma we immediately obtain the following (cf. the natural isomorphism $\theta_s\theta_s \cong \theta_s \oplus \theta_s$).

Theorem 6.9 *Suppose $W = \{1,s\}$ and $\mathfrak{h}$ is a 1-dimensional realization of W over a field $\Bbbk$ with $\operatorname{char}\Bbbk \neq 2$. Then the split Grothendieck ring $[\mathcal{D}]$ of $\mathcal{D}$ (i.e. the* *ring of isomorphism classes of objects of* $\mathcal{D}$*) is isomorphic to the following:*

$$\begin{aligned}[\mathcal{D}] &\longrightarrow \mathcal{H}(S_2) = \mathbb{Z}[v^{\pm 1}][b_s]/(b_s^2 - (v+v^{-1})b_s) \\ [\mathbf{1}(1)] &\longmapsto v \\ [B_s] &\longmapsto b_s.\end{aligned}$$

Remark There is a generalization of Theorem 6.9 to all Coxeter systems known as *Soergel's categorification theorem.* It states that (under mild assumptions on the realization $\mathfrak{h}$) the split Grothendieck ring $[\mathcal{D}]$ is isomorphic to the Iwahori–Hecke algebra $\mathcal{H}(W)$. In the setting of classical Soergel bimodules, this result was proven by Soergel in [17, Satz 1.10] for suitably 'nice' realizations, and in the diagrammatic setting it was proven more generally by Elias–Williamson [9, Corollary 6.27].

We conclude with some applications and references.

1. The original motivating application for Soergel was the Kazhdan–Lusztig conjectures, which describe the characters of the simple modules of $\mathcal{O}_\lambda$ in terms of Kazhdan–Lusztig polynomials. This was originally proven in the 1980s by Beilinson–Bernstein [2] (and independently by Brylinski–Kashiwara [4]) using highly geometric techniques. In the 1990s Soergel suggested an alternative proof based on decomposing $B_s \otimes B_t \otimes \cdots \otimes B_u$ into a direct sum of indecomposable Soergel bimodules [15]. Soergel's proof was substantially more algebraic, but relied crucially on an important geometric result called the Decomposition Theorem. In [8] Elias–Williamson removed this dependence to produce an entirely algebraic proof (for a more readable introduction, see also [7, 18]).
2. A similar character-theoretic conjecture in modular representation theory is Lusztig's conjecture, which describes the characters of simple modules for a semisimple algebraic group G over a field of characteristic $p > 0$. Soergel showed that Soergel bimodules for the Weyl group in characteristic p give an analogous description of "modular category $\mathcal{O}$" [16], a subquotient of the category of rational G-modules. In the celebrated paper [19] Williamson used this framework to show that Lusztig's conjecture is in fact false, except when p is extremely large!
3. Soergel's categorification theorem provides another way to think about the above results wholly within the context of Soergel bimodules. To be more precise, Soergel showed in [15] that the Kazhdan–Lusztig conjectures hold if and only if a statement known as Soergel's conjecture holds. Soergel's conjecture states that the indecomposable Soergel bimodules correspond to the Kazhdan–Lusztig basis of the corresponding Hecke algebra. This is

difficult to prove because the Kazhdan–Lusztig basis is defined 'combinatorially' with no reference to the morphisms in $\mathcal{D}$. Elias–Williamson [8] proved Soergel's conjecture algebraically in characteristic 0, while Williamson [19] found counterexamples to Soergel's conjecture in positive characteristic. These counterexamples suggest defining the *p-canonical basis* or *p-Kazhdan–Lusztig basis* to be the basis of the Hecke algebra corresponding to the indecomposable Soergel bimodules in characteristic p [12]. Unlike the ordinary Kazhdan–Lusztig basis, the p-Kazhdan–Lusztig basis is *not* combinatorial and requires understanding of the morphisms in $\mathcal{D}$ in general.

4 Achar *et al.* have shown that the p-Kazhdan–Lusztig basis for the corresponding affine Weyl group in characteristic p gives the characters of tilting modules (another class of G-modules parametrized by highest weight) [1]. This fits in with a conjectured categorical equivalence involving the functors $\{\theta_s\}$ in characteristic p [14], similar to Theorem 6.5. In type A these decompositions also give the simple characters of the symmetric group. More recently the author (together with Chris Bowman and Anton Cox) has given an alternative, more direct proof of the symmetric group result [3].

Bibliography

[1] Achar, Pramod N., Makisumi, Shotaro, Riche, Simon, and Williamson, Geordie. 2017 (Mar.). *Free-monodromic mixed tilting sheaves on flag varieties*. arXiv: 1703.05843.

[2] Beĭlinson, A., and Bernstein, J. 1981. Localisation de $\mathfrak{g}$-modules. *C. R. Acad. Sci. Paris Sér. I Math.*, **292**(1), 15–18.

[3] Bowman, Chris, Cox, Anton, and Hazi, Amit. 2020 (May). *Path isomorphisms between quiver Hecke and diagrammatic Bott-Samelson endomorphism algebras*. arXiv:2005.02825.

[4] Brylinski, J.-L. and Kashiwara, M. 1981. Kazhdan-Lusztig conjecture and holonomic systems. *Invent. Math.*, **64**(3), 387–410.

[5] Elias, Ben. 2016. The two-color Soergel calculus. *Compos. Math.*, **152**(2), 327–398.

[6] Elias, Ben, and Khovanov, Mikhail. 2010. Diagrammatics for Soergel categories. *Int. J. Math. Math. Sci.*, Art. ID 978635, 58.

[7] Elias, Ben, and Williamson, Geordie. 2013 (Mar.). *Soergel bimodules and Kazhdan-Lusztig conjectures*. www.maths.usyd.edu.au/u/geordie/aarhus/.

[8] Elias, Ben, and Williamson, Geordie. 2014. The Hodge theory of Soergel bimodules. *Ann. of Math. (2)*, **180**(3), 1089–1136.

[9] Elias, Ben, and Williamson, Geordie. 2016. Soergel calculus. *Represent. Theory*, **20**, 295–374.

[10] Elias, Ben, Makisumi, Shotaro, Thiel, Ulrich, and Williamson, Geordie. 2020. *Introduction to Soergel bimodules*. RSME Springer Series, vol. 5. Springer, Cham.

[11] Humphreys, James E. 2008. *Representations of semisimple Lie algebras in the BGG category* $\mathcal{O}$. Graduate Studies in Mathematics, vol. 94. American Mathematical Society, Providence, RI.

[12] Jensen, Lars Thorge, and Williamson, Geordie. 2017. The p-canonical basis for Hecke algebras. Pages 333–361 of: *Categorification and higher representation theory*. Contemp. Math., vol. 683. Amer. Math. Soc., Providence, RI.

[13] Libedinsky, Nicolas. 2008. Sur la catégorie des bimodules de Soergel. *J. Algebra*, **320**(7), 2675–2694.

[14] Riche, Simon, and Williamson, Geordie. 2018. Tilting modules and the p-canonical basis. *Astérisque*, ix+184.

[15] Soergel, Wolfgang. 1990. Kategorie $\mathcal{O}$, perverse Garben und Moduln über den Koinvarianten zur Weylgruppe. *J. Amer. Math. Soc.*, **3**(2), 421–445.

[16] Soergel, Wolfgang. 2000. On the relation between intersection cohomology and representation theory in positive characteristic. *J. Pure Appl. Algebra*, **152**(1–3), 311–335. Commutative algebra, homological algebra and representation theory (Catania/Genoa/Rome, 1998).

[17] Soergel, Wolfgang. 2007. Kazhdan-Lusztig-Polynome und unzerlegbare Bimoduln über Polynomringen. *J. Inst. Math. Jussieu*, **6**(3), 501–525.

[18] Williamson, Geordie. 2012 (Dec.). *Soergel bimodules and representation theory*. www.maths.usyd.edu.au/u/geordie/sydney/.

[19] Williamson, Geordie. 2017. Schubert calculus and torsion explosion. *J. Amer. Math. Soc.*, **30**(4), 1023–1046. With a joint appendix with Alex Kontorovich and Peter J. McNamara.

7
A Companion to Quantum Groups

Bart Vlaar

7.1 Introduction

Note that the category of modules of a bialgebra has a tensor product structure. Given two modules V, W, in general it is not clear that $V \otimes W$ is isomorphic to $W \otimes V$ as modules. If the bialgebra possesses a quasitriangular structure, i.e. a universal R-matrix, then there exists a natural isomorphism which is compatible with the tensor structure. Quantum groups form a rich family of (non-commutative and non-cocommutative) quasitriangular bialgebras which in some sense deform associative algebras naturally associated with certain groups and Lie algebras.

Quantum groups were discovered in the 1980s in the context of quantum integrability (simultaneous diagonalizability of commuting Hamiltonians via solutions of the Yang–Baxter equation), initially in [27]. Their theory is a vast topic that has developed immensely in the last four decades. There are various reasons why quantum groups are interesting: there are connections with low-dimensional topology (e.g. representations of braid groups and construction of quasi-invariants for knots, links, etc.), q-deformed harmonic analysis (closely connected to the older theory of special functions depending on a deformation parameter) and non-commutative geometry (the study of deformed algebras of functions on algebraic groups).

It is difficult to give a precise definition of a quantum group that encompasses the various classes of examples which are known as such. Focusing on deformations of *co*commutative bialgebras, one may propose the following "soft" definition, which will be our guide in these notes:

A quantum group is a non-commutative bialgebra depending on a parameter that is
(1) quasitriangular for all values of the parameter and
(2) cocommutative only for special values of the parameter.

We will showcase a special class of quantum groups: quantized enveloping algebras $U_q\mathfrak{g}$, also known as Drinfeld–Jimbo quantum groups, and explain the construction of the universal R-matrix. The representation theory of $U_q\mathfrak{g}$, if q is not a root of unity, stays close to that of $U\mathfrak{g}$, i.e. that of $\mathfrak{g}$, which is well known. Unfortunately it is beyond the scope of this account to discuss in detail other types of quantum groups (e.g. Yangians and RTT algebras, compact quantum groups, bicrossproduct quantum groups) or delve very deep into particular advanced branches and applications of this theory such as: Lie bialgebra quantization; diagonalization of commuting transfer matrices; the definition of knot (quasi-)invariants; root-of-unity phenomena; canonical bases (crystal bases) and the $q \to 0$ limit; Knizhnik–Zamolodchikov equations. Standard textbook resources focusing on quantum groups are for instance [7, 17, 23, 24, 30, 32]. For an account tailored to the application of quantum groups to quantum integrability. We recommend the lecture notes [36] and the book [19]. Note that quantum groups have applications in mathematical physics beyond integrability; one can in particular point out their role in quantum gravity (see [31]).

We hope that these notes provide the reader with a basic working knowledge of quantum groups and spur them on to a deeper investigation in this rich variety of topics. For most of these lecture notes, we assume fairly little background knowledge beyond abstract linear algebra and basic notions of group theory and representation theory, although some familiarity with Lie theory and category theory is helpful.

7.1.1 Outline

First of all we will review bialgebras, Hopf algebras and their representations (and introduce notation that we will use throughout this chapter) in Section 7.2. We discuss quasitriangular bialgebras and the induced braiding on their categories of representations in Section 7.3. In Section 7.4 we discuss their quantizations, Drinfeld–Jimbo quantum groups $U_q\mathfrak{g}$, and in Section 7.5 the quasitriangular structure of $U_q\mathfrak{g}$; in these sections we pay particular attention to the special case $U_q\mathfrak{sl}_2$. We end the notes with a brief investigation in Section 7.6 into more recent developments involving quasitriangularity for special types of subalgebras of bialgebras, and the associated cylindrically braided structure on the category of modules.

7.1.2 Acknowledgements

The author was supported by EPSRC grant EP/R009465/1. The lecture series associated with these notes, part of the LMS Autumn Algebra School 2020, was supported by the London Mathematical Society and the European Research Council and hosted by the International Centre for Mathematical Sciences. The author is grateful to Stefan Kolb for useful comments.

7.2 Bialgebras

In this section we recall some basic theory surrounding bialgebras and monoidal categories, setting the stage for the next section which deals with quasitriangular bialgebras and braided monoidal categories. For more background on some of this material the reader can consult for instance [7, Sec. 4.1] or the lecture notes [39].

7.2.1 Notation

We fix an arbitrary[1] field k; linear structures will always be with respect to k, which we may suppress from the notation. In this way $\otimes = \otimes_k$ is the tensor product over k, $\mathrm{Hom}(V,W) = \mathrm{Hom}_k(V,W)$ is the vector space of k-linear maps from V to W and $\mathrm{End}(V) = \mathrm{End}_k(V)$ is the algebra of k-linear maps on V.

Let V,W be vector spaces. We denote the identity map on V by id_V. We denote by $\sigma_{V,W}$ the unique linear map from $V \otimes W$ to $W \otimes V$ which sends $v \otimes w$ to $w \otimes v$ for all $v \in V, w \in W$. If there is no cause for confusion, we will simply write id instead of id_V and σ instead of $\sigma_{V,V}$.

7.2.2 Algebras

We consider an algebra A over k, i.e. a vector space that possesses a bilinear multiplication map: $A \times A \to A$ which is compatible with scalar multiplication: $\lambda(ab) = (\lambda a)b = a(\lambda b)$ for all $\lambda \in k$ and $a, b \in A$. Note that since the multiplication map is bilinear we can view it as a linear map $m : A \otimes A \to A$. We will always[2] assume that A is *associative*, i.e.

$$m \circ (m \otimes \mathsf{id}_A) = m \circ (\mathsf{id}_A \otimes m) \in \mathrm{Hom}(A, A \otimes A \otimes A), \tag{7.2.1}$$

and *unital*, i.e. there is a linear map $\eta : k \to A$ such that

$$m \circ (\eta \otimes \mathsf{id}_A) = \mathsf{id}_A = m \circ (\mathsf{id}_A \otimes \eta) \in \mathrm{End}(A, A), \tag{7.2.2}$$

where we have used that $k \otimes A \cong A \cong A \otimes k$. Note that such η must be injective and hence we can, and shall, identify k with $\eta(k) \subseteq A$. In particular, there is an

[1] Later we will assume that k is of characteristic zero and algebraically closed.

[2] Note that Lie algebras are not algebras using this restricted definition.

element $\eta(1) \in A$, which we simply denoted by 1, with the property $1a = a1 = a$ for all $a \in A$ (conversely, given such an element, there is a unique linear map $\eta : k \to A$ sending $1 \in k$ to $1 \in A$).

Let A and B be algebras with multiplications m_A, m_B and unit maps η_A, η_B, respectively. An *algebra morphism* from A to B is a linear map $f : A \to B$ such that

$$m_B \circ (f \otimes f) = f \circ m_A \in \mathrm{Hom}(A \otimes A, B), \qquad \eta_B = f \circ \eta_A \in \mathrm{Hom}(k, B).$$

Also, $A \otimes B$ is an algebra in a natural way, with multiplication

$$\begin{aligned} m_{A\otimes B} := (m_A \otimes m_B) \circ (\mathsf{id}_A \otimes \sigma_{A,B} \otimes \mathsf{id}_B) \\ \in \mathrm{Hom}(A \otimes B \otimes A \otimes B, A \otimes B) \end{aligned} \tag{7.2.3}$$

(note that the swap σ is really necessary here) and unit map

$$\eta_{A\otimes B} := \eta_A \otimes \eta_B \in \mathrm{Hom}(k, A \otimes B). \tag{7.2.4}$$

Naturally associated to an algebra A are two groups: the subset $A^\times$ of invertible elements and the set of algebra automorphisms $\mathrm{Aut}_{\mathsf{alg}}(A)$ (invertible algebra morphisms from A to itself). For any $x \in A^\times$ we denote by $\mathsf{Ad}(x)$ the automorphism of A given by conjugation by x: $\mathsf{Ad}(x)(a) = xax^{-1}$ for all $a \in A$; thus we obtain a group morphism Ad from $A^\times$ to $\mathrm{Aut}_{\mathsf{alg}}(A)$.

7.2.3 Algebra Representations

In general, a good way to study (or "test") A is by looking at representations of A. A *representation* of A on V is an algebra morphism $\pi_V : A \to \mathrm{End}(V)$ (more loosely, we also say that V carries a representation of A if such a π_V exists). For all $a \in A$ and $v \in V$ the element $\pi_V(a)(v)$ depends linearly on both a and v and is thus a linear map on the tensor product $A \otimes V$. Accordingly, we say that V has a *(left) A-module structure* consisting of the left action map $\lambda_V : A \otimes V \to V$ defined by $\lambda_V(a \otimes v) = \pi_V(a)(v)$, which is sometimes denoted $a \cdot v$ if the representation or the module structure is clear from the context.

If V, W are left A-modules, then we call a linear map $\varphi : V \to W$ an *A-intertwiner* (or A-module morphism) if φ commutes with the action of A, i.e. if the following diagram commutes for all $a \in A$:

$$\begin{array}{ccc} V & \xrightarrow{\varphi} & W \\ {\scriptstyle \pi_V(a)}\downarrow & & \downarrow{\scriptstyle \pi_W(a)} \\ V & \xrightarrow[\varphi]{} & W. \end{array} \tag{7.2.5}$$

7.2.4 Tensor Products of Modules

If V and W are left A-modules, then $V \otimes W$ is not automatically an A-module, but merely an $A \otimes A$-module, with representation map $\pi_V \otimes \pi_W : A \otimes A \to \mathrm{End}(V) \otimes \mathrm{End}(W) \subseteq \mathrm{End}(V \otimes W)$.

Example 7.1 Let G be a group and consider the *group algebra* kG (the k-linear space with basis given by the group elements, which we turn into an algebra by extending the group multiplication linearly). Note that group representations of G are in a natural 1-to-1 correspondence with algebra representations of kG; if $\pi : G \to \mathrm{GL}(V)$ is a group representation then the corresponding algebra representation from kG on V is denoted by the same symbol. If we have two group representations $\pi_V : G \to \mathrm{GL}(V)$ and $\pi_W : G \to \mathrm{GL}(W)$ then $V \otimes W$ automatically carries a representation $\pi_{V \otimes W} : G \to \mathrm{GL}(V \otimes W)$ defined by

$$\pi_{V \otimes W}(g)(v \otimes w) := \pi_V(g)(v) \otimes \pi_W(g)(w) \tag{7.2.6}$$

for all $g \in G$, $v \in V$ and $w \in W$. Viewing π_V, π_W and $\pi_{V \otimes W}$ as algebra representations, note that we have defined $\pi_{V \otimes W} = (\pi_V \otimes \pi_W) \circ \Delta$, where Δ is the algebra morphism form kG to $kG \otimes kG$ uniquely determined by $\Delta(g) = g \otimes g$ for all $g \in G$.

In general, a natural framework for constructing representations of A from tensor products of representations of A arises whenever there is a distinguished algebra morphism $\Delta : A \to A \otimes A$: in this case we immediately see that $\pi_{V \otimes W} := (\pi_V \otimes \pi_W) \circ \Delta$ is an algebra morphism from A to $\mathrm{End}(V \otimes W)$. In this way we have related an additional structure on the representations of A to an additional structure on A itself.

The simplest meaningful representations of A are 1-dimensional representations or *characters*, i.e. algebra morphisms from A to $\mathrm{End}(k) \cong k$. To guarantee the existence of such a representation, it is convenient to further extend the structure on A given by Δ by stipulating that we also have a distinguished algebra morphism $\varepsilon : A \to k$.

Note that the new structure maps Δ and ε are similar to m and η, respectively, but go in the reverse direction. Therefore we call them *coproduct* (or *comultiplication*) and *counit (map)*.

7.2.5 Monoidal Categories

The notion of A-module is naturally bolted onto a more basic notion of vector space, to which is associated a notion of taking tensor products and a special vector space k, which acts as a neutral element when taking tensor products.

We want to constrain the structure maps Δ, ε so that tensor products of A-modules and the special module k behave as tensor products of vector spaces and the special vector space k. So we need to capture which properties of vector spaces we wish to preserve.

Note that for all vector spaces U,V,W we have[3]

$$(U\otimes V)\otimes W \cong U\otimes (V\otimes W), \qquad k\otimes V\cong V\otimes k\cong V. \tag{7.2.7}$$

These properties are reminiscent of the definition of a monoid, and we call a category $\mathcal{C}$ a *monoidal category* (or *tensor category*) if there exists a bifunctor $\otimes: \mathcal{C}\times\mathcal{C}\to\mathcal{C}$ and a special object $1_{\mathcal{C}}$ that are abstractions of the tensor product operation on vector spaces and the special vector space k. To flesh this out precisely requires a bit more work; the precise definition of monoidal category[4] is given for instance in [7, Sec. 5.1B]. In particular, the collection of k-linear spaces together with the k-linear maps between them constitutes a monoidal category, called Vect, with the monoidal structure given by the usual tensor product of vector spaces and the vector space k.

The representations of our algebra A together with their intertwiners also form a category, which we denote by $\mathsf{Rep}(A)$. Note that we have a forgetful function For from $\mathsf{Rep}(A)$ to Vect, mapping each module to the underlying vector space and each intertwiner to the underlying linear map. It is natural to require of A that the isomorphisms in (7.2.7) are preserved when we interpret them as statements about $\mathsf{Rep}(A)$. More precisely, we want For to become a *monoidal functor* (i.e. it maps the tensor product of A-modules to the tensor product of vector spaces). It gives rise to the following definition.

7.2.6 Bialgebras

Definition 7.2 An algebra A is called a *bialgebra* if there exist algebra morphisms $\Delta: A\to A\otimes A$ and $\varepsilon: A\to k$ satisfying *coassociativity* and *counit axioms*:

$$(\Delta\otimes \mathsf{id}_A)\circ\Delta = (\mathsf{id}_A\otimes\Delta)\circ\Delta \qquad \in \mathrm{Hom}(A, A\otimes A\otimes A), \tag{7.2.8}$$

$$(\varepsilon\otimes \mathsf{id}_A)\circ\Delta = \mathsf{id}_A = (\mathsf{id}_A\otimes\varepsilon)\circ\Delta \qquad \in \mathrm{End}(A). \tag{7.2.9}$$

Remark A vector space A which possesses linear maps $\Delta: A\to A\otimes A$ and $\varepsilon: A\to k$ satisfying (7.2.8–7.2.9) is called a *coalgebra*. Bialgebras are at the

[3] Here we are careful in writing isomorphisms instead of identities, since tensor products of vector spaces are only defined up to isomorphism.

[4] Often the terms *monoidal category* and *tensor category* are used interchangeably but in [7] the terminology *tensor category* and *quasitensor category* correspond to what is commonly known as a *symmetric monoidal category* and *braided monoidal category*, respectively.

same time algebras and coalgebras in such a way that the two types of additional structures are compatible: the coalgebra structure maps Δ, ε are algebra morphisms (equivalently, the algebra structure maps m, η are coalgebra morphisms).

Before we study examples of bialgebras, we will develop the basic theory further. If A is a bialgebra, the properties (7.2.8) and (7.2.9) guarantee that the identities in (7.2.7) are identities of A-modules, so $\mathsf{Rep}(A)$ forms a monoidal category. In fact, we have a somewhat stronger statement:

Theorem 7.3 ([e.g. 39, Prop. 1.1]) *Let A be an algebra with multiplication map m. Let $\Delta : A \to A \otimes A$ and $\varepsilon : A \to k$ be algebra maps. Let $\otimes : \mathsf{Rep}(A) \times \mathsf{Rep}(A) \to \mathsf{Rep}(A)$ be the functor which associates to a pair of A-modules (V, W) a module, uniquely defined by stipulating that the underlying vector space is the usual tensor product $V \otimes W$ and the representation is*

$$\pi_{V\otimes W} = (\pi_V \otimes \pi_W) \circ \Delta : A \to \operatorname{End}(V) \otimes \operatorname{End}(W) \subseteq \operatorname{End}(V \otimes W). \quad (7.2.10)$$

Also let k be an A-module with representation $\pi_k = \varepsilon : A \to \operatorname{End}(k) \cong k$. Then $(\mathsf{Rep}(A), \otimes, k)$ is a monoidal category with the same isomorphisms as Vect *in (7.2.7) if and only if (Δ, ε) satisfies (7.2.8–7.2.9).*

If $B \subseteq A$ is a subalgebra and $\Delta(B) \subseteq B \otimes B$ then we call B a *subbialgebra of* A (as a consequence, B is a bialgebra in its own right). If A and B are bialgebras with coproducts Δ_A, Δ_B and counits ε_A, ε_B, respectively, then a *bialgebra morphism* is an algebra morphism $f : A \to B$ with the additional property

$$(f \otimes f) \circ \Delta_A = \Delta_B \circ f, \qquad \varepsilon_A = \varepsilon_B \circ f. \quad (7.2.11)$$

Let A be a bialgebra. In order to describe explicitly the A-module structure of a tensor product of any (finite) number of A-modules, we can recursively define *iterated coproducts* $\Delta^{(n)} \in \operatorname{Hom}(A, A^{\otimes n})$ for $n \in \mathbb{Z}_{\geq 0}$ as follows:

$$\Delta^{(0)} = \varepsilon, \qquad \Delta^{(n+1)} = (\Delta^{(n)} \otimes \mathsf{id}) \circ \Delta \quad (7.2.12)$$

so that $\Delta^{(1)} = \mathsf{id}$ by (7.2.9) and hence $\Delta^{(2)} = \Delta$. By virtue of (7.2.8), replacing the recursion in (7.2.12) by $\Delta^{(n+1)} = (\mathsf{id} \otimes \Delta^{(n)}) \circ \Delta$ for any or all n produces the same linear maps $\Delta^{(n)}$.

Remark Since $\Delta^{(n)}$ maps into $A^{\otimes n}$, for $a \in A$ and $n \in \mathbb{Z}_{>0}$ there exist (non-unique) $a_i^{(1)}, a_i^{(2)}, \ldots, a_i^{(n)} \in A$ such that

$$\Delta^{(n)}(a) = \sum_i a_i^{(1)} \otimes a_i^{(2)} \otimes \cdots \otimes a_i^{(n)}. \quad (7.2.13)$$

This may be abbreviated to *Sweedler notation*:

$$\Delta^{(n)}(a) = \sum a^{(1)} \otimes a^{(2)} \otimes \cdots \otimes a^{(n)}. \tag{7.2.14}$$

For example, we may write (7.2.9) as $\sum \varepsilon(a^{(1)})a^{(2)} = a = \sum a^{(1)}\varepsilon(a^{(2)})$.

7.2.7 Commutativity and Cocommutativity

Let A be a bialgebra. The *opposite bialgebra* A^{op} is the bialgebra obtained from A by replacing the multiplication map, say $m : A \otimes A \to A$ by $m^{\mathrm{op}} := m \circ \sigma$. We call A *commutative* if $A^{\mathrm{op}} = A$, i.e. if $m^{\mathrm{op}} = m$ (in other words the underlying algebra is commutative).

We will be more interested in the *co-opposite bialgebra* A^{cop}. It is the bialgebra obtained from A by replacing Δ by $\Delta^{\mathrm{op}} := \sigma \circ \Delta$. We call A *cocommutative* if $A^{\mathrm{cop}} = A$, i.e. if $\Delta^{\mathrm{op}} = \Delta$. Note that if A is cocommutative then the monoidal category $\mathsf{Rep}(A)$ is *symmetric*: for all $V, W \in \mathsf{Rep}(A)$ there is an A-intertwiner $c_{V,W}$ from the object $V \otimes W$ to the object $W \otimes V$ such that $c_{W,V} c_{V,W} = \mathsf{id}_{V \otimes W}$. It is given by $c_{V,W} = \sigma_{V,W}$. The fact that $\sigma_{V,W}$ is an intertwiner is equivalent to $\Delta = \Delta^{\mathrm{op}}$.

7.2.8 Antipodes and Hopf Algebras

Many bialgebras appearing "in nature" have an additional structure map called *antipode*, which enriches the category of representations of such a bialgebra. To define it, first let A and B be bialgebras with multiplications m_A, m_B, unit maps η_A, η_B, coproducts Δ_A, Δ_B and counit maps ε_A, ε_B. The set of linear maps from A to B, $\mathrm{Hom}(A, B)$, possesses a natural product structure called *convolution product*, sending $f, g : A \to B$ to

$$f * g := m_B \circ (f \otimes g) \circ \Delta_A : A \to B. \tag{7.2.15}$$

It follows from the definition of bialgebra that $(\mathrm{Hom}(A, B), *)$ is a monoid with neutral element $\eta_B \circ \varepsilon_A : A \to B$ called the *convolution monoid*. Setting $B = A$ and suppressing subscripts, an *antipode* is a map $S \in \mathrm{Hom}(A, A)$ which is a $*$-inverse of $\mathsf{id} \in \mathrm{Hom}(A, A)$. In other words, an antipode S satisfies

$$m \circ (S \otimes \mathsf{id}) \circ \Delta = \eta \circ \varepsilon = m \circ (\mathsf{id} \otimes S) \circ \Delta. \tag{7.2.16}$$

One can now combine the convolution monoid construction with uniqueness of inverses to prove a slew of basic properties of antipodes. For proofs of the following we refer for instance to [10, Sec. 4.2].

Lemma 7.4 *Let A be a bialgebra with antipode S. Then*

1 *S is unique;*
2 *S is a bialgebra morphism from A to $(A^{\mathsf{op}})^{\mathsf{cop}}$ (and hence S^2 is a bialgebra endomorphism of A);*
3 *if S is invertible (with respect to composition) then S^{-1} is an antipode for the bialgebras A^{op} and A^{cop};*
4 *if A is commutative or cocommutative then S is involutive;*
5 *if B is another bialgebra with antipode S' and $f : A \to B$ is a bialgebra morphism, then $S' \circ f = f \circ S$.*

Since we are interested in "deformations" of (co)commutative bialgebras, considering Lemma 7.4 (4), it is natural to require that S is invertible (although we have to relinquish involutiveness).

Definition 7.5 We call a bialgebra A a *Hopf algebra* if it has an antipode which is invertible (with respect to composition).

7.2.9 Representations of Bialgebras and Hopf Algebras

We review some more standard terminology of the representation theory of bialgebras. Let A be a bialgebra. Any vector space V automatically becomes a left A-module if we set $a \cdot v = \varepsilon(a)v$ for all $a \in A$, $v \in V$. This is called the *trivial* A-module structure on V.

The *(left) regular representation* of A is the A-module structure on A itself given by left multiplication. If A is additionally a Hopf algebra, the antipode also allows us to define the *(left) adjoint representation* of A on itself. Namely, for all $a, b \in A$ set, in terms of Sweedler notation,

$$\mathsf{ad}(a)(b) := \sum a^{(1)} b S(a^{(2)}). \tag{7.2.17}$$

It is a nice exercise to show that the map $\mathsf{ad} : A \to \mathrm{End}_k(A)$ defined by this assignment is indeed an algebra morphism.

If A is a Hopf algebra then the dual $V^* = \mathrm{Hom}(V, k)$ of $V \in \mathsf{Rep}(A)$ becomes an A-module by setting

$$(a \cdot f)(v) = f(S(a) \cdot v) \qquad \text{for all } a \in A,\ f \in V^*,\ v \in V. \tag{7.2.18}$$

As a consequence of the first equation of (7.2.16), this action of A on V^* implies that the canonical linear map: $V^* \otimes V \to k$ is an A-intertwiner.

Note that S can be replaced by S^{-1} in (7.2.18), requiring us to distinguish between the right-dual V^* and the left-dual ${}^*V$ of A-modules. A natural condition on a Hopf algebra A which implies that $V^* \cong {}^*V$ as A-modules is that the square of the antipode is inner, i.e. if $S^2 = \mathsf{Ad}(u)$ for some $u \in A^\times$. This claim follows from the observation that $\varphi_u : V^* \to {}^*V$ defined by $f \mapsto u^{-1} \cdot f$ is an A-intertwiner.

7.2.10 Key Example 1: Group Algebras

We discuss some important families of examples of bialgebras which can be defined in terms of a group G or a Lie algebra $\mathfrak{g}$, both algebraic structures with a well-defined notion of representations.

Let G be a group. Restating the key observation of Example 7.1, the group algebra kG becomes a Hopf algebra if we set

$$\Delta(g) = g \otimes g, \qquad \varepsilon(g) = 1, \qquad S(g) = g^{-1} \tag{7.2.19}$$

for all $g \in G$ and extend linearly. The adjoint representation of kG on itself extends the conjugation action of G on itself given by $g \cdot h = ghg^{-1}$ for all $g, h \in G$. More generally, we can let G be a monoid and the same assignments for Δ and ε define a bialgebra structure on kG, which extends to a Hopf algebra structure if and only if G is a group.

We can "dualize" this example. Consider the commutative algebra k^G of functions $f : G \to k$ (with pointwise addition and multiplication). Note that $k^G \otimes k^G$ naturally embeds into $k^{G \times G}$. If G is finite then this is an algebra isomorphism, so we may identify these algebras, and in that case the following definitions make sense:

$$\Delta(f)(g,h) = f(gh), \qquad \varepsilon(f) = f(1_G), \qquad S(f)(g) = f(g^{-1}) \tag{7.2.20}$$

for all $f \in k^G$ and $g, h \in G$. This endows k^G with a Hopf algebra structure. There are also infinite groups and associated function algebras $F(G)$ where we can make the identification $F(G) \otimes F(G) \cong F(G \times G)$ so that the same construction endows $F(G)$ with a bialgebra structure, for instance:

- let G be an algebraic group over k and replace $F(G)$ by the algebra of regular functions;
- let G be a compact topological group, set $k = \mathbb{R}$ or $k = \mathbb{C}$ and replace $F(G)$ by the algebra generated by the matrix entries of all finite-dimensional representations of G.

7.2.11 Key Example 2: Universal Enveloping Algebras

Let $\mathfrak{g}$ be a Lie algebra. Consider its tensor algebra

$$T\mathfrak{g} := k \oplus \mathfrak{g} \oplus \mathfrak{g}^{\otimes 2} \oplus \cdots, \tag{7.2.21}$$

a free algebra with multiplication given by the tensor product. The *universal enveloping algebra* is the algebra $U\mathfrak{g} := T\mathfrak{g}/I$, where I is the two-sided ideal of $T\mathfrak{g}$ generated by all elements of the form $x \otimes y - y \otimes x - [x,y]$ for $x, y \in \mathfrak{g}$. Before we give the bialgebra structure, we highlight two key properties of universal enveloping algebras (see e.g. [5]).

1 The *universal property* of $U\mathfrak{g}$ and the canonical embedding $\iota : \mathfrak{g} \to U\mathfrak{g}$ is the following statement. For any Lie algebra morphism $\varphi : \mathfrak{g} \to A$ (where A is any algebra) there exists a unique algebra morphism $\widehat{\varphi} : U\mathfrak{g} \to A$ such that $\varphi = \widehat{\varphi} \circ \iota$. In particular we may take $A = \mathrm{End}(V)$ with V a vector space and obtain that Lie algebra representations of $\mathfrak{g}$ correspond 1-to-1 to algebra representations of $U\mathfrak{g}$.
2 The *Poincaré–Birkhoff–Witt theorem* states that given a totally ordered k-basis X of $\mathfrak{g}$, a k-basis of $U\mathfrak{g}$ is given by the set

$$\bigcup_{n \geq 0} \{\iota(x_1) \cdots \iota(x_n) \,|\, x_1, \ldots, x_n \in X, x_1 \leqslant \cdots \leqslant x_n\}. \tag{7.2.22}$$

The natural Hopf algebra structure on $U\mathfrak{g}$ is uniquely determined by

$$\Delta(x) = x \otimes 1 + 1 \otimes x, \qquad \varepsilon(x) = 0, \qquad S(x) = -x \tag{7.2.23}$$

for all $x \in \mathfrak{g}$. This map Δ corresponds precisely to the standard action of Lie algebras on tensor products of their representations: if $\pi_V : \mathfrak{g} \to \mathrm{End}(V)$ and $\pi_W : \mathfrak{g} \to \mathrm{End}(W)$ are Lie algebra representations then $\pi_{V\otimes W}$ defined by $\pi_{V\otimes W}(x) := \pi_V(x) \otimes \mathsf{id}_W + \mathsf{id}_V \otimes \pi_W(x)$ for all $x \in \mathfrak{g}$ is a representation of $\mathfrak{g}$ on $V \otimes W$. Also, in this case the adjoint representation of $U\mathfrak{g}$ on itself corresponds to the usual adjoint representation of $\mathfrak{g}$ given by $x \cdot y = [x,y]$ for all $x, y \in \mathfrak{g}$.

7.2.12 Grouplike and Skew-primitive Elements

Let A be a bialgebra. Inspired by the examples above we highlight two important types of elements of A. We call an element $a \in A$ *grouplike* if

$$\Delta(a) = a \otimes a, \qquad a \neq 0. \tag{7.2.24}$$

The set of grouplike elements is denoted by $\mathrm{Gr}(A)$. From (7.2.9) it follows that $\varepsilon(a) = 1$ for all $a \in A$ and hence $\mathrm{Gr}(A) \cap A^\times$ is a group. If A has an antipode S then $\mathrm{Gr}(A)$ is a subgroup of $A^\times$ and S acts on $\mathrm{Gr}(A)$ by inversion.

We call $a \in A$ *skew-primitive* if there exist $g, h \in \mathrm{Gr}(A)$ such that

$$\Delta(a) = a \otimes g + h \otimes a. \tag{7.2.25}$$

The set of such elements, denoted $\mathsf{Pri}_{g,h}(A)$, is a subspace of $\mathsf{Ker}(\varepsilon)$. If A has an antipode S then for all $g, h \in \mathrm{Gr}(A)$, S acts on $\mathsf{Pri}_{g,h}(A)$ as $a \mapsto -h^{-1}ag^{-1}$.

Elements of $\mathsf{Pri}_{1,1}(A)$ are called *primitive*. We have an inclusion of Lie algebras (with Lie bracket given by the commutator) $\mathsf{Pri}_{1,1}(A) \subseteq \mathsf{Ker}(\varepsilon) \subset A$.

Note that $U\mathfrak{g}$ is generated by primitive elements. More precisely, the embedding ι maps $\mathfrak{g}$ into $\mathsf{Pri}_{1,1}(U\mathfrak{g}) \subset U\mathfrak{g}$ and the Poincaré–Birkhoff–Witt theorem can be used to deduce that, if k is of characteristic 0, $\mathsf{Pri}_{1,1}(U\mathfrak{g}) = \mathfrak{g}$. A result due to Kostant [26] gives a wide-ranging converse to this observation: if k is algebraically closed of characteristic zero then any cocommutative Hopf algebra A such that $G(A) = \{1\}$ is isomorphic to the universal enveloping algebra of $\mathsf{Pri}_{1,1}(A)$. One can in fact remove the constraint on $G(A)$ and show that A is isomorphic to a particular type of semidirect product $U(\mathsf{Pri}_{1,1}(A)) \rtimes kG(A)$ known as *Hopf smash product*. The Hopf algebras we will highlight in these notes will be generated by grouplike and skew-primitive elements.

7.3 Quasitriangular Bialgebras and Braided Monoidal Categories

Note that the bialgebras discussed so far are all either cocommutative or commutative. We will see that quantum groups arise in a certain way as non-(co)commutative variations of them. Note especially that in algebraic geometry one studies algebraic groups via their (commutative) algebras of regular functions; it is natural to consider a "modified" or "deformed" algebraic group by making the algebra of regular functions non-commutative. This is the origin of the name "quantum group".

7.3.1 Generalizing Cocommutativity

To provide a context for this, we discuss a generalization of cocommutativity. It starts with the following idea. Let A be a bialgebra with coproduct Δ. Suppose there exists $\mathcal{R} \in (A \otimes A)^\times$ such that

$$\Delta^{\mathsf{op}} = \mathsf{Ad}(\mathcal{R}) \circ \Delta, \tag{7.3.1}$$

i.e. $\mathcal{R}\Delta(a) = \Delta^{\mathsf{op}}(a)\mathcal{R}$ for all $a \in A$. Note that (7.2.9) implies that

$$u^{(1)} := (\varepsilon \otimes \mathsf{id})(\mathcal{R}) \in A^\times, \qquad u^{(2)} := (\mathsf{id} \otimes \varepsilon)(\mathcal{R}) \in A^\times \tag{7.3.2}$$

are central elements of A and hence the element

$$\widetilde{\mathcal{R}} := (\varepsilon \otimes \varepsilon)(\mathcal{R}) \cdot (u^{(2)} \otimes u^{(1)})^{-1} \cdot \mathcal{R} \in (A \otimes A)^\times \tag{7.3.3}$$

satisfies both (7.3.1) and $(\varepsilon \otimes \mathrm{id})(\widetilde{\mathcal{R}}) = (\mathrm{id} \otimes \varepsilon)(\widetilde{\mathcal{R}}) = 1$. Hence without loss of generality we may assume that $\mathcal{R}$ satisfies

$$(\varepsilon \otimes \mathrm{id})(\mathcal{R}) = (\mathrm{id} \otimes \varepsilon)(\mathcal{R}) = 1. \tag{7.3.4}$$

Let us explore how (7.3.1) constrains the element R. Both A and A^{cop} are bialgebras and hence (7.2.8) is satisfied both as-is and with Δ replaced by Δ^{op}. Note that the coassociativity axiom (7.2.8) is a statement about linear maps from A to $A \otimes A \otimes A$. It is therefore convenient to identify the three canonical linear embeddings of $A \otimes A$ into $A \otimes A \otimes A$ and introduce notation for them. For all $a, b \in A$ we will write

$$(a \otimes b)_{12} = a \otimes b \otimes 1, \quad (a \otimes b)_{13} = a \otimes 1 \otimes b, \quad (a \otimes b)_{23} = 1 \otimes a \otimes b \tag{7.3.5}$$

and extend this notation linearly, so that e.g. $\mathcal{R}_{12} = \mathcal{R} \otimes 1$. Now note that $(\Delta^{\mathrm{op}} \otimes \mathrm{id}) \circ \Delta^{\mathrm{op}} = (\mathrm{id} \otimes \Delta^{\mathrm{op}}) \circ \Delta^{\mathrm{op}}$ together with (7.3.1) implies

$$\mathrm{Ad}\big(\mathcal{R}_{12}(\Delta \otimes \mathrm{id})(\mathcal{R})\big) \circ (\Delta \otimes \mathrm{id}) \circ \Delta = \mathrm{Ad}\big(\mathcal{R}_{23}(\mathrm{id} \otimes \Delta)(\mathcal{R})\big) \circ (\mathrm{id} \otimes \Delta) \circ \Delta.$$

Now using (7.2.8) for Δ itself we obtain that the element

$$X := \big(\mathcal{R}_{23} \cdot (\mathrm{id} \otimes \Delta)(\mathcal{R})\big)^{-1} \cdot \mathcal{R}_{12} \cdot (\Delta \otimes \mathrm{id})(\mathcal{R}) \in (A \otimes A \otimes A)^{\times} \tag{7.3.6}$$

centralizes the image of $\Delta^{(3)}$ in $A \otimes A \otimes A$. Owing to (7.3.4) we have

$$(\varepsilon \otimes \mathrm{id} \otimes \mathrm{id})(X) = (\mathrm{id} \otimes \varepsilon \otimes \mathrm{id})(X) = (\mathrm{id} \otimes \mathrm{id} \otimes \varepsilon)(X) = 1 \otimes 1. \tag{7.3.7}$$

The simplest possible X satisfying these constraints is $X = 1 \otimes 1 \otimes 1$. If we assume this, $\mathcal{R}$ satisfies the *cocycle condition*

$$\mathcal{R}_{12}(\Delta \otimes \mathrm{id})(\mathcal{R}) = \mathcal{R}_{23}(\mathrm{id} \otimes \Delta)(\mathcal{R}). \tag{7.3.8}$$

Without loss of generality we may write $(\Delta \otimes \mathrm{id})(\mathcal{R}) = \mathcal{R}_{13}\mathcal{R}_{23}Y$ for some $Y \in (A \otimes A \otimes A)^{\times}$. Taking counits again, we obtain

$$(\varepsilon \otimes \mathrm{id} \otimes \mathrm{id})(Y) = (\mathrm{id} \otimes \varepsilon \otimes \mathrm{id})(Y) = (\mathrm{id} \otimes \mathrm{id} \otimes \varepsilon)(Y) = 1, \tag{7.3.9}$$

and hence again we assume the simplest possible solution: $Y = 1 \otimes 1 \otimes 1$, so that

$$(\Delta \otimes \mathrm{id})(\mathcal{R}) = \mathcal{R}_{13}\mathcal{R}_{23}. \tag{7.3.10}$$

Combining (7.3.8) and (7.3.10) and applying (7.3.1), we obtain

$$\mathcal{R}_{12}\mathcal{R}_{13}\mathcal{R}_{23} = \mathcal{R}_{12}(\Delta \otimes \mathrm{id})(\mathcal{R}) = \mathcal{R}_{23}(\mathrm{id} \otimes \Delta)(\mathcal{R}) = (\mathrm{id} \otimes \Delta^{\mathrm{op}})(\mathcal{R})\mathcal{R}_{23},$$

so that $(\mathrm{id} \otimes \Delta^{\mathrm{op}})(\mathcal{R}) = \mathcal{R}_{12}\mathcal{R}_{13}$. Left-multiplying by $\mathrm{id} \otimes \sigma$, we obtain

$$(\mathrm{id} \otimes \Delta)(\mathcal{R}) = \mathcal{R}_{13}\mathcal{R}_{12}. \tag{7.3.11}$$

7.3.2 Quasitriangularity: Definition and Basic Properties

The above analysis motivates the following generalization of cocommutativity based on (7.3.1), originally due to Drinfeld [12].

Definition 7.6 Let A be a bialgebra and $\mathcal{R} \in (A \otimes A)^\times$. The pair $(A, \mathcal{R})$ is called *quasitriangular* and $\mathcal{R}$ a *(universal) R-matrix* for A if (7.3.1) and (7.3.10–7.3.11) hold.

If $(A, \mathcal{R})$ and $(B, \mathcal{S})$ are quasitriangular bialgebras then a bialgebra morphism $\psi : A \to B$ is called a *quasitriangular morphism* if $(\psi \otimes \psi)(\mathcal{R}) = \mathcal{S}$.

Lemma 7.7 *Let* $(A, \mathcal{R})$ *be a quasitriangular bialgebra.*

1 The (universal) Yang–Baxter equation *is satisfied:*

$$\mathcal{R}_{12}\mathcal{R}_{13}\mathcal{R}_{23} = \mathcal{R}_{23}\mathcal{R}_{13}\mathcal{R}_{12} \qquad \in A \otimes A \otimes A. \tag{7.3.12}$$

2 The bialgebras $(A, \sigma(\mathcal{R})^{-1})$, $(A^{\mathrm{op}}, \sigma(\mathcal{R}))$ *and* $(A^{\mathrm{cop}}, \sigma(\mathcal{R}))$ *are quasitriangular.*
3 The counit condition (7.3.4) *is satisfied.*
4 If A is a Hopf algebra then

$$S^2 = \mathrm{Ad}(u), \quad \text{where } u = \big(m \circ (S \otimes \mathrm{id}) \circ \sigma\big)(\mathcal{R}) \in A^\times \tag{7.3.13}$$

$$(S \otimes \mathrm{id})(\mathcal{R}) = \mathcal{R}^{-1} = (\mathrm{id} \otimes S^{-1})(\mathcal{R}), \qquad (S \otimes S)(\mathcal{R}) = \mathcal{R}. \tag{7.3.14}$$

For the proofs see e.g. [7, Props. 4.2.3 and 4.2.7]. Here we reproduce the proof of (7.3.12), which relies on (7.3.1) and (7.3.10):

$$\begin{aligned} \mathcal{R}_{12}\mathcal{R}_{13}\mathcal{R}_{23} &= \mathcal{R}_{12}(\Delta \otimes \mathrm{id})(\mathcal{R}) &&= (\Delta^{\mathrm{op}} \otimes \mathrm{id})(\mathcal{R})\mathcal{R}_{12} \\ &= (\sigma \otimes \mathrm{id})(\mathcal{R}_{13}\mathcal{R}_{23})\mathcal{R}_{12} &&= \mathcal{R}_{23}\mathcal{R}_{13}\mathcal{R}_{12}. \end{aligned} \tag{7.3.15}$$

Note that (7.3.12) can also be deduced, in a very similar way, from (7.3.1) and the other coproduct formula (7.3.11). Indeed, it is natural for a given quasi-triangular bialgebra to possess a symmetry interchanging the two coproduct formulas. In the following lemma we identify such a symmetry.

Lemma 7.8 *Let A be a bialgebra with a bialgebra morphism $\omega : A \to A^{\mathrm{cop}}$. If $\mathcal{R} \in (A \otimes A)^\times$ is fixed by $\sigma \circ (\omega \otimes \omega)$ then conditions* (7.3.10) *and* (7.3.11) *are equivalent.*

Proof This follows from $\mathrm{id} \otimes \Delta = (\sigma \otimes \mathrm{id}) \circ (\mathrm{id} \otimes \sigma) \circ (\Delta \otimes \mathrm{id}) \circ \sigma$ and $(\omega \otimes \omega) \circ \sigma = \sigma \circ (\omega \otimes \omega)$. □

Remark Another condition on quasitriangular bialgebras guaranteeing the equivalence of (7.3.10) and (7.3.11) is $\sigma(\mathcal{R}) = \mathcal{R}^{-1}$; such quasitriangular bialgebras are called *triangular*.

7.3.3 Braided Monoidal Categories

The main point of having a quasitriangular structure on a bialgebra is that the category of (left) A-modules is not just a monoidal category, but that the two possible tensor products of A-modules V and W, namely $V \otimes W$ and $W \otimes V$, are naturally isomorphic as A-modules, thereby preserving a key property of symmetric monoidal categories. Moreover the category of A-modules carries a natural braided structure.

More precisely, let V and W be A-modules with corresponding representations $\pi_V : A \to \mathrm{End}(V)$, $\pi_W : A \to \mathrm{End}(W)$. Denote $R_{V,W} = (\pi_V \otimes \pi_W)(\mathcal{R})$ (the linear map on $V \otimes W$ corresponding to the action of $\mathcal{R}$). Recall the linear map $\sigma_{V,W} : V \otimes W \to W \otimes V$ and define

$$\check{R}_{V,W} := \sigma_{V,W} \circ R_{V,W} \in \mathrm{Hom}(V \otimes W, W \otimes V). \tag{7.3.16}$$

Note that, since $\mathcal{R}$ is invertible, $\check{R}_{V,W}$ is invertible.

Lemma 7.9 *The map $\check{R}_{V,W}$ intertwines the modules $V \otimes W$ and $W \otimes V$:*

$$\check{R}_{V,W}\pi_{V\otimes W}(a) = \pi_{W\otimes V}(a)\check{R}_{V,W} \qquad \text{for all } a \in A. \tag{7.3.17}$$

In particular, $V \otimes W$ and $W \otimes V$ are isomorphic as A-modules.

Proof The axiom (7.3.1) implies

$$R_{V,W}(\pi_V \otimes \pi_W)(\Delta(a)) = (\pi_V \otimes \pi_W)(\Delta^{\mathrm{op}}(a))R_{V,W} \tag{7.3.18}$$

for all $a \in A$. Left-multiplying by $\sigma_{V,W}$ we obtain (7.3.17). □

It is possible to represent the category $\mathsf{Rep}(A)$ using a diagrammatical calculus developed in [37]. Here A-intertwiners from $U_1 \otimes \cdots \otimes U_m$ to $V_1 \otimes \cdots \otimes V_n$ correspond to diagrams with m incoming arrows and n outgoing arrows, labelled by the corresponding modules. Furthermore taking tensor products corresponds to horizontal juxtaposition, and composition of intertwiners corresponds to vertical juxtaposition; we use the convention that composition is downward, which is also indicated by arrows. In particular, the intertwiners id_U and $\check{R}_{V,W}$ are represented by a single strand and a braiding:

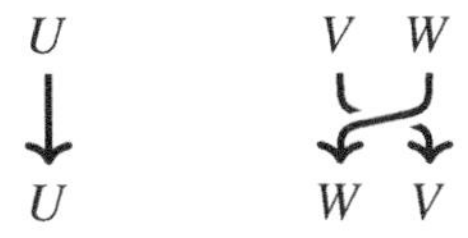

U V W
U W V
(7.3.19)

We also represent the action of $a \in A$ on $V \in \mathsf{Rep}(A)$ by a decoration, marked by a, of the strand labelled by V:

(7.3.20)

In particular, (7.3.17) corresponds to

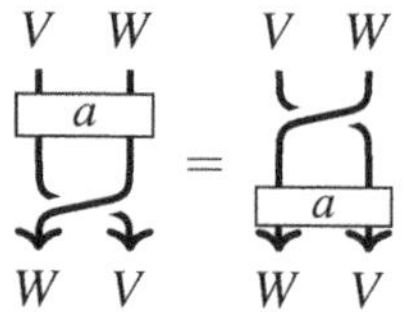

(7.3.21)

Also the coproduct axioms (7.3.10–7.3.11) correspond to natural conditions. Namely, let $U, V, W \in \mathsf{Rep}(A)$. Applying $\pi_U \otimes \pi_V \otimes \pi_W$ to (7.3.10) yields

$$R_{U \otimes V, W} = (R_{U,W})_{13}(R_{V,W})_{23}. \tag{7.3.22}$$

Left-multiplying by $(\sigma_{U,W})_{12}(\sigma_{V,W})_{23} = \sigma_{U \otimes V, W}$, we obtain

$$\check{R}_{U \otimes V, W} = (\check{R}_{U,W} \otimes \mathsf{id}_V)(\mathsf{id}_U \otimes \check{R}_{V,W}), \tag{7.3.23}$$

an equation in $\mathrm{Hom}(U \otimes V \otimes W, W \otimes U \otimes V)$. In the same way, from (7.3.11) we obtain

$$\check{R}_{U, V \otimes W} = (\mathsf{id}_V \otimes \check{R}_{U,W})(\check{R}_{U,V} \otimes \mathsf{id}_W), \tag{7.3.24}$$

an equation in $\mathrm{Hom}(U \otimes V \otimes W, V \otimes W \otimes U)$. In terms of the diagrammatical calculus, (7.3.23–7.3.24) correspond to the topological identities

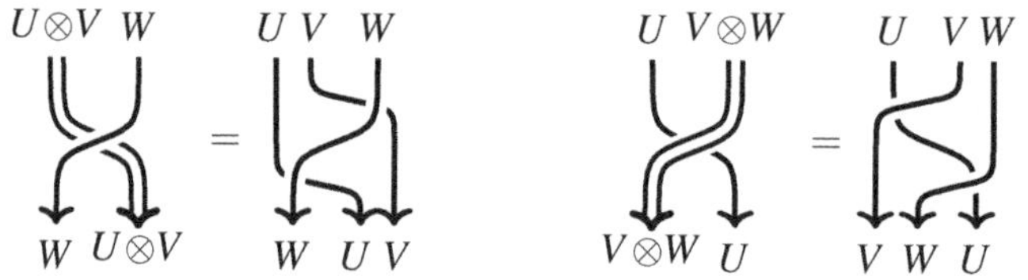

(7.3.25)

This means that the monoidal category $\mathsf{Rep}(A)$ is *braided*, see [20].

Remark If $(A, \mathcal{R})$ is triangular then, for all $V, W \in \mathsf{Rep}(A)$, we have $\check{R}_{W,V} = \check{R}_{V,W}^{-1}$ so that $\mathsf{Rep}(A)$ is a symmetric monoidal category.

To complete the description of the braided structure on $\mathsf{Rep}(A)$, suppose $U, V, W \in \mathsf{Rep}(A)$ and apply $\pi_U \otimes \pi_V \otimes \pi_W$ to (7.3.12). We obtain the (matrix) Yang–Baxter equation

$$\begin{aligned}(R_{U,V})_{12}(R_{U,W})_{13}(R_{V,W})_{23} \\ = (R_{V,W})_{23}(R_{U,W})_{13}(R_{U,V})_{12} \in \mathrm{End}(U \otimes V \otimes W),\end{aligned} \tag{7.3.26}$$

or, equivalently,

$$\begin{aligned}(\mathrm{id}_W \otimes \check{R}_{U,V})(\check{R}_{U,W} \otimes \mathrm{id}_V)(\mathrm{id}_U \otimes \check{R}_{V,W}) \\ = (\check{R}_{V,W} \otimes \mathrm{id}_U)(\mathrm{id}_V \otimes \check{R}_{U,W})(\check{R}_{U,V} \otimes \mathrm{id}_W),\end{aligned} \tag{7.3.27}$$

an equation in $\mathrm{Hom}(U \otimes V \otimes W, W \otimes V \otimes U)$. This corresponds diagrammatically to

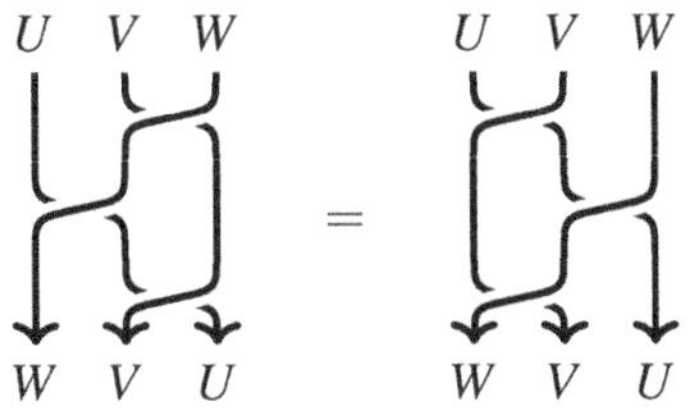

U V W   U V W

=   (7.3.28)

W V U   W V U

Finally, for arbitrary $L \in \mathbb{Z}_{>0}$ consider the braid group

$$\mathsf{Br}_L := \left\langle b_1, \ldots, b_{L-1} \,\middle|\, b_i b_{i+1} b_i = b_{i+1} b_i b_{i+1}, b_i b_j = b_j b_i \text{ if } |i-j| > 1 \right\rangle$$

(the fundamental group of the L-th unordered configuration space of the disk). For all $V \in \mathsf{Rep}(A)$, we obtain a representation of Br_L on $V^{\otimes L}$, given by

$$b_i \mapsto \mathrm{id}_V^{\otimes(i-1)} \otimes \check{R}_{V,V} \otimes \mathrm{id}_V^{\otimes(L-i-1)}. \tag{7.3.29}$$

7.3.4 Sweedler's Hopf Algebra – A Warm-up Exercise

We discuss a finite-dimensional quasitriangular Hopf algebra with a nontrivial R-matrix found in [41]. It does not depend on a parameter (and so is not a quantum group following our soft definition from the introduction). Consider the algebra A generated by symbols f and g subject to the relations

$$f^2 = 0, \qquad g^2 = 1, \qquad fg = -gf. \tag{7.3.30}$$

Note that $\{1, f, g, fg\}$ is a k-basis for A. Straightforward checks on generators show that the assignments

$$\begin{aligned} &\Delta(f) = f \otimes g + 1 \otimes f, \qquad &&\varepsilon(f) = 0, \qquad &&S(f) = gf, \\ &\Delta(g) = g \otimes g, &&\varepsilon(g) = 1, &&S(g) = g \end{aligned} \tag{7.3.31}$$

define a Hopf algebra structure on A with $g \in \mathrm{Gr}(A)$ and $f \in \mathsf{Pri}_{g,1}(A)$. The algebra A is the smallest noncommutative non-cocommutative Hopf algebra. Some of its properties foreshadow similar properties of Drinfeld–Jimbo quantum groups.

Now assume that $\mathrm{char}(k) \neq 2$. We will show that, for all $\beta \in k$, the following expression determines a quasitriangular structure on A

$$\mathcal{R}_\beta = \tfrac{1}{2}(1\otimes 1 + 1\otimes g + g\otimes 1 - g\otimes g)(1\otimes 1 + \beta f\otimes gf) \in A\otimes A. \quad (7.3.32)$$

See [7, Sec. 4.2F] for a somewhat different approach. Consider

$$\widetilde{\mathcal{R}}_\beta := \mathcal{R}_0^{-1}\mathcal{R}_\beta = 1\otimes 1 + \beta f\otimes gf \quad (7.3.33)$$

and note that $\widetilde{\mathcal{R}}_\beta\widetilde{\mathcal{R}}_{-\beta} = 1\otimes 1$. By writing $\mathcal{R}_0 = 1\otimes 1 - 2\frac{1-g}{2}\otimes\frac{1-g}{2}$ and noting that $\frac{1-g}{2}$ is an idempotent, we deduce that $\mathcal{R}_0$ is an involution. From the invertibility of $\mathcal{R}_0$ and $\widetilde{\mathcal{R}}_\beta$ we deduce that $\mathcal{R}_\beta$ is invertible. Moreover, by a direct computation we obtain

$$\mathcal{R}_0(f\otimes g) = (f\otimes 1)\mathcal{R}_0, \qquad \mathcal{R}_0(1\otimes f) = (g\otimes f)\mathcal{R}_0. \quad (7.3.34)$$

Lemma 7.10 *The involutive linear map $\omega : A \to A$ is uniquely determined by $\omega(1) = 1$, $\omega(g) = g$ and $\omega(f) = fg$ is a bialgebra morphism from A to A^{cop}. Furthermore $\mathcal{R}_\beta$ is fixed by $\sigma\circ(\omega\otimes\omega)$.*

Proof The first statement follows directly from (7.3.30–7.3.31). The second statement is a consequence of $\sigma(\mathcal{R}_0) = \mathcal{R}_0 = (\omega\otimes\omega)(\mathcal{R}_0)$ and $\sigma(\widetilde{\mathcal{R}}_\beta) = (\omega\otimes\omega)(\widetilde{\mathcal{R}}_\beta)$. □

Theorem 7.11 *For all $\beta \in k$, $(A, \mathcal{R}_\beta)$ is quasitriangular.*

Proof This is essentially a computation, but it is instructive to highlight some salient points. For the axiom (7.3.1), it suffices[5] to prove

$$\widetilde{\mathcal{R}}_\beta\Delta(a) = \Delta(a)\widetilde{\mathcal{R}}_\beta, \qquad \mathcal{R}_0\Delta(a) = \Delta^{\mathsf{op}}(a)\mathcal{R}_0 \qquad \text{for all } a\in A. \quad (7.3.35)$$

In turn, it suffices to verify these statements for $a \in \{f, g\}$, which is a straightforward consequence of (7.3.34). By Lemma 7.8 it now suffices to prove the axiom (7.3.10). A direct computation shows that

$$(\Delta\otimes\mathsf{id})(\mathcal{R}_0) = (\mathcal{R}_0)_{13}(\mathcal{R}_0)_{23}, \quad (7.3.36)$$

so that it remains to prove that

$$(\Delta\otimes\mathsf{id})(\widetilde{\mathcal{R}}_\beta) = (\mathcal{R}_0)_{23}(\widetilde{\mathcal{R}}_\beta)_{13}(\mathcal{R}_0)_{23}(\widetilde{\mathcal{R}}_\beta)_{23}. \quad (7.3.37)$$

[5] This is the main reason for introducing the factorization $\mathcal{R}_\beta = \mathcal{R}_0\widetilde{\mathcal{R}}_\beta$. We will approach the quasitriangularity of $U_q\mathfrak{g}$ in a similar way.

We expand with respect to powers of β. It now suffices to prove:

$$(\Delta \otimes \mathrm{id})(1 \otimes 1) = (\mathcal{R}_0)_{23}(\mathcal{R}_0)_{23}, \tag{7.3.38}$$

$$(\Delta \otimes \mathrm{id})(f \otimes gf) = (\mathcal{R}_0)_{23}(f \otimes 1 \otimes gf)(\mathcal{R}_0)_{23} + 1 \otimes f \otimes gf, \tag{7.3.39}$$

$$0 = (\mathcal{R}_0)_{23}(f \otimes 1 \otimes gf)(\mathcal{R}_0)_{23}(1 \otimes f \otimes gf). \tag{7.3.40}$$

Note that the first equation is trivial. The second equation follows by combining (7.3.34) with the coproduct formulas for f and g. Finally, the third equation follows by combining (7.3.34) with $f^2 = 0$. □

Remark The quasitriangular bialgebra $(A, \mathcal{R})$ is in fact triangular, since $\sigma(\mathcal{R}_\beta) = \mathcal{R}_\beta^{-1}$. This follows from the identity $\mathcal{R}_0 \sigma(\widetilde{\mathcal{R}}_\beta) = \widetilde{\mathcal{R}}_{-\beta}\mathcal{R}_0$, a direct consequence of (7.3.34).

To illustrate how nontrivial solutions of (7.3.26) arise in tensor products of modules over a quasitriangular bialgebra, consider the following two non-isomorphic indecomposable representations of A on a 2-dimensional vector space V. They are defined by:

$$\pi^{\pm}(f) = \begin{pmatrix} 0 & 0 \\ 1 & 0 \end{pmatrix}, \qquad \pi^{\pm}(g) = \begin{pmatrix} \pm 1 & 0 \\ 0 & \mp 1 \end{pmatrix} \tag{7.3.41}$$

with respect to a fixed ordered basis (v_1, v_2). With respect to the ordered basis $(v_1 \otimes v_1, v_1 \otimes v_2, v_2 \otimes v_1, v_2 \otimes v_2)$ of $V \otimes V$, we have

$$(\pi^{\pm} \otimes \pi^{\pm})(\mathcal{R}_\beta) = \begin{pmatrix} \pm 1 & 0 & 0 & 0 \\ 0 & 1 & 0 & 0 \\ 0 & 0 & 1 & 0 \\ \beta & 0 & 0 & \mp 1 \end{pmatrix}, \tag{7.3.42}$$

a nontrivial solution of the Yang–Baxter equation (7.3.26). Unfortunately, the representation theory of A is not very rich:

Theorem 7.12 *If k is algebraically closed, A has exactly four isomorphism classes of indecomposable modules. More precisely, up to isomorphism there are two 1-dimensional modules, given by $\pm\varepsilon$ and the two 2-dimensional modules defined by* (7.3.41).

Proof This follows from the fact that we may assume that g acts as a diagonalizable map on the module, which must therefore split up as a direct sum of ± 1-eigenspaces. For more details, see [7, 4.2F(g)]. □

On the other hand, semisimple Lie algebras $\mathfrak{g}$ have a very rich category of modules. Their enveloping algebras $U\mathfrak{g}$ are naturally cocommutative (and hence quasitriangular) Hopf algebras; their quantizations $U_q\mathfrak{g}$ inherit the category of modules and are, up to a technicality, quasitriangular Hopf algebras themselves.

7.4 Drinfeld–Jimbo Quantum Groups

In Section 7.4 we will study a deformation of the universal enveloping algebra of a Lie algebra associated to a (connected, complex, semisimple) Lie group, called quantized universal enveloping algebras or Drinfeld–Jimbo quantum groups. From now on we assume that k is algebraically closed and $\mathsf{char}(k)=0$ (in particular $\mathbb{Q} \subset k$). First we deal with the $\mathfrak{sl}_2$ case.

7.4.1 $\mathfrak{sl}_2$ and $U\mathfrak{sl}_2$

Let us first study the basic case of $\mathfrak{sl}_2$, the Lie algebra of traceless 2×2-matrices over k. It has a basis given by

$$e = \begin{pmatrix} 0 & 1 \\ 0 & 0 \end{pmatrix}, \qquad f = \begin{pmatrix} 0 & 0 \\ 1 & 0 \end{pmatrix}, \qquad h = \begin{pmatrix} 1 & 0 \\ 0 & -1 \end{pmatrix}. \tag{7.4.1}$$

In this case it is not hard to see that we have only the following Lie bracket relations between the basis elements:

$$[h,e] = 2e, \qquad [h,f] = -2f, \qquad [e,f] = h. \tag{7.4.2}$$

Hence, it follows immediately that $U\mathfrak{sl}_2$ is obtained from the free algebra over the symbols E,F,H by imposing the relations

$$HE - EH = 2E, \qquad HF - HF = -2F, \qquad EF - FE = H. \tag{7.4.3}$$

The canonical embedding $\iota : \mathfrak{sl}_2 \to U\mathfrak{sl}_2$ is given by $e \mapsto E$, $f \mapsto F$ and $h \mapsto H$. In the quantum deformed version we will "keep" E and F and "replace" H by a well-chosen linear combination of a grouplike element and its inverse.

7.4.2 Quantum $\mathfrak{sl}_2$

Let q be an indeterminate[6] and consider the algebra $U_q\mathfrak{sl}_2$ generated over $k(q)$ by symbols E, F, t and t^{-1} subject to the relations

[6] In a related formalism, q is not an indeterminate but a scalar. Typically it is imposed that q is nonzero and not a root of unity. The study of quantum groups for root-of-unity values of q is very interesting but outside the scope of this introductory course.

$$\begin{aligned} tE &= q^2Et, & tF &= q^{-2}Ft, \\ [E,F] &= \frac{t-t^{-1}}{q-q^{-1}}, & tt^{-1} &= t^{-1}t = 1. \end{aligned} \tag{7.4.4}$$

We will use the following convention for the additional structure maps:

$$\begin{aligned} \Delta(E) &= E\otimes 1 + t\otimes E, & \varepsilon(E) &= 0, & S(E) &= -t^{-1}E, \\ \Delta(F) &= F\otimes t^{-1} + 1\otimes F, & \varepsilon(F) &= 0, & S(F) &= -Ft, \\ \Delta(t^{\pm 1}) &= t^{\pm 1}\otimes t^{\pm 1}, & \varepsilon(t^{\pm 1}) &= 1, & S(t^{\pm 1}) &= t^{\mp 1}. \end{aligned} \tag{7.4.5}$$

It is a standard check to see that this endows $U_q\mathfrak{sl}_2$ with a Hopf algebra structure. Note that the subalgebras $\langle E,t,t^{-1}\rangle$ and $\langle F,t,t^{-1}\rangle$ are Hopf subalgebras, whereas $\langle E,t\rangle$ and $\langle F,t^{-1}\rangle$ are subbialgebras which are not Hopf subalgebras.

7.4.3 The Topological Quantum Group $U_{[[h]]}\mathfrak{sl}_2$

Morally, sending $q\to 1$ should recover the defining relations and Hopf algebra structure of $U\mathfrak{sl}_2$. By making the formal substitution $t = q^H$ this can indeed be done. For instance, in the right-hand side of the relation $[E,F] = \frac{t-t^{-1}}{q-q^{-1}}$ one may take the formal limit $q\to 1$ and immediately obtain H, as required. Writing $t = q^H$ and $q^2t = q^{H+2}$ as formal power series in $\log(q)$, the relation $tE = q^2Et$ is equivalent to

$$\sum_{r\geq 0}\frac{1}{r!}\log(q)^r H^r E = \sum_{r\geq 0}\frac{1}{r!}\log(q)^r E(H+2)^r. \tag{7.4.6}$$

Since q is an indeterminate, this should be true on the level of the coefficients, yielding the $U\mathfrak{sl}_2$-relations $H^rE = E(H+2)^r$. This suggests a connection between $U_q\mathfrak{sl}_2$ and $U\mathfrak{sl}_2[[\log(q)]]$.

To make this rigorous, choose a new indeterminate h. The h-adic topology on a vector space V over $k[[h]]$ is defined by stipulating that

1 $\{h^nV \mid n\in\mathbb{Z}_{\geq 0}\}$ is a base of the neighbourhoods of 0 in V,

2 translations in V are continuous.

It follows then that $k[[h]]$-linear maps are continuous. A *topological Hopf algebra* over $k[[h]]$ is an h-adic complete $k[[h]]$-module A equipped with $k[[h]]$-linear structure maps η, m, ε, Δ and S satisfying the Hopf algebra axioms discussed in Section 2, but with algebraic tensor products replaced by h-adic completions. We then may study the topological Hopf algebra $U_{[[h]]}\mathfrak{sl}_2$, defined as follows.

Namely, consider the free algebra $\mathcal{P} := k\langle E,F,H\rangle$ and the algebra of power series $\mathcal{P}[[h]]$. Consider the two-sided ideal I of $\mathcal{P}[[h]]$ generated by

$$[H,E]-2E, \qquad [H,F]+2F, \qquad [E,F]-\frac{\mathrm{e}^{hH}-\mathrm{e}^{-hH}}{\mathrm{e}^{h}-\mathrm{e}^{-h}} \tag{7.4.7}$$

and let I^{cl} be its closure in the h-adic topology. Then we can define $U_{[[h]]}\mathfrak{sl}_2 := \mathcal{P}[[h]]/I^{\mathrm{cl}}$. One then can deduce that $U_{[[h]]}\mathfrak{sl}_2 \cong (U\mathfrak{sl}_2)[[h]]$ as algebras over $k[[h]]$ (see [7, Cor. 6.5.4]).

7.4.4 Some Representations of $U_q\mathfrak{sl}_2$

It is easy to explicitly construct finite-dimensional representations of $U_q\mathfrak{sl}_2$, which simplify to $\mathfrak{sl}_2$-representations if we let q go to 1. We denote, for $m \in \mathbb{Z}$,

$$[m]_q = \frac{q^m - q^{-m}}{q-q^{-1}} \in k(q). \tag{7.4.8}$$

Since $q^{1-m}[m]_q$ is a power series in $q-1$ with constant term m, in the formal limit $q \to 1$, we recover the integer m. Consider, for $n \in \mathbb{Z}_{>0}$, the n-dimensional vector space

$$V^{(n)} = k(q)v_1^{(n)} \oplus \cdots \oplus k(q)v_n^{(n)} \cong k(q)^n. \tag{7.4.9}$$

Consider the assignments

$$\begin{aligned} \pi^{(n)}(E)(v_i^{(n)}) &= [i-1]_q v_{i-1}^{(n)}, \\ \pi^{(n)}(F)(v_i^{(n)}) &= [n-i]_q v_{i+1}^{(n)}, \\ \pi^{(n)}(t^{\pm 1})(v_i^{(n)}) &= q^{\pm(n-2i+1)} v_i^{(n)} \end{aligned} \tag{7.4.10}$$

for $i \in \{1,2,\ldots,n\}$, where we have set $v_i^{(n)} := 0$ if $i<1$ or $i>n$. By straightforward checks it follows that $\pi^{(n)}$ extends to a representation of $U_q(\mathfrak{sl}_2)$ on $V^{(n)}$. Note that $\pi^{(1)} = \varepsilon$. Also note that

$$\pi^{(n)}\left(\frac{t-t^{-1}}{q-q^{-1}}\right)(v_i^{(n)}) = [n-2i+1]_q v_i^{(n)}.$$

Hence, formally letting $q \to 1$, we obtain the representation of $\mathfrak{sl}_2$ on $kv_1^{(n)} \oplus \cdots \oplus kv_n^{(n)}$ given by

$$\begin{aligned} e \cdot v_i^{(n)} &= (i-1)v_{i-1}^{(n)}, \\ f \cdot v_i^{(n)} &= (n-i)v_{i+1}^{(n)}, \\ h \cdot v_i^{(n)} &= (n-2i+1)v_i^{(n)}. \end{aligned} \tag{7.4.11}$$

7.4.5 Chevalley–Serre Presentation of Finite-dimensional Semisimple Lie Algebras

Likewise we are interested in constructing quantum groups $U_q\mathfrak{g}$ for arbitrary finite-dimensional semisimple Lie algebras $\mathfrak{g}$. This is most conveniently done using the Chevalley–Serre presentation of $\mathfrak{g}$ in terms of its Cartan matrix. More precisely, let $C = (c_{ij})_{i,j\in I}$ be an arbitrary Cartan matrix, i.e. $c_{ii} = 2$, $c_{ij} \in \mathbb{Z}_{\geq 0}$, $c_{ij} = 0$ if and only if $c_{ji} = 0$ and finally all submatrices $(c_{ij})_{i,j\in J}$ with $J \subseteq I$ have positive determinant (we briefly discuss the Kac–Moody generalization in Section 7.4.7). There exist positive setwise-coprime integers d_i such that $d_ic_{ij} = d_jc_{ji}$ for all $i, j \in I$. Then each semisimple finite-dimensional Lie algebra arises as follows. Consider the Lie algebra $\mathfrak{g} = \mathfrak{g}(C)$ generated by the subalgebras

$$\mathfrak{sl}_{2,i} := \langle e_i, f_i, h_i \rangle \tag{7.4.12}$$

for all $i \in I$, subject to the $\mathfrak{sl}_2$-relations (7.4.2) with e, f, h replaced by e_i, f_i, h_i, respectively, and, for $i \neq j$, the cross relations

$$\begin{gathered}[h_i, h_j] = 0, \quad [h_i, e_j] = c_{ij}e_j, \quad [h_i, f_j] = -c_{ij}f_j, \quad [e_i, f_j] = 0, \\ [e_i, [e_i, \ldots, [e_i, e_j] \cdots]] = [f_i, [f_i, \ldots, [f_i, f_j] \cdots]] = 0,\end{gathered} \tag{7.4.13}$$

where there are $1 - c_{ij}$ nested Lie brackets in the last two relations (Serre relations). Consider the subalgebras

$$\mathfrak{n}^+ = \langle e_i \,|\, i \in I \rangle, \qquad \mathfrak{h} = \langle h_i \,|\, i \in I \rangle \qquad \mathfrak{n}^- = \langle f_i \,|\, i \in I \rangle. \tag{7.4.14}$$

Like any other Lie algebra, $\mathfrak{g}$ acts on itself by the adjoint action. It is particularly useful to study the adjoint action of $\mathfrak{h}$ on $\mathfrak{g}$, with respect to which we have the *triangular decomposition*

$$\mathfrak{g} = \mathfrak{n}^+ \oplus \mathfrak{h} \oplus \mathfrak{n}^- \qquad \text{as } \mathfrak{h}\text{-modules}. \tag{7.4.15}$$

More generally, we are interested in representations $V \in \mathsf{Rep}(\mathfrak{g})$ with a weight decomposition with respect to $\mathfrak{h}$:

$$V = \bigoplus_{\lambda \in \mathfrak{h}^*} V_\lambda, \qquad V_\lambda = \{v \in V \,|\, h \cdot v = \lambda(h)v \text{ for all } h \in \mathfrak{h}\}. \tag{7.4.16}$$

Any $\lambda \in \mathfrak{h}^*$ for which V_λ is nontrivial is called a ($\mathfrak{h}$-)*weight* for V. Consider the *weight lattice*

$$P = \{\lambda \in \mathfrak{h}^* \,|\, \lambda(h_i) \in \mathbb{Z} \text{ for all } i \in I\}. \tag{7.4.17}$$

A weight for the adjoint action is called a *root* and the *root system* Φ is the set of nonzero roots. Then $\mathfrak{n}^+ = \bigoplus_{\alpha\in\Phi^+} \mathfrak{g}_\alpha$ for some $\Phi^+ \subset \Phi$. For $j \in I$, define the simple root $\alpha_j \in \mathfrak{h}^*$ by $\alpha_j(h_i) = c_{ij}$ so that $\mathfrak{g}_{\alpha_j} = \mathbb{C}e_j$ and hence

$\alpha_j \in \Phi^+$. Now define a symmetric bilinear form $(\,,\,)$ on $\mathfrak{h}^*$ by $(\alpha_i, \alpha_j) = d_i c_{ij}$ for all $i, j \in I$; it satisfies $(\lambda, \alpha_i) = \lambda(d_i h_i)$ for all $i \in I$.

The category $\mathcal{O}$ is the full subcategory of $\mathsf{Rep}(\mathfrak{g}) \cong \mathsf{Rep}(U\mathfrak{g})$ whose objects are $\mathfrak{g}$-modules V with the decomposition (7.4.16) with all V_λ finite-dimensional, such that $U\mathfrak{n}^+$ acts locally finitely, i.e. for all $v \in V$ the $U\mathfrak{n}^+$-module generated by v is finite-dimensional. The category $\mathcal{O}$ is monoidal[7].

A subcategory called $\mathcal{O}_{\mathsf{int}}$ is obtained by additionally assuming that for each $i \in I$ the subalgebra $U\mathfrak{sl}_{2,i}$ acts locally finitely; by the triangular decomposition for this subalgebra, this is equivalent to $E_i = \iota(e_i)$ and $F_i = \iota(f_i)$ acting locally nilpotently: for all $v \in V$ there exists $m \in \mathbb{Z}_{\geq 0}$ such that $E_i^m \cdot v = F_i^m \cdot v = 0$. Then $\mathcal{O}_{\mathsf{int}}$ is a monoidal category and a semisimple category, with the simple objects given by irreducible highest-weight representations (more precisely, the associated highest weight λ is dominant and integral: $\lambda(h_i) \geq 0$ for all $i \in I$ and $\lambda \in P$). In fact, $\mathcal{O}_{\mathsf{int}}$ corresponds to the category of finite-dimensional $\mathfrak{g}$-representations, with, after a suitable choice of basis, each e_i acting as a strict upper triangular matrix and f_i as its transpose.

7.4.6 Drinfeld–Jimbo Quantum Groups

The definition of the quantum deformation of $U\mathfrak{g}$, independently due to Drinfeld [12] and Jimbo [18], is as follows. For a given Cartan matrix C, let q be an indeterminate and set $q_i := q^{d_i}$. The *Drinfeld–Jimbo quantum group* is the algebra $U_q\mathfrak{g}$ generated over $k(q)$ by subalgebras

$$U_q\mathfrak{sl}_{2,i} := \langle E_i, F_i, t_i, t_i^{-1} \rangle \tag{7.4.18}$$

for $i \in I$, subject to the $U_q\mathfrak{sl}_2$ relations (7.4.4) with $E, F, t^{\pm}, q$ replaced by $E_i, F_i, t_i^{\pm}, q_i$, respectively,[8] and, for $i \neq j$, the cross relations

$$\begin{gathered} t_i E_j = q_i^{c_{ij}} E_j t_i, \quad t_i F_j = q_i^{-c_{ij}} F_j t_i, \quad [t_i, t_j] = 0, \quad [E_i, F_j] = 0, \\ [E_i, [E_i, \ldots, [E_i, E_j]_{q_i^{c_{ij}}} \cdots]_{q_i^{-c_{ij}-2}}]_{q_i^{-c_{ij}}} = 0, \\ [F_i, [F_i, \ldots, [F_i, F_j]_{q_i^{-c_{ij}}} \cdots]_{q_i^{c_{ij}+2}}]_{q_i^{c_{ij}}} = 0, \end{gathered} \tag{7.4.19}$$

where we have used the notation $[x, y]_p := xy - pyx$ for the deformed commutator. Denote the root lattice by $Q = \sum_{i \in I} \mathbb{Z}\alpha_i$. For $\mu = \sum_{i \in I} m_i \alpha_i \in Q$ we denote $t_\mu = \prod_{i \in I} t_i^{m_i}$. The *(quantum) Cartan subalgebra* is the commutative subalgebra

$$U^0 = \langle t_i, t_i^{-1} \,|\, i \in I \rangle = \mathrm{Sp}_k\{t_\mu \,|\, \mu \in Q\}. \tag{7.4.20}$$

[7] Note that we are using the version of category O favoured by Kac, see [21]. The category $\mathcal{O}$ as originally defined in [4] is not a monoidal category.

[8] In the spirit of Section 7.4.3, we may think of t_i as $q_i^{H_i} = q^{d_i H_i}$.

Remark To be precise, we have given the so-called *adjoint form* of $U_q\mathfrak{g}$, where the Cartan subalgebra is defined in terms of the root lattice $\mathbb{Z}\Phi \subseteq P$. More generally, we may take any sublattice Λ of the weight lattice P, yielding a larger Cartan subalgebra generated by t_λ for $\lambda \in \Lambda$ satisfying e.g. $t_\lambda E_j = q^{(\lambda,\alpha_j)} E_j t_\lambda$. The *simply connected form* of $U_q\mathfrak{g}$ is obtained when we choose $\Lambda = P$. In the case $\mathfrak{g} = \mathfrak{sl}_2$, $P = \frac{1}{2}Q$ and the simply connected and adjoint forms are the only relevant forms of $U_q\mathfrak{g}$, with the simply connected form obtained from the adjoint form by adjoining square roots of the generators t_i and t_i^{-1}.

By a straightforward check on generators (see e.g. [17, Lem. 4.8]), one has the following result.

Lemma 7.13 *$U_q\mathfrak{g}$ is a (non-cocommutative) Hopf algebra with the additional structure maps given by*

$$\begin{aligned} \Delta(E_i) &= E_i \otimes 1 + t_i \otimes E_i, & \varepsilon(E_i) &= 0, & S(E_i) &= -t_i^{-1}E_i, \\ \Delta(F_i) &= F_i \otimes t_i^{-1} + 1 \otimes F_i, & \varepsilon(F_i) &= 0, & S(F_i) &= -F_i t_i, \\ \Delta(t_i^{\pm 1}) &= t_i^{\pm 1} \otimes t_i^{\pm 1}, & \varepsilon(t_i^{\pm 1}) &= 1, & S(t_i^{\pm 1}) &= t_i^{\mp 1}. \end{aligned} \tag{7.4.21}$$

For later convenience we record the explicit formulas for S^{-1}:

$$S^{-1}(E_i) = -E_i t_i^{-1}, \quad S^{-1}(F_i) = -t_i F_i, \quad S^{-1}(t_i^{\pm 1}) = t_i^{\mp 1}. \tag{7.4.22}$$

The triangular decomposition induced on $U\mathfrak{g}$ by the multiplication map, namely $U\mathfrak{g} \cong U\mathfrak{n}^+ \otimes U\mathfrak{h} \otimes U\mathfrak{n}^-$, lifts directly to

$$U_q\mathfrak{g} \cong U^+ \otimes U^0 \otimes U^-, \tag{7.4.23}$$

where we have introduced the subalgebras

$$U^+ = \langle E_i \,|\, i \in I \rangle, \qquad U^- = \langle F_i \,|\, i \in I \rangle. \tag{7.4.24}$$

For $V \in \mathsf{Rep}(U_q\mathfrak{g})$ and $\lambda \in P$, denote the (quantum) weight space

$$V_\lambda = \{ v \in V \,|\, t_i \cdot v = q_i^{\lambda(h_i)} v = q^{(\alpha_i,\lambda)} v \text{ for all } i \in I \}. \tag{7.4.25}$$

In particular, as part of the adjoint action of $U_q\mathfrak{g}$ on itself, the t_i act by conjugation, and we have the root space decompositions

$$U^\pm = \bigoplus_{\lambda \in Q^+} U^\pm_{\pm\lambda}, \qquad \text{where } Q^+ := \sum_{i \in I} \mathbb{Z}_{\geq 0}\alpha_i. \tag{7.4.26}$$

The category $\mathcal{O}_q$ is defined as the full subcategory of $\mathsf{Rep}(U_q\mathfrak{g})$ whose objects are modules V such that

$$V = \bigoplus_{\lambda \in P} V_\lambda \tag{7.4.27}$$

with all V_λ finite-dimensional and such that U^+ acts locally finitely. As before, $\mathcal{O}_q$ is a monoidal category. Note that the E_i-action and F_i-action on $V \in \mathcal{O}_q$ satisfy

$$E_i(V_\lambda) \subseteq V_{\lambda+\alpha_i}, \qquad F_i(V_\lambda) \subseteq V_{\lambda-\alpha_i}. \tag{7.4.28}$$

The subcategory $\mathcal{O}_{q,\mathsf{int}}$ is obtained by additionally assuming that each subalgebra $U_q\mathfrak{sl}_{2,i}$ acts locally finitely. Then $\mathcal{O}_{q,\mathsf{int}}$ is the category of finite-dimensional representations such that each t_i acts diagonalizably with integer powers of q_i as eigenvalues (so-called type-$\mathbf{1}$ representations). As in the $(q \to 1)$-limit, $\mathcal{O}_{q,\mathsf{int}}$ is a monoidal category and a semisimple category, whose simple objects are irreducible highest-weight representations with dominant integral highest weight, see e.g. [30, Cor. 6.2.3] or [7, Sec. 10.1].

In the case $\mathfrak{g} = \mathfrak{sl}_2$, the weight lattice is $P = \frac{\mathbb{Z}}{2}\alpha$, where α is the unique simple root and the representation $\pi^{(n)}$ defined in (7.4.10) defines a simple object in $\mathcal{O}_{q,\mathsf{int}}$. It is an irreducible highest-weight representation with highest weight vector $v_1^{(n)}$ and highest weight $\frac{n-1}{2}\alpha$.

7.4.7 Kac–Moody Generalization

The definition of the Drinfeld–Jimbo quantum group can straightforwardly be extended to the case where $C = (c_{ij})_{i,j\in I}$ is a symmetrizable generalized Cartan matrix, thereby quantum-deforming universal enveloping algebras of Kac–Moody Lie algebras [21]. This means we require $c_{ii} = 2$, $c_{ij} \in \mathbb{Z}_{\geq 0}$, $c_{ij} = 0$ if and only if $c_{ji} = 0$ and the existence of a set of positive setwise-coprime integers d_i such that $d_i c_{ij} = d_j c_{ji}$ for all $i, j \in I$. As in the classical $(q \to 1)$ case, the Cartan subalgebra is larger: U^0 is defined in terms of a lattice which as a free abelian group has rank $|I| + \mathsf{cork}(C)$. The category $\mathcal{O}_q$ and the subcategory $\mathcal{O}_{q,\mathsf{int}}$ can be defined as above and are still monoidal categories. Moreover $\mathcal{O}_{q,\mathsf{int}}$ is still semisimple with simple objects given by irreducible highest-weight representations with dominant integral highest weight. However, neither category contains nontrivial finite-dimensional representations.

We say that C is of *affine type* if $\det(C) = 0$ and all *proper* submatrices $(c_{ij})_{i,j\in J}$ with $J \subset I$ have positive determinant, see e.g. [6, 15]. If C is of affine type, then $\mathfrak{g}' := \langle e_i, f_i, h_i \,|\, i \in I\rangle$ and similarly $U_q\mathfrak{g}' := \langle E_i, F_i, t_i^{\pm 1} \,|\, i \in I\rangle$ (but not $\mathfrak{g}$ and $U_q\mathfrak{g}$ themselves) have finite-dimensional representations that arise from the identification of $\mathfrak{g}'$ as a central extension of a loop algebra $\mathfrak{g}_0 \otimes k[z, z^{-1}]$ of a finite-dimensional simple Lie algebra $\mathfrak{g}_0$ (or a twisted loop algebra), see e.g. [6, 15] for details. These affine quantum groups are the most relevant in quantum integrability.

7.5 Quasitriangularity for $U_q\mathfrak{g}$

Let $\mathfrak{g}$ be a finite-dimensional semisimple Lie algebra over k. We will construct the universal R-matrix for $U_q\mathfrak{g}$, roughly following the approach from [17, Ch. 6 and 7] which is based on the approaches by [30, Ch. 4] and [42]. A closely related construction is that via the quantum double construction due to Drinfeld, see [13] and cf. [7, Sec. 4.2D]. We complement the fairly technical arguments by explicit formulas for the special case $\mathfrak{g} = \mathfrak{sl}_2$.

From now on we work over the larger field $k(q^{1/d})$ for a suitable positive integer d, since we want to allow linear maps acting on objects in $\mathcal{O}_{q,\mathrm{int}}$ by multiplication by such scalars in certain weight spaces. Let us set the stage.

7.5.1 The Bar Involution

The *bar involution* is an involutive algebra automorphism of $U_q\mathfrak{g}$ denoted by $\overline{\cdot}$ which acts nontrivially on the base field $k(q^{1/d})$: it sends $q^{1/d}$ to $q^{-1/d}$. On the generators it is defined as follows:

$$\overline{E_i} = E_i, \qquad \overline{F_i} = F_i, \qquad \overline{t_i^{\pm 1}} = t_i^{\mp 1}. \tag{7.5.1}$$

It is straightforward to check that these assignments preserve the defining relations of $U_q\mathfrak{g}$, as required.

We will give a construction of the universal R-matrix by considering, in addition to Δ and Δ^{op}, a third coproduct $\overline{\Delta} := (\overline{\cdot} \otimes \overline{\cdot}) \circ \Delta \circ \overline{\cdot}$. Explicitly, we have

$$\begin{gathered}\overline{\Delta}(E_i) = E_i \otimes 1 + t_i^{-1} \otimes E_i, \qquad \overline{\Delta}(F_i) = F_i \otimes t_i + 1 \otimes F_i, \\ \overline{\Delta}(t_i^{\pm 1}) = t_i^{\pm 1} \otimes t_i^{\pm 1}.\end{gathered} \tag{7.5.2}$$

Our plan is to construct an invertible element $\mathcal{R}$ that intertwines Δ with Δ^{op} as follows:

$$\mathcal{R}\Delta(u) = \Delta^{\mathrm{op}}(u)\mathcal{R} \qquad \text{for all } u \in U_q\mathfrak{g}. \tag{7.5.3}$$

It turns out that $\overline{\Delta}$ is convenient in an intermediate stage of the proof of this. Namely, we will establish (7.5.3) by constructing two elements $\widetilde{\mathcal{R}}$ and κ that intertwine Δ with $\overline{\Delta}$, and $\overline{\Delta}$ with Δ^{op}, respectively:

$$\widetilde{\mathcal{R}}\Delta(x) = \overline{\Delta}(x)\widetilde{\mathcal{R}}, \qquad \kappa\overline{\Delta}(x) = \Delta^{\mathrm{op}}(x)\kappa \qquad \text{for all } x \in U_q\mathfrak{g}. \tag{7.5.4}$$

From these two equations (7.5.3) readily follows if we set $\mathcal{R} = \kappa\widetilde{\mathcal{R}}$. Compare this with the proof of Theorem 7.11 for Sweedler's Hopf algebra.

7.5.2 The Chevalley Involution

Define the Chevalley involution of $U_q\mathfrak{g}$ on its generators as follows:

$$\omega(E_i) = -F_i, \qquad \omega(F_i) = -E_i, \qquad \omega(t_i^{\pm 1}) = t_i^{\mp 1}. \tag{7.5.5}$$

It corresponds to the matrix Lie algebra automorphism $x \mapsto -x^{\mathrm{t}}$ in finite-dimensional representations of $\mathfrak{g}$. By straightforward checks we obtain the following result:

Lemma 7.14 *The map ω is a bialgebra morphism from $U_q\mathfrak{g}$ to $U_q\mathfrak{g}^{\mathrm{cop}}$.*

7.5.3 The Completions $\widehat{U}$ and $\widehat{U}^{(2)}$

In order to construct the universal R-matrix $\mathcal{R}$ for $U_q\mathfrak{g}$, we will consider an algebra properly containing $U_q\mathfrak{g} \otimes U_q\mathfrak{g}$. The fact that $\mathcal{R}$ does not lie in $U_q\mathfrak{g} \otimes U_q\mathfrak{g}$ is the only obstacle for $U_q\mathfrak{g}$ being quasitriangular, so we say that $U_q\mathfrak{g}$ is quasitriangular "up to completion". It means that $\mathcal{R}$ has a well-defined action on a proper subcategory of $\mathsf{Rep}(U_q\mathfrak{g})$, namely $\mathcal{O}_{q,\mathsf{int}}$.

We discuss now one possible definition of the completion, closely following [2, Sec. 3.1], but also see [e.g. 38, Sec. 1.3]. Since $\mathcal{O}_{q,\mathsf{int}}$ is a subcategory of $\mathsf{Rep}(U_q\mathfrak{g})$, we have a forgetful functor $\mathsf{For} : \mathcal{O}_{q,\mathsf{int}} \to \mathsf{Vect}$, which is a monoidal functor (it preserves tensor products). Consider now the algebra $\widehat{U}$ of all natural transformations from For to itself. A *natural transformation* of For is a tuple (φ_V), where V runs through $\mathcal{O}_{q,\mathsf{int}}$, consisting of linear maps $\varphi_V : \mathsf{For}(V) \to \mathsf{For}(V)$ such that the following diagram in Vect commutes:

$$\begin{array}{ccc} \mathsf{For}(V) & \xrightarrow{\varphi_V} & \mathsf{For}(V) \\ {\scriptstyle \mathsf{For}(f)}\downarrow & & \downarrow{\scriptstyle \mathsf{For}(f)} \\ \mathsf{For}(W) & \xrightarrow[\varphi_W]{} & \mathsf{For}(W) \end{array} \tag{7.5.6}$$

for all $V, W \in \mathcal{O}_{q,\mathsf{int}}$ and for all $U_q\mathfrak{g}$-intertwiners $f : V \to W$. Note that $\widehat{U}$ naturally has the structure of an algebra over $k(q^{1/d})$ since we can add, scalar-multiply and compose such tuples entrywise. Furthermore, the $U_q\mathfrak{g}$-action on objects of $\mathcal{O}_{q,\mathsf{int}}$ produces an algebra morphism $U_q\mathfrak{g} \to \widehat{U}$. Indeed, compare the definition of natural transformation with the definition of intertwiner (defined by means of another commuting diagram of linear maps (7.2.5)). This morphism is injective, see [Lu94, Prop. 3.5.4], and henceforth we will view $U_q\mathfrak{g}$ as a subalgebra of $\widehat{U}$.

We also have a functor $\mathsf{For}^{(2)} : \mathcal{O}_{q,\mathsf{int}} \times \mathcal{O}_{q,\mathsf{int}} \to \mathsf{Vect}$, sending pairs of modules (V, W) to $\mathsf{For}(V \otimes W)$ and pairs of intertwiners (f, g) to $\mathsf{For}(f \otimes g)$. We define $\widehat{U}^{(2)} = \mathsf{End}(\mathsf{For}^{(2)})$, which is an algebra for the same reasons as $\widehat{U}$, and we may view $\widehat{U} \otimes \widehat{U} \subset \widehat{U}^{(2)}$ via $(\varphi_V)_V \otimes (\psi_W)_W \mapsto (\varphi_V \otimes \psi_W)_{(V,W)}$. Analogously we can define a completion $\widehat{U}^{(n)}$ for any $n \in \mathbb{Z}_{\geq 0}$ with natural algebra embeddings $\widehat{U}^{(m)} \otimes \widehat{U}^{(n)} \to \widehat{U}^{(m+n)}$.

Any $\varphi \in \widehat{U}$ can be restricted to $\mathsf{For}(V \otimes W)$ for all $V, W \in \mathcal{O}_{q,\mathsf{int}}$; since restriction is compatible with composition and linearity of natural transformations, we obtain an algebra morphism

$$\Delta : \widehat{U} \to \widehat{U}^{(2)}, \qquad (\varphi_V) \mapsto (\varphi_{V \otimes W}). \tag{7.5.7}$$

It restricts to the usual coproduct of the embedded subalgebra $U_q\mathfrak{g} \subset \widehat{U}$, motivating the notation. The algebra maps $\Delta \otimes \mathsf{id}, \mathsf{id} \otimes \Delta$ from $\widehat{U} \otimes \widehat{U}$ to $\widehat{U}^{(3)}$ extend to algebra maps from $\widehat{U}^{(2)}$ to $\widehat{U}^{(3)}$.

7.5.4 The Element κ

For $V, W \in \mathcal{O}_{q,\mathsf{int}}$ a linear map $\kappa_{V,W} \in \mathrm{End}(V \otimes W)$ is uniquely determined by the condition

$$\kappa_{V,W}(v \otimes w) = q^{(\mu,\nu)} v \otimes w \text{ for all } \mu, \nu \in P, v \in V_\mu, w \in W_\nu. \tag{7.5.8}$$

The tuple $\kappa := (\kappa_{V,W})$ lies in $\widehat{U}^{(2)}$ (but not in $\widehat{U} \otimes \widehat{U}$).

Lemma 7.15 *The map* $\mathsf{Ad}(\kappa)$ *preserves* $U_q\mathfrak{g} \otimes U_q\mathfrak{g}$; *more precisely*

$$\begin{aligned} \mathsf{Ad}(\kappa)(E_i \otimes 1) &= E_i \otimes t_i, & \mathsf{Ad}(\kappa)(1 \otimes E_i) &= t_i \otimes E_i, \\ \mathsf{Ad}(\kappa)(F_i \otimes 1) &= F_i \otimes t_i^{-1}, & \mathsf{Ad}(\kappa)(1 \otimes F_i) &= t_i^{-1} \otimes F_i, \\ \mathsf{Ad}(\kappa)(t_i^{\pm 1} \otimes 1) &= t_i^{\pm 1} \otimes 1, & \mathsf{Ad}(\kappa)(1 \otimes t_i^{\pm 1}) &= 1 \otimes t_i^{\pm 1}. \end{aligned} \tag{7.5.9}$$

Proof Note that $U_q\mathfrak{g} \otimes U_q\mathfrak{g}$ is generated by $E_i \otimes 1$, $F_i \otimes 1$, $t_i^{\pm 1} \otimes 1$, $1 \otimes E_i$, $1 \otimes F_i$ and $1 \otimes t_i^{\pm 1}$. Let $V, W \in \mathcal{O}_{q,\mathsf{int}}$ be arbitrary and let $\mu, \nu \in P$. Owing to (7.4.28), we have

$$\begin{aligned} \mathsf{Ad}(\kappa)(E_i \otimes 1)|_{V_\mu \otimes W_\nu} &= \kappa(E_i \otimes 1)\kappa^{-1}|_{V_\mu \otimes W_\nu} \\ &= q^{-(\mu,\nu)}\kappa(E_i \otimes 1)|_{V_\mu \otimes W_\nu} \\ &= q^{(\mu+\alpha_i,\nu)-(\mu,\nu)}(E_i \otimes 1)|_{V_\mu \otimes W_\nu} \\ &= q^{(\alpha_i,\nu)}(E_i \otimes 1)|_{V_\mu \otimes W_\nu} \\ &= (E_i \otimes t_i)|_{V_\mu \otimes W_\nu}, \end{aligned}$$

as required. The computations for $F_i \otimes 1$, $1 \otimes E_i$ and $1 \otimes F_i$ are entirely similar. Finally, since $t_i^{\pm 1} \otimes 1$ and $1 \otimes t_i^{\pm 1}$ preserve the weight summands of objects in $\mathcal{O}_{q,\mathsf{int}}$, they are fixed by conjugation by κ. □

From (7.5.9) we obtain the desired intertwining property of κ:

Lemma 7.16 *We have* $\mathrm{Ad}(\kappa)\circ\overline{\Delta}=\Delta^{\mathrm{op}}$.

We continue our study of the element κ with the following result:

Lemma 7.17 *We have* $(\Delta\otimes\mathrm{id})(\kappa)=\kappa_{13}\kappa_{23}$.

Proof Let $U,V,W\in\mathcal{O}_{q,\mathrm{int}}$ and $\lambda,\mu,\nu\in P$. From the definition of the coproduct map $\widehat{U}\to\widehat{U}^{(2)}$ and the embedding $\widehat{U}^{(2)}\to\widehat{U}^{(3)}$, we obtain

$$\begin{aligned}(\Delta\otimes\mathrm{id})(\kappa)_{U,V,W}|_{U_\lambda\otimes V_\mu\otimes W_\nu} &= \kappa_{U\otimes V,W}|_{U_\lambda\otimes V_\mu\otimes W_\nu}\\ &= \text{multiplication by } q^{(\lambda+\mu,\nu)}|_{U_\lambda\otimes V_\mu\otimes W_\nu}\\ &= \text{multiplication by } q^{(\lambda,\nu)}q^{(\mu,\nu)}|_{U_\lambda\otimes V_\mu\otimes W_\nu}\\ &= (\kappa_{U,W})_{13}(\kappa_{V,W})_{23}|_{U_\lambda\otimes V_\mu\otimes W_\nu},\end{aligned}$$

as required. Here we have used that $U_\lambda\otimes V_\mu\subseteq(U\otimes V)_{\lambda+\mu}$, which follows directly from the definition of weight space. □

7.5.5 The Algebra $\widehat{U}^{-+}$

We also consider the algebra $\widehat{U}^{+}:=\prod_{\mu\in Q^+}U^+_\mu$. Let $(x_\mu)_{\mu\in Q^+}\in\widehat{U}^+$ be arbitrary. Note that for all $V\in\mathcal{O}_{q,\mathrm{int}}$ and all $v\in V$, there are only finitely many $\mu\in Q^+$ such that $x_\mu\cdot v$ is nonzero. Hence the expression $\sum_{\mu\in Q^+}x_\mu\cdot v$ is a well-defined element of V. It can be checked that $(x_\mu)_{\mu\in Q^+}$ defines an element of $\widehat{U}$, so that we may consider $\widehat{U}^+$ as a subalgebra of $\widehat{U}$. Considering the inclusion $U^+\subseteq\widehat{U}^+$, it is safe to write elements of $\widehat{U}^+$ additively as $x=\sum_{\mu\in Q^+}x_\mu$.

Owing again to the finiteness of the U^+-action, elements of the form $\sum_{\mu,\nu\in Q^+}y_\nu\otimes x_\mu$ with $x_\mu\in U^+_\mu$, $y_\nu\in U^-_{-\nu}$ have a well-defined action on $V\otimes W$ for all $V,W\in\mathcal{O}^+_q$, and lie in $\widehat{U}^{(2)}$. The subalgebra of $\widehat{U}^{\otimes 2}$ generated by such elements is denoted $U^-\widehat{\otimes}U^+$.

Remark We can define subalgebras $U^+\widehat{\otimes}U^+,U^+\widehat{\otimes}U^-\subset\widehat{U}^{(2)}$ in a similar way, but not $U^-\widehat{\otimes}U^-$: its putative elements do not have a well-defined action on objects in $\mathcal{O}_{q,\mathrm{int}}$.

We now claim that the desired element $\mathcal{R}$ can be chosen in the subalgebra

$$\widehat{U}^{-+}:=\langle U_q\mathfrak{g}\otimes U_q\mathfrak{g},U^-\widehat{\otimes}U^+,\kappa\rangle\subset\widehat{U}^{(2)}. \tag{7.5.10}$$

We can extend the composition $\omega^{(2)}:=\sigma\circ(\omega\otimes\omega)$ from an involutive algebra automorphism of $U_q\mathfrak{g}\otimes U_q\mathfrak{g}$ to an involutive algebra automorphism of $\widehat{U}^{-+}$; we simply stipulate that the extension fix κ and act on $U^-\widehat{\otimes}U^+$ as follows:

$$\sum_{\mu,\nu\in Q^+}c_{\mu,\nu}y_\nu\otimes x_\mu\leftrightarrow\sum_{\mu,\nu\in Q^+}c_{\mu,\nu}\omega(x_\mu)\otimes\omega(y_\nu). \tag{7.5.11}$$

This is consistent with the relations (7.5.9) and the natural relations involving series, and hence defines an algebra automorphism of $\widehat{U}^{-+}$.

7.5.6 Bialgebra Pairings

We now start the construction of the desired element $\widetilde{\mathcal{R}} \in U^{-}\widehat{\otimes}U^{+}$. Suppose A, B are two bialgebras (with coproducts Δ_A, Δ_B and counits ε_A, ε_B, respectively). A *bialgebra pairing* between A and B (see [7, 4.1D]), is a k-linear map $\langle \cdot, \cdot \rangle : A \otimes B \to k$ with the properties

$$\begin{aligned} \langle \Delta_A(a), b \otimes b' \rangle &= \langle a, bb' \rangle, & \langle a \otimes a', \Delta_B(b) \rangle &= \langle aa', b \rangle, \\ \varepsilon_A(a) &= \langle a, 1 \rangle, & \varepsilon_B(b) &= \langle 1, b \rangle \end{aligned} \tag{7.5.12}$$

for all $a, a' \in A$ and $b, b' \in B$. Here we denote by the same symbol the canonical extension of the pairing of A and B to a k-linear map: $(A \otimes A) \otimes (B \otimes B) \to k$ defined by $\langle a \otimes a', b \otimes b' \rangle = \langle a, b \rangle \langle a', b' \rangle$ for all $a, a' \in A$ and $b, b' \in B$. In particular, $\langle a \otimes a', b \otimes b' \rangle = \langle a' \otimes a, b' \otimes b \rangle$ and we automatically obtain a bialgebra pairing between A^{cop} and B^{op} and between A^{op} and B^{cop}.

The quantum analogues of the standard Borel subalgebras, viz.

$$U_q\mathfrak{b}^{+} = \langle E_i, t_i, t_i^{-1} \,|\, i \in I \rangle, \qquad U_q\mathfrak{b}^{-} = \langle F_i, t_i, t_i^{-1} \,|\, i \in I \rangle \tag{7.5.13}$$

are subbialgebras of $U_q\mathfrak{g}$ over $k(q^{1/d})$. Then the following assignments define a unique bialgebra pairing between $U_q\mathfrak{b}^{-,\mathrm{cop}}$ and $U_q\mathfrak{b}^{+}$:

$$\begin{aligned} \langle t_\lambda, t_\mu \rangle &= q^{-(\lambda,\mu)}, & \langle F_i, E_j \rangle &= \frac{\delta_{ij}}{q_i^{-1} - q_i}, \\ \langle t_\lambda, E_j \rangle &= 0, & \langle F_i, t_\mu \rangle &= 0 \end{aligned} \tag{7.5.14}$$

for all $i, j \in I$ and $\lambda, \mu \in Q$ (see e.g. [17, 6.12]), where it is presented as a pairing between $U_q\mathfrak{b}^{-}$ and $U_q\mathfrak{b}^{+,\mathrm{op}}$. This is a nondegenerate pairing (see [17, 6.21]); moreover its restriction to $U^{-}_{-\nu} \times U^{+}_{\mu}$ vanishes if $\mu \neq \nu$ and is nondegenerate if $\mu = \nu$: if for some $x \in U^{+}_{\mu}$ we have $\langle y, x \rangle = 0$ for all $y \in U^{-}_{-\mu}$ then $x = 0$ (and vice versa).

7.5.7 Skew Derivations

In order to construct the desired element $\widetilde{\mathcal{R}}$ and establish its key properties we introduce linear maps on U^{+} called *right and left skew derivations* due to Lusztig (see [30, Sec. 1.2 and 3.1]). More precisely, for each $i \in I$ there exist $D_i^{(r,\ell)} \in \mathrm{End}_{k(q^{1/d})}(U^{+})$ uniquely determined by stipulating that $D_i^{(r,\ell)}(E_j) = \delta_{ij}$ for all $j \in I$ and

$$
\begin{aligned}
D_i^{(r)}(xx') &= D_i^{(r)}(x)\mathsf{Ad}(t_i)(x') + xD_i^{(r)}(x'),\\
D_i^{(\ell)}(xx') &= D_i^{(\ell)}(x)x' + \mathsf{Ad}(t_i)(x)D_i^{(\ell)}(x')
\end{aligned}
\tag{7.5.15}
$$

and $x, x' \in U^+$. Clearly, the two maps $D_i^{(r,\ell)}$ send U_μ^+ to $U_{\mu-\alpha_i}^+$ for all $\mu \in Q^+$. By [17, 6.14], for all $\mu \in Q^+$, $x \in U_\mu^+$, we have

$$
\begin{aligned}
\Delta(x) - x \otimes 1 - \sum_{i \in I} D_i^{(r)}(x)t_i \otimes E_i &\in \sum_{\substack{\nu \in Q^+\\ \nu \neq \alpha_j, \nu \neq 0}} U_{\mu-\nu}^+ t_\nu \otimes U_\nu^+,\\
\Delta(x) - t_\mu \otimes x - \sum_{i \in I} E_i t_{\mu-\nu_i} \otimes D_i^{(\ell)}(x) &\in \sum_{\substack{\nu \in Q^+\\ \nu \neq \alpha_j, \nu \neq 0}} U_\nu^+ t_{\mu-\nu} \otimes U_{\mu-\nu}^+.
\end{aligned}
\tag{7.5.16}
$$

By [17, 6.15 (5)], for all $x \in U^+$, $y \in U^-$ and $i \in I$, we have

$$
\langle F_i y, x\rangle = \langle F_i, E_i\rangle \langle y, D_i^{(\ell)}(x)\rangle, \quad \langle yF_i, x\rangle = \langle F_i, E_i\rangle \langle y, D_i^{(r)}(x)\rangle. \tag{7.5.17}
$$

By [17, 6.17], for all $x \in U^+$ and $i \in I$, we have

$$
[x, F_i] = (q_i - q_i^{-1})^{-1}\left(D_i^{(r)}(x)t_i - t_i^{-1}D_i^{(\ell)}(x)\right). \tag{7.5.18}
$$

In fact, each of (7.5.16–7.5.18) can be used to define the linear maps $D_i^{(r,\ell)}$ uniquely. By [30, Lem. 1.2.15 (a)] we have

$$
\forall i \in I\, D_i^{(r)}(x) = 0 \quad \Leftrightarrow \quad \forall i \in I\, D_i^{(\ell)}(x) = 0 \quad \Leftrightarrow \quad x = 0 \tag{7.5.19}
$$

for all $x \in U^+$.

Note that ω restricts to a bialgebra morphism from $U_q\mathfrak{b}^+$ to $U_q\mathfrak{b}^{-,\mathrm{cop}}$ interchanging U_μ^+ and $U_{-\mu}^-$ for all $\mu \in Q^+$. One can similarly define skew derivations for U^- in the natural way, namely via the compositions $\omega \circ D_i^{(r,\ell)} \circ \omega$. As a consequence we have $\langle \omega(x), \omega(y)\rangle = \langle y, x\rangle$ for all $x \in U_q\mathfrak{b}^+$, $y \in U_q\mathfrak{b}^-$, see [17, 6.16].

7.5.8 The Element Θ

For arbitrary $\mu \in Q^+$, choose a basis $(x_{\mu,r})_r$ for the finite-dimensional $k(q^{1/d})$-vector space U_μ^+ and let $(y_{\mu,r})_r$ be the dual basis of $U_{-\mu}^-$ with respect to the bilinear pairing $\langle\, ,\rangle$. Consider the element defined by

$$
\Theta = \sum_{\mu \in Q^+} \Theta_\mu \in \widehat{U}^{(2)}, \quad \Theta_\mu = \sum_r y_{\mu,r} \otimes x_{\mu,r} \in U_{-\mu}^- \otimes U_\mu^+. \tag{7.5.20}
$$

In other words, Θ is the "canonical element" of the restriction of $\langle\, ,\rangle$ to $U^- \times U^+$. Note that $U_0^- = U_0^+ = k$ so that $\Theta_0 = 1 \otimes 1$.

Since Θ_μ is independent of the choice of basis for U_μ^+ and ω is invertible, each Θ_μ is fixed by $\omega^{(2)}$. Straightaway we obtain

Lemma 7.18 *The element Θ satisfies $\omega^{(2)}(\Theta) = \Theta$.*

We stay close to the approach in [17, Ch. 7] in establishing key properties of Θ.

Theorem 7.19 *The linear space*

$$\{X \in U^- \widehat{\otimes} U^+ \,|\, \mathsf{Ad}(X) \circ \overline{\Delta} = \Delta\} \tag{7.5.21}$$

is 1-dimensional and spanned by Θ.

Proof Suppose $X \in U^- \widehat{\otimes} U^+$ is such that $X\overline{\Delta}(u) = \Delta(u)X$ for all $u \in U_q\mathfrak{g}$. Since $\Delta(t_i^{\pm 1}) = \overline{\Delta}(t_i^{\pm 1}) = t_i^{\pm 1} \otimes t_i^{\pm 1}$ for all $i \in I$, it follows that $X = \sum_{\mu \in Q^+} X_\mu$ with $X_\mu \in U_{-\mu}^- \otimes U_\mu^+$. Assuming X is of this form, the condition $\mathsf{Ad}(X) \circ \overline{\Delta} = \Delta$ is equivalent to the following identities:

$$[X_\mu, E_i \otimes 1] = (t_i \otimes E_i)X_{\mu-\alpha_i} - X_{\mu-\alpha_i}(t_i^{-1} \otimes E_i), \tag{7.5.22}$$

$$[X_\mu, 1 \otimes F_i] = (F_i \otimes t_i^{-1})X_{\mu-\alpha_i} - X_{\mu-\alpha_i}(F_i \otimes t_i) \tag{7.5.23}$$

for all $i \in I$, where $X_\mu := 0$ if $\mu \notin Q^+$. By applying $\sigma \circ (\omega \otimes \omega)$ one sees that these identities are equivalent. By (7.5.18) and linear independence, (7.5.23) is equivalent to the system

$$\begin{aligned} (F_i \otimes 1)X_{\mu-\alpha_i} &= (q_i^{-1} - q_i)^{-1}(\mathsf{id} \otimes D_i^{(\ell)})(X_\mu), \\ X_{\mu-\alpha_i}(F_i \otimes 1) &= (q_i^{-1} - q_i)^{-1}(\mathsf{id} \otimes D_i^{(r)})(X_\mu) \end{aligned} \tag{7.5.24}$$

for all $i \in I$. It suffices to prove that the solution set of (7.5.24) are precisely the scalar multiples of Θ. The computation in [17, 7.1], which relies on (7.5.17), shows that $X = \Theta$ satisfies these conditions. To show uniqueness up to scalar multiples, we may follow the proof of [30, Thm. 4.1.2], which relies on (7.5.19). □

Theorem 7.20 *We have*

$$(\Delta \otimes \mathsf{id})(\Theta) = \Theta_{23}\mathsf{Ad}(\kappa_{23}^{-1})(\Theta_{13}). \tag{7.5.25}$$

Proof Considering (7.5.9), it suffices to prove

$$(\Delta \otimes \mathsf{id})(\Theta_\mu) = \sum_{\nu \in Q^+} (\Theta_{\mu-\nu})_{23}(1 \otimes t_\nu^{-1} \otimes 1)(\Theta_\nu)_{13}. \tag{7.5.26}$$

Fix $y \in U^-_{-\mu}$ with $\mu = \sum_{i\in I} m_i\alpha_i$ with $m_i \in \mathbb{Z}_{\geq 0}$. By induction with respect to the height of μ, viz. $\sum_{i\in I} m_i$, we obtain from the coproduct formula for F_i in (7.4.21) that, for all

$$\Delta(y) \in \sum_{\nu\in Q^+} U^-_{-\nu} \otimes U^-_{\nu-\mu} t_\nu^{-1} \tag{7.5.27}$$

and hence

$$\Delta(y) = \sum_{\nu\in Q^+} \sum_{r,s} c^\nu_{r,s} y_{\nu,r} \otimes y_{\mu-\nu,s} t_\nu^{-1} \tag{7.5.28}$$

for some $c^\nu_{r,s} \in k(q^{1/d})$. Since the basis $\{y_{\mu,r}\}_r$ is dual to $\{x_{\mu,r}\}_r$ with respect to $\langle,\rangle$, we have

$$c^\nu_{r,s} = \langle \Delta(y), x_{s,\nu} \otimes x_{r,\mu-\nu} \rangle = \langle \Delta^{\mathrm{op}}(y), x_{r,\mu-\nu} \otimes x_{s,\nu} \rangle = \langle y, x_{r,\mu-\nu} x_{s,\nu} \rangle.$$

Now (7.5.26) follows by recalling the definition of Θ_μ in terms of the basis elements $y_{\mu,r}$ and $x_{\mu,r}$. We refer to the proof of [17, Lem. 7.4] for the remaining computation. □

7.5.9 The Quasi R-matrix $\widetilde{\mathcal{R}}$ and the Universal R-matrix $\mathcal{R}$

We now simply define $\widetilde{\mathcal{R}} = \Theta^{-1}$. Immediately we obtain from Lemma 7.18, Theorem 7.19 and Theorem 7.20 the following result for $\widetilde{\mathcal{R}}$:

Theorem 7.21 *The element $\widetilde{\mathcal{R}}$ is fixed by $\omega^{(2)}$, has the intertwining property $\mathrm{Ad}(\widetilde{\mathcal{R}}) \circ \Delta = \overline{\Delta}$ and the coproduct formula $(\Delta \otimes \mathrm{id})(\widetilde{\mathcal{R}}) = \mathrm{Ad}(\kappa_{23}^{-1})(\widetilde{\mathcal{R}}_{13})\widetilde{\mathcal{R}}_{23}$.*

The desired element $\mathcal{R}$ is now given by

$$\mathcal{R} = \kappa\widetilde{\mathcal{R}} \in \widehat{U}^{-+}. \tag{7.5.29}$$

As a consequence of Theorem 7.21 and the various properties of κ from Section 7.5.4 we obtain that $\mathcal{R}$ satisfies the properties of a universal R-matrix:

Theorem 7.22 *The element $\mathcal{R}$ is fixed by $\omega^{(2)}$, has the intertwining property $\mathrm{Ad}(\mathcal{R}) \circ \Delta = \Delta^{\mathrm{op}}$ and the coproduct formula $(\Delta \otimes \mathrm{id})(\mathcal{R}) = \mathcal{R}_{13}\mathcal{R}_{23}$.*

Recall that the other coproduct formula $(\mathrm{id} \otimes \Delta)(\mathcal{R}) = \mathcal{R}_{13}\mathcal{R}_{12}$ follows from Lemma 7.8.

7.5.10 The R-matrix for $U_q\mathfrak{sl}_2$

For the quantum group $U_q\mathfrak{sl}_2$ it is possible to make the formula for $\widetilde{\mathcal{R}}$ rather explicit. It leads to the following result.

Theorem 7.23 *The subspace of elements $\widetilde{\mathcal{R}} \in U^{-}\widehat{\otimes}U^{+}$ satisfying*

$$\mathsf{Ad}(\widetilde{\mathcal{R}}) \circ \Delta = \overline{\Delta} \tag{7.5.30}$$

is 1-dimensional. The unique such element with $(\varepsilon \otimes \varepsilon)(\widetilde{\mathcal{R}}) = 1$ is

$$\widetilde{\mathcal{R}} = \sum_{r=0}^{\infty} c_r (F \otimes E)^r, \tag{7.5.31}$$

where for $r \in \mathbb{Z}_{\geq 0}$ we have

$$c_r := \frac{(q-q^{-1})^r}{[r]_q!} q^{r(r-1)/2}, \quad \text{with } [r]_q! := [r]_q[r-1]_q \cdots [2]_q[1]_q. \tag{7.5.32}$$

Additionally, it satisfies

$$(\Delta \otimes \mathsf{id})(\widetilde{\mathcal{R}}) = \mathsf{Ad}(\kappa_{23}^{-1})(\widetilde{\mathcal{R}}_{13})\widetilde{\mathcal{R}}_{23}. \tag{7.5.33}$$

It can be proven directly from the relations and the coproduct formulas for the generators of $U_q\mathfrak{g}$. We refer to [17, Ch. 3] for details, but if tempted the reader might want to take it on as a useful exercise. As a hint towards the solution, it is helpful to prove the following relation

$$[E, F^{r+1}] = [r+1]_q \frac{q^r t - q^{-r} t^{-1}}{q - q^{-1}} F^r, \tag{7.5.34}$$

and the following formula for the coproduct

$$\Delta(F^r) = \sum_{s=0}^{r} q^{s(s-r)} \binom{r}{s}_q F^{r-s} \otimes t^{s-r} F^s, \tag{7.5.35}$$

where for $r \in \mathbb{Z}_{\geq 0}$, $s \in \mathbb{Z}$, we have

$$\binom{r}{s}_q := \begin{cases} \frac{[r]_q!}{[s]_q![r-s]_q!} & \text{if } 0 \leq s \leq r, \\ 0 & \text{otherwise.} \end{cases} \tag{7.5.36}$$

Note, in the formal limit $q \to 1$, both κ and $\widetilde{\mathcal{R}}$, and hence also $\mathcal{R} = \kappa\widetilde{\mathcal{R}}$, go to $1 \otimes 1$.

A large range of matrix solutions to the Yang–Baxter equation (7.3.26) in representations of $U_q(\mathfrak{sl}_2)$ now arises naturally. Recall the n-dimensional representation $(\pi^{(n)}, V^{(n)})$ of $U_q(\mathfrak{sl}_2)$ defined in (7.4.10). By evaluating $(\pi^{(m)} \otimes \pi^{(n)})(\mathcal{R})$ for various m, n, we obtain linear maps on $V^{(m)} \otimes V^{(n)}$ which satisfy (7.3.26) in $V^{(l)} \otimes V^{(m)} \otimes V^{(n)}$ for various l, m, n.

To make this explicit as well, with respect to the basis $(v_1^{(2)}, v_2^{(2)})$, the 2-dimensional representation $\pi^{(2)}$ can be written as

$$\pi^{(2)}(E) = \begin{pmatrix} 0 & 1 \\ 0 & 0 \end{pmatrix}, \ \pi^{(2)}(F) = \begin{pmatrix} 0 & 0 \\ 1 & 0 \end{pmatrix}, \ \pi^{(2)}(t) = \begin{pmatrix} q & 0 \\ 0 & q^{-1} \end{pmatrix}. \tag{7.5.37}$$

With respect to the basis $(v_1^{(2)} \otimes v_1^{(2)}, v_1^{(2)} \otimes v_2^{(2)}, v_2^{(2)} \otimes v_1^{(2)}, v_2^{(2)} \otimes v_2^{(2)})$ of $V^{(2)} \otimes V^{(2)}$, we obtain

$$(\pi^{(2)} \otimes \pi^{(2)})(\kappa) = \begin{pmatrix} q^{1/2} & 0 & 0 & 0 \\ 0 & q^{-1/2} & 0 & 0 \\ 0 & 0 & q^{-1/2} & 0 \\ 0 & 0 & 0 & q^{1/2} \end{pmatrix},$$
$$(\pi^{(2)} \otimes \pi^{(2)})(\widetilde{\mathcal{R}}) = \begin{pmatrix} 1 & 0 & 0 & 0 \\ 0 & 1 & 0 & 0 \\ 0 & q-q^{-1} & 1 & 0 \\ 0 & 0 & 0 & 1 \end{pmatrix} \tag{7.5.38}$$

and hence the following nontrivial solution of the Yang–Baxter equation:

$$R := (\pi^{(2)} \otimes \pi^{(2)})(\mathcal{R}) = q^{-1/2} \begin{pmatrix} q & 0 & 0 & 0 \\ 0 & 1 & 0 & 0 \\ 0 & q-q^{-1} & 1 & 0 \\ 0 & 0 & 0 & q \end{pmatrix}. \tag{7.5.39}$$

7.5.11 The Dual Quantum Group $F_q(\mathsf{SL}(2))$

We mention here also the dual object $F_q(\mathsf{SL}(2))$, the quantized algebra of scalar-valued functions on $\mathsf{SL}(2)$. We refer to [7, Ch. 7] for a more in-depth discussion. The algebra $F_q(\mathsf{SL}(2))$ is generated over $k(q)$ by elements a, b, c, d subject to

$$\begin{gathered} ab = qba, \quad bd = qdb, \quad ac = qca, \quad cd = qdc, \quad bc = cb, \\ ad - qbc = 1 = da - q^{-1}cb. \end{gathered} \tag{7.5.40}$$

The Hopf algebra structure on $F_q(\mathsf{SL}(2))$ is as follows:

$$\begin{aligned} \Delta(a) &= a \otimes a + b \otimes c, & \varepsilon(a) &= 1, & S(a) &= d, \\ \Delta(b) &= a \otimes b + b \otimes d, & \varepsilon(b) &= 0, & S(b) &= -q^{-1}b, \\ \Delta(c) &= c \otimes a + d \otimes c, & \varepsilon(c) &= 0, & S(c) &= -qc, \\ \Delta(d) &= c \otimes b + d \otimes d, & \varepsilon(d) &= 1, & S(d) &= a. \end{aligned} \tag{7.5.41}$$

As $q \to 1$, we formally recover the commutative algebra of functions on $\mathsf{SL}(2)$, where a corresponds to the function returning the $(1,1)$-entry, b to the function returning the $(1,2)$-entry, etc., with the standard Hopf algebra structure on k-valued functions on $\mathsf{SL}(2)$, given by (7.2.20).

Since the generators correspond to matrix entries, it is natural to form the matrix

$$T = \begin{pmatrix} a & b \\ c & d \end{pmatrix} \in \mathrm{End}(k(q)^2) \otimes F_q(\mathsf{SL}(2)). \tag{7.5.42}$$

Then the Hopf algebra structure maps are given simply as

$$\Delta(T) = T \otimes T, \quad \varepsilon(T) = \begin{pmatrix} 1 & 0 \\ 0 & 1 \end{pmatrix}, \quad S(T) = T^{-1}, \tag{7.5.43}$$

where the $\otimes$ in $T \otimes T$ means: use ordinary matrix multiplication and take tensor products at the level of the matrix entries. Also, recall the matrix $R \in \mathrm{End}(k(q)^2 \otimes k(q)^2)$ from (7.5.39); then the *RTT-relation*

$$R_{12}T_{13}T_{23} = T_{23}T_{13}R_{12} \in \mathrm{End}(k(q)^2 \otimes k(q)^2) \otimes F_q(\mathsf{SL}(2)), \tag{7.5.44}$$

together with the *q-determinant* condition $ad - qbc = 1$, is equivalent to the relations (7.5.40). Here we see a nice aspect of duality at play: the object $\mathcal{R}$ controls the failure of cocommutativity for the algebra $U_q\mathfrak{sl}_2$ and the failure of commutativity for the algebra $F_q(\mathsf{SL}(2))$.

7.6 Coideal Subalgebras and Cylinder Braiding

The material in this supplementary section deals with more recent developments in the field of braided monoidal categories with cylinder twists [1, 2, 16, 43], for which the algebraic counterpart is a coideal subalgebra of the quasi-triangular bialgebra, which is itself endowed with a quasitriangular structure. Various aspects of quantum group theory have been extended to this setting such as q-deformed harmonic analysis [29, 33, 34] and canonical bases and q-analogues of Schur–Weyl duality [3, 14].

7.6.1 Coideal Subalgebras

First, we return to the setting where k is any field. Let A be a bialgebra and $B \subseteq A$ a subalgebra. We call B a *two-sided coideal subalgebra*, *right coideal subalgebra* or *left coideal subalgebra* if

$$\begin{gathered} \Delta(B) \subseteq B \otimes A + A \otimes B, \\ \Delta(B) \subseteq B \otimes A, \qquad\qquad \Delta(B) \subseteq A \otimes B, \end{gathered} \tag{7.6.1}$$

respectively. Given a bialgebra A, all subbialgebras of A are right and left coideal subalgebras of A. Also, all right coideal subalgebras of A and all left coideal subalgebras of A are two-sided coideal subalgebras. If A is cocommutative then these four concepts coincide.

Example 7.24 Consider a bialgebra A, an element $g \in \mathrm{Gr}(A)$ and an element $b \in \mathsf{Pri}_{g,1}(A)$. Consider the subalgebra B generated by b (its elements are polynomial expressions in b). Then B is a right coideal, since $\Delta(b) = b \otimes g + 1 \otimes b \in B \otimes A$. Also, B is graded by the degree in b and $B \otimes B$ is graded by the sum of the degrees. Using this it is straightforward to see that $B \cap \mathrm{Gr}(A) = \{1\}$, so that B is not a subbialgebra unless $g = 1$. The assumptions on A are indeed met if, for instance, A is equal to Sweedler's Hopf algebra or $U_q\mathfrak{sl}_2$.

7.6.2 Cylinder Braiding

We saw in Section 7.3.3 that the monoidal category $\mathsf{Rep}(A)$, if A is a quasitriangular bialgebra, possesses a braided structure. In particular, there is an action of the braid group Br_L on $V^{\otimes L}$, for any $V \in \mathsf{Rep}(A)$, given by (7.3.29). Let us adjoin a generator b_0 to Br_L to obtain a larger group Br_L^0, subject to the relations

$$b_0 b_1 b_0 b_1 = b_1 b_0 b_1 b_0, \qquad b_0 b_i = b_i b_0 \text{ if } i > 1. \tag{7.6.2}$$

This is known as the Artin–Tits braid group of type B_L (the subgroup Br_L is the Artin–Tits braid group of type A_{L-1}) and is the fundamental group of the L-th unordered configuration space of the *punctured* disk. Given the representation of Br_L in terms of $\check{R}_{V,V}$, see (7.3.29), it is natural to require that b_0 acts as follows:

$$b_0 \mapsto K_V \otimes \mathsf{id}_V^{\otimes(L-1)} \tag{7.6.3}$$

for some invertible $K_V \in \mathrm{End}(V)$ since then automatically the relations $b_0 b_i = b_i b_0$ for $i > 1$ are preserved. In order to preserve the quartic relation, we must have

$$(K_V \otimes \mathsf{id}_V)\check{R}_{V,V}(K_V \otimes \mathsf{id}_V)\check{R}_{V,V} = \check{R}_{V,V}(K_V \otimes \mathsf{id}_V)\check{R}_{V,V}(K_V \otimes \mathsf{id}_V),$$

or, equivalently,

$$\begin{aligned}(K_V \otimes \mathsf{id}_V)\sigma(R_{V,V})(\mathsf{id}_V \otimes K_V)R_{V,V} \\ = \sigma(R_{V,V})(\mathsf{id}_V \otimes K_V)R_{V,V}(K_V \otimes \mathsf{id}_V),\end{aligned} \tag{7.6.4}$$

which is also known as the *(constant) reflection equation.*

Remark The study of this tensorial version of the quartic braid relation originated in mathematical physics, more precisely quantum integrability in the presence of a boundary, see [8, 40] and for a more general type of reflection equation cf. [9]. Note, however, that in this application the objects $R_{V,V}$ and K_V depend on an additional parameter, called the *spectral parameter*, which varies in the equation:

$$(K_V(y) \otimes \mathsf{id}_V)\sigma(R_{V,V}(yz))(\mathsf{id}_V \otimes K_V(z))R_{V,V}(z/y) \\ = \sigma(R_{V,V}(z/y))(\mathsf{id}_V \otimes K_V(z))R_{V,V}(yz)(K_V(y) \otimes \mathsf{id}_V). \tag{7.6.5}$$

This spectral parameter roughly corresponds to the loop parameter appearing in the definition of a loop algebra.

It is now natural to ask what the additional structure on the braided monoidal category $\mathsf{Rep}(A)$ is, which allows for the action of Br_L to extend to an action of Br_L^0. Because of the embedding $\mathsf{Br}_L < \mathsf{Br}_L^0$, in terms of the graphical calculus from Section 7.3.3 topologically we are adding an obstacle, requiring us to interpret the generator b_0 as the interaction of one of the L strands with the obstacle. This extended calculus was given in [43] and discussed further in [16]. Let us indicate the obstacle by a vertical grey bar to the left of the strands and parallel to their direction of travel and the linear map K_V by a winding around this obstacle:

(7.6.6)

V

The pictorial version of (7.6.4) (or rather its generalization to the case where the two modules are distinct) is as follows:

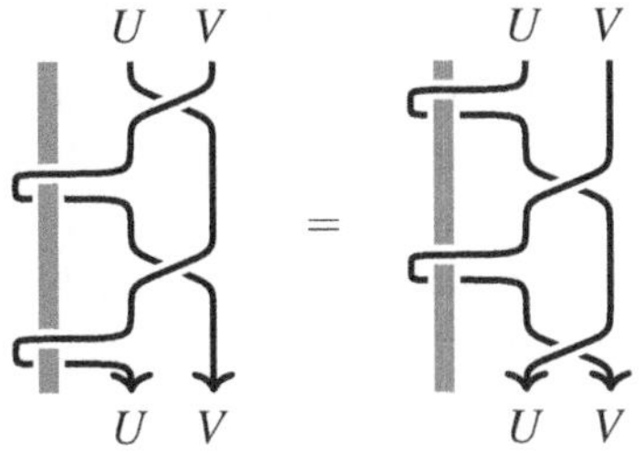

U V U V

= (7.6.7)

U V U V

Let us identify further natural conditions on the objects K_V. Given an A-module V, let $\pi_V : A \to \mathrm{End}(V)$ be the corresponding algebra morphism. Let us take inspiration from the situation for $\check{R}_{U,V}$. By construction, K_V is an intertwiner for the subalgebra

$$B = \{b \in A \,|\, K_V \pi_V(b) = \pi_V(b) K_V\}, \tag{7.6.8}$$

which typically will be a proper subalgebra of A. Note furthermore that, as a consequence of the intertwining property of $\check{R}^{V,V}$, the action of $\Delta^{(L)}(A)$ and the action of Br_L mutually commute in $\mathrm{End}(V^{\otimes L})$, for all $L \in \mathbb{Z}_{>0}$. It is natural to impose that the action of $\Delta^{(L)}(B)$ and the action of Br_L^0 mutually commute for all $L \in \mathbb{Z}_{>0}$, since it is already satisfied for $L = 1$. The most general assumption on B that allows us to perform recursion and obtain the mutual commutativity for all $L \in \mathbb{Z}_{>0}$ is that $\Delta(b) \in B \otimes A$ for all $b \in B$, in other words that B is a right coideal subalgebra; in particular, we can define B to be the largest right coideal subalgebra of the algebra defined in (7.6.8).

We also need a rule for assigning a value to $K_{U\otimes V}$. A topologically natural condition is for instance given by

$$K_{U\otimes V} = \check{R}_{U,V}^{-1}(K_V \otimes \mathrm{id}_U)\check{R}_{U,V}(K_U \otimes \mathrm{id}_V) \tag{7.6.9}$$

or, diagrammatically,

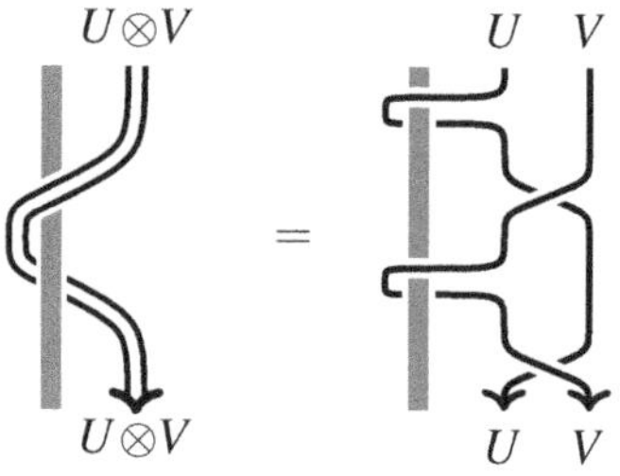

(7.6.10)

Such a collection of linear maps $(K_V)_{V\in\mathrm{Rep}(A)}$ is called a *cylinder twist* on the category $\mathrm{Rep}(A)$.

7.6.3 Cylindrical Quasitriangularity

We now consider an additional structure on a quasitriangular bialgebra $(A, \mathcal{R})$ such that its category of A-modules is braided with a cylinder twist. We allow a generalization, which appeared in [2], namely twisting by a quasitriangular automorphism. Given a quasitriangular bialgebra $(A, \mathcal{R})$ with a quasitriangular automorphism ψ, we denote $\mathcal{R}^\psi := (\psi \otimes \mathrm{id})(\mathcal{R})$. Recall that $\mathcal{R}^{\psi\psi} := (\psi \otimes \psi)(\mathcal{R})$ equals $\mathcal{R}$.

Definition 7.25 ([2, 43]) Let $(A, \mathcal{R})$ be a quasitriangular bialgebra, ψ a quasitriangular automorphism of A and $\mathcal{K} \in A^\times$. We call $(A, \mathcal{R}, \psi, \mathcal{K})$ *cylindrically quasitriangular* and $\mathcal{K}$ a *ψ-twisted universal K-matrix* for A if

$$\Delta(\mathcal{K}) = \mathcal{R}^{-1}(1 \otimes \mathcal{K})\mathcal{R}^\psi(\mathcal{K} \otimes 1). \tag{7.6.11}$$

Furthermore, we call a subalgebra $B \subseteq A$ *ψ-cylindrically invariant* if

$$\mathrm{Ad}(\mathcal{K})|_B = \psi|_B. \tag{7.6.12}$$

Theorem 7.26 *If $(A, \mathcal{R}, \psi, \mathcal{K})$ is cylindrically quasitriangular then $\mathcal{K}$ satisfies the (universal) ψ-twisted reflection equation:*

$$(\mathcal{K} \otimes 1)\sigma(\mathcal{R}^\psi)(1 \otimes \mathcal{K})\mathcal{R} = \sigma(\mathcal{R})(1 \otimes \mathcal{K})\mathcal{R}^\psi(\mathcal{K} \otimes 1). \tag{7.6.13}$$

Proof This follows from (7.3.1) and (7.6.11):

$$\begin{aligned}(\mathcal{K}\otimes 1)\sigma(\mathcal{R}^{\psi})(1\otimes\mathcal{K})\mathcal{R} &= \sigma(\mathcal{R})\Delta^{\mathrm{op}}(\mathcal{K})\mathcal{R}\\ &= \sigma(\mathcal{R})\mathcal{R}\Delta(\mathcal{K})\\ &= \sigma(\mathcal{R})(1\otimes\mathcal{K})\mathcal{R}^{\psi}(\mathcal{K}\otimes 1)\\ &= \sigma(\mathcal{R})(1\otimes\mathcal{K})\mathcal{R}^{\psi}(\mathcal{K}\otimes 1).\end{aligned}$$ □

7.6.4 The Construction of a Universal K-matrix for $U_q\mathfrak{sl}_2$

Given a finite-dimensional semisimple Lie algebra $\mathfrak{g}$ over $\mathbb{C}$ and an involutive Lie algebra automorphism $\theta : \mathfrak{g} \to \mathfrak{g}$, Letzter gave a general theory of the quantization of symmetric pairs $(\mathfrak{g}, \mathfrak{s})$, where $\mathfrak{s} = \mathfrak{g}^{\theta}$ is the fixed-point Lie subalgebra, see [29] and cf. [33, 34]. In this theory $U_q\mathfrak{g}$ is the usual Drinfeld–Jimbo quantum group and $U_q\mathfrak{s}$ is a coideal subalgebra of $U_q\mathfrak{g}$. In [2] a construction is given of a universal K-matrix associated to such a pair $(U_q\mathfrak{g}, U_q\mathfrak{s})$, generalizing a construction first given in [3]. This can be seen as a coideal version of the construction of the universal R-matrix due to Lusztig [30].

Here we follow the approach of [1], which gives a further generalization of the above construction. We discuss this by illustrating a special case, which can be seen in parallel to the discussion of the universal R-matrix for $U_q\mathfrak{sl}_2$ in Section 7.5.10. Essentially the same object appeared previously in a more ad-hoc setting in [28, 43].

Recall the Hopf algebra $U_q\mathfrak{sl}_2$ with generators E, F, t, t^{-1} subject to relations (7.4.4) and with the additional structure maps given by (7.4.5). For $\gamma \in k(q^{1/d})^{\times}$, let $H_\gamma \in \widehat{U}$ be the element acting on objects V of $\mathcal{O}_{q,\mathrm{int}}$ as

$$H_\gamma(v) = q^{-(\mu,\mu)/2}\chi_\gamma(\mu)^{-1}v, \qquad v \in V_\mu,\ \mu \in P, \tag{7.6.14}$$

where $\chi_\gamma : P \to k(q^{1/d})$ is any group homomorphism with the property $\chi_\gamma(\alpha) = \gamma$. Then the algebra automorphism $\theta_\gamma = \omega \circ \mathrm{Ad}(H_\gamma)$ of $U_q\mathfrak{sl}_2$ satisfies

$$\theta_\gamma(E) = -q\gamma^{-1}tF, \quad \theta_\gamma(F) = -q^{-1}\gamma Et^{-1}, \quad \theta_\gamma(t^{\pm 1}) = t^{\mp 1}. \tag{7.6.15}$$

By a direct check, we see that the subalgebra B_γ of $U_q\mathfrak{sl}_2$ generated by $F + \theta_\gamma(F) = F - q^{-1}\gamma Et^{-1}$ is a right coideal subalgebra. It is the Letzter quantization of the fixed-point subalgebra of the Chevalley involution of $\mathfrak{sl}_2$ with an extra parameter dependence. We are interested in an element $\mathcal{K} \in \widehat{U}$ satisfying the coproduct formula (7.6.11) and the intertwining property (7.6.12) with $\psi = \mathrm{id}$.

There exists (see [22]) an element $\mathcal{T} \in \widehat{U}$ which resolves the R-matrix in the sense that $\mathcal{R} = (\mathcal{T}^{-1} \otimes \mathcal{T}^{-1})\Delta(\mathcal{T})$ and satisfies $\mathrm{Ad}(\mathcal{T}) = \omega$; such an element can be constructed as a Cartan modification of an element $\widetilde{\mathcal{T}} \in \widehat{U}$ which satisfies

$\widetilde{\mathcal{R}} = (\widetilde{\mathcal{T}}^{-1} \otimes \widetilde{\mathcal{T}}^{-1})\Delta(\widetilde{\mathcal{T}})$ (see [30, 5.2.1]). Hence it suffices to construct an element $\mathcal{K}' = \mathcal{T}\mathcal{K}$ such that

$$\begin{aligned} \mathsf{Ad}(\mathcal{K}')|_{B_\gamma} &= \omega|_{B_\gamma}, \\ \Delta(\mathcal{K}') &= (1 \otimes \mathcal{K}')(\omega \otimes \mathsf{id})(\mathcal{R})(\mathcal{K}' \otimes 1) \end{aligned} \tag{7.6.16}$$

(note that the coproduct formula is now an expression of three factors). An algebra morphism $f : B_\gamma \to \overline{B_{\overline{\gamma}}}$ is uniquely defined by

$$f(F - q^{-1}\gamma E t^{-1}) = \overline{F - q^{-1}\overline{\gamma} E t^{-1}} = F - q\gamma E t. \tag{7.6.17}$$

It can be checked by a direct computation that

$$\mathsf{Ad}(H_\gamma) \circ f|_{B_\gamma} = \omega|_{B_\gamma}, \qquad \Delta(H_\gamma) = H_\gamma \otimes H_\gamma \kappa^{-1}, \tag{7.6.18}$$

in terms of the element κ defined by (7.5.8). Now from (7.6.16) and (7.6.18) one deduces that it suffices to construct an element $\widetilde{K} = H_\gamma^{-1} K'$ such that

$$\begin{aligned} &\mathsf{Ad}(\widetilde{\mathcal{K}})|_{B_\gamma} = f|_{B_\gamma}, \\ &\Delta(\widetilde{\mathcal{K}}) = \mathsf{Ad}(\kappa)(1 \otimes \widetilde{\mathcal{K}}) \cdot (\mathsf{Ad}(H_\gamma^{-1})\omega \otimes \mathsf{id})(\widetilde{\mathcal{R}}) \cdot (\widetilde{\mathcal{K}} \otimes \mathsf{id}). \end{aligned} \tag{7.6.19}$$

In [2, Sec. 6] and, without constraints on γ, in [1, Sec. 7] it is shown that the element $\widetilde{\mathcal{K}}$ exists in $\widehat{U}^+$ and is unique if we impose $\varepsilon(\widetilde{\mathcal{K}}) = 1$. We can give an explicit formula.

Lemma 7.27 ([11]) *The system* (7.6.19) *is satisfied by the following element* $\widetilde{\mathcal{K}} \in \widehat{U}^+$:

$$\widetilde{\mathcal{K}} = \sum_{r \in \mathbb{Z}_{\geq 0}} q^{2r(r-1)} [2r]_q!! (\gamma(q-q^{-1})E^2)^r \tag{7.6.20}$$

where $[2r]_q!! := [2r]_q [2(r-1)]_q \cdots [2]_q [0]_q$.

We obtain that the element $\mathcal{K} = \mathcal{T}^{-1} H_\gamma \widetilde{\mathcal{K}}$ satisfies (7.6.12) and (7.6.11) and hence the reflection equation (7.6.13) (all with $\psi = \mathsf{id}$).

The representations $\pi^{(n)}$ are objects in category $\mathcal{O}_{q,\mathsf{int}}$ and hence we can evaluate $\mathcal{K}$ for instance by applying $\pi^{(2)}$, obtaining an element of $\mathrm{End}(V^{(2)})$. We recall the matrices defined in (7.5.37). The action on $V^{(2)}$ of the three constituent factors of $\mathcal{K}$ is as follows:

$$\begin{aligned} \pi^{(2)}(\mathcal{T}^{-1}) &= q^{-3/4} \begin{pmatrix} 0 & -1 \\ 1 & 0 \end{pmatrix}, \\ \pi^{(2)}(H_\gamma) &= q^{3/4} \begin{pmatrix} 1 & 0 \\ 0 & \gamma \end{pmatrix}, \\ \pi^{(2)}(\widetilde{\mathcal{K}}) &= \begin{pmatrix} 1 & 0 \\ 0 & 1 \end{pmatrix}. \end{aligned} \tag{7.6.21}$$

There is an indeterminacy in H_γ (more precisely, in χ_γ) corresponding to the choice of an overall scalar. We choose the scalar so that H_γ acts in the way above. It follows that

$$K := \pi^{(2)}(\mathcal{K}) = \begin{pmatrix} 0 & -\gamma \\ 1 & 0 \end{pmatrix} \tag{7.6.22}$$

is a solution to the matrix reflection equation

$$(K \otimes \mathsf{id}_{V(2)})\check{R}(K \otimes \mathsf{id}_{V(2)})\check{R} = \check{R}(K \otimes \mathsf{id}_{V(2)})\check{R}(K \otimes \mathsf{id}_{V(2)}), \tag{7.6.23}$$

where $\check{R} = \sigma \circ R$ with R given by (7.5.39).

7.6.5 Cylinder Quasitriangularity Twisted by an Algebra Automorphism

The setup outlined in Section 7.6.3 can be generalized as follows, which allows us to apply it in the context of Drinfeld–Jimbo quantum groups of non-finite type, see [1]. It can be directly motivated from the setup in Definition 7.25 by noting that also the coproduct formula $\Delta(\mathcal{K}) = \sigma(\mathcal{R})(1 \otimes \mathcal{K})\mathcal{R}^\psi(\mathcal{K} \otimes 1)$, which is the one actually used in [2, 25], also leads to the reflection equation (7.6.13), and so a more general ansatz $\Delta(\mathcal{K}) = \mathcal{F}^{-1}(1 \otimes \mathcal{K})\mathcal{R}^\psi(\mathcal{K} \otimes 1)$ with $\mathcal{F} \in (A \otimes A)^\times$ is tempting.

To make it precise, assume that $\psi : A \to A$ is merely an algebra automorphism (so not necessarily a quasitriangular or even a bialgebra automorphism) and recall the notations $\mathcal{R}^\psi := (\psi \otimes \mathsf{id})(\mathcal{R})$, $\mathcal{R}^{\psi\psi} := (\psi \otimes \psi)(\mathcal{R})$. We can generalize Definition 7.25 and call $(A, \mathcal{R}, \psi, \mathcal{K})$ *cylindrically quasitriangular* and $\mathcal{K}$ a *ψ-twisted universal K-matrix* for A if there exists $\mathcal{F} \in (A \otimes A)^\times$ such that

$$\Delta(\mathcal{K}) = \mathcal{F}^{-1}(1 \otimes \mathcal{K})\mathcal{R}^\psi(\mathcal{K} \otimes 1), \tag{7.6.24}$$

$$\sigma(\mathcal{R}^{\psi\psi}) = \sigma(\mathcal{F})\mathcal{R}\mathcal{F}^{-1}. \tag{7.6.25}$$

(The notion of a ψ-cylindrically invariant subalgebra remains the same as in Definition 7.25.)

Note that the element $\mathcal{F}$ appearing in (7.6.11) has to satisfy certain constraints. By applying coassociativity (7.2.8) and the counit axiom (7.2.9), we obtain

$$\begin{aligned}
&\big(\mathcal{F}_{12}(\Delta \otimes \mathsf{id})(\mathcal{F})\big)^{-1}(1 \otimes 1 \otimes \mathcal{K})\big((\mathsf{Ad}(\mathcal{F}) \circ \Delta \circ \psi) \otimes \mathsf{id}\big)(\mathcal{R}) \\
&= \big(\mathcal{F}_{23}(\mathsf{id} \otimes \Delta)(\mathcal{F})\big)^{-1}(1 \otimes 1 \otimes \mathcal{K})\big(((\psi \otimes \psi) \circ \Delta^{\mathsf{op}}) \otimes \mathsf{id}\big)(\mathcal{R}), \\
&\varepsilon(\mathcal{K}) = (\mathsf{id} \otimes \varepsilon)(\mathcal{F}), \\
&\mathcal{K}\big((\varepsilon \circ \psi) \otimes \mathsf{id}\big)(\mathcal{R})\varepsilon(\mathcal{K}) = (\varepsilon \otimes \mathsf{id})(\mathcal{F})\mathcal{K},
\end{aligned}$$

which is most easily satisfied by making two assumptions. First of all, recalling (7.3.8), we assume that $\mathcal{F}$ satisfies the cocycle condition

$$\mathcal{F}_{12}(\Delta \otimes \mathsf{id})(\mathcal{F}) = \mathcal{F}_{23}(\mathsf{id} \otimes \Delta)(\mathcal{F}), \tag{7.6.26}$$

so that by (7.2.9) we obtain

$$(\mathsf{id} \otimes \varepsilon)(\mathcal{F}) = (\varepsilon \otimes \mathsf{id})(\mathcal{F}) = \varepsilon(\mathcal{K}) \in k^\times, \tag{7.6.27}$$

which we may as well set equal to 1 (by rescaling $\mathcal{F}$ and $\mathcal{K}$ in a compatible manner). Such $\mathcal{F}$ are called *Drinfeld twists* (for bialgebras), see [13, 35]. Secondly, we assume that ψ and $\mathcal{F}$ twist the bialgebra structure on A in related ways:

$$(\psi \otimes \psi) \circ \Delta^{\mathsf{op}} \circ \psi^{-1} = \mathsf{Ad}(\mathcal{F}) \circ \Delta, \qquad \varepsilon \circ \psi^{-1} = \varepsilon. \tag{7.6.28}$$

This extends the assumption made on $\mathcal{F}$ in (7.6.25), yielding related twists of quasitriangular structures on A.

The proof of Theorem 7.26 now produces the following generalization of the ψ-twisted reflection equation:

$$(\mathcal{K} \otimes 1)\sigma(\mathcal{R}^{\psi})(1 \otimes \mathcal{K})\mathcal{R} = \sigma(\mathcal{R}^{\psi\psi})(1 \otimes \mathcal{K})\mathcal{R}^{\psi}(\mathcal{K} \otimes 1). \tag{7.6.29}$$

It is argued in [1, Sec. 9] that, if $A = U_q(\mathfrak{g})$ with $\mathfrak{g}$ a Kac–Moody algebra of affine type then, for suitable choices of ψ, the image of (7.6.29) in finite-dimensional modules recovers the matrix equation (7.6.5), or rather a generalized version of it, see [9, Eq. (4.15)].

Bibliography

[1] Appel, A., and Vlaar, B. 2020. Universal k-matrices for quantum Kac-Moody algebras. *arXiv preprint arXiv:2007.09218.*

[2] Balagović, M., and Kolb, S. 2019. Universal K-matrix for quantum symmetric pairs. *Journal für die reine und angewandte Mathematik*, **2019**(747), 299–353.

[3] Bao, H, and Wang, W. 2018. *A new approach to Kazhdan-Lusztig theory of type B via quantum symmetric pairs.* Société mathématique de France Paris.

[4] Bernstein, I N, Gelfand, I M, and Gelfand, S I. 1975. Differential operators on the base affine space and a study of $\mathfrak{g}$-modules. *Lie groups and their representations (Proc. Summer School, Bolyai János Math. Soc., Budapest, 1971)*, 21–64.

[5] Bourbaki, N. 1994. Lie groups and Lie algebras. Pages 247–267 of: *Elements of the History of Mathematics.* Springer.

[6] Chari, V, and Pressley, A. 1991. Quantum affine algebras. *Communications in Mathematical Physics*, **142**(2), 261–283.

[7] Chari, V, and Pressley, A. 1995. *A guide to quantum groups.* Cambridge University Press.

[8] Cherednik, I. 1984. Factorizing particles on a half-line and root systems. *Theoretical and Mathematical Physics*, **61**(1), 977–983.
[9] Cherednik, I. 1992. Quantum Knizhnik-Zamolodchikov equations and affine root systems. *Communications in Mathematical Physics*, **150**(1), 109–136.
[10] Dascalescu, S, Nastasescu, C, and Raianu, S. 2001. Hopf Algebras: an Introduction. *Marcel Decker Inc., New York.*
[11] Dobson, L, and Kolb, S. 2019. Factorisation of quasi K-matrices for quantum symmetric pairs. *Selecta Mathematica*, **25**(4), 1–55.
[12] Drinfeld, V G. 1985. Hopf algebras and the quantum Yang-Baxter equation. *Soviet Math. Dokl.*, **32**(6), 254–258.
[13] Drinfeld, V G. 1987. Quantum Groups. Pages 798–820 of: Gleason, A M (ed), *Proceedings of the International Congress of Mathematicians*. American Mathematical Society.
[14] Ehrig, M, and Stroppel, C. 2018. Nazarov–Wenzl algebras, coideal subalgebras and categorified skew Howe duality. *Advances in Mathematics*, **331**, 58–142.
[15] Frenkel, I B, and Reshetikhin, N Yu. 1992. Quantum affine algebras and holonomic difference equations. *Communications in Mathematical Physics*, **146**(1), 1–60.
[16] Häring-Oldenburg, R. 2001. Actions of tensor categories, cylinder braids and their Kauffman polynomial. *Topology and its Applications*, **112**(3), 297–314.
[17] Jantzen, J C. 1996. *Lectures on quantum groups*. Vol. 6. American Mathematical Soc.
[18] Jimbo, M. 1986. A q-analogue of $U(\mathfrak{gl}(N+1))$, Hecke algebra, and the Yang-Baxter equation. *Letters in Mathematical Physics*, **11**(3), 247–252.
[19] Jimbo, M, and Miwa, T. 1994. *Algebraic analysis of solvable lattice models*. Vol. 85. Amer. Math. Soc.
[20] Joyal, A, and Street, R. 1993. Braided tensor categories. *Advances in Mathematics*, **102**(1), 20–78.
[21] Kac, V G. 1990. *Infinite-dimensional Lie algebras*. Cambridge University Press.
[22] Kamnitzer, J, and Tingley, P. 2009. The crystal commutor and Drinfeld's unitarized R-matrix. *Journal of Algebraic Combinatorics*, **29**(3), 315–335.
[23] Kassel, C. 1995. Quantum Groups. *Grad. Texts in Math.*
[24] Klimyk, A, and Schmüdgen, K. 1997. *Quantum Groups and their Representations (1997)*. Springer Texts and Monographs in Physics.
[25] Kolb, S. 2020. Braided module categories via quantum symmetric pairs. *Proceedings of the London Mathematical Society*, **121**(1), 1–31.
[26] Kostant, B. 1977. Graded manifolds, graded Lie theory, and prequantization. Pages 177–306 of: *Differential Geometrical Methods in Mathematical Physics*. Springer.
[27] Kulish, P P, and Reshetikhin, N Yu. 1983. Quantum linear problem for the sine-Gordon equation and higher representations. *Journal of Soviet Mathematics*, **23**(4), 2435–2441.
[28] Kulish, P P, Sasaki, R, and Schwiebert, C. 1993. Constant solutions of reflection equations and quantum groups. *Journal of Mathematical Physics*, **34**(1), 286–304.
[29] Letzter, G. 1999. Symmetric pairs for quantized enveloping algebras. *Journal of Algebra*, **220**(2), 729–767.
[30] Lusztig, G. 1994. *Introduction to quantum groups*. Birkhäuser, Boston.

[31] Majid, S. 1988. Hopf algebras for physics at the Planck scale. *Classical and Quantum Gravity*, **5**(12), 1587.
[32] Majid, S. 2002. *A quantum groups primer*. Cambridge University Press.
[33] Noumi, M, and Sugitani, T. 1995. Quantum symmetric spaces and related q-orthogonal polynomials. Pages 28–40 of: Arima, A *et al.* (ed), *Group theoretical methods in physics*. World Scientific.
[34] Noumi, M, Dijkhuizen, M S, and Sugitani, T. 1997. Multivariable Askey-Wilson polynomials and quantum complex Grassmannians. *AMS Field Inst. Commun.*, **14**, 167–177.
[35] Reshetikhin, N. 1990. Multiparameter quantum groups and twisted quasitriangular Hopf algebras. *Letters in Mathematical Physics*, **20**(4), 331–335.
[36] Reshetikhin, N. 2010. Lectures on the integrability of the six-vertex model. *Exact methods in low-dimensional statistical physics and quantum computing*, 197–266.
[37] Reshetikhin, N Yu, and Turaev, V G. 1990. Ribbon graphs and their invariants derived from quantum groups. *Communications in Mathematical Physics*, **127**(1), 1–26.
[38] Saito, Y. 1994. PBW basis of quantized universal enveloping algebras. *Publications of the Research Institute for Mathematical Sciences*, **30**(2), 209–232.
[39] Schneider, H-J. 1995. Lectures on Hopf algebras. *Trabajos de Matemática*, **31**, 95.
[40] Sklyanin, E K. 1988. Boundary conditions for integrable quantum systems. *Journal of Physics A: Mathematical and General*, **21**(10), 2375.
[41] Sweedler, M E. 1969. *Hopf algebras*. Benjamin, New York.
[42] Tanisaki, T. 1992. Killing forms, Harish-Chandra isomorphisms, and universal R-matrices for quantum algebras. *International Journal of Modern Physics A*, **7**(supp01b), 941–961.
[43] tom Dieck, T, and Häring-Oldenburg, R. 1998. Quantum groups and cylinder braiding. Pages 619–639 of: *Forum Math.*, vol. 10.

8

Infinite-dimensional Lie Algebras and Their Multivariable Generalizations

Brian Williams

The loop algebra $L\mathfrak{g} = \mathfrak{g}[z, z^{-1}]$, consisting of Laurent polynomials valued in a Lie algebra $\mathfrak{g}$, admits a nontrivial central extension $\widehat{\mathfrak{g}}$ for each choice of invariant pairing on $\mathfrak{g}$. This affine Lie algebra and its cousin the Virasoro algebra are foundational objects in representation theory and conformal field theory. A natural question then arises: do there exist multivariable, or higher dimensional, generalizations of the affine algebra?

Overview and Prerequisites

In §8.1 we give a rapid, biased review of the theory of infinite-dimensional Lie algebras. In §8.2 we introduce higher dimensional, multivariable generalizations of the Lie algebras from §8.1. In the final section we give a less formal discussion of further topics related to the theory of multivariable Lie algebras. The goal of §8.3.1 is to provide a geometric perspective on infinite-dimensional Lie algebras and modules based on the theory of factorization algebras. Finally, §8.3.2 introduces some modest enhancements of multivariable algebras analogous to the theory of affine algebras associated to super Lie algebras.

In these notes we assume some familiarity with Lie algebras, including the notion of a module. We assume that the reader is familiar with the notion of a cochain complex (especially in §8.2 and §8.3), and some other rudimentary notions in homological algebra. Some basic exposure to algebraic geometry will also be helpful for following the later sections.

Acknowledgements

I first learned about the multivariable algebras discussed here from Kevin Costello's suggestion to study symmetries in certain higher dimensional quantum field theories. Most importantly for these notes I'd like to acknowledge the work of Faonte, Hennion and Kapranov in [10], which develops the general theory of higher dimensional Kac–Moody algebras and whose presentation we closely follow in §8.2. I'm also extremely grateful for valuable collaborations with Owen Gwilliam and Ingmar Saberi whose results are discussed in §8.2 and §8.3.

8.1 Infinite-dimensional Lie algebras

The first infinite-dimensional Lie algebra we focus on arises from the algebra of Laurent polynomials $\mathbb{C}[z,z^{-1}]$, which is the algebra of functions on the punctured affine line $\mathbb{A}^{\times}$.

The space of derivations of any commutative algebra always forms a Lie algebra where the bracket is simply the *commutator* of the endomorphisms defining the derivations. The *Witt algebra* is the Lie algebra of derivations of $\mathbb{C}[z,z^{-1}]$. As a vector space, the Witt algebra admits a presentation in terms of vector fields on the punctured disk:

$$\mathsf{witt} \overset{\text{def}}{=} \left\{ f(z)\frac{\mathrm{d}}{\mathrm{d}z} \mid f(z) \in \mathbb{C}[z,z^{-1}] \right\} \cong \mathbb{C}[z,z^{-1}]\frac{\mathrm{d}}{\mathrm{d}z}.$$

The Lie bracket is simply the Lie bracket of vector fields

$$\left[f(z)\frac{\mathrm{d}}{\mathrm{d}z}, g(z)\frac{\mathrm{d}}{\mathrm{d}z} \right] = \left(f(z)g'(z) - f'(z)g(z) \right) \frac{\mathrm{d}}{\mathrm{d}z}.$$

It is standard to choose a basis $\{L_n \overset{\text{def}}{=} -z^{n+1}\frac{\mathrm{d}}{\mathrm{d}z}\}_{n\in\mathbb{Z}}$ for witt so that the commutator takes the form $[L_n, L_m] = (n-m)L_{n+m}$.

8.1.1 Central Extensions

We will turn to examples of modules for infinite-dimensional Lie algebras momentarily. The Witt algebra, for example, has a much richer theory of modules if instead of plain modules, one considers "projective" modules. This means that it is not the Witt algebra that acts but rather a certain *central extension* of the Witt algebra. We begin with some examples of central extensions.

The Heisenberg Algebra

There is another Lie algebra associated to the commutative algebra $\mathbb{C}[z,z^{-1}]$. In a trivial way, we can view it as an *abelian* Lie algebra. This is a rather boring infinite-dimensional Lie algebra, but there is a particular extension of it that will be characteristic of many of the constructions to come.

Suppose V is a vector space[1] equipped with a bilinear pairing $(\cdot,\cdot)\colon V \times V \to \mathbb{C}$. Let $\widehat{V}$ denote the vector space $V \oplus \mathbb{C} = V \oplus \operatorname{span}\{K\}$, and denote elements of this vector space by $v + \lambda K$, where $v \in V$ and $\lambda \in \mathbb{C}$. Then, we ask whether the formulas $[v,w] = (v,w)K$ and $[v,K] = 0$ for $v,w \in V$ define the structure of a Lie algebra on $\widehat{V}$. The Jacobi identity for the bracket $[\cdot,\cdot]$ just defined is trivially satisfied. Thus, the only condition we must consider is the skew-symmetry of the bracket. Clearly, skew-symmetry holds if and only if the original bracket $(\cdot,\cdot)$ is anti-symmetric: we need $(v,w) = -(w,v)$ for all $v,w \in \mathbb{C}$.

Thus, these formulas define a Lie algebra structure on $\widehat{V} = V \oplus \mathbb{C}K$ if and only if the pairing $(\cdot,\cdot)$ is anti-symmetric. Let us return to the algebra $\mathbb{C}[z,z^{-1}]$. As a vector space $V = \mathbb{C}[z,z^{-1}]$ is equipped with an anti-symmetric pairing $(z^n,z^m) = m\delta_{n,-m}K$, where $\delta_{n,-m}$ is the function that is 1 when $n = -m$ and zero otherwise.

Definition 8.1 The *Heisenberg algebra* is the Lie algebra $\mathsf{h} = \mathbb{C}[z,z^{-1}] \oplus \mathbb{C}K$ with brackets defined by

$$[z^n,z^m] = m\delta_{n,-m}K$$

and $[K,z^n] = 0$ for all $n,m \in \mathbb{Z}$.

The Heisenberg algebra is an example of the following more general notion (which is meaningful even for finite-dimensional algebras, of course):

Definition 8.2 A *central extension* of a Lie algebra $\mathfrak{g}$ by a Lie algebra $\mathfrak{c}$ is a Lie algebra $\widetilde{\mathfrak{g}}$ that sits in an exact sequence of Lie algebras

$$\mathfrak{c} \to \widetilde{\mathfrak{g}} \to \mathfrak{g}$$

with the property that $\mathfrak{c}$ lies in the center of $\widetilde{\mathfrak{g}}$.

In particular, $\mathfrak{c}$ must be an abelian Lie algebra. There is a classification of equivalences classes of central extensions based on the *Lie algebra cohomology* of $\mathfrak{g}$ with coefficients in $\mathfrak{c}$. We take a brief detour in this direction.

Lie Algebra Cohomology

We define the Lie algebra cohomology of a Lie algebra $\mathfrak{g}$ with coefficients in a $\mathfrak{g}$-module. Categorically speaking, Lie algebra cohomology of a module is a certain "derived" replacement for a very classical notion in Lie theory: the $\mathfrak{g}$-invariants of the module. There is a particular cochain complex modeling this replacement that we recall.

[1] We will work over $\mathbb{C}$ unless otherwise stated.

The commutator in an associative algebra is a Lie bracket. Given any Lie algebra $\mathfrak{g}$, we let $U\mathfrak{g}$ denote its universal enveloping algebra. This is an associative algebra equipped with a canonical map $i\colon \mathfrak{g} \to U\mathfrak{g}$ of Lie algebras. It is universal in the sense that if A is any other algebra which admits a Lie algebra map $f : \mathfrak{g} \to A$, there is a unique map $\widetilde{f} : U\mathfrak{g} \to A$, for which $f = \widetilde{f} \circ i$. Explicitly, one obtains $U\mathfrak{g}$ as a quotient of the tensor algebra $T(\mathfrak{g}) = \oplus_{n \geqslant 0} \mathfrak{g}^{\otimes n}$. The famous Poincaré–Birkhoff–Witt theorem identifies $U\mathfrak{g}$ with $\mathrm{Sym}(\mathfrak{g})$ as vector spaces (but not algebras!).

Given a Lie algebra $\mathfrak{g}$ and a module M, the Lie algebra cohomology is the derived functor

$$H^n(\mathfrak{g};M) = \mathrm{Ext}^n_{U\mathfrak{g}}(\mathbb{C},M).$$

Here, we view $\mathbb{C}$ as a trivial $\mathfrak{g}$-module. One can use a particular resolution for the trivial $\mathfrak{g}$-module to come up with the following cochain model for Lie algebra cohomology.

Definition 8.3 Let $\mathfrak{g}$ be a Lie algebra and M a $\mathfrak{g}$-module. The *Chevalley–Eilenberg* cochain complex $\mathrm{C}^\bullet(\mathfrak{g};M)$ computing Lie algebra cohomology is the cochain complex whose underlying graded vector space is $\mathrm{Hom}\left(\mathrm{Sym}\left(\mathfrak{g}[1]\right),\right.$ $\left.M\right) = \mathrm{Hom}\left(\oplus_{k \geqslant 0}(\wedge^k \mathfrak{g})[k]),M\right)$. The differential is defined as follows. Given a k-cochain $\varphi : \wedge^k \mathfrak{g} \to M$, the $(k+1)$-cochain $\mathrm{d}(\varphi)$ is defined by

$$\begin{aligned} \mathrm{d}(\varphi)(x_1,\dots,x_{k+1}) &= \sum_{i=1}^{k+1} (-1)^{i+1} x_i \cdot \varphi(x_1,\dots,\widehat{x_i},\dots,x_{k+1}) \\ &+ \sum_{1 \leqslant i<j \leqslant k+1} (-1)^{i+j} \varphi([x_i,x_j],x_1,\dots,\widehat{x_i},\dots,\widehat{x_j},\dots,x_{k+1}). \end{aligned} \tag{8.1.1}$$

It is a tedious but straightforward exercise to verify that $\mathrm{d} \circ \mathrm{d} = 0$, so that we have defined a cochain complex. The Chevalley–Eilenberg cochain complex computes the Lie algebra cohomology $H^\bullet(\mathfrak{g};M) = H^\bullet(\mathrm{C}^\bullet(\mathfrak{g};M))$.

Remark The linear dual cochain complex $\mathrm{C}_\bullet(\mathfrak{g};M)$ computes Lie algebra *homology*. As a graded vector space $\mathrm{C}_\bullet(\mathfrak{g};M)$ is $\mathrm{Sym}(\mathfrak{g}[1]) \otimes_{\mathbb{C}} M = \left(\oplus_{k \geqslant 0}(\wedge^k \mathfrak{g})[k])\right) \otimes_{\mathbb{C}} M$. We leave it as an exercise to write down the differential of this cochain complex which is linear dual to the one above.

Lemma 8.1 *Central extensions of $\mathfrak{g}$ by a $\mathfrak{c}$ are in one-to-one correspondence with the second cohomology group $H^2(\mathfrak{g}\ ;\ \mathfrak{c})$.*

Proof We will use our model for Lie algebra cohomology. Suppose φ is a 2-cocycle representing a class in $H^2(\mathfrak{g};\mathfrak{c})$. Then, define the Lie algebra $\widetilde{\mathfrak{g}}$, which as a vector space is $\mathfrak{g} \oplus \mathfrak{c}$ with Lie brackets $[x,x'] = [x,x']_{\mathfrak{g}} + \varphi(x,x')$, and $[x,c] = [c,c'] = 0$ for $x,x' \in \mathfrak{g}$ and $c,c' \in \mathfrak{c}$. Here $[\cdot,\cdot]_{\mathfrak{g}}$ denotes the original Lie bracket on $\mathfrak{g}$.

Conversely, suppose $\widetilde{\mathfrak{g}}$ is such a central extension with Lie bracket $[\cdot,\cdot]$. Then, define the 2-cochain $\varphi(x,x') = [x,x'] - [x,x']_{\mathfrak{g}}$. It is immediate to check that φ is a 2-cocycle. We leave it as an exercise to formulate the appropriate notion of "equivalence" of two central extensions and to show how cohomologous cocycles give rise to it. □

Example 8.4 Consider the algebra $\mathbb{C}[z,z^{-1}]$ thought of as an abelian Lie algebra. Then, the bilinear map $\mathbb{C}[z,z^{-1}] \times \mathbb{C}[z,z^{-1}] \to \mathbb{C}$ defined by $(z^n,z^m) = m\delta_{n,-m}$ defines a 2-cocycle in $\mathrm{C}^2(\mathbb{C}[z,z^{-1}])$. This is rather trivial in this abelian case, as the only condition to check is that it is antisymmetric.

Gelfand–Fuks Cohomology

We now turn to classifying central extensions of the Witt algebra. The Lie algebra witt is infinite dimensional, so one must use caution when defining the Chevalley–Eilenberg cochain complex.

The following result is well known and can be found in [14].

Proposition 8.2 *The cohomology $H^2(\mathsf{witt})$ is one-dimensional spanned by a single class. This class may be represented by the cocycle $\varphi_{\mathrm{Vir}}(L_m,L_n) = \frac{1}{12}\delta_{m,-n}(m^3-m)$. Then φ_{Vir}.*

Remark Let $\mathrm{Res} : \mathbb{C}((z))\mathrm{d}z \to \mathbb{C}$ be the formal residue map, which sends a Laurent 1-form $\sum a_n z^n \mathrm{d}z$ to a_{-1}. In terms of vector fields on the formal punctured disk, one can rewrite this cocycle as $\frac{1}{12}\mathrm{Res}_z(f'(z)\mathrm{d}g'(z))$. Here, d is the formal de Rham differential $\mathrm{d}(h(z)) = h'(z)\mathrm{d}z$. The factor of $\frac{1}{12}$ is conventional and can be traced back to the Grothendieck–Riemann–Roch theorem and string theory.

Definition 8.5 The *Virasoro algebra* vir is the one-dimensional central extension of the Witt algebra witt defined by the 2-cocycle φ_{Vir}.

Remark The Witt algebra describes the infinitesimal symmetries of the punctured disk. The Virasoro algebra is closely tied to the moduli space of Riemann surfaces. The Lie algebra of conformal transformations on a punctured domain in a Riemann surface is given by two copies of the Witt algebra

$$\mathsf{witt} \oplus \overline{\mathsf{witt}} = \mathbb{C}[z,z^{-1}]\partial_z \oplus \mathbb{C}[\bar{z},\bar{z}^{-1}]\partial_{\bar{z}}.$$

The Virasoro algebra (associated to the holomorphic copy of the Witt algebra) corresponds to a certain *line bundle* over the moduli space of Riemann surfaces.

8.1.2 Affine Algebras

Algebraically, affine algebras are constructed from generalized Cartan matrices of a particular type [22]. Geometrically, and closer to the perspective we take here, affine algebras arise infinitesimally from *loop groups* – smooth maps from a circle S^1 into a Lie group [31]. We will not work directly with loop groups in these notes, rather we pass straight to the level of the Lie algebra.

Suppose that $\mathfrak{g}$ is a Lie algebra. For applications in representation theory, one often restricts to the case that $\mathfrak{g}$ is simple, but for now it will make no difference. We can tensor the commutative algebra of Laurent polynomials $\mathbb{C}[z,z^{-1}]$ with the Lie algebra $\mathfrak{g}$ to obtain a new Lie algebra

$$\mathfrak{g}\otimes\mathbb{C}[z,z^{-1}]=\mathfrak{g}[z,z^{-1}].$$

Write elements of this Lie algebra as $x\otimes f(z)$. Explicitly, the Lie bracket is defined by

$$[x\otimes f(z),y\otimes g(z)]=[x,y]\otimes(f\cdot g)(z).$$

We refer to $\mathfrak{g}[z,z^{-1}]$ as the *current algebra* associated to $\mathfrak{g}$. Equivalently, if we think of $\mathfrak{g}$ as an affine variety, this Lie algebra is the same as maps from the punctured affine space $\mathbb{A}^\times$ to $\mathfrak{g}$.

Just as in the case of the Virasoro algebra, there is a one-dimensional central extension of this Lie algebra. According to Lemma 8.1 we know to look for such central extensions in the Lie algebra cohomology $H^2(\mathfrak{g}[z,z^{-1}])$. To describe the relevant piece of this cohomology we introduce some terminology.

The algebra of polynomials on a vector space V is $\mathbb{C}[V]=\mathrm{Sym}(V^*)$. When $V=\mathfrak{g}$ we note that $\mathfrak{g}$ acts on its polynomials $\mathbb{C}[\mathfrak{g}]$ by the adjoint representation, which we will denote on elements by ad_x. By definition, an *invariant polynomial* of $\mathfrak{g}$ is a polynomial P on $\mathfrak{g}$ such that $\mathrm{ad}_x(P)=0$ for all $x\in\mathfrak{g}$. One has the following class of 2-cocycles on the current algebra $\mathfrak{g}[z,z^{-1}]$.

Definition 8.6 Suppose κ is an invariant quadratic polynomial of $\mathfrak{g}$. Define the 2-cochain $\varphi_\kappa\in C^2(\mathfrak{g}[z,z^{-1}])$ by the formula $\varphi_\kappa(f(z)\otimes x,g(z)\otimes y)=\mathrm{Res}_z(f\mathrm{d}g)\kappa(x,y)$.

Remark More concisely, if we view $x(z)\in\mathfrak{g}[z,z^{-1}]$ as a $\mathfrak{g}$-valued function on the punctured disk, then we can write the 2-cochain as $\mathrm{Res}_z\kappa(x(z)\mathrm{d}y(z))$.

One can calculate that φ_κ is a cocycle. In this way, there is an embedding from invariant quadratic polynomials into the second cohomology:

$$\mathrm{Sym}^2(\mathfrak{g}^*)^{\mathfrak{g}}\hookrightarrow H^2(\mathfrak{g}[z,z^{-1}]).$$

For every such invariant quadratic, we obtain a central extension.

Definition 8.7 The *Kac–Moody affine algebra*, or simply *affine algebra*, $\widehat{\mathfrak{g}}_\kappa$ associated to an invariant quadratic polynomial κ is the central extension of $\mathfrak{g}[z,z^{-1}]$ defined by the 2-cocycle φ_κ.

Example 8.8 There is a natural quadratic invariant polynomial associated to any Lie algebra. The *Killing form* is the invariant polynomial $\kappa_{\mathrm{Kill}}(x,y) = \mathrm{Tr}(\mathrm{ad}_x \circ \mathrm{ad}_y)$ where the trace on the right-hand side is in the adjoint representation. When $\mathfrak{g}$ is simple, this is the unique invariant quadratic polynomial up to scale. Furthermore, $\mathfrak{g}$ is semisimple if and only if κ_{Kill} is nondegenerate.

8.1.3 Modules

The theory of modules for the infinite-dimensional Lie algebras is extremely rich and we make no assertions of giving a complete account here. Instead, we opt to give the idea of the flavor of modules we are interested in generalizing to the multivariable case. Let us first examine an example called the *Fock module*. This is a module for the Heisenberg Lie algebra h.

Fock Module

As a vector space the Heisenberg Lie algebra admits a presentation

$$\mathrm{span}_{\mathbb{C}}\{b[n]\}_{n\in\mathbb{Z}} \oplus \mathbb{C}\cdot K,$$

where in our previous notation $b[n]$ corresponds to the homogenous Laurent polynomial z^n. The commutator is simply $[b[n],K] = 0$ and $[b[n],b[m]] = m\delta_{n,-m}K$.

Consider the following polynomial algebra on an infinite number of generators:

$$V = \mathbb{C}[x_{-1},x_{-2},\ldots].$$

We will describe the structure of an h-module on this space.

The definition of this module is somewhat motivated by a physical system: one views the Heisenberg algebra h as some algebra of operators and V as the space of "states" of the system. In this case, one should imagine V as describing the configurations of some number of particles. The unit element $|0\rangle \overset{\mathrm{def}}{=} 1 \in V$ plays the role of the "vacuum", whereby no particles are present.

The usual convention is that the operator $b[-n]$, for $n > 0$, plays the role of the "creation" operator. That is, from the vacuum $b[-n]$ creates the nth excited state $b[-n]|0\rangle = x_{-n}$. Similarly, if $F(x_1,x_2,\ldots) \in V$ is any combination of multiparticle states it is natural to define

$$b[-n]F(x_{-1},x_{-2}\ldots) \overset{\mathrm{def}}{=} x_{-n}F(x_{-1},x_{-2},\ldots).$$

This determines part of the module structure. Now, how do the $b[n]$ act for $n \geqslant 0$?

In the Heisenberg algebra, $b[n]$ participates in the bracket $[b[n], b[-n]] = nK$. Since K is central, we can choose that it acts on V diagonally by some fixed scalar. For now, we will assume it acts by the identity.

In order to have an action, then, we see that $b[n]$, for $n > 0$, must act on linear homogenous polynomials by $b[n]x_m = n\delta_{m,-n}|0\rangle$. Similarly, if $F(x_1, x_2, \ldots) \in V$ is any combination of multiparticle states one defines

$$b[n]F(x_{-1}, x_{-2} \ldots) \stackrel{\text{def}}{=} n \frac{\partial}{\partial x_{-n}} F(x_{-1}, x_{-2} \ldots).$$

It remains to define the operator b_0. One obvious choice is to declare it acts by zero, but in fact we obtain a module structure by declaring that it acts diagonally by multiplication by an arbitrary complex number $\mu \in \mathbb{C}$. The μ is called the *weight*.

To summarize, we have postulated an action of h on V where K acts by the identity and $b[n]$ acts by

$$b[n] = \begin{cases} x_n, & n < 0 \\ n \dfrac{\partial}{\partial x_{-n}}, & n > 0 \\ \mu, & n = 0 \end{cases}.$$

It is immediate to check that this endows V with an h-module structure. We arrive at the following definition. From now on, we refer to $F(\mu) = \mathbb{C}[x_{-1}, x_{-2} \ldots]$, with this module structure, as the *Fock module* of weight $\mu \in \mathbb{C}$.

In principle, we can generalize this definition slightly by declaring that K act diagonally by some arbitrary scalar $k \in \mathbb{C}$ rather than by the identity. There are essentially two cases. First, if $k \neq 0$ one can show that the reparametrization $b[n] \mapsto \frac{1}{\sqrt{k}} b[n]$ defines an isomorphism of the resulting module with $F(\mu)$ as defined above. Second, if $k = 0$ then $\mathbb{C}[x_{-1}, x_{-2}, \ldots]$ descends to a module for the abelian Lie algebra $\mathfrak{h}/(\mathbb{C}K)$ spanned by $\{b[n]\}_{n \in \mathbb{Z}}$.

One of the key properties of the module $F(\mu)$ is that it is irreducible, meaning it has no proper submodules. In fact, something stronger is true.

Proposition 8.3 *$F(\mu) \simeq F(\nu)$ if and only if $\mu = \nu$. Furthermore, if V is any irreducible h-module such that K acts by 1 and b_0 acts by μ, then $V \cong F(\mu)$.*

Here is a more invariant way to present this module. The Heisenberg algebra has an abelian subalgebra $\mathbb{C}[z] \oplus \mathbb{C} \cdot K \subset \mathsf{h}$ which is spanned by $\{b[n], K\}$ for $n \geqslant 0$. Define the one-dimensional module $\mathbb{C}(\mu) \simeq \mathbb{C}$ for this subalgebra by the rule that $b[0]$ acts by μ, K acts by 1 and $b[n]$ acts by zero for $n > 0$.

The enveloping algebra $U(\mathfrak{h})$ of the Heisenberg algebra is naturally an $\mathfrak{h}$-module. In particular, it is a module for the subalgebra $\mathbb{C}[z] \oplus \mathbb{C} \cdot K$.

Proposition 8.4 *There is an isomorphism of* $\mathfrak{h}$*-modules* $F(\mu) \cong U(h) \otimes_{U(\mathbb{C}[z] \oplus K)} \mathbb{C}(\mu)$. *In other words,* $F(\mu)$ *is the* $\mathfrak{h}$*-module* **induced** *from the* $\mathbb{C}[z] \oplus K$ *module* $\mathbb{C}(\mu)$.

Vacuum Modules

The other class of modules we introduce are modules over the affine algebra $\widehat{\mathfrak{g}}_\kappa$. Like the Fock module, it is induced from a representation of a subalgebra. Consider the subalgebra

$$\mathfrak{g}[z] \oplus \mathbb{C}K \subset \widehat{\mathfrak{g}}_\kappa.$$

Define the one-dimensional module $\mathbb{C}_1 \cong \mathbb{C}$ for this subalgebra where $\mathfrak{g}[z]$ acts trivially and K acts by the identity.

Definition 8.9 The *vacuum module* of level κ is the $\widehat{\mathfrak{g}}_\kappa$ module

$$V_\kappa(\mathfrak{g}) \stackrel{\text{def}}{=} U(\widehat{\mathfrak{g}}_\kappa) \otimes_{U(\mathfrak{g}[z] \oplus K)} \mathbb{C}_1. \tag{8.1.2}$$

While we do not touch on many details here, this class of modules for the affine Kac–Moody algebra appears in areas of physics, representation theory, and number theory. Perhaps most importantly, the vacuum module carries the structure of a *vertex algebra* which we will briefly touch on later in §8.3.1. The description as a vertex algebra has led to progress in the context of Langlands duality, specifically in the characterization of the center of the affine algebra in terms of "opers" on the Langlands dual group; see [11] and references therein for formulations of this perspective. Many of these applications rest on the relationship between the affine algebra and the Virasoro algebra. The famous Segal–Sugawara construction endows $V_\kappa(\mathfrak{g})$ with the structure of a $\mathsf{vir}_{c(\kappa)}$-module (for some value of $c(\kappa)$) for all but a single value of κ, namely when $\kappa_c = -\frac{1}{2}\kappa_{\text{Kill}}$ called the *critical level.*

8.2 Multivariable Generalizations

We return to the algebraic situation to discuss possible multivariable generalizations of the algebra of Laurent polynomials $\mathbb{C}[z, z^{-1}]$, which we recall is the algebra of functions on punctured affine space $\mathbb{A}^\times$. For $n > 1$, there are essentially two obvious candidates:

- Replace punctured affine space by the n-fold product

$$(\mathbb{A}^\times)^n = \mathbb{A}^\times \times \cdots \times \mathbb{A}^\times.$$

The algebra of functions is generated by n invertible algebraic parameters

$$\mathcal{O}((\mathbb{A}^\times)^n) = \mathbb{C}[z_1^\pm, \dots, z_n^\pm].$$

- Replace punctured affine space $\mathbb{A}^\times = \mathbb{A}^1 \setminus 0$ by punctured n-space $\mathbb{A}^n \setminus 0$. By an algebraic version of Hartogs' theorem, one finds that the algebra of functions on $\mathbb{A}^n \setminus 0$ is the algebra of polynomials in n-variables

$$\mathcal{O}(\mathbb{A}^n \setminus 0) \simeq \mathbb{C}[z_1, \dots, z_n].$$

  In particular, $\mathcal{O}(\mathbb{A}^n \setminus 0) = \mathcal{O}(\mathbb{A}^n)$.

There are intermediate versions of these two extreme cases where one takes some number of punctured m-spaces for $1 \leqslant m \leqslant n$, e.g. $\mathbb{A}^\times \times (\mathbb{A}^2 \setminus 0)$ when $n = 3$.

Central extensions play a crucial role in the theory of infinite-dimensional Lie algebras. The most basic example is the Heisenberg algebra which is the one-dimensional central extension of the abelian Lie algebra of Laurent polynomials $\mathbb{C}[z, z^{-1}]$ defined by the 2-cocycle $(f, g) \mapsto \mathrm{Res}(f dg)$. We ask for analogs of the Heisenberg algebra starting with the multivariable algebras above. In other words, do there exist analogs of this 2-cocycle defined in terms of some residue pairing?

The algebra $\mathbb{C}[z_1^\pm, \dots, z_n^\pm]$ is the building block for the theory of toroidal algebras. There is no obvious analog of the one-dimensional central extension, as in the Heisenberg algebra. However, there is a "universal" central extension which takes values in the space $\Omega^1_B / d_B \mathcal{O}$, of Kähler differentials modulo exact Kähler differentials. We will not be concerned with this extension.

Let us turn to punctured n-space. For $n > 1$, we recalled that $\mathcal{O}(\mathbb{A}^n \setminus 0) = \mathcal{O}(\mathbb{A}^n)$. This result might suggest that $\mathbb{A}^n \setminus 0$ is an unnatural place to seek a generalization of the loop algebra. Such pessimism is misplaced because of the fundamental difference with the one-dimensional case: $\mathbb{A}^n \setminus 0$ *is not an affine scheme* for $n > 1$. In other words, its sheaf of holomorphic functions has non-trivial cohomology. From the point of view of algebraic geometry it is more natural to use a *derived* algebra of functions modeling the space of derived global sections $\mathbb{R}\Gamma(\mathbb{A}^n \setminus 0, \mathcal{O})$.

It will be convenient to reference the cohomology of higher dimensional punctured affine space. A standard computation reveals the following description

$$H^i(\mathbb{A}^n \setminus 0, \mathcal{O}) = \begin{cases} 0, & i \neq 1, \dots, n-2, n \\ \mathbb{C}[z_1, \dots, z_n], & i = 0 \\ z_1^{-1} \cdots z_n^{-1} \mathbb{C}[z_1^{-1}, \dots, z_n^{-1}], & i = n-1 \end{cases}.$$

This description arises from the calculation of the Čech cohomology with respect to the cover by the affine open sets $\mathbb{A}^n \setminus \{z_i = 0\}$. The algebra structure on the cohomology is easy to describe. In degree zero, one finds the usual polynomial ring. The piece in degree $(n-1)$ is a module for this ring and the full graded algebra structure is a square-zero extension of this ring by this module.

In the first section we recall a particular model following [10] at the level of cochain complexes for the derived algebra of global sections of punctured affine space.

8.2.1 Differential Graded Algebras

The problem is to effectively probe the "non-affineness" of $\mathbb{A}^n \setminus 0$. A characterizing property is the existence of nontrivial cohomology classes in degrees other than zero. One of the core principles of "derived mathematics" is that one should keep track not of the cohomology, but of a particular *model* for it as a cochain complex.

As an example of a "model," consider the sheaf cohomology $H^\bullet(X, \mathcal{F})$ for $\mathcal{F}$ a sheaf on X. The Čech complex associated to an open cover $\mathcal{U}$ of X is a particular cochain complex $(\check{C}^\bullet(\mathcal{U}, \mathcal{F}), \delta)$ which computes the cohomology of $\mathcal{F}$:

$$H^\bullet(X, \mathcal{F}) \simeq H^\bullet\Big(\check{C}^\bullet(\mathcal{U}, \mathcal{F}), \delta\Big).$$

Any such model for sheaf cohomology of the pair $(X, \mathcal{F})$ is denoted $\mathbb{R}\Gamma(X, \mathcal{F})$ and is referred to as the "derived" global section of $\mathcal{F}$. This object is well defined only up to quasi-isomorphism of cochain complexes. For instance, if X is a smooth complex manifold and $\mathcal{F}$ is the global section of a holomorphic vector bundle F, another model is given by the so-called "Dolbeault complex"

$$\Big(\Omega^{0,\bullet}(X, F)\,,\, \overline{\partial}\Big).$$

The Dolbeault complex is analogous to the ordinary de Rham complex of differential forms. The Dolbeault complex is concentrated in degrees $0, \ldots, n = \dim_{\mathbb{C}}(X)$. In degree k, the space $\Omega^{0,k}(X, F)$ is the smooth sections of the vector bundle $\wedge^k \mathrm{T}_X^{*0,1} \otimes F$, where $\mathrm{T}_X^{*0,1}$ stands for the anti-holomorphic piece of the complexified cotangent bundle of X. The differential $\overline{\partial}$ is a component of the de Rham differential, which is extracted using the complex structure of both X and the bundle F. We refer to [21] for a more detailed exposition of Dolbeault cohomology.

Let us spell out what this looks like locally. We assume that $X = \mathbb{C}^n$ and that F is the trivial bundle, for simplicity. Introduce a holomorphic coordinate system $z = (z_1, \dots, z_n)$. The zeroth piece of the Dolbeault complex $\Omega^{0,0}(\mathbb{C}^n)$ consists simply of smooth functions $f(z,\bar{z})$ on $\mathbb{C}^n$. In fact, the full Dolbeault complex is a graded module over smooth functions freely generated by n elements $\mathrm{d}\bar{z}_1, \dots, \mathrm{d}\bar{z}_n$ providing a frame for $\mathrm{T}^{*0,1}$; they have cohomological degree $+1$. That is, as a graded vector space the Dolbeault complex is the graded polynomial algebra over smooth functions

$$\Omega^{0,\bullet}(\mathbb{C}^n) = \mathbb{C}^\infty(\mathbb{C}^n)[\mathrm{d}\bar{z}_1, \dots, \mathrm{d}\bar{z}_n]. \tag{8.2.1}$$

Here, "graded" means that we impose the relations $\mathrm{d}\bar{z}_i \mathrm{d}\bar{z}_j = -\mathrm{d}\bar{z}_j \mathrm{d}\bar{z}_i$[2] since $\mathrm{d}\bar{z}_i$ have cohomological degree $+1$. The $\bar{\partial}$ operator is defined by the formula

$$\bar{\partial} \stackrel{\text{def}}{=} \sum_i \mathrm{d}\bar{z}_i \frac{\partial}{\partial \bar{z}_i}, \tag{8.2.2}$$

which is easily seen to satisfy $\bar{\partial}^2 = 0$.

There are two very special properties of the Dolbeault complex that we'd like to emphasize:

- First, the Dolbeault complex extends to a *sheaf* of cochain complexes. In particular, if $U \subset \mathbb{C}^n$ is an open set then the Dolbeault complex $\Omega^{0,\bullet}(U)$ is defined and there is a natural restriction map $\Omega^{0,\bullet}(\mathbb{C}^n) \to \Omega^{0,\bullet}(U)$. Explicitly, $\Omega^{0,\bullet}(U)$ is the free polynomial algebra over $\mathbb{C}^\infty(U)$ on the same generators $\mathrm{d}\bar{z}_i$ as above. Restriction is simply the restriction of smooth functions.
- At least for the trivial bundle, the Dolbeault complex has the structure of a *differential graded algebra.*

Definition 8.10 A *differential graded (dg) algebra* is a cochain complex (A, d) together with a multiplication $\cdot \colon A \times A \to A$ such that

$$\mathrm{d}(a \cdot b) = (\mathrm{d}a) \cdot b + (-1)^{|a|} a \cdot (\mathrm{d}b).$$

A commutative dg algebra is one that additionally satisfies $a \cdot b = (-1)^{|a||b|} b \cdot a$ for all a, b.

In particular, for any open $U \subset \mathbb{C}^n$, the Dolbeault complex $\Omega^{0,\bullet}(U)$ is a commutative dg algebra model for the derived sections of the structure sheaf $\mathbb{R}\Gamma(U, \mathcal{O}_U)$, where $\mathcal{O}_U$ stands for the sheaf of holomorphic functions on U.

Take $U = \mathbb{C}^n \setminus 0$. There is a small difference between the cohomology of the Dolbeault complex of $\mathbb{C}^n \setminus 0$ and the cohomology of $\mathbb{A}^n \setminus 0$, as we recalled in the introduction to this section. When we write $\mathbb{C}^n$, we are working in the category

[2] We drop the conventional $\wedge$ symbol in these notes.

of *complex manifolds*, whereas $\mathbb{A}^n$ is an *algebraic* variety. In this section we want to work algebraically as an attempt to follow the setting in the first chapter as closely as possible.

An Algebraic Model for Punctured Affine Space

All polynomials are holomorphic functions, so there is an embedding of algebras

$$\mathcal{O}(\mathbb{A}^n) = \mathbb{C}[z_1, \dots, z_n] \hookrightarrow \mathcal{O}(\mathbb{C}^n).$$

We are interested in an *algebraic* analog of the Dolbeault complex which gives a particular model for $\mathbb{R}\Gamma\left(\mathbb{A}^n \setminus 0, \mathcal{O}_{\mathbb{A}^n \setminus 0}\right)$.

For affine space, one way to do this is to formally replace $\mathbb{C}^\infty(\mathbb{C}^n)$ in (8.2.1) with polynomials that have both holomorphic and anti-holomorphic dependence

$$\mathbb{C}[z_1, \dots, z_n, \bar{z}_1, \dots, \bar{z}_n][\mathrm{d}\bar{z}_1, \dots, \mathrm{d}\bar{z}_n].$$

This is, in fact, a subalgebra of $\Omega^{0,\bullet}(\mathbb{C}^n)$ and the differential $\bar{\partial}$ clearly preserves it. So, this gives us an algebraic version of $\Omega^{0,\bullet}(\mathbb{C}^n)$. A useful way to characterize this subalgebra is as the sum of eigenspaces for the natural action by the group $\mathrm{U}(1)^{\times n}$, which rotates each coordinate independently.

What about an algebraic version of $\Omega^{0,\bullet}(\mathbb{C}^n \setminus 0)$? The model we review here was proposed first in [10] motivated by the Jouanolou torsor. First, consider the localized ring

$$\mathsf{R}_n \overset{\text{def}}{=} \mathbb{C}[z_1, \dots, z_n, \bar{z}_1, \dots, \bar{z}_n]\left[\frac{1}{z\bar{z}}\right],$$

where $z\bar{z} = z_1\bar{z}_1 + \cdots + z_n\bar{z}_n$. One should think of R_n as the ring of smooth polynomial functions on punctured affine space.

Next, consider the graded algebra freely generated over R_n by degree $+1$ elements $\mathrm{d}\bar{z}_i$

$$\mathsf{R}_n[\mathrm{d}\bar{z}_1, \dots, \mathrm{d}\bar{z}_n]. \tag{8.2.3}$$

Assign a "$\bar{z}$-weight" to this ring by declaring that $\bar{z}_i$ and $\mathrm{d}\bar{z}_i$ have $\bar{z}$-weight $+1$ and $(z\bar{z})^{-1}$ has $\bar{z}$-weight -1.

Definition 8.11 Let A_n be the graded subalgebra of (8.2.3) consisting of elements $\alpha = \alpha(z, \bar{z}, \mathrm{d}\bar{z}, (z\bar{z})^{-1})$ satisfying the following two conditions:

- the $\bar{z}$-weight of α is zero;
- the contraction of α with the anti-holomorphic Euler vector field $\bar{E} = \sum_i \bar{z}_i \partial_{\bar{z}_i}$ vanishes.

The graded algebra A_n equipped with the $\overline{\partial}$ operator (8.2.2) provides a commutative dg algebra model for $\mathbb{R}\Gamma\left(\mathbb{A}^n \setminus 0, \mathcal{O}_{\mathbb{A}^n\setminus 0}\right)$.

Proposition 8.5 ([10]) *The $\overline{\partial}$ differential acting on the graded ring* (8.2.3) *restricts to a differential on A_n. Furthermore, this endows $(\mathsf{A}_n, \overline{\partial})$ with the structure of a commutative dg algebra whose cohomology is isomorphic to $H^\bullet\left(\mathbb{A}^n \setminus 0, \mathcal{O}_{\mathbb{A}^n\setminus 0}\right)$.*

We will unpack this model in a few low-dimensional cases. First, notice that A_1 is concentrated in degree zero. Indeed $\overline{z}(z\overline{z})^{-1} = z^{-1}$ thus $\mathsf{R}_1 = \mathbb{C}[z, z^{-1}, \overline{z}, \overline{z}^{-1}]$ and $\mathsf{A}_1 = \mathbb{C}[z, z^{-1}]$ as well.

Let's look at the case $n = 2$. The graded algebra A_2 is only concentrated in degrees zero and one:

$$\mathsf{A}_2 : \quad \left(\mathsf{A}_2^0 \xrightarrow{\overline{\partial}} \mathsf{A}_2^1\right).$$

The degree zero piece A_2^0 is the algebra on generators z_1, z_2 and $\overline{z}_1/(z\overline{z}), \overline{z}_2/(z\overline{z})$ subject to the relation

$$z_1(\overline{z}_1/(z\overline{z})) + z_2(\overline{z}_2/(z\overline{z})) = 1.$$[3]

In particular, an element can be written as a linear combination of monomials of the form

$$z_1^{k_1} z_2^{k_2} \frac{\overline{z}_1^{\ell_1}\overline{z}_2^{\ell_2}}{(z\overline{z})^{\ell_1+\ell_2}}, \quad k_i, \ell_i \geqslant 0,$$

subject to the relation. Of course, the differential $\overline{\partial}$ annihilates z_1, z_2. Moreover,

$$\overline{\partial} \frac{\overline{z}_1^{\ell_1}\overline{z}_2^{\ell_2}}{(z\overline{z})^{\ell_1+\ell_2}} = \mathrm{d}\overline{z}_1 \frac{\overline{z}_1^{\ell_1-1}\overline{z}_2^{\ell_2}}{(z\overline{z})^{\ell_1+\ell_2+1}}(\ell_1 z_2\overline{z}_2 - \ell_2 z_1\overline{z}_1) + \mathrm{d}\overline{z}_2 \frac{\overline{z}_1^{\ell_1}\overline{z}_2^{\ell_2-1}}{(z\overline{z})^{\ell_1+\ell_2+1}}(\ell_2 z_1\overline{z}_1 - \ell_1 z_2\overline{z}_2). \tag{8.2.4}$$

This is zero if and only if $\ell_1 = \ell_2 = 0$. Thus, we see that the cohomology in degree zero is simply polynomials in z_1, z_2: $H^0(\mathsf{A}_2) = \mathbb{C}[z_1, z_2]$.

In degree one, A_2 consists of elements of the form $g_1\,\mathrm{d}\overline{z}_1 + g_2\,\mathrm{d}\overline{z}_2, g_i \in \mathsf{R}_2$ subject to the condition that the $\overline{z}$-weight of g_i is -1 and $g_1\overline{z}_1 + g_2\overline{z}_2 = 0$. Similarly as in the previous paragraph, we find that A_2^1 is equal to $\mathsf{A}_2^0 \cdot \omega$ where ω is

$$\omega = \frac{\overline{z}_2\,\mathrm{d}\overline{z}_1 - \overline{z}_1\,\mathrm{d}\overline{z}_2}{(z\overline{z})^2}. \tag{8.2.5}$$

[3] Generally, these equations describe a quadric in weighted projective space.

This is the *Bochner–Martinelli* kernel in complex dimension two, which we will discuss in more detail in the next subsection.

To identify the cohomology in degree one, we can use (8.2.4) to show that non-exact elements can be identified with $h\omega$ where h is a polynomial in $\bar{z}_1/(z\bar{z}), \bar{z}_2/(z\bar{z})$. Thus

$$H^1(\mathsf{A}_2) \simeq \mathbb{C}\left[\frac{\bar{z}_1}{z\bar{z}}, \frac{\bar{z}_2}{z\bar{z}}\right]\omega.$$

There is an abstract isomorphism between this space and the presentation we gave for $H^1(\mathbb{A}^2 \setminus 0, \mathcal{O})$ as $z_1^{-1}z_2^{-1}\mathbb{C}[z_1^{-1}, z_2^{-1}]$ in the introduction of this section. We will see momentarily how the higher residue implements this isomorphism. We turn to that in the next section.

Before moving on, we want to observe that there is an embedding of cochain complexes from the algebraic model A_n of punctured affine space to the analytic one provided by the Dolbeault complex of $\mathbb{C}^n \setminus 0$

$$\mathsf{A}_n \hookrightarrow \Omega^{0,\bullet}(\mathbb{C}^n \setminus 0), \tag{8.2.6}$$

which is simply the inclusion $z, \bar{z}, \mathrm{d}\bar{z} \mapsto z, \bar{z}, \mathrm{d}\bar{z}$. In fact, this defines an embedding into the Dolbeault complex of $D^n \setminus 0$ for any punctured disk.

8.2.2 Interlude: Residues

Cauchy's integral formula for a disk $D \subset \mathbb{C}$ states that for any smooth function $f: D \to \mathbb{C}$ and point $z \in \mathbb{C}$,

$$2\pi\sqrt{-1}f(w) = \oint_{\partial D} \frac{\mathrm{d}z}{z-w} f(z) + \mathsf{int}_D \frac{\mathrm{d}z}{z-w} \wedge \bar{\partial} f.$$

If f is holomorphic the last term drops out and we obtain the more familiar formula $f(w) = \frac{1}{2\pi\sqrt{-1}} \oint \frac{\mathrm{d}z}{z-w} f(z)$. One uses the integral formula to prove the famous residue formula: for any holomorphic one-form ω defined on a punctured disk $D \setminus \{w\}$ one has $\mathrm{Res}_w(\omega) = \frac{1}{2\pi\sqrt{-1}} \oint_C \omega$ where C is a closed Jordan curve contained in $D \setminus \{w\}$. Note that a special role in the Cauchy integral formula is played by the "Cauchy kernel" $\frac{\mathrm{d}z}{z-w}$, which is a holomorphic one-form defined on $D \setminus \{w\}$ satisfying $\mathrm{Res}_w\left(\frac{\mathrm{d}z}{z-w}\right) = \frac{1}{2\pi\sqrt{-1}}$.

There are higher dimensional versions of the Cauchy kernel, which involve integrals over products of punctured disks. We will be more interested in integrals over the once punctured n-disk.

Let $w=(w_1,\dots,w_n)\in\mathbb{C}^n$. Define the *Bochner–Martinelli* kernel to be the differential form of type $(0,n-1)$ defined on $\mathbb{C}^n\setminus w$ by

$$\omega_{\mathrm{BM}}(z,w)=(-1)^{n-1}\frac{(n-1)!}{(2\pi\sqrt{-1})^n}\frac{1}{|z-w|^{2n}}$$
$$\sum_{1\leqslant i\leqslant n}(\overline{z}_i-\overline{w}_i)\wedge \mathrm{d}\overline{z}_1\wedge\cdots\wedge\widehat{\mathrm{d}\overline{z}_i}\wedge\cdots\wedge \mathrm{d}\overline{z}_n.$$

What will often appear in integral expressions is the differential form of type $(n,n-1)$ given by $\mathrm{d}^n z\wedge\omega_{\mathrm{BM}}(z,w)$.

The Bochner–Martinelli kernel appears in the higher dimensional generalization of the Cauchy integral formula. Consider an n-disk D^n around a point $w\in\mathbb{C}^n$ and suppose f is a smooth function $f\colon D^n\to\mathbb{C}$. Then

$$f(w)=\oint_{\partial D^n}\mathrm{d}^n z\wedge\omega_{\mathrm{BM}}(z,w)f(z)+\mathsf{int}_{D^n}\mathrm{d}^n z\wedge\omega_{\mathrm{BM}}(z,w)\wedge\overline{\partial}f(z). \tag{8.2.7}$$

In (8.2.6) we have seen how the dg algebra A_n embeds in $\Omega^{0,\bullet}(D^n\setminus 0)$ for any disk D^n centered at the origin. Use this embedding to define the following "higher residue."

Definition 8.12 The n-dimensional *residue* at $z=0$ is the linear map $\mathrm{Res}_{z=0}\colon \mathsf{A}_n[n-1]\to\mathbb{C}$ defined by the formula

$$\mathrm{Res}_{z=0}(\alpha)=\oint_{\partial D^n}\mathrm{d}^n z\wedge\alpha,$$

where D^n is any closed n-disk centered at zero.

The residue can be extended to points other than $z=0$ in a standard way. We want to point out the peculiar cohomological shift in the definition of the residue. The reason for this is that the integral in question is over a $2n-1$-dimensional sphere and so $\mathrm{d}^n z\wedge\alpha$ must be a form of total degree $2n-1$, which means that α must be a Dolbeault form of type $(0,n-1)$. Also, we observe that the residue is trivially a cochain map: A_n is concentrated in degrees $[0,n-1]$ and the residue is only nonzero on the piece in top degree. Slightly more nontrivial is the following lemma.

Lemma 8.6 *For any* $\alpha\in\mathsf{A}_n$ *one has* $\mathrm{Res}_{z=0}(\overline{\partial}\alpha)=0$.

Proof By Stokes' theorem $\mathrm{Res}_{z=0}(\overline{\partial}\alpha)=\oint\mathrm{d}^n z\wedge\overline{\partial}\alpha=\oint\overline{\partial}(\mathrm{d}^n z\wedge\alpha)=0$. □

This lemma implies the residue descends to a map on cohomology $\mathrm{Res}_{z=0}\colon H^{n-1}(\mathsf{A}_n)\to\mathbb{C}$. The residue provides a useful characterization of the cohomology. Consider the Bochner–Martinelli kernel $\omega\overset{\mathrm{def}}{=}\omega_{\mathrm{BM}}(z,0)$. First, we observe that ω is an element in A_n of degree $(n-1)$. Also, notice that when $n=1$ this is simply the function $\frac{1}{z}$. When $n=2$ this is the $(0,1)$ form defined in (8.2.5). Notice that we verified part of this proposition in the previous section, by hand, without explicitly using the residue.

Proposition 8.7 *The class $[\omega] \in H^{n-1}(A_n)$ is nontrivial in cohomology. In fact, every element in the $(n-1)$st cohomology can be written as linear combinations of classes of elements of the form*

$$\partial_{z_1}^{k_1} \cdots \partial_{z_n}^{k_n} \omega \in A_n^{n-1}.$$

Proof By the integral formula (8.2.7) we have $\mathrm{Res}_{z=0}(\omega) = 1$. To see that the remaining classes in the proposition are nontrivial, we simply apply integration by parts in the integral formula. □

By this proposition we have the following description of the cohomology of A_n: in degree zero it is $\mathbb{C}[z_1, \ldots, z_n]$ and in degree $(n-1)$ it is

$$H^{n-1}(\mathsf{A}_n) \simeq \mathbb{C}[\partial_{z_1}, \ldots, \partial_{z_n}]\omega.$$

8.2.3 Higher Current Algebras

A familiar generalization of the notion of a Lie algebra in the context of supersymmetry and certain topics in representation theory is a *super* Lie algebra. A super Lie algebra is a $\mathbb{Z}/2$-graded vector space $\mathfrak{h} = \mathfrak{h}^{\mathrm{even}} \oplus \Pi\mathfrak{h}^{\mathrm{odd}}$ equipped with a bracket that preserved the parity. Here, the $\Pi(-)$ denotes parity shift. This bracket is required to satisfy $\mathbb{Z}/2$ graded versions of the antisymmetry and Jacobi relations as well. A natural further lift of a super Lie algebra is the notion of a $\mathbb{Z}$-graded Lie algebra, henceforth referred to as just a graded Lie algebra. We will be interested in graded Lie algebras equipped with the further data of a differential.

Definition 8.13 A *dg Lie algebra* is the data of a graded vector space $\mathfrak{h} = \oplus \mathfrak{h}^i[-i]$, a linear map $\mathrm{d}\colon \mathfrak{h}^\bullet \to \mathfrak{h}^{\bullet+1}$ and a bracket $[\cdot,\cdot]\colon \mathfrak{h}^\bullet \times \mathfrak{h}^\bullet \to \mathfrak{h}^\bullet$ such that the following conditions hold.

- d turns $\mathfrak{h} = \oplus \mathfrak{h}^i[-i]$ into a cochain complex (so $\mathrm{d}^2 = 0$) and the graded Leibniz rule is satisfied

$$\mathrm{d}[x,y] = [\mathrm{d}x, y] + (-1)^{|x|}[x, \mathrm{d}y].$$

- Graded anti-symmetry $[x,y] = (-1)^{|x||y|+1}[y,x]$.

- Graded Jacobi identity

$$(-1)^{|x||z|}[x,[y,z]]+(-1)^{|x||y|}[y,[z,x]]+(-1)^{|y||z|}[z[x,y]]=0$$

is satisfied.

The starting point for affine algebras is the loop algebra $\mathfrak{g}[z,z^{-1}]=\mathfrak{g}\otimes_{\mathbb{C}}\mathbb{C}[z,z^{-1}]$. The Lie algebra structure on this loop algebra is induced from the Lie bracket on $\mathfrak{g}$ together with the commutative product on Laurent polynomials. A similar construction holds in the setting of dg Lie algebras.

Lemma 8.8 *Suppose (A,d) is a commutative dg algebra and $(\mathfrak{h},[\cdot,\cdot]_{\mathfrak{h}})$ is an ordinary Lie algebra. Then $A\otimes\mathfrak{h}$ is endowed with the natural structure of a dg Lie algebra. The differential is $d\otimes\mathrm{id}_{\mathfrak{h}}$ and the bracket is $[a\otimes X,b\otimes Y]=(ab)\otimes[X,Y]_{\mathfrak{h}}$.*

We arrive at the precise notion of the "higher current algebras" we referenced in the introduction. Recall the dg model A_n for the derived global sections of punctured affine space $\mathbb{A}^n\setminus 0$.

Definition 8.14 The *n-current algebra* of a Lie algebra $\mathfrak{g}$ is the dg Lie algebra $\mathfrak{g}\otimes\mathsf{A}_n$. This dg Lie algebra is concentrated in degrees $[0,n-1]$ and the differential is given by the $\overline{\partial}$ operator.

If $(\mathfrak{h},\mathrm{d},[\cdot,\cdot])$ is any dg Lie algebra then the d-cohomology $H^\bullet(\mathfrak{h})$ inherits the structure of a graded Lie algebra from the original bracket.

Proposition 8.9 *The graded Lie algebra $H^\bullet(\mathfrak{g}\otimes\mathsf{A}_n)$ is concentrated in degrees zero and $n-1$ and admits the presentation:*

- *In degree zero the cohomology is $\mathfrak{g}\otimes\mathbb{C}[z_1,\dots,z_n]$. We use the notation for elements:*

$$X[k_1,\dots,k_n]\overset{\text{def}}{=}X\otimes z_1^{k_1}\cdots z_n^{k_n}\in\mathfrak{g}\otimes\mathbb{C}[z_1,\dots,z_n],$$

where $X\in\mathfrak{g}$, $k_1,\dots,k_n\geqslant 0$.
- *In degree $n-1$ the cohomology is $\mathfrak{g}\otimes\mathbb{C}[\partial_{z_1},\dots,\partial_{z_n}]\omega$. We use the notation for elements:*

$$X[\ell_1,\dots,\ell_n]\overset{\text{def}}{=}X\otimes\left(\prod_{j=1}^{n}\frac{(-1)^{-\ell_j-1}}{(-\ell_j-1)!}\partial_{z_j}^{-\ell_j-1}\right)\omega,$$

where $X\in\mathfrak{g}$, $\ell_1,\dots,\ell_n<0$.[4]

[4] The complicated-looking normalization makes for simpler formulas, which will appear later on.

The (graded) Lie bracket is described by

$$\Big[X[k_1,\dots,k_n],Y[k_1',\dots,k_n']\Big] = [X,Y]_{\mathfrak{g}}[k_1+k_1',\dots,k_n+k_n'], \quad k_i,k_j' \geqslant 0,$$

$$\Big[X[k_1,\dots,k_n],Y[\ell_1,\dots,\ell_n']\Big] = [X,Y]_{\mathfrak{g}}[k_1+\ell_1,\dots,k_n+\ell_n], \quad 0 < -\ell_j \leqslant k_j.$$

Remaining brackets are determined by graded skew symmetry.

We point out a property of the graded Lie bracket in cohomology which is different than the ordinary (one-dimensional) current algebra. Consider the case $n=2$, and for simplicity take the Lie algebra $\mathfrak{g}=\mathfrak{sl}(2)$ with standard basis $\{e,f,h\}$. The cohomology $\mathfrak{sl}(2)\otimes H^\bullet(\mathsf{A}_2)$ is spanned by elements $\{e[k,\ell],f[k,\ell],h[k,\ell]\}$ for $k,\ell\in\mathbb{Z}$.

Consider, for instance, the elements $e[k,\ell]$ and $f[r,s]$. The elements with $k=\ell=r=s\geqslant 0$ satisfy the $\mathfrak{sl}(2)$-type relation $[e[k,\ell],f[r,s]]=h[k+r,\ell+s]$. However, notice that

$$\big[e[1,0],f[-1,-1]\big]=0.$$

This property of the bracket is quite different than in the $n=1$ case. Indeed, if $e[k],f[r]\in\mathfrak{sl}(2)\otimes\mathsf{A}_1$ then

$$\big[e[k],f[r]\big]=h[k+r]$$

for *all* $k,r\in\mathbb{Z}$. In other words, it appears that for $n>1$, taking cohomology seems to result in the loss of some amount of structure that one might expect to be present in the current algebra. This reflects the existence of "higher order" operations present in the level of cohomology, reminiscent of Massey products in the de Rham cohomology of a non-formal space. It leads us to the theory of L_∞ algebras, which we will also need in order to describe the centrally extended algebras.

8.2.4 Interlude: L_∞ Algebras

The most explicit description of central extensions of higher dimensional current algebras will force us to leave the world (ever so slightly) of dg Lie algebras. An L_∞ *algebra* is a precise weakening of the notion of a dg Lie algebra. We give a brief synopsis here, but refer to [20, 26, 27] for further background and further explanations of the formulas given below. These algebras are a lot like dg Lie algebras: there is a differential d and bracket $[\cdot,\cdot]$, and d is required to be a derivation for the bracket. The key difference is: *the Jacobi identity may not hold*. But, it is required to hold *up to homotopy*. Precisely, this means that we have to prescribe the data of a new 3-ary "bracket"

$$[\cdot,\cdot,\cdot]_3 \colon \mathfrak{g}^\bullet \times \mathfrak{g}^\bullet \times \mathfrak{g}^\bullet \to \mathfrak{g}^\bullet[-1]$$

of cohomological degree -1 which satisfies an identity of the form

$$\begin{aligned}&[[x_1,x_2],x_3] \pm [[x_2,x_3],x_1] \pm [[x_3,x_1],x_2] \\ &\quad = \mathrm{d}[x_1,x_2,x_3]_3 \pm [\mathrm{d}x_1,x_2,x_3]_3 \pm [x_1,\mathrm{d}x_2,x_3]_3 \pm [x_1,\mathrm{d}x_2,x_3]_3. \quad (8.2.8)\end{aligned}$$

This identity says that the left-hand side, while not zero, is trivial up to homotopy – the 3-bracket provides this homotopy. In particular, the Jacobi identity holds in cohomology. In addition, there are higher compatibilities that the 3-linear bracket $[\cdot]_3$ must satisfy which also potentially involve 4-ary brackets, and so on.

Before giving the precise definition, we set up some notation which is known as the "Koszul sign rule." Consider a collection of homogenous elements $x_1,\dots,x_n$. The *Koszul sign* $\varepsilon(x_1,\dots,x_n;\sigma)$ of a permutation $\sigma \in \Sigma_n$ is defined by the following relation

$$x_1 \cdots x_n = \varepsilon(x_1,\dots,x_n;\sigma) x_{\sigma(1)} \cdots x_{\sigma(n)}.$$

We will abbreviate this by $\varepsilon(\sigma)$ in the definition below.

Definition 8.15 Let $\mathfrak{g}^\bullet = \oplus_{j\in\mathbb{Z}} \mathfrak{g}^j[-j]$ be a $\mathbb{Z}$-graded vector space. An *L_∞ algebra* structure on V is a collection of multilinear maps

$$[\cdot]_k \colon \mathfrak{g}^{\times k} \to \mathfrak{g}[2-k]$$

for $k \geqslant 1$ such that the following conditions hold.

(i) Graded skew symmetry. For all $\sigma \in S_k$, $x_i \in \mathfrak{g}$ one has

$$[x_{\sigma(1)},\dots,x_{\sigma(k)}]_k = (-1)^\sigma \varepsilon(\sigma)[x_1,\dots,x_k]_k.$$

(ii) Higher Jacobi identities. For all $x_i \in \mathfrak{g}$ one has

$$\sum_{i+j=k+1} \sum_{\sigma} (-1)^{i(j-1)}(-1)^\sigma \varepsilon(\sigma)\Big[\big[x_{\sigma(1)},\dots,x_{\sigma(i)}\big]_i, x_{\sigma(i+1)},\dots,x_{\sigma(k)}\Big]_j,$$

where σ ranges over $(i,k-i)$ unshuffles.

We point out that Lie, graded Lie and dg Lie algebras are special cases: when $g^j = 0$ for $j \neq 0$ this returns the definition of an ordinary Lie algebra; a graded Lie algebra is an L_∞ algebra with $[\cdot]_k = 0$ for $k \neq 2$; a dg Lie algebra is an L_∞ algebra with $[\cdot]_k = 0$ for $k > 2$.

Transferred Structure

Before moving towards central extensions, we follow up the discussion following Proposition 8.9 about the "lack" of structure present in the cohomology of the current algebra $\mathfrak{g}\otimes \mathsf{A}_n$.

Given any dg Lie algebra $(\mathfrak{g},\mathrm{d},[\cdot,\cdot])$ its d-cohomology $H^\bullet(\mathfrak{g},\mathrm{d})$ has the structure of a graded Lie algebra (a dg Lie algebra with zero differential). This theorem follows from a general result about operads [28, 10.3.15].

Theorem 8.16 *Suppose that $(\mathfrak{g},d,[\cdot,\cdot])$ is a dg Lie algebra. Then, there exists an L_∞ structure $\{[\cdot]_k\}$ on $H = H^\bullet(\mathfrak{g},d)$ such that $H \simeq \mathfrak{g}$ as L_∞ algebras.*

Generally speaking, the transferred L_∞ structure on $H^\bullet(\mathfrak{g}\otimes \mathsf{A}_n)$ has nontrivial higher operations. We will not fully characterize the transferred L_∞ algebra in these notes. We state a nontrivial higher bracket for the example $n=2$ and $\mathfrak{g}=\mathfrak{sl}(2)$. Recall that the 2-ary bracket of both $e[1,0]$ and $e[0,1]$ with $f[-1,-1]$ is trivial in $\mathfrak{sl}(2)\otimes H^\bullet(\mathsf{A}_2)$. However, there is a higher 3-ary operation of the form

$$\big[e[1,0],e[0,1],f[-1,-1]\big]_3 = e[0,0].$$

The existence of these higher operations implies that $\mathfrak{sl}(2)\otimes \mathsf{A}_2$ is not formal as a dg Lie algebra.

Remark In general, the L_∞ structure on $H^\bullet(\mathfrak{g}\otimes \mathsf{A}_n) \simeq \mathfrak{g}\otimes H^\bullet(\mathsf{A}_n)$ is obtained from an A_∞ structure on $H^\bullet(\mathsf{A}_n)$. This A_∞ algebra has higher operations for $n>1$.

8.2.5 Central Extensions, Redux

Recall that one-dimensional central extensions of an ordinary Lie algebra $\mathfrak{h}$ are on bijective correspondence with the cohomology group $H^2(\mathfrak{h}\,;\,\mathbb{C})$. There is a very similar correspondence for dg Lie algebras (in fact, all L_∞ algebras).

The Lie algebra cohomology of a dg Lie algebra $(\mathfrak{h},\mathrm{d},[\cdot,\cdot])$ is defined very similarly as in the ordinary case. Explicitly, it is cohomology of the cochain complex

$$\mathrm{C}^\bullet(\mathfrak{h}) \overset{\mathrm{def}}{=} \Big(\mathrm{Sym}(\mathfrak{h}^*[-1])\,,\,\mathrm{d}_1+\mathrm{d}_2\Big),$$

where

- d_1 stands for the "internal" differential of the cochain complex $(\mathfrak{h},\mathrm{d})$. On $\mathfrak{h}^*[-1]$ it is the linear dual of d. It is extended to the full symmetric algebra by the graded Leibniz rule. In particular, it preserves symmetric degree

$$\mathrm{d}_1 : \mathrm{Sym}^k(\mathfrak{h}^*[-1]) \to \mathrm{Sym}^k(\mathfrak{h}^*[-1]).$$

- d_2 is the familiar Chevalley–Eilenberg differential associated to the bracket $[\cdot,\cdot]$. It is defined exactly as in the case of Lie algebra cohomology of an ordinary Lie algebra; see Definition 8.3.

Proposition 8.10 *Let $(\mathfrak{h}, d, [\cdot,\cdot])$ be a dg Lie algebra. Suppose that $\varphi \in C^\bullet(\mathfrak{h}\,;\mathbb{C})$ is a cocycle of total cohomological degree N. Moreover, assume that φ admits a decomposition of the form $\varphi = \varphi^{(1)} + \cdots + \varphi^{(k)} + \cdots$, where $\varphi^{(k)} : \mathrm{Sym}^k(\mathfrak{g}[1]) \to \mathbb{C}[N]$ is the kth Taylor component of φ. Then, φ defines an L_∞ algebra $\widetilde{\mathfrak{h}}_\varphi$ which as a vector space is*

$$\widetilde{\mathfrak{h}}_\varphi = \mathfrak{h} \oplus \mathbb{C}[N-2]\cdot K$$

and whose brackets are defined by the rules:

- *the element K is central;*
- *the remaining brackets $[\cdot]_k : \mathfrak{h}^{\times k} \to \widetilde{\mathfrak{h}}_\varphi$ are defined by*

$$[\cdot]_1 = d + K\varphi^{(1)}$$
$$[\cdot]_2 = [\cdot,\cdot] + K\varphi^{(2)}$$
$$[\cdot]_3 = K\varphi^{(3)}, \quad etc.$$

This result generalizes Lemma 8.1 – *every* cocycle in $C^\bullet(\mathfrak{h})$ represents an L_∞ central extension. We turn to the simplest central extension in the context of multivariable infinite-dimensional Lie algebras – the higher dimensional Heisenberg algebra.

The Higher Dimensional Heisenberg Algebra

Consider the cochain complex A_n modeling punctured affine space considered as an abelian dg Lie algebra. The following lemma is a rephrasing of observations made in §8.2.2.

Lemma 8.11 *The multilinear map $\varphi_n : (A_n)^{\times n+1} \to \mathbb{C}$ defined by*

$$\varphi_n(\alpha_0,\ldots,\alpha_n) \stackrel{\text{def}}{=} \mathrm{Res}_{z=0}(\alpha_0\,\partial\alpha_1 \cdots \partial\alpha_n)$$

defines a nontrivial degree +2 cocycle $\varphi_n \in C^\bullet(A_n\,;\mathbb{C})$.

Using Proposition 8.10 φ_n determines a one-dimensional L_∞ central extension of A_n.

Definition 8.17 The *n-Heisenberg algebra* is the L_∞ central extension of the abelian dg Lie algebra A_n by the degree +2 cocycle φ_n.

The underlying cochain complex of h_n is

$$\left(\mathsf{A}_n \oplus \mathbb{C}K\,,\,\overline{\partial}\right),$$

where $\overline{\partial}$ acts on A_n in the usual way and $\overline{\partial}(K) = 0$. The only nontrivial higher bracket is $(n+1)$-ary and is given in terms of the higher residue

$$[\alpha_1, \ldots, \alpha_{n+1}]_{n+1} = \mathrm{Res}_{z=0}\left(\alpha_1 \partial \alpha_2 \cdots \partial \alpha_{n+1}\right).$$

The cohomology of the Heisenberg algebra admits a simple presentation. Recall, by Theorem 8.16 that the cohomology of any L_∞ algebra has the inherited structure of an L_∞ algebra with $[\cdot]_1 = 0$.

Proposition 8.12 *The transferred L_∞ structure present in the cohomology $H^\bullet = H^\bullet(\mathsf{h}_n)$ has trivial k-ary operation $[\cdot]_k$ for $k \neq n+1$. The $(n+1)$-ary operation is given by the residue*

$$\begin{array}{rccl} [\cdot]_{n+1}\colon & (H^0)^{\times n} \times H^{n-1} & \to & \mathbb{C}\cdot K, \\ & (f_1, \ldots, f_n; \alpha) & \mapsto & K \cdot \mathrm{Res}_{z=0}\left(f_1 \partial f_2 \cdots \partial f_n \partial \alpha\right). \end{array}$$

Let's unravel this proposition in the case $n = 2$. We will use the holomorphic coordinates $(z, w) = (z_1, z_2)$ for $\mathbb{C}^2$ in this section to avoid clutter. The cohomology of h_2 is nontrivial only in degrees zero and one. There is the following presentation for the degree zero cohomology $\mathrm{span}_{\mathbb{C}}\left\{b[n,m]\,,\, K \mid n, m \in \mathbb{Z}_{\geqslant 0}\right\}$. In terms of holomorphic polynomials in two variables, $b[n,m]$ corresponds to $b[n,m] = z^n w^m$ for $n, m \geqslant 0$, just as in Proposition 8.9 (where $\mathfrak{g}$ is the one-dimensional Lie algebra $\mathbb{C}\cdot b$.) In degree one, the cohomology is $\mathrm{span}_{\mathbb{C}}\left\{b[n,m] \mid n, m \in \mathbb{Z}_{<0}\right\}$. The generators $b[n,m]$ for $n, m < 0$ are given by the $\overline{\partial}$-class of derivatives of the Bochner–Martinelli kernel, see the formulas in Proposition 8.9.

The only nontrivial higher bracket is a 3-ary bracket on $H^\bullet(\mathsf{h}_2)$ of the form

$$[\cdot,\cdot,\cdot]_3\colon H^1(\mathsf{h}_2) \times H^0(\mathsf{h}_2) \times H^0(\mathsf{h}_2) \to \mathbb{C}K$$

and permutations thereof. In terms of these generators, one can read off this 3-ary bracket using the explicit formula involving the residue class

$$\left[b[n,m]\,,\, b[k,\ell]\,,\, b[r,s]\right]_3 = (ks - \ell r)\delta_{-n,k+r}\delta_{-m,\ell+s}K. \tag{8.2.9}$$

It is instructive to check that the higher Jacobi identity is satisfied using this combinatorial description of the higher bracket. (One may find the identity useful: $(ks - \ell r)\delta_{-n,k+r}\delta_{-m,\ell+s} = (mr - ns)\delta_{-n,k+r}\delta_{-m,\ell+s}$.)

Higher Dimensional Kac–Moody Algebra

We now consider central extensions of the n-current algebra $\mathfrak{g}\otimes \mathsf{A}_n$, where $\mathfrak{g}$ is an ordinary Lie algebra. Recall that in the case $n=1$, we considered central extensions built from invariant quadratic polynomials on $\mathfrak{g}$. One of the main results of [10] is a concrete relationship of higher order invariant polynomials with central extensions of the n-current algebra.

Theorem 8.18 ([10]) *Let $n\geqslant 1$ and suppose that $\mathfrak{g}$ is semisimple. Then there is an embedding of vector spaces*

$$\mathrm{Sym}^{n+1}(\mathfrak{g}^*)^{\mathfrak{g}} \hookrightarrow H^2(\mathfrak{g}\otimes A_n),$$

which sends an invariant polynomial θ to a class $[\varphi_\theta]$.

A particular representative for the class φ_θ can be constructed as follows. First, like the n-Heisenberg algebra the representative φ_θ is $(n+1)$-linear. If $x_0\otimes\alpha_0,\dots,x_n\otimes\alpha_n$ are n-currents then we can consider the differential form $\theta(x_0,\dots,x_n)\alpha_0\partial\alpha_1\cdots\partial\alpha_n$ which is necessarily of type $(n,\bullet)$. The cochain φ_θ is the residue of this class

$$\varphi_\theta(x_0\otimes\alpha_0,\dots,x_n\otimes\alpha_n)=\theta(x_0,\dots,x_n)\mathrm{Res}_{z=0}\big(\alpha_0\partial\alpha_1\cdots\partial\alpha_n\big).$$

It is an instructive exercise to verify that φ_θ is indeed of degree $+2$ and is closed for the Chevalley–Eilenberg differential on $\mathrm{C}^\bullet(\mathfrak{g}\otimes\mathsf{A}_n)$.

Definition 8.19 Let θ be a degree $(n+1)$ invariant polynomial of $\mathfrak{g}$. The *n-Kac–Moody algebra* associated to θ is the central extension $\widehat{\mathfrak{g}}_{n,\theta}$ of the n-current algebra associated to the class $[\varphi_\theta]$.

8.2.6 Derived Fock Modules

To begin our discussion of modules, we attempt to parallel the construction of the Fock module $F(\mu)$ for the ordinary Heisenberg algebra h (see §8.1.3) to modules for the higher dimensional Heisenberg algebra. We will find that already in the case $n=2$, such modules are inherently *derived* in the sense that they are "L_∞ modules" for the 2-Heisenberg algebra h_2.

We try to mimic the definition of the ordinary Fock module as closely as possible. We immediately run into a discrepancy with the ordinary situation: the "creation operators" $b[n,m]$, $n,m<0$ lie in a nontrivial cohomological degree. Thus, any generalization of the Fock module to higher dimensions must have components in nontrivial cohomological degree.

Consider the following $\mathbb{Z}$-graded vector space given by a polynomial algebra on a (doubly) infinite number of generators:

$$V = \mathbb{C}[x_{i,j}]_{i,j \leqslant -1}.$$

The cohomological $\mathbb{Z}$-grading is determined by declaring that $x_{i,j}$ has cohomological degree $+1$ for all i,j.

We define the action of the "creation operators" by

$$b[n,m]|0\rangle \stackrel{\text{def}}{=} x_{n,m}, \quad n,m < 0.$$

More generally, if $F(x_{i,j}) \in V$ is any state, define

$$b[n,m]F(x_{i,j}) \stackrel{\text{def}}{=} x_{n,m}F(x_{i,j}), \quad n,m < 0.$$

We will assume that the central element K acts by the identity on V, just as in the case of the ordinary Fock module. Analogous to the ordinary Fock module, for $n,m > 0$ the elements $b[n,m]$ will be "annihilation operators". However, their action on V is more subtle. Notice that there are no elements in h_2 of cohomological degree -1. So, the naive definition of annihilation

$$b[k,\ell]x_{n,m} \stackrel{?}{=} C(k,\ell,m,n)\delta_{k,n}\delta_{\ell,m}|0\rangle,$$

where $C(k,\ell,n,m)$ is some constant, is not sensible. One, perhaps disappointing, definition is to declare that $b[n,m]$ act trivially for $n,m > 0$. This would certainly be compatible with the fact that h_2 has no nontrivial 2-ary brackets. But, what about compatibility with the 3-ary bracket $[\cdot,\cdot,\cdot]_3$ which we know is present in cohomology?

The key is that we do not obtain the structure of a *strict* $H^\bullet(\mathsf{h}_2)$-module on V. Instead, one finds the structure of an L_∞*-module*.

Interlude: L_∞ Modules

The notion of an L_∞ module is a weakening of the notion of a module for a Lie algebra in the same way that an L_∞ algebra weakens the notion of a Lie algebra. Let $\mathfrak{h}$ be a Lie algebra. An $\mathfrak{h}$-module is prescribed by the data of a vector space M and a map $\rho\colon \mathfrak{h} \times M \to M$, $(x,y) \mapsto \rho(x;y)$ which satisfies $\rho([x,x'];y) = \rho(x;\rho(x';y)) - \rho(x';\rho(x;y))$.

Suppose that $\mathfrak{h} = (\mathfrak{h}, \mathrm{d}_{\mathfrak{h}})$ is a dg Lie algebra. A dg $\mathfrak{h}$-module is a cochain complex (M, d_M) together with a grading-preserving map ρ as above which additionally satisfies $\rho(\mathrm{d}_{\mathfrak{h}}x;y) = \mathrm{d}_M\rho(x;y) - \rho(x;\mathrm{d}_M y)$. In the case of an L_∞ algebra, we have the following generalization of this definition.

Definition 8.20 Let $(\mathfrak{h}, [\cdot]_k)$ be an L_∞ algebra and (M, d_M) a cochain complex. An L_∞ $\mathfrak{h}$-*module structure* on M is a collection of graded skew-symmetric multilinear maps

$$\rho^{(k)} \colon \mathfrak{h}^{\times k} \times M \to M[1-k]$$

such that

$$0 = \sum_{i+j=k+2} \sum_{\sigma} \chi(\sigma)(-1)^{i(j-1)} \rho^{(j)}(\rho^{(i)}(x_{\sigma(1)}, \ldots, x_{\sigma(i)}), x_{\sigma(i+1)}, \ldots, x_{\sigma(k)}),$$

where σ ranges over $(i, n-i)$-unshuffles, $x_1, \ldots, x_{n-1} \in \mathfrak{h}$, $x_n \in M$. In the formulas above we set $\rho^{(0)} \overset{\text{def}}{=} \mathrm{d}_M$ and $\rho^{(n)}(y_1, \ldots, y_m) = [x_1, \ldots, x_m]_n$ if $y_1, \ldots, y_m \in \mathfrak{h}$.

Let's unpack this in the simple case where $\mathfrak{h}$ is a graded Lie algebra. Then, we see that $\rho^{(1)}$ does *not* define an $\mathfrak{h}$-module structure, but rather

$$\rho^{(1)}(x; \rho^{(1)}(y; m)) - \rho^{(1)}(y; \rho^{(1)}(x; m)) = \mathrm{d}_M \rho^{(2)}(x, y; m), \quad x, y \in \mathfrak{h}, \quad m \in M.$$

In other words, the failure for $\rho^{(1)}$ to be an $\mathfrak{h}$-module is homotopically trivializable, with homotopy given by $\rho^{(2)}$.

Example 8.21 To any Lie algebra $\mathfrak{h}$, one can associate the module $\mathrm{Sym}(\mathfrak{h})$. Similarly, if $\mathfrak{h}$ is an L_∞ algebra, then $\mathrm{Sym}(\mathfrak{h})$ has the canonical structure of an L_∞ $\mathfrak{h}$-module. If $[\cdot]_k \colon \mathfrak{h}^{\times k} \to \mathfrak{h}$ is the k-ary operation of $\mathfrak{h}$ then this module structure is defined on generators $\mathfrak{h} \subset \mathrm{Sym}(\mathfrak{h})$ by the collection of maps $\rho^{(k)}$ with

$$\rho^{(k)}(x_1, \ldots, x_k; y) = [x_1, \ldots, x_k, y]_{k+1}.$$

The L_∞ Fock Module

The Fock module $V = \mathbb{C}[x_{i,j}]$ has the structure of an L_∞ module for the L_∞ algebra $H^\bullet(\mathsf{h}_2)$. This module structure has a nontrivial linear $\rho^{(1)}$ and 2-linear $\rho^{(2)}$ component which we now define.

For $n, m > 0$ the operators $b[n, m]$ participate in the L_∞-module structure as follows. First, the linear action of $b[n, m]$ on V is trivial

$$\rho^{(1)}(b[n, m]; F(x_{i,j})) = 0, \quad n, m > 0$$

for all $F(x_{i,j}) \in V$. Next, there is a 2-ary component to the L_∞-module structure defined by

$$\rho^{(2)}(b[k, \ell], b[r, s]; x_{n,m}) = (ks - \ell r)\delta_{-n,k+r}\delta_{-m,\ell+s}|0\rangle.$$

More generally, for any $F(x_{i,j}) \in V$ we define

$$\rho^{(2)}(b[k, \ell], b[r, s]; F(x_{i,j})) = (ks - \ell r)\frac{\partial}{\partial x_{-k-r,-\ell-s}} F(x_{i,j}).$$

Notice that this formula is sensible for all k,ℓ,r,s such that the elements $b[k,\ell],b[r,s]$ are defined.

Proposition 8.13 *These formulas for* $\{\rho^{(1)},\rho^{(2)}\}$ *endow* V *with the structure of an* L_∞ $H^\bullet(\mathsf{h}_2)$*-module.*

Proof Let $F = F(x_{i,j}) \in \mathbb{C}[x_{i,j}]$ and suppose that $n,m<0$ and $r,s,k,\ell \geqslant 0$. We check the following ∞-module relation:

$$\begin{aligned}\rho^{(2)}(b[k,\ell],b[r,s];\rho^{(1)}(b[n,m];F(x_{i,j})))\\ +\rho^{(1)}(b[n,m];\rho^{(2)}(b[k,\ell],b[r,s];F(x_{i,j})))\\ \stackrel{?}{=}\rho^{(1)}(\left[b[k,\ell],b[r,s],b[n,m]\right]_3;F(x_{i,j})).\end{aligned} \tag{8.2.10}$$

Since $n,m<0$, the first term on the left-hand side is

$$\rho^{(2)}(b[k,\ell],b[r,s];x_{n,m}F(x_{i,j})) = (ks-\ell r)\left(F - x_{n,m}\frac{\partial F}{\partial x_{-k-r,-\ell-s}}\right).$$

The second term on the left-hand side is

$$\rho^{(1)}\left(b[n,m];(ks-\ell r)\frac{\partial F}{\partial x_{-k-r,-\ell-s}}\right) = (ks-\ell r)x_{n,m}\frac{\partial F}{\partial x_{-k-r,-\ell-s}}.$$

Using Equation (8.2.9) we see that the right-hand side is

$$(ks-\ell r)\rho^{(1)}(K;F(x_{i,j}) = (ks-\ell r)F(x_{i,j})$$

thus verifying the relation. The proofs for other values of indices are analogous and we leave them to the interested reader. □

Derived Creation/Annihilation

We have just seen how to construct a class of modules for the cohomology of the higher dimensional Heisenberg algebra. In this section, we show how this can be lifted to the cochain level. We return to the case of a general dimension $n>1$.

Recall that the complex A_n is concentrated in degrees $0,\dots,n-1$. In particular, there is a quotient map

$$\mathsf{A}_n[n-1] \to H^{n-1}(\mathsf{A}_n),$$

which sends a class $\alpha \in \mathsf{A}_n^{n-1}$ (which is automatically a cocycle) to its cohomology class $[\alpha]$.

Definition 8.22 Define the vector space $\mathsf{A}_{n,-}$ to be the top cohomology $H^{n-1}(\mathsf{A}_n)$. Define $\mathsf{A}_{n,+}$ to be the short exact sequence of dg algebras

$$\mathsf{A}_{n,+} \hookrightarrow \mathsf{A}_n \to \mathsf{A}_{n,-}.$$

Notice that like A_n, the algebra $\mathsf{A}_{n,+}$ is concentrated in degrees $[0, n-1]$. However, its cohomology is concentrated in degree zero.

Let's return to the n-dimensional Heisenberg L_∞ algebra h_n. Notice that the higher residue pairing is trivial when restricted to the subalgebra $\mathsf{A}_{n,+}$. In particular, the *abelian* graded Lie algebra $\mathsf{A}_{n,+} \oplus \mathbb{C}K$ is a subalgebra of the full Heisenberg algebra

$$\mathsf{A}_{n,+} \oplus \mathbb{C}K \subset \mathsf{h}_n.$$

With this, we can proceed to a family of modules for h_n by induction just as in the ordinary case.

Fix complex numbers $\mu, k \in \mathbb{C}$. Let $\mathbb{C}(\mu, k)$ denote the one-dimensional $(\mathsf{A}_{n,+} \oplus \mathbb{C}K)$-module where:

- K acts by k;
- $b[0,0] = 1 \in \mathsf{A}^0_{n,+}$ acts by μ;
- the remaining elements act trivially.

The universal enveloping construction can be extended to L_∞ algebras (see [3]). In this way, any L_∞ algebra $\mathfrak{g}$ determines an A_∞ algebra $U(\mathfrak{g})$. In the definition below, recall that $U(\mathsf{h}_n) \simeq \mathrm{Sym}(\mathsf{h}_n)$ as complexes, and hence has the natural structure of a left L_∞ h_n-module.

Definition 8.23 The *derived Fock module* for the n-Heisenberg algebra h_n associated to the pair of numbers (μ, k) is the L_∞ module

$$\mathsf{F}_n(\mu, k) \overset{\text{def}}{=} U(\mathsf{h}_n) \otimes_{U(\mathsf{A}_{n,+} \oplus \mathbb{C}K)} \mathbb{C}(\mu, k).$$

The reader will recognize the similarities of this definition with formula (8.1.2). The cohomology of the derived Fock module is simple to describe based on previous calculations.

Proposition 8.14 *As a graded vector space, the higher residue identifies $H^\bullet(F_n(\mu,k))$ with the graded symmetric algebra on $A_{n,-}[1-n]$. In particular:*

- *When n is even, the cohomology is a symmetric algebra*

$$H^\bullet(F_n(\mu,k)) \simeq \bigoplus_{k \geqslant 0} \mathrm{Sym}^k(\mathsf{A}_{n,-})\,[k(1-n)].$$

- *When n is odd, the cohomology is an exterior algebra*

$$H^\bullet(F_n(\mu,k)) \simeq \bigoplus_{k \geqslant 0} \wedge^k(A_{n,-})\,[k(1-n)].$$

Remark This statement is valid even when $n = 1$. Indeed, we recover the usual presentation of the Fock module since $\mathsf{A}_{1,-} \cong z^{-1}\mathbb{C}[z^{-1}]$ by convention. For any n, the cohomology of the Fock module is the graded symmetric algebra of a vector space concentrated in degree $n-1$. In particular, when $n > 1$, the Fock module always carries nontrivial cohomological degree.

When $n = 2$, we see that the cohomology of the derived Fock module is the graded symmetric algebra on a vector space in concentrated in degree $+1$. Thus, there is some chance for $H^\bullet(\mathsf{F}_2(\mu,k))$ to agree with the module we constructed by hand in the introduction to this section. This is indeed the case. The transferred L_∞ $H^\bullet(\mathsf{h}_2)$-module structure on $H^\bullet(\mathsf{F}_2(\mu,1))$ is equivalent to the one we described in Proposition 8.13.

Just like for the ordinary Fock module, the behavior of $\mathsf{F}_n(\mu,k)$ is not very sensitive to the number $k \in \mathbb{C}$. Indeed, for $k,k' \neq 0$ there is an equivalence of modules $\mathsf{F}_n(\mu,k) \simeq \mathsf{F}_n(\mu,k')$. Furthermore, $\mathsf{F}(\mu,0)$ is a trivial h_n-module.

We leave further questions pertaining to the higher dimensional Fock modules that we do not attempt to answer in these lectures.

- Does $\mathsf{F}_n(\mu,k) \simeq \mathsf{F}_n(\mu',k)$ imply $\mu = \mu'$?
- In what sense (if at all) is $\mathsf{F}_n(\mu,k)$ "irreducible" as an L_∞-module for h_n? In what sense (if at all) is $H^\bullet\mathsf{F}_n(\mu,k)$ "irreducible" as an L_∞-module for $H^\bullet\mathsf{h}_n$?

Vacuum Modules

We define a class of modules for the higher dimensional Kac–Moody modules. Fix an invariant polynomial θ of degree $n+1$ and consider the associated higher Kac–Moody algebra $\widehat{\mathfrak{g}}_{n,\theta}$. Then, there is the subalgebra

$$\mathfrak{g} \otimes \mathsf{A}_{n,+} \oplus \mathbb{C}K \subset \widehat{\mathfrak{g}}_{n,\theta}.$$

One can see this is a subalgebra by checking that the restriction of the higher residue vanishes. Define the module $\mathbb{C}_1 \cong \mathbb{C}$ for this subalgebra where $\mathfrak{g} \otimes \mathsf{A}_{n,+}$ acts trivially and K acts by 1.

Definition 8.24 The *n-vacuum module* of level θ is the L_∞ $\widehat{\mathfrak{g}}_{n,\theta}$ module

$$V_{n,\theta}(\mathfrak{g}) \overset{\text{def}}{=} U(\widehat{\mathfrak{g}}_{n,\theta}) \otimes_{U(\mathfrak{g}\otimes\mathsf{A}_{n,+}\oplus K)} \mathbb{C}_1. \tag{8.2.11}$$

We will not develop any theory behind these vacuum modules, but will raise some questions pertaining to them at the end of the next section. Also, we offer one geometric interpretation in terms of factorization algebras in §8.3.1.

8.2.7 Higher Virasoro Algebra

So far we have not touched upon multivariable versions of the Virasoro algebra. For the Kac–Moody algebra based on $\mathfrak{g}$, one starts by simply tensoring with the dg algebra A_n to obtain the n-current algebra $\mathfrak{g} \otimes \mathsf{A}_n$.[5] When $n = 1$ the Virasoro algebra is a central extension of the Witt algebra. So, a more basic question is what the n-dimensional version of the Witt algebra should be.

Recall that A_n is a derived model for the sheaf of functions on punctured affine space. For the n-dimensional version of the Witt algebra, one should consider derivations of the dg algebra A_n. Geometrically, this corresponds to looking at a subalgebra of the derived global sections of the tangent bundle $\mathrm{T}_{\mathbb{A}^n \setminus 0}$ on $\mathbb{A}^n \setminus 0$. Recall that vector fields form a Lie algebra under the commutator; we require any model for the derived global sections to have the structure of a dg Lie algebra which extends this. Since the tangent bundle of $\mathbb{A}^n \setminus 0$ admits a framing, we obtain the following model for this dg Lie algebra.

Definition 8.25 Let witt_n denote the cochain complex

$$\mathsf{A}_n \otimes \mathbb{C}\{\partial_{z_1}, \ldots, \partial_{z_n}\} \cong \mathsf{A}_n \otimes \mathbb{C}^n,$$

where the differential is defined by $\overline{\partial}(\alpha \otimes \partial_{z_i}) = (\overline{\partial}\alpha) \otimes \partial_{z_i}$, for $\alpha \in \mathsf{A}_n$. The bracket

$$\left[\alpha \otimes \partial_{z_i}, \beta \otimes \partial_{z_j}\right] = \alpha L_{\partial_{z_i}}(\beta) \otimes \partial_{z_j} - (-1)^{|\alpha||\beta|} \beta L_{\partial_{z_j}}(\alpha) \otimes \partial_{z_i}$$

endows $(\mathsf{witt}_n, \overline{\partial}, [\cdot,\cdot])$ with the structure of a dg Lie algebra. Furthermore $H^\bullet(\mathsf{witt}_n)$ agrees with the graded Lie structure on $H^\bullet(\mathbb{A}^n \setminus 0, \mathrm{T})$.

From the embedding of dg algebras $\mathsf{A}_n \hookrightarrow \Omega^{0,\bullet}(\mathbb{C}^n \setminus 0)$ we obtain an embedding of dg Lie algebras $\mathsf{witt}_n \hookrightarrow \Omega^{0,\bullet}(\mathbb{C}^n \setminus 0, \mathrm{T})$. Again, this is almost a quasi-isomorphism of dg Lie algebras; at the level of cohomology, this embedding is dense.

We address the problem of classifying central extensions of the n-Witt algebra. A key difference with the case $n = 1$, and the ordinary Witt algebra, is that there is no longer a one-dimensional space of central extensions.

Higher Central Charges

In [18], Hennion and Kapranov have used methods of factorization homology to partially characterize central extensions of the n-Witt algebra in terms of the Lie algebra cohomology of *formal vector fields* w_n. The Lie algebra w_n is

[5] One could think of this as derived sections of the adjoint bundle associated to some principal G-bundle on punctured affine space.

defined as the ∞-jets of sections of the tangent bundle on (non-punctured) affine space and its cohomology has been extensively studied by Fuks [13].

Theorem 8.26 ([18]) *There is a map*

$$H^{2n+1}(\mathsf{w}_n) \to H^2(\mathsf{witt}_n). \tag{8.2.12}$$

In other words, any class $c \in H^{2n+1}(\mathsf{w}_n)$ defines a central extension of the n-Witt algebra that we denote by $\mathrm{vir}_{n,c}$.

In upcoming work [35] we offer another construction of central extensions of witt_n from cohomology classes of formal vector fields based on the method of "descent" for the de Rham cohomology of ∞-jets of the tangent bundle; for some details see the final chapter of [34]. This construction will divert from the content of this series, but we mention a particular example to get a sense of these central extensions.

Consider the 2-Witt algebra $\mathsf{witt}_2 = \mathsf{A}_2\{\partial_{z_1}, \partial_{z_2}\}$. The Jacobian of a vector field on $\mathbb{C}^2$ is a 2×2 matrix whose entries are functions. This definition extends to the 2-Witt algebra; the Jacobian of an element $\xi = \sum \alpha_i \partial_{z_i}$ of witt_2 is the 2×2 matrix with values in A_2 whose ij entry is

$$[\mathrm{Jac}(\xi)]^i_j = \partial_{z_i}\alpha_j \in \mathsf{A}_2.$$

Consider the following two cochains of witt_2

$$\psi_1(\xi_0,\xi_1,\xi_2) = \mathrm{Res}_{z=0}\mathrm{Tr}\left(\mathrm{Jac}(\xi_0)\partial\mathrm{Jac}(\xi_1)\partial\mathrm{Jac}(\xi_2)\right),$$
$$\psi_2(\xi_0,\xi_1,\xi_2) = \mathrm{Res}_{z=0}\mathrm{Tr}\left(\mathrm{Jac}(\xi_0)\right)\mathrm{Tr}\left(\partial\mathrm{Jac}(\xi_1)\partial\mathrm{Jac}(\xi_2)\right).$$

It is an involved exercise to verify directly that both ψ_1, ψ_2 are cocycles of cohomological degree 2. We expect that these two cocycles are inequivalent. This problem is related to the question of the injectivity of the map (8.2.12).

We end this section with a series of questions and goals that we hope can be attacked in the near future.

- Is there a "free field" realization of the higher dimensional Virasoro algebra on the Fock module introduced in §8.2.6? See [16] for a simple case of free field realization for n-Kac–Moody algebras.
- For which invariant polynomials $\theta \in \mathrm{Sym}^{n+1}(\mathfrak{g}^*)^{\mathfrak{g}}$ (if any) does the vacuum module $\mathbb{V}_{n,\theta}(\mathfrak{g})$ have an action by the higher dimensional Virasoro algebra? In other words, formulate a higher dimensional version of the Segal–Sugawara construction.
- In [33] we began an effort to compute characters of higher dimensional Kac–Moody and Virasoro algebras. Analogs of Weyl–Kac character formula have yet to be determined.

- These algebras describe symmetries of higher dimensional *holomorphic* field theories. Progress towards this has been developed in [16, 33, 34]. Recently, Kapranov has utilized similar techniques to characterize an infinite-dimensional conformal algebra extending the conformal algebra on $\mathbb{R}^n$ for all n [25]; it would be interesting to see what role such algebras play in non-holomorphic theories.

8.3 Further Topics and Applications

8.3.1 Factorization Algebras and a Geometric Perspective

As we've mentioned above, the Virasoro algebra describes the symmetries in conformal field theory. The so-called "observables" of a conformal field theory are described by a mathematical object called a vertex algebra [7, 12, 23]. Algebro-geometrically, vertex algebras are closely related to chiral algebras developed by Beilinson and Drinfeld [5]. More recently, the Costello–Gwilliam theory of *factorization algebras* has provided another geometric formulation of vertex algebras [9]. The rigorous definition of a vertex algebra is quite complicated at first glance. The main advantage of the perspective that factorization algebras offer is in their intrinsic link to the observables of a conformal field theory. One connection between factorization algebras and CFT is given in [9, Theorem 2.2.1], which provides a construction of a vertex algebra from a factorization algebra. We also point to work of Bruegmann who has further elucidated the connection between vertex algebras and factorization algebras [8].

(Pre)factorization Algebras

We will not give a complete account of factorization algebras. Rather, we extract some geometric intuition and sketch how it connects to the algebraic situation that we have developed in the previous sections. Additionally, we will only glance at the theory of *pre*factorization algebras; the "gluing" axiom will play no role for us. Following [9], a prefactorization algebra is an algebra over a certain *colored* operad. A colored operad is similar to an operad except that there is a set of colors which can be fed into the multi-operations. It sometimes goes by the name "(symmetric) multicategory" and is also closely related to the notion of a pseudotensor category, see [2]. For a nice review of colored operads (equivalently, multicategories) we refer to [9, §A.2].

Recall that the collection of open sets on a manifold M forms a poset. This can be enhanced to the structure of a colored operad $\mathcal{C}_M$ as follows. For a collection of open sets $\{U_i, V\}$ define the set of multi-operations

$$\mathcal{C}_M\left(\boxtimes_{i\in I} U_i, V\right)$$

by:

- the singleton set $\{\star\}$ if the collection of open sets $\{U_i\}$ is mutually disjoint and each U_i is contained in V;
- the empty set, otherwise.

Definition 8.27 A *prefactorization algebra* $\mathcal{F}$ on a manifold M is a $\mathcal{C}_M$-algebra. Unpacking, this is an assignment

$$\mathcal{F}\colon U \subset M \mapsto A(U)$$

together with a "multiplication rule"

$$m_{\{U_i\},V} : \otimes_{i \in I} \mathcal{F}(U_i) \to \mathcal{F}(V)$$

whenever the open sets $\{U_i\}$ are mutually disjoint and contained in V. These multiplications are required to satisfy the natural associativity axiom.

We have not explicitly written out the associativity axiom; we refer to [9, §1.1] for a more precise definition. To get a feel for this notion of associativity, we mention a special case. A prefactorization algebra is called *locally constant* if for every embedding of open balls $B \hookrightarrow B'$ in M the induced map $A(B) \xrightarrow{\simeq} A(B')$ is an equivalence.

Theorem 8.28 (Lurie) *There is an equivalence of categories between locally constant (pre)factorization algebras on $\mathbb{R}_{>0}$ (or any homeomorphic space) and associative algebras.*

This theorem is valid only at the level of ∞-categories. The notion of an associative algebra should be taken up to homotopy which can be modeled by the category of A_∞ algebras or algebras over the operad of little 1-disks. Extending this, Lurie [30, §5.4.5] identifies locally constant factorization algebras on $\mathbb{R}^n$, $n \geqslant 1$ with algebras over the operad of little n-disks introduced by Boardman and Vogt [6].

Current Algebras

To connect with the style of infinite-dimensional algebras we have encountered in this note we will not be directly concerned with locally constant factorization algebras (though they will play some role). Instead, we are interested in the concept of a factorization algebra which depends *holomorphically* on the manifold; in particular the manifold must come equipped with a complex structure. Since the full structure of a holomorphic factorization algebra will not play a significant role in our discussion we divert the interested reader to the textbook [9, §5] or the survey [17].

In physics, "conserved currents" refer to observables of a physical system whose integrals over cycles lead to conserved quantities. We will find that the algebras we have introduced in §8.2 arise from the value of holomorphic factorization algebras on codimension one spheres. The structure of the factorization algebras allows one to multiply currents which we will see corresponds to taking the enveloping algebra of the underlying Lie algebra of currents.

Let us begin by asking about the "factorization model" of the simplest example, the Heisenberg algebra. We have already encountered the Dolbeault complex $\Omega^{0,\bullet}(X)$ of a complex manifold X. If X is not compact, one defines the compactly supported forms as a subalgebra $\Omega^{0,\bullet}_c(X) \subset \Omega^{0,\bullet}(X)$.

Suppose $U \subset X$ is an open set in the complex manifold. Then, consider the cochain complex $\mathrm{Sym}(\Omega^{0,\bullet}_c(U)[1])$. This is the graded symmetric algebra on the shifted compactly supported Dolbeault forms. The assignment $U \mapsto \mathrm{Sym}(\Omega^{0,\bullet}_c(U)[1])$ actually extends to the structure of a prefactorization algebra on X. The multiplication map $m_{U,V;W}$ for the configuration $i : U \sqcup V \hookrightarrow W$ is defined by the following composition

$$\begin{array}{ccc} \mathrm{Sym}\left(\Omega^{0,\bullet}_c(U)[1]\right) \otimes \mathrm{Sym}\left(\Omega^{0,\bullet}_c(V)[1]\right) & \xrightarrow{\cong} & \mathrm{Sym}\left(\Omega^{0,\bullet}_c(U)[1] \oplus \Omega^{0,\bullet}_c(V)[1]\right) \\ & \searrow^{m_{U,V;W}} & \downarrow i_* \\ & & \mathrm{Sym}\left(\Omega^{0,\bullet}_c(W)[1]\right). \end{array}$$

The shift by $[1]$ may seem arbitrary, but leads to an interesting consequence as we will now see. For the remainder of this section, let $\mathcal{F}$ stand for this prefactorization algebra.

We restrict to the complex manifold $X = \mathbb{C}^n$ and consider the following types of open sets. Choose holomorphic coordinates $(z_1, \ldots, z_n)$. The radius $\mathrm{rad} = |z_1|^2 + \cdots + |z_n|^2$ defines a map $\mathrm{rad} \colon \mathbb{C}^n \setminus 0 \to \mathbb{R}_{>0}$. For $r < R$ the open set $\mathrm{rad}^{-1}(r < R) \subset \mathbb{C}^n$ is in the neighborhood of a $(2n-1)$-sphere in $\mathbb{C}^n$ centered at zero of radius $(r+R)/2$.

Recall the dg algebra A_n modeling punctured affine space that we introduced in §8.2. We realized this as a subalgebra of (non-compactly supported) Dolbeault forms $\Omega^{0,\bullet}(\mathbb{C}^n \setminus 0)$. We relate compactly supported Dolbeault forms to A_n using a version of Serre duality. Indeed, let ρ be a smooth function on $\mathbb{C}^n \setminus 0$ which takes value 1 near the outer boundary of $\mathrm{rad}^{-1}(r < R)$ and value zero near the inner boundary. We further normalize ρ so that

$$\mathsf{int}_{\mathbb{C}^n} \overline{\partial} \rho \, \mathrm{d}^n z = 1.$$

The assignment $\alpha \mapsto \alpha \wedge \overline{\partial}(\rho)$ defines a cochain map

$$i_\rho : \mathsf{A}_n \hookrightarrow \Omega_c^{0,\bullet}\left(\mathrm{rad}^{-1}(r<R)\right)[1].$$

Notice that the map is defined by wedging with a particular $(0,1)$-form, hence the cohomological shift. One can use Stokes' theorem to see that this map intertwines the $\overline{\partial}$ operators. This map has a very special property, which is proved using Serre duality: at the level of cohomology the map is *dense*. Applying the symmetric algebra functor $\mathrm{Sym}(-)$, we see that i_ρ gives a map of cochain complexes

$$i_\rho : \mathrm{Sym}\left(\mathsf{A}_n\right) \to \mathcal{F}\left(\mathrm{rad}^{-1}(r<R)\right).$$

Instead of considering fixed values $r<R$ we can ask about the *pushforward* of the factorization algebra $\mathrm{rad}_*\mathcal{F}$ along the radius map. This one-dimensional factorization algebra on $\mathbb{R}_{>0}$ is *almost* locally constant. Consider the following construction.

- There is a natural action of the torus T^n on $\mathbb{C}^n \setminus 0$ given by rotations $(e^{2\pi i\theta_1}z_1, \ldots, e^{2\pi i\theta_n}z_n)$. This action extends to the factorization algebra $\mathrm{rad}_*\mathcal{F}$. Let $\mathcal{F}(\mathrm{rad}^{-1}(r<R))^{(\vec{N})}$ denote the subspace of $\mathrm{rad}_*(\mathcal{F})(r<R)$ where T^n acts by weight $\vec{N} \in \mathbb{Z}^n$. Then, the assignment

$$(r<R) \subset \mathbb{R}_{>0} \mapsto \oplus_{N\in\mathbb{Z}}\mathcal{F}(\mathrm{rad}^{-1}(r<R))^{(N)}$$

  defines a *locally constant* factorization algebra on $\mathbb{R}_{>0}$.
- By Theorem 8.28 this construction defines an associative dg algebra that we denote by $\oint\mathcal{F}$. The map i_ρ defines an isomorphism of associative dg algebras

$$i_\rho : \mathrm{Sym}(\mathsf{A}_n) \xrightarrow{\cong} \oint\mathcal{F}.$$

This case was somewhat boring as we used the data of the factorization algebra $\mathcal{F}$ to extract a *commutative* dg algebra $\mathrm{Sym}(\mathsf{A}_n)$. Recall, the Heisenberg algebra is a central extension of the (abelian) Lie algebra of algebraic functions on the punctured disk. So, a natural question to ask is where the central extension is hiding at the level of the factorization algebra construction. The key idea is that at the level of factorization algebras, this extension appears as a *deformation* of the differential $\overline{\partial}$.

Consider the (familiar looking) multi-linear functional ψ on the compactly supported forms $\Omega_c^{0,\bullet}(\mathbb{C}^n \setminus 0)$:

$$\psi(\omega_0, \ldots, \omega_n) = \mathsf{int}_{\mathbb{C}^n\setminus 0}\omega_0\partial\omega_1 \cdots \partial\omega_n.$$

Notice that instead of taking the residue, as in Definition 8.17 we are integrating along all of $\mathbb{C}^n \setminus 0$. Because of this, we see that ψ is actually of *cohomological degree* $+1$ in the sense that for the value $\psi(\omega_0, \dots, \omega_n)$ to be nonzero one must have $|\omega_0| + \cdots + |\omega_n| = n$.

Using ψ, we can perform the following deformation of the factorization algebra $\mathcal{F}$. To an open set $U \subset \mathbb{C}^n$ consider the cochain complex

$$\widetilde{\mathcal{F}}(U) = \left(\mathrm{Sym}\left(\Omega_c^{0,\bullet}(U)[1]\right) \, , \, \overline{\partial} + \psi \right).$$

Notice that the term in the differential is the one used in the definition of $\mathcal{F}$, the functional ψ is the deformation.

Proceeding just as we did above by looking at codimension one spheres we can build a locally constant one-dimensional factorization algebra out of $\widetilde{\mathcal{F}}$ and hence an A_∞ algebra $\oint \widetilde{\mathcal{F}}$.

Proposition 8.15 *Let h_n be the n-Heisenberg algebra. Then, there is a quasi-isomorphism of A_∞ algebras $i_\rho : U(\mathsf{h}_n) \to \oint \widetilde{\mathcal{F}}$.*

The key to the proof of this proposition is the observation that the functional ψ is compatible with the cocycle φ_n of Lemma 8.11 under the map i_ρ defined above. We refer to [16] for more details and for the following generalization. Recall that $\mathfrak{g} \otimes \Omega^{0,\bullet}(X)$ has the structure of a dg Lie algebra; the same is true if we replace forms by compactly supported forms.

Theorem 8.29 ([16]) *Let $\mathfrak{g}$ be a Lie algebra and $\theta \in \mathrm{Sym}^{n+1}(\mathfrak{g}^*)^{\mathfrak{g}}$ an invariant polynomial. Let $\mathcal{F}_\theta(\mathfrak{g})$ be the factorization algebra which assigns to an open set $U \subset \mathbb{C}^n$ the cochain complex*

$$\widetilde{C}_\bullet^\theta \left(\mathfrak{g} \otimes \Omega_c^{0,\bullet}(U)\right) = \left(\mathrm{Sym}(\mathfrak{g} \otimes \Omega_c^{0,\bullet}(U)[1] \, , \, \overline{\partial} + \mathrm{d}_{\mathrm{CE}} + \psi_\theta \right).$$

Then, there is an equivalence of A_∞ algebras

$$U(\widehat{\mathfrak{g}}_{n,\theta}) \to \oint \mathcal{F}_\theta(\mathfrak{g}).$$

In this proposition, ψ_θ is the multi-linear functional on $\mathfrak{g} \otimes \Omega_c^{0,\bullet}$ defined by $\mathrm{int}\theta(\omega_0 \partial \omega_1 \cdots \partial \omega_n)$. The complex $\widetilde{\mathrm{C}}_\bullet$ is a deformation of the Chevalley–Eilenberg complex computing the Lie algebra *homology* of $\mathfrak{g} \otimes \Omega_c^{0,\bullet}$. There is a completely analogous result for the Virasoro algebra, and its higher dimensional versions. We refer to [34, Chapter 3] for more details.

To summarize, we have used the "radial" part of the structure of factorization algebras on $\mathbb{C}^n$ to extract the current algebras, and central extensions thereof. Of course, there is much more data present in the factorization algebra than just its radial part. In complex dimension one, the factorization product of disks, for instance, is the geometric underpinning of the operator product expansion in the context of vertex algebras. We will not directly involve ourselves with the product of disks, but we discuss another configuration which will lead us to a familiar theory of modules for infinite-dimensional Lie algebras.

Vacuum Modules

We've already mentioned the passage from associative algebras to locally constant factorization algebras on $\mathbb{R}_{>0}$. This equivalence is best thought of as an equivalence between locally constant factorization algebras and the operad of little intervals $\{(a,b) \mid a<b\}$ in $\mathbb{R}_{>0}$. The underlying vector space (or cochain complex) of the associative algebra is the value of the prefactorization algebra on such an interval $A = \mathcal{F}((a,b))$.

What happens when we look at a factorization algebra $\mathcal{F}$ on $\mathbb{R}_{\geqslant 0}$ instead? The condition of being locally constant is replaced by the notion of a "constructible" factorization algebra, we refer to [1] for a full development. At the operadic level, this has the effect of introducing a new type of interval of the form $[0,a)$ together with operations corresponding to configurations of intervals

$$(b,c) \sqcup (d,e) \hookrightarrow (a,f),$$
$$[0,c) \sqcup (d,e) \hookrightarrow [0,f),$$

where $0<a<b<c<d<e<f$. The first type of configuration gives rise to an algebra structure on $A = \mathcal{F}((b,c))$, just as it did in the case of $\mathbb{R}_{>0}$. The second type of configuration endows $M = \mathcal{F}([0,c))$ with the structure of an A-module. The following result is a very special case of the general formalism developed in [1]. See also [9, Chapter 8].

Theorem 8.30 *A constructible factorization algebra on $\mathbb{R}_{\geqslant 0}$ is equivalent to the data of an associative algebra A together with an A-module M.*

We have shown how factorization algebras on $\mathbb{C}^n$ give rise to associative algebras by restricting to the "radial" part $\oint \mathcal{F}$. Extending this, we consider the radial projection map as $\mathrm{rad}\colon \mathbb{C}^n \to \mathbb{R}_{\geqslant 0}$. Restricting to $\mathbb{R}_{>0}$ we know how to recover $\oint \mathcal{F}$ by looking at S^1-eigenspaces. Now, we look at the S^1-eigenspaces of $\mathcal{F}([0,r))$, this is simply the value of $\mathcal{F}$ on the n-disk $D_0(r)$ of radius r centered at zero. Denote the direct sum of all of the eigenspaces by $V_{\mathcal{F}}$.

Proposition 8.16 *Suppose that $\mathcal{F}$ is a holomorphic factorization algebra (such as $\mathcal{F} = \mathrm{Sym}(\Omega_c^{0,\bullet}[1])$). Then the subspace $V_{\mathcal{F}} \subset \mathcal{F}(D_r^n(0))$ is a module for the A_∞ algebra $\oint \mathcal{F}$.*

Let's work out an example of this in the case of the n-Heisenberg algebra. We have already seen that the radial part of the factorization algebra $\mathcal{F} = \mathrm{Sym}(\Omega_c^{0,\bullet}[1])$, where the differential $\overline{\partial} + \psi$ recovers the enveloping algebra of the n-Heisenberg algebra h_n. We will argue that $V_{\mathcal{F}}$ is the vacuum module $\mathsf{F}_n(0,1)$ for h_n constructed in §8.2.6.

To see that the cohomology of $V_{\mathcal{F}}$ is the right thing is not difficult. Indeed, by Serre duality $H^\bullet(\Omega_c^{0,\bullet}(D_0(r)))$ can be identified with the continuous linear dual of the space of holomorphic n-forms $\Omega^{n,hol}(D_0(r))$ concentrated in degree $+n$. The residue defines a linear embedding

$$\mathsf{A}_{n,-} \hookrightarrow \Omega^{n,hol}(D_0(r))^\vee$$

sending α to the functional $\omega \mapsto \mathrm{Res}_{z=0}(\alpha \wedge \omega)$, which one can show agrees precisely with the subspace of S^1-eigenspaces. Applying the symmetric algebra one then finds that $H^\bullet V_{\mathcal{F}} \cong \mathrm{Sym}(\mathsf{A}_{n,-}[1-n])$, which we compare to Proposition 8.14.

We consider the module structure in the case $n = 1$. The characterizing property of the vacuum module involved the existence of a vector $|0\rangle \in V_{\mathcal{F}}$ such that $b[n]|0\rangle = 0$ for all $n \geqslant 0$. The vacuum vector is simply $|0\rangle = 1 \in \mathrm{Sym}^0$. To represent $b[n] = z^n \in \mathrm{Sym}(\mathbb{C}[z,z^{-1}]) \cong \oint \mathcal{F}$ in the value of the factorization algebra on an annulus we use the bump function $\overline{\partial}\rho$ of the previous section. The element $\widetilde{b}[n] = z^n \overline{\partial}\rho \in \mathcal{F}(\mathbb{A})$ represents $b[n]$ in cohomology.

We want to compute the value of $\widetilde{b}[n] \otimes 1$ under the factorization product

$$m : \mathcal{F}(D_0(r)) \otimes \mathcal{F}(\mathrm{rad}^{-1}(R < R')) \to \mathcal{F}(D_0(r')),$$

where $r < R < R' < r'$. It suffices to show that when viewed as an element of $\mathcal{F}(D_0(r'))$, the class $\widetilde{b}[n]$ is cohomologically trivial for $n \geqslant 0$. Indeed, when $n \geqslant 0$, the element ρz^n defines an element of $\mathcal{F}(D_0(r'))$ which satisfies $\overline{\partial}(\rho z^n) = \widetilde{b}[n]$.

Proposition 8.17 *Consider the factorization algebra* $\mathcal{F} = \left(\mathrm{Sym}(\Omega_c^{0,\bullet}[1]), \overline{\partial} + \psi\right)$ *on* $\mathbb{C}^n$. *The module* $V_{\mathcal{F}}$ *for the* A_∞ *algebra* $\oint \mathcal{F} \simeq U(h_n)$ *is equivalent to the vacuum Fock module* $F_n(0,1)$.

A very similar result holds for n-Kac–Moody algebras $\widehat{\mathfrak{g}}_{n,\theta}$ and their associated vacuum modules that we introduced in §8.2.6. Associated to any Lie algebra $\mathfrak{g}$ and degree $n+1$ invariant polynomial θ there is the factorization algebra $\mathcal{F}_{\mathfrak{g},\theta}$ on $\mathbb{C}^n$ described in Theorem 8.29. Its radial part $\oint \mathcal{F}_{\mathfrak{g},\theta}$ returns the enveloping algebra of $\widehat{\mathfrak{g}}_{n,\theta}$. The module $V_{\mathcal{F}_{\mathfrak{g},\theta}}$ is equivalent to the level θ vacuum module $V_{n,\theta}(\mathfrak{g})$. We refer to [16] for more details.

8.3.2 Super Enhancements

Super Lie algebras play a role in symmetries of physical systems which have objects of different parity. Most common examples of such systems arise from *supersymmetry* which involve fields or particles that have both "bosonic" and "fermionic" statistics.

The blend of supersymmetry with infinite-dimensional Lie algebras is rich from both a mathematical and physical perspective. Physically, super enhancements of the (ordinary) Virasoro algebra, for instance, play a significant role in superstring theory in an analogous way that the Virasoro algebra appears in conformal field theory or in the bosonic string. We will see an example of one such enhancement below. Mathematically, Kac has classified all finite-dimensional super Lie algebras which naturally leads to the theory of affine super Lie algebras [22]. The connection between the character theory of affine super Lie algebras and number theory is a deep and fascinating subject on its own [24].

In this section we will follow a similar trend and consider super enhancements of higher dimensional Kac–Moody and Virasoro algebras. These algebras appear as symmetries in higher dimensional supersymmetric field theories. We do not stress that realization here, but refer to [16, 32, 33] for accounts of this.

These enhancements are interesting in their own right and most certainly lead to a rich theory of representations extending what we have dipped our toes into throughout this note. One aspect of the theory that we focus on is how certain super enhancements of the multivariable algebras that we have introduced interpolate between ordinary (single-variable) infinite-dimensional Lie algebras, like affine algebras.

Example: the "Topological" Virasoro Algebra

We begin with an example of a super enhancement of a familiar infinite-dimensional Lie algebra. Recall that the Virasoro algebra is linearly generated by elements $\{L_n\}_{n\in\mathbb{Z}}$. The super Lie algebra $\mathsf{vir}_{1|1}$ spanned by even elements $\{L_n, J_n\}_{n\in\mathbb{Z}}$ and odd elements $\{G_n, Q_n\}_{n\in\mathbb{Z}}$ satisfies the commutation relations

$$\begin{aligned}
[L_n, L_m] &= (n-m)L_{n+m} - \frac{c}{2}n(n^2-1)\delta_{n+m,0},\\
[L_n, G_m] &= -mG_{n+m}, \quad [L_n, Q_m] = (n-m)Q_{n+m},\\
[G_n, Q_m] &= L_{n+m} + nJ_{n+m} + \frac{c}{2}(n^2-n)\delta_{n+m,0},\\
[J_n, J_m] &= cn\delta_{n+m,0},\\
[J_n, G_m] &= G_{n+m}, \quad [J_n, Q_m] = -Q_{n+m}.
\end{aligned}$$

This super Lie algebra admits a lift to a $\mathbb{Z}$-graded Lie algebra by declaring that the elements $\{Q_n\}$ have degree $+1$ and the elements $\{G_n\}$ have degree -1. We will assume this grading in what follows. Notice that the ordinary Virasoro algebra appears as a subalgebra. This algebra is referred to as the "topological Virasoro" algebra [15] due to its connection with $\mathcal{N} = (2,2)$ supersymmetry.

With this example we can detail a particular construction that will be useful in the next section. Recall, a Maurer–Cartan element of a dg Lie algebra $\mathfrak{g}$ is an element $\alpha \in \mathfrak{g}$ satisfying the Maurer–Cartan equation

$$\mathrm{d}\alpha + \frac{1}{2}[\alpha, \alpha] = 0.$$

Such elements connect dg Lie algebras with the rich subject of formal deformation theory (see [19, 29], for instance). For us, it will be important to note that a Maurer–Cartan element allows one to *deform* the dg Lie algebra via

$$(\mathfrak{g}, \mathrm{d}) \rightsquigarrow (\mathfrak{g}, \mathrm{d} + [\alpha, -])$$

while keeping the bracket the same. More generally, if D is a Maurer–Cartan element in the dg Lie algebra of derivations of $\mathfrak{g}$ one can consider the deformation $\mathrm{d} + D$; in this case α determines the inner derivation $D = [\alpha, -]$.

The element $Q_0 \in \mathsf{vir}_{1|1}$ is a Maurer–Cartan element; indeed the differential is trivial and $[Q_0, Q_0] = 0$. In particular we obtain a deformed algebra $(\mathsf{vir}_{1|1}, [Q, -])$. It is actually not difficult to show that this deformation completely collapses on to the trivial Lie algebra.

Higher Dimensional Enhancements

We focus on super enhancements of the multivariable current algebra. Any Lie algebra $\mathfrak{g}$ acts on itself by the adjoint representation. Using this, one can define a super Lie algebra structure on $\mathfrak{g} \oplus \Pi\mathfrak{g}$ where $\mathfrak{g} \times \mathfrak{g} \to \mathfrak{g}$ is the usual bracket, $\mathfrak{g} \times \Pi\mathfrak{g} \to \Pi\mathfrak{g}$ is the adjoint action and $\Pi\mathfrak{g} \times \Pi\mathfrak{g} \to \mathfrak{g}$ is the zero map. If $\mathbb{C}[\varepsilon]$ denotes the "odd dual numbers," the free graded algebra generated by a single *odd* element ε, then this super Lie algebra is identical to $\mathfrak{g}[\varepsilon] = \mathfrak{g} \otimes \mathbb{C}[\varepsilon]$. In the case $\mathfrak{g} = \mathfrak{gl}(N)$, note that $\mathfrak{gl}(N)[\varepsilon]$ is a subalgebra of the matrix super Lie algebra $\mathfrak{gl}(N|N)$.

The definition of the n-current algebra makes sense for a super or graded Lie algebra. For the example at hand, we consider the n-current algebra associated to $\mathfrak{g}[\varepsilon]$ which is $\mathfrak{g}[\varepsilon] \otimes \mathsf{A}_n$. Concretely, elements are of the form $\alpha + \varepsilon\alpha'$ where $\alpha, \alpha' \in \mathsf{A}_n$. There are two gradings at hand: there is a $\mathbb{Z}$-grading arising from the $\mathbb{Z}$-grading on A_n and there is a $\mathbb{Z}/2$-grading arising from the parity of ε. In fact, this $\mathbb{Z}/2$ grading actually lifts to a $\mathbb{Z}$-grading by declaring that ε has degree -1. In this way we obtain a dg Lie algebra structure on $\mathfrak{g}[\varepsilon] \otimes \mathsf{A}_n$ whereby A_n carries its usual $\mathbb{Z}$-grading and ε has degree -1. We will use this totalized $\mathbb{Z}$-grading in what follows.

In the last section we saw how one can deform a dg Lie algebra; the same is true for the current algebra $\mathfrak{g}[\varepsilon] \otimes \mathsf{A}_n$. Given any derivation $D \in \mathrm{Der}(\mathfrak{g}[\varepsilon] \otimes \mathsf{A}_n)$ of cohomological degree $+1$ which satisfies

$$[\overline{\partial}, D] + \frac{1}{2}[D, D] = 0, \tag{8.3.1}$$

we can consider the deformed current algebra

$$\left(\mathfrak{g}[\varepsilon] \otimes \mathsf{A}_n\,,\, \overline{\partial}\right) \rightsquigarrow \left(\mathfrak{g}[\varepsilon] \otimes \mathsf{A}_n\,,\, \overline{\partial} + D\right).$$

Recall that A_n is a model for punctured affine space whose coordinates are $(z_1, \ldots, z_n)$. We define the following derivation of $\mathfrak{g}[\varepsilon] \otimes \mathsf{A}_n$

$$D \overset{\text{def}}{=} z_n \frac{\partial}{\partial \varepsilon}.$$

It is immediate to see that D satisfies (8.3.1) and so defines a deformed algebra $\left(\mathfrak{g}[\varepsilon] \otimes \mathsf{A}_n\,,\, \overline{\partial} + z_n \frac{\partial}{\partial \varepsilon}\right)$.

Proposition 8.18 *Let* $\iota\colon \mathbb{A}^{n-1} \to \mathbb{A}^n$ *be the map* $\iota(z_1, \ldots, z_{n-1}) \mapsto (z_1, \ldots, z_{n-1}, 0)$. *The map*

$$\left(\mathfrak{g}[\varepsilon] \otimes A_n\,,\, \overline{\partial} + z_n \frac{\partial}{\partial \varepsilon}\right) \xrightarrow{\simeq} \mathfrak{g} \otimes \mathsf{A}_{n-1},$$

which sends $X \otimes \alpha$ *to* $X \otimes \iota^* \alpha$, *is a quasi-isomorphism of dg Lie algebras.*

Proof This statement follows from showing that $\iota^*\colon \mathsf{A}_n[\varepsilon] \to \mathsf{A}_{n-1}$ is a quasi-isomorphism of dg algebras. At the level of coordinates ι^* sends $\varepsilon \mapsto 0$, $z_i \mapsto z_i$ for $i \neq n$ and $z_n \mapsto 0$. It is not difficult to see that the map intertwines the differentials. To see that ι^* is a quasi-isomorphism one can write down an explicit retraction, see [32]. We remark that the key to describing the cohomology is to observe that there is a degree -1 operator on $\mathsf{A}_2[\varepsilon]$ defined by $\varepsilon \partial_{z_2}$ which satisfies $[\overline{\partial} + z_n \partial_\varepsilon, \varepsilon \partial_{z_2}] = z_n \partial_{z_n} + \varepsilon \partial_\varepsilon$. Thus, $\varepsilon \partial_{z_n}$ can be used to define a homotopy between $\mathsf{A}_n[\varepsilon]$ and the complex obtained by setting ε and z_n to zero.

The explicit formula for the retraction in the case $n = 2$ is

$$s(z^{-n}) = \frac{\overline{z}_1^n}{(z_1\overline{z}_1 + z_2\overline{z}_2)^n} - \varepsilon n \frac{\overline{z}_1^{n-1}}{(z_1\overline{z}_1 + z_2\overline{z}_2)^{n-1}} \omega,$$

where $\omega \in \mathsf{A}_2^1$ is the Bochner–Martinelli kernel as defined in (8.2.5). A direct calculation shows

$$\overline{\partial} s(z^{-n}) = z_2 n \frac{\overline{z}_1^{n-1}}{(z_1\overline{z}_1 + z_2\overline{z}_2)^{n-1}} \omega,$$

which implies $\left(\overline{\partial}+z_2\frac{\partial}{\partial\varepsilon}\right)s(z_1^{-n})=0$. Thus showing that s is a cochain map. □

Applied to the case $n=2$, this proposition implies that there is a deformation of the two-dimensional current algebra $\mathfrak{g}\otimes \mathsf{A}_2[\varepsilon]$ which is equivalent to the ordinary current algebra $\mathfrak{g}[z,z^{-1}]$. To see the affine algebra $\widehat{\mathfrak{g}}_\kappa$ as a deformation one must introduce a central extension of this 2-current algebra. For an invariant quadratic polynomial κ on $\mathfrak{g}$ consider the bilinear map

$$\widetilde{\varphi}_\kappa(\alpha+\varepsilon\alpha',\beta+\varepsilon\beta')=\mathrm{Res}^{(2)}_{z=0}\left(\kappa(\alpha,\partial\beta')\mathrm{d}z_2+\kappa(\alpha',\partial\beta)\mathrm{d}z_2\right).$$

Here, we emphasize that the righthand side denotes the two-dimensional residue. It is an exercise to see that $\widetilde{\varphi}_\kappa$ is a two-cocycle of the dg Lie algebra $(\mathfrak{g}\otimes\mathsf{A}_2[\varepsilon],\overline{\partial}+z_2\partial_\varepsilon)$ and hence defines a central extension that we denote by $\widehat{\mathfrak{g}}_{2|1,\kappa}$.

Proposition 8.19 ([32]) *The quasi-isomorphism of the above proposition extends to a quasi-isomorphism between $\widehat{\mathfrak{g}}_{2|1,\kappa}$ with differential $\overline{\partial}+z_2\partial_\varepsilon$ and the affine algebra $\widehat{\mathfrak{g}}_\kappa$.*

We remark that there are similar results at the level of vacuum, and Verma, modules for the higher dimensional Kac–Moody algebra. The above algebraic results are closely related to the work of [4] where they give a construction of vertex algebras from the observables of four-dimensional supersymmetric field theory. We refer to [32] for a further discussion of this and an extension of these results to factorization algebras.

Bibliography

[1] Ayala, D., Francis, J. and Tanaka, H. L. 2017. Factorization homology of stratified spaces. *Selecta Math. (N.S.)*, **23**(1), 293–362.

[2] Bakalov, Bojko, D'Andrea, Alessandro, and Kac, Victor G. 2001. Theory of finite pseudoalgebras. *Adv. Math.*, **162**(1), 1–140.

[3] Baranovsky, Vladimir. 2008. A universal enveloping for L_∞-algebras. *Math. Res. Lett.*, **15**(6), 1073–1089.

[4] Beem, Christopher, Lemos, Madalena, Liendo, Pedro, Peelaers, Wolfger, Rastelli, Leonardo, and van Rees, Balt C. 2015. Infinite chiral symmetry in four dimensions. *Comm. Math. Phys.*, **336**(3), 1359–1433.

[5] Beilinson, Alexander, and Drinfeld, Vladimir. 2004. *Chiral algebras*. American Mathematical Society Colloquium Publications, vol. 51. Providence, RI: American Mathematical Society.

[6] Boardman, J. M. and Vogt, R. M. 1973. *Homotopy invariant algebraic structures on topological spaces*. Lecture Notes in Mathematics, Vol. 347. Springer-Verlag, Berlin-New York.

[7] Borcherds, Richard E. 1986. Vertex algebras, Kac-Moody algebras, and the Monster. *Proc. Nat. Acad. Sci. U.S.A.*, **83**(10), 3068–3071.

[8] Bruegmann, Daniel. 2020. Vertex Algebras and Costello-Gwilliam Factorization Algebras.

[9] Costello, Kevin, and Gwilliam, Owen. 2017. *Factorization algebras in quantum field theory. Vol. 1*. New Mathematical Monographs, vol. 31. Cambridge University Press, Cambridge.

[10] Faonte, Giovanni, Hennion, Benjamin, and Kapranov, Mikhail. 2019. Higher Kac–Moody algebras and moduli spaces of G-bundles. *Advances in Mathematics*, **346**, 389–466.

[11] Frenkel, Edward. 2007. *Langlands correspondence for loop groups*. Cambridge Studies in Advanced Mathematics, vol. 103. Cambridge University Press, Cambridge.

[12] Frenkel, I. B. 1985. Representations of Kac-Moody algebras and dual resonance models. Pages 325–353 of: *Applications of group theory in physics and mathematical physics (Chicago, 1982)*. Lectures in Appl. Math., vol. 21. Amer. Math. Soc., Providence, RI.

[13] Fuks, D. B. 1986. *Cohomology of infinite-dimensional Lie algebras*. Contemporary Soviet Mathematics. Consultants Bureau, New York. Translated from the Russian by A. B. Sosinskiĭ.

[14] Gelfand, I. M. and Fuks, D. B. 1968. Cohomologies of the Lie algebra of vector fields on the circle. *Funkcional. Anal. i Priložen.*, **2**(4), 92–93.

[15] Getzler, E. 1994. Two-dimensional topological gravity and equivariant cohomology. *Comm. Math. Phys.*, **163**(3), 473–489.

[16] Gwilliam, Owen, and Williams, Brian R. 2018. Higher Kac–Moody algebras and symmetries of holomorphic field theories.

[17] Gwilliam, Owen, and Williams, Brian R. 2021. A survey of holomorphic field theory. *To appear*.

[18] Hennion, Benjamin, and Kapranov, Mikhail. 2018. Gelfand–Fuchs cohomology in algebraic geometry and factorization algebras.

[19] Hinich, Vladimir. 2001. DG coalgebras as formal stacks. *J. Pure Appl. Algebra*, **162**(2-3), 209–250.

[20] Hinich, Vladimir, and Schechtman, Vadim. 1993. Homotopy Lie algebras. Pages 1–28 of: *I. M. Gelfand Seminar*. Adv. Soviet Math., vol. 16. Amer. Math. Soc., Providence, RI.

[21] Huybrechts, Daniel. 2005. *Complex geometry*. Universitext. Springer-Verlag, Berlin. An introduction.

[22] Kac, V. G. 1977. Lie superalgebras. *Advances in Math.*, **26**(1), 8–96.

[23] Kac, Victor. 1998. *Vertex algebras for beginners*. Second edn. University Lecture Series, vol. 10. American Mathematical Society, Providence, RI.

[24] Kac, Victor G. and Wakimoto, Minoru. 1994. Integrable highest weight modules over affine superalgebras and number theory. Pages 415–456 of: *Lie theory and geometry*. Progr. Math., vol. 123. Birkhäuser Boston, Boston, MA.

[25] Kapranov, Mikhail. 2021. Conformal maps in higher dimensions and derived geometry. 2.

[26] Lada, Tom, and Markl, Martin. 1995. Strongly homotopy Lie algebras. *Comm. Algebra*, **23**(6), 2147–2161.

[27] Lada, Tom, and Stasheff, Jim. 1993. Introduction to SH Lie algebras for physicists. *Internat. J. Theoret. Phys.*, **32**(7), 1087–1103.
[28] Loday, Jean-Louis, and Vallette, Bruno. 2012. *Algebraic operads*. Grundlehren der Mathematischen Wissenschaften [Fundamental Principles of Mathematical Sciences], vol. 346. Springer, Heidelberg.
[29] Lurie, Jacob. 2011. *Derived Algebraic Geometry X: Formal Moduli Problems*.
[30] Lurie, Jacob. 2017. *Higher Algebra*.
[31] Pressley, Andrew, and Segal, Graeme. 1986. *Loop groups*. Oxford Mathematical Monographs. The Clarendon Press, Oxford University Press, New York. Oxford Science Publications.
[32] Saberi, Ingmar, and Williams, Brian R. 2019. Superconformal algebras and holomorphic field theories. *arXiv:1910.04120*.
[33] Saberi, Ingmar, and Williams, Brian R. 2020. Twisted characters and holomorphic symmetries. *Lett. Math. Phys.*, **110**(10), 2779–2853.
[34] Williams, Brian R. *Holomorphic sigma-models and their symmetries*. Thesis (Ph.D.)–Northwestern University.
[35] Williams, Brian R. 2021. On the local cohomology of holomorphic vector fields. *To appear*.

9

An Introduction to Crowns in Finite Groups

Gareth Tracey

Notation and Conventions

The following is a list of notation and conventions that will be used throughout the chapter. In what follows, G is a group.

- The notation $H \leq G$ means that H is a subgroup of G; while $H \trianglelefteq G$ means that H is a normal subgroup of G.
- For a subgroup H of G, $H \backslash G$ denotes the set of right cosets of H in G.
- $C_G(H)$ denotes the centraliser of the subgroup H in G.
- $Z(G)$ denotes the centre of G, while $\Phi(G)$ denotes the Frattini subgroup of G (see Definition 9.9).
- $\mathrm{Aut}(G)$ denotes the automorphism group of G.
- For elements x and g of G, we write $x^g = g^{-1}xg$.
- More generally, group actions will always be written on the right. So if the group G acts on the set Ω, we will write ω^g for the image of $\omega \in \Omega$ under the action of $g \in G$.
- For a positive integer k, we will write G^k for the direct product of k copies of G. That is, G^k is the group which, as a set, is the cartesian product of k copies of G, equipped with pointwise multiplication.
- $\mathrm{core}_G(H) := \bigcap_{g \in G} H^g$ denotes the *core* of the subgroup H in G (i.e. the largest normal subgroup of G contained in H).
- We will write Z_n for the cyclic group of order n, and $\mathbb{F}_p$ for the finite field of order p, for p prime.
- Alt_n and Sym_n will denote the alternating and symmetric groups of degree n, respectively.
- We will write $\mathrm{SL}_n(\mathbb{F})$ and $\mathrm{GL}_n(\mathbb{F})$ for the special and general linear groups of dimension n over the field $\mathbb{F}$.
- Abelian groups will always be written multiplicatively.

- The term *minimal normal subgroup* will always refer to a non-trivial minimal normal subgroup.
- If $f\colon X \to Y$ is a map between sets X and Y, and $A \subseteq X$, we will write $f\downarrow_A$ for the restriction of f to A.

9.1 An Introduction to the Theory of Crowns

Roughly speaking, crowns are certain quotients of finite groups which have a "large" normal subgroup isomorphic to a direct product of simple groups. In order to define crowns rigorously, a number of basic notions from group and representation theory are required. In this section, we note the definitions and results necessary. We then conclude (see Subsection 9.1.3) by defining an equivalence relation on the set of chief factors of a finite group. This will set us up to define and study the notion of a crown (see Section 9.2).

9.1.1 Chief Factors in Finite Groups

Recall that for finite groups G and H, $H \leq G$ means that H is a subgroup of G, and $H \trianglelefteq G$ means that H is a normal subgroup of G. We begin by defining sections and normal sections in G.

Definition 9.1 Let G be a finite group. A *section* of G is a group X/Y, where $X, Y \leq G$ with $Y \trianglelefteq X$. If X and Y are both normal in G, then we say that X/Y is a *normal section* of G.

Thus, the composition factors in a finite group G are all sections of G, but are not necessarily normal sections. To study crowns in finite groups, we will be interested in the normal sections in G, and specifically the "minimal normal sections". These are called the chief factors of G, and their formal definition is as follows:

Definition 9.2 Let G be a finite group. A *chief factor* of G is a normal section X/Y of G with the property that if $Y \leq Z \leq X$ with $Z \trianglelefteq G$, then either $Z = X$ or $Z = Y$.

The most common (and some of the most important) examples of chief factors of a finite group G are the minimal normal subgroups of G. That is, those normal subgroups N of G with the property that if $Z \leq N$ with $Z \trianglelefteq G$, then $Z = 1$ or $Z = N$. These can be seen as chief factors of G by taking $Y := 1$ and $X := N$ in Definition 9.2. These groups are particularly important for inductive arguments in finite group theory, and they have a very particular structure:

Lemma 9.3 *Let G be a finite group, and let N be a minimal normal subgroup of G. Then $N \cong S^t$ is isomorphic to a direct product of t copies of a finite simple group S.*

Proof We prove the lemma by induction on $|G|$. The socle $\mathrm{soc}(X)$ of a finite group X is the product of its minimal normal subgroups, and is clearly a non-trivial characteristic subgroup of X. Thus, $\mathrm{soc}(N)$, being characteristic in $N \trianglelefteq G$, is normal in G. Hence, $\mathrm{soc}(N) = N$, since N is a minimal normal subgroup of G.

Now, let N_1 be a minimal normal subgroup of N. Then $\prod_{g \in G} N_1^g$ is a normal subgroup of G contained in N, so must be equal to N, by the minimality of N (we caution the reader that the product $\prod_{g \in G} N_1^g$ here is not necessarily a direct product). Choose a set $\{N_1, \ldots, N_r\}$ of G-conjugates N_i of N_1 which is minimal with the property that $N = \prod_{i=1}^r N_i$. Then $N_i \not\leq \prod_{j \neq i} N_j$, for each i. Since N_i is a minimal normal subgroup of N, it follows that N_i intersects $\prod_{j \neq i} N_j$ trivially, for each i. Hence $N = N_1 \times \ldots \times N_r$. If $N = G$, then G is simple, and the result follows. So assume that $N < G$. Then the inductive hypothesis implies that each N_i is a direct product of isomorphic simple groups: $N_i \cong T_i^{k_i}$. But all N_i are G-conjugate, so $T_i \cong T_j$ for all i, j. This completes the proof. □

Since a chief factor X/Y of G is a minimal normal subgroup of G/Y, the following is immediate.

Corollary 9.4 *Let G be a finite group, and let X/Y be a chief factor of G. Then $X/Y \cong S^t$ is isomorphic to a direct product of t copies of a finite simple group S.*

We finish this section by noting that one can inductively define a series of subgroups of a finite group G as follows: Set $X_0 := 1$, and for $i \geqslant 1$, let X_i/X_{i-1} be a minimal normal subgroup of the group G/X_{i-1}. We then have a series:

$$1 = X_0 < X_1 < \ldots < X_t = G. \tag{9.1.1}$$

This is a so-called *normal series* (i.e. every group X_i in the series is normal in G, not just in X_{i+1}).

Definition 9.5 Let G be a finite group. A series (9.1.1) in G is called a chief series for G.

Like a composition series for G, a chief series for G is unique in the following sense: if $1 = X_0 < X_1 < \ldots < X_t$ and $1 = Y_0 < Y_1 < \ldots < Y_s$ are two chief series for G, then $s = t$ and there is a bijection f from $\{X_i/X_{i-1} : 1 \leq i \leq t\}$ to $\{Y_i/Y_{i-1} : 1 \leq i \leq s\}$ such that $X_i/X_{i-1} \cong f(X_i/X_{i-1})$ for all i. Thus, we may speak of t as the chief length of G, and the set $\{X_i/X_{i-1}: 1 \leq i \leq t\}$ as the set of chief factors of G.

9.1.2 Representations and the Action of a Finite Group on Its Chief Factors

Suppose that G and A are finite groups, and that G acts on A via $a \to a^g$, $a \in A$, $g \in G$. We say that G acts on A *via automorphisms* if $(ab)^g = a^g b^g$ for all $a, b \in A$, and all $g \in G$. In this case, the map $\theta_g\colon A \to A$, $a \to a^g$, is an automorphism of A. The associated map $g \to \theta_g$ is a homomorphism from G to $\mathrm{Aut}(A)$ with kernel denoted $C_G(A) = \{g \in G\colon a^g = a \text{ for all } a \in A\}$.

For example, a finite group G acts via automorphisms (by conjugation) on any normal section of G. In particular, if X/Y is a chief factor of G and $X/Y \cong S^t$, for a simple group S, we get a well-defined map $G \to \mathrm{Aut}(S^t)$ with kernel denoted $C_G(X/Y)$. The group $G/C_G(X/Y)$ is called the group *induced* by G on X/Y. Since $G/C_G(X/Y)$ is isomorphic to a subgroup of $\mathrm{Aut}(X/Y)$, we will abuse notation and write $G/C_G(X/Y) \leq \mathrm{Aut}(X/Y)$.

We would now like to garner more information on the groups induced by a finite group on its chief factors. Before doing so, we need the following definition:

Definition 9.6 Let A be a finite group, and let T be a subgroup of the symmetric group Sym_t of degree $t \geqslant 1$. Then the *(permutational) wreath product* of A by T is the group $A \wr T := A^t \rtimes T$, where the action of T on A^t is defined by

$$(a_1, a_2 \ldots, a_t)^x = (a_{1^{x^{-1}}}, a_{2^{x^{-1}}} \ldots, a_{t^{x^{-1}}})$$

for $x \in T$, $a_i \in A$. The subgroups A^t and T are called the *base group* and *top group* of $A \wr T$, respectively.

Definition 9.6 will be useful not only for our next lemma, but also for examples throughout the chapter.

Now let G be a finite group, and let X/Y be a chief factor of G. By Corollary 9.4, X/Y is isomorphic to a direct product S^t of t copies of a finite simple group S. Then S is either abelian (i.e. $S \cong Z_p$, for a prime p), or S is a non-abelian simple group. Since the induced group $G/C_G(X/Y)$ is a subgroup of $\mathrm{Aut}(X/Y) \cong \mathrm{Aut}(S^t)$, it will be useful to have information on the automorphism group of a direct product of isomorphic simple groups.

Lemma 9.7 *Let S be a finite simple group, $t \geqslant 1$.*

1 *If S is abelian (i.e. $S \cong Z_p$ for a prime p), then* $\mathrm{Aut}(S^t) \cong \mathrm{GL}_t(p)$.
2 *If S is non-abelian, then* $\mathrm{Aut}(S^t) \cong \mathrm{Aut}(S) \wr \mathrm{Sym}_t$.

Proof For part (i), note that an elementary abelian p-group is simply a vector space over $\mathbb{F}_p$, and that an automorphism is an invertible linear map. Part (ii) is straightforward, but requires a bit more effort. We refer the interested reader to [3, Theorem 3.1] for the details. □

Recall that a *representation* of a finite group G over a field $\mathbb{F}$ is a homomorphism from G into $\mathrm{GL}_n(\mathbb{F})$. We call n the degree of the representation, and the vector space $\mathbb{F}^n$ is called the *natural module* for G. We remark that we view matrices as acting on row vectors, so all modules considered here are right modules. Lemma 9.7(i) then states that each abelian chief factor $X/Y \cong Z_p^t$ of a finite group G yields a t-dimensional representation for G over the field $\mathbb{F}_p$ of p elements. Similarly, a permutation representation of a finite group G is a homomorphism from G into Sym_n, for some $n \geqslant 1$. The natural number n is called the *degree* of the permutation representation. Lemma 9.7(ii) states that each non-abelian chief factor of a finite group G yields a permutation representation for G of degree t.

The following lemma gives more information on the groups induced by a finite group on its chief factors.

Lemma 9.8 *Let G be a finite group, and let X/Y be a chief factor of G so that X/Y is isomorphic to a direct product, S^t, of t isomorphic copies of a non-abelian finite simple group S.*

(i) If S is abelian (i.e. $S \cong Z_p$ for a prime p), then $G/C_G(X/Y) \leq GL_t(\mathbb{F}_p)$ acts irreducibly on the natural module $\mathbb{F}_p^t$.
(ii) If S is non-abelian, then consider the projection $\pi\colon \mathrm{Aut}(S^t) \cong \mathrm{Aut}(S) \wr \mathrm{Sym}_t \to \mathrm{Sym}_t$. Then $\pi(G/C_G(X/Y))$ is a transitive subgroup of Sym_t.

Proof The proof follows immediately from the fact that if A is normal in G with $Y \leq A \leq X$, then $A = Y$ or $A = X$. □

9.1.3 An Equivalence Relation on a Special Set of Chief Factors of a Finite Group

Recall that our aim in the first section of these notes is to define an equivalence relation on the set of chief factors in a finite group G. We are almost ready to do so. But first, we require a standard definition.

Definition 9.9 Let G be a finite group.

(a) The *Frattini subgroup*, written $\Phi(G)$, of G is the intersection of all maximal subgroups of G. Thus, $\Phi(G) := \bigcap_{M <_{\max} G} M$.

(b) A chief factor X/Y of G is called *non-Frattini* if X/Y is not a subgroup of $\Phi(G/Y)$.

Recall that a finite group G is *nilpotent* if all Sylow subgroups of G are normal in G. The following lemma states, in particular, that $\Phi(G)$ is nilpotent. Its proof can be found in any standard textbook in finite group theory (for example, see [6, Chapter 1]).

Lemma 9.10 *Let G be a finite group.*

(i) The subgroup $\Phi(G)$ is nilpotent.
(ii) G is nilpotent if and only if $G/\Phi(G)$ is abelian.
(iii) $\Phi(G)$ is the set of "non-generators" of G. More precisely,
$\Phi(G) = \{x \in G \colon A \subseteq G \text{ and } \langle x, A\rangle = G \text{ if and only if } \langle A\rangle = G\}$.

Notice that Lemma 9.10(i) implies that every non-abelian chief factor of G is non-Frattini. Suppose, then, that X/Y is an abelian chief factor of G. If X/Y is non-Frattini, then either $G = X$ or G/Y has the form $G/Y = X/Y \rtimes H/Y$, for some subgroup H of G containing Y. Indeed, X/Y being non-Frattini implies that there exists a maximal subgroup H/Y of G/Y not containing X/Y. Then $G/Y = (X/Y)(H/Y)$. Moreover, $(X/Y) \cap (H/Y)$ is a normal subgroup of G/Y (we leave the proof of this fact as an exercise). Thus, $(X/Y) \cap (H/Y)$ must be either trivial or equal to X/Y. Thus, either $G/Y = X/Y$ or $G/Y = X/Y \rtimes H/Y$, as claimed. For this reason, the non-Frattini chief factors in a finite group are often also called the *complemented* chief factors of G (a complement of a subgroup H in a finite group G is a subgroup K such that $HK = G$ and $H \cap K = 1$).

We would now like to define an equivalence relation on the set of non-Frattini chief factors in a finite group G. We begin with a definition.

Definition 9.11 A finite group L is called *monolithic* if L has a unique minimal normal subgroup N. If in addition N is not contained in $\Phi(L)$, then L is called a *monolithic primitive group*.

The reason for the terminology "primitive" in Definition 9.11 is that if $N \not\leq \Phi(L)$, then there exists a maximal subgroup M of L which does not contain N. It follows that M is core-free in L (i.e. $\mathrm{core}_L(M) = 1$), and hence that L has a faithful primitive permutation action on the cosets of M (we will discuss this in more depth in Subsection 9.2.1).

Our next definition introduces the "crown" terminology:

Definition 9.12 Let L be a monolithic primitive group and let N be its unique minimal normal subgroup. For each positive integer k, let L^k be the k-fold direct power of L. The *crown-based power of L of length k* is the subgroup L_k of L^k defined by

$$L_k = \{(l_1, \ldots, l_k) \in L^k \mid l_1 \equiv \cdots \equiv l_k \bmod N\}.$$

Equivalently, $L_k = N^k \operatorname{diag}(L^k)$, where $\operatorname{diag}(L^k) = \{(l_1, \ldots, l_k) \in L^k \mid l_i = l_j$ for all $i, j\}$.

Example 9.13 Let $A := Z_p$ be a cyclic group of order p, with p an odd prime. Let $H = \langle h \rangle$ be a cyclic group of order 2, and define an action of H on the k-fold direct power A^k by $a^h = a^{-1}$ for all $a \in A^k$. Let $L := A \rtimes H$. Then $A^k \rtimes H \cong L_k$ is the crown-based power of L of length k.

We are now almost ready to define the equivalence relation on chief factors in finite groups mentioned at the beginning of the section. First, recall that if a group G acts on a group A via automorphisms, then we say that A is a *G-group*. Some of the most widely studied G-groups are the groups of the form $A = \mathbb{F}^n$, for some field $\mathbb{F}^n$: these are the $\mathbb{F}[G]$-modules, and the associated maps $G \to \operatorname{Aut}(A) \cong \mathrm{GL}_n(\mathbb{F})$ are the $\mathbb{F}[G]$-representations. Our next definition generalises some basic notions in representation theory to arbitrary G-groups.

Definition 9.14 Let G be a finite group, and let A and B be G-groups.

(a) If G does not stabilise (set-wise) any non-trivial subgroup of A, then A is called an *irreducible* G-group.
(b) If there exists an isomorphism $f\colon A \to B$ such that $f(a)^g = f(a^g)$ for all $g \in G$, then A and B are said to be *G-isomorphic*.

We are now ready to define G-equivalent G-groups:

Definition 9.15 Let G be a finite group. We say that two G-groups A_1 and A_2 are *G-equivalent* and we put $A_1 \sim_G A_2$, if there are isomorphisms $\varphi\colon A_1 \to A_2$ and $\Phi\colon A_1 \rtimes G \to A_2 \rtimes G$ such that the following diagram commutes:

$$\begin{array}{ccccccccc}
1 & \longrightarrow & A_1 & \longrightarrow & A_1 \rtimes G & \longrightarrow & G & \longrightarrow & 1 \\
 & & \downarrow{\scriptstyle \varphi} & & \downarrow{\scriptstyle \Phi} & & \downarrow{\scriptstyle id} & & \\
1 & \longrightarrow & A_2 & \longrightarrow & A_2 \rtimes G & \longrightarrow & G & \longrightarrow & 1.
\end{array} \tag{9.1.2}$$

Note that the two rows in the diagram (9.1.2) represent split short exact sequences. Moreover, the map $A_i \to A_i \rtimes G$ is the usual inclusion map, while $A_i \rtimes G \to G$ is the quotient map $(a_i, g) \mapsto g$ for $a_i \in A_i$, $g \in G$.

The following lemma shows that being G-equivalent is weaker than being G-isomorphic.

Lemma 9.16 *Let G be a finite group, and let A_1 and A_2 be G-groups.*

(i) If A_1 and A_2 are G-isomorphic, then A_1 and A_2 are G-equivalent.
(ii) In the particular case where A_1 and A_2 are abelian the converse is true: if A_1 and A_2 are abelian and G-equivalent, then A_1 and A_2 are also G-isomorphic.

Proof For Part (i), let $f\colon A_1 \to A_2$ be a G-isomorphism. Then define the isomorphisms φ and Φ from Definition 9.15 by $\varphi := f$ and $\Phi\colon A_1 \rtimes G \to A_2 \rtimes G$, $a_1 g \to f(a_1)g$. It is straightforward to check that these maps are indeed isomorphisms. It is then trivial to see that the diagram (9.1.2) commutes.

For part (ii), assume that A_1 and A_2 are abelian, and that A_1 and A_2 are G-equivalent. Let φ and Φ be the maps from Definition 9.15. We claim that $\varphi\colon A_1 \to A_2$ is in fact a G-isomorphism. Indeed, fix $g \in G$, and let a_1 be an element of A_1. Then using the commuting diagram (9.1.2), we have

$$\varphi(a_1^g) = \Phi(a_1^g) = \Phi(a_1)^{\Phi(g)}, \tag{9.1.3}$$

where the last equality follows since Φ is a group homomorphism. The diagram (9.1.2), however, implies that $\Phi(g) = ug$, for some $u \in A_2$. Since A_2 is abelian, we deduce from equation (9.1.3) that $\varphi(a_1^g) = \Phi(a_1)^g$. Since $\Phi\downarrow_{A_1} = \varphi$, it follows that φ is a G-isomorphism, as claimed. □

Remark Note that if two G-groups A_1 and A_2 are G-isomorphic, then it follows from the definition of G-isomorphism that $C_G(A_1) = C_G(A_2)$. This is often a quick and easy way to show that two G-groups are not G-isomorphic.

The following is an example where two G-groups are G-equivalent, but not G-isomorphic:

Example 9.17 Let $G = \mathrm{Alt}_5 \times \mathrm{Alt}_5$, and let A_1 and A_2 be the normal subgroups $A_1 := \mathrm{Alt}_5 \times 1$, $A_2 := 1 \times \mathrm{Alt}_5$. Using the conjugation actions of G on A_1 and A_2, we can construct the (external) semidirect products $A_1 \rtimes G$ and $A_2 \rtimes G$.

Now, $C_G(A_1) = A_2$ and $C_G(A_2) = A_1$, so A_1 and A_2 are not G-isomorphic (see Remark 9.1.3).

On the other hand, define $\varphi\colon A_1 \to A_2$ by $\varphi((x,1)) = (1,x)$, and $\Phi\colon A_1 \rtimes G \to A_2 \rtimes G$ by $\Phi(((x,1),(g,h))) = ((1, xgh^{-1}),(g,h))$. Then it is a routine exercise to check that φ and Φ are isomorphisms, and the associated diagram as at (9.1.2) commutes.

9.2 Equivalence Classes of Non-Frattini Chief Factors

In this section, our aim is to build on our work in Section 9.1 to define the *set of crowns* of a finite group G (see Definition 9.30). We begin with a necessary discussion on permutation group theory.

9.2.1 Primitive Permutation Groups

As Definition 9.11 suggests, primitive permutation groups play an important role in the theory of crowns in a finite group. In this subsection, we will briefly recall some important notions from permutation group theory and, in particular, from the theory of primitive groups.

First, recall from Subsection 9.1.2 that a *permutation group on a set* Ω is a subgroup G of the symmetric group $\mathrm{Sym}(\Omega)$. In this case, Ω is called a *G-set*. If Ω is finite of cardinality n, then we say that G is a *permutation group of degree n*. Recall also that if G is a finite group acting on a set Ω, then the associated homomorphism $G \to \mathrm{Sym}(\Omega)$ is called a *permutation representation* of G.

As with G-groups, we have a notion of isomorphism between G-sets.

Definition 9.18 Let G be a finite group. Two G-sets Ω_1 and Ω_2 are said to be *G-isomorphic* if there is a bijection $f\colon \Omega_1 \to \Omega_2$ such that $f(\omega_1^g) = f(\omega_1)^g$ for all $\omega_1 \in \Omega_1, g \in G$.

The following are special types of permutation representations.

Definition 9.19 Let G be a finite group acting on a finite set Ω.

(a) G is said to act *transitively* on Ω if, for all $\omega_1, \omega_2 \in \Omega$, there exists $g \in G$ such that $\omega_1^g = \omega_2$.
(b) G is said to act *primitively* on Ω if G acts transitively on Ω and a point stabiliser $G_\omega = \{g \in G\colon \omega^g = \omega\}$ is a maximal subgroup of G.

The following are basic, but important, remarks about transitive and primitive permutation representations of a finite group. See [6, Chapter 8] for a more detailed discussion.

Remark Suppose that G is a finite group acting transitively on a finite set Ω.

(1) All point stabilisers are G-conjugate, so if one point stabiliser is maximal in G, then they all are.
(2) Consider the G-set $G_\omega \backslash G$ (i.e. the set of right G-cosets of G_ω, acted upon by G by right multiplication). Then the G-sets Ω and $G_\omega \backslash G$ are G-isomorphic. Thus, each transitive permutation representation of a finite

group G may be viewed as an action on the set of right cosets of a subgroup. In particular, each primitive permutation representation of G can be viewed as an action on the set of right cosets of a maximal subgroup of G.

(3) The *kernel of the action of* G on Ω is the set $\{g\in G\colon \omega^g=\omega \text{ for all } \omega\in\Omega\}$. When H is a subgroup of G, the kernel of the action of G on $H\backslash G$ is precisely the core of H in G.

A famous result, due independently to O'Nan and Scott, characterises the primitive permutation groups into types, usually based on geometric considerations. In this chapter, we will only be concerned with two of these types, which we now define. Recall that an *almost simple* group is a finite group L such that $S \leq L \leq \mathrm{Aut}(S)$ for some finite simple group S.

Let L be an almost simple group, $S \leq L \leq \mathrm{Aut}(S)$, such that L/S is cyclic of prime order. Consider the crown-based power $G := L_2 = \{(l_1,l_2) \in L^2\colon a_1 \equiv a_2 \bmod S\}$. Fix $\alpha \in \mathrm{Aut}(S)$ with the property that α centralises the group L/S (that is, $l^\alpha S{=}lS$ for all $l\in L$). Then $H := \{(l,l^\alpha)\colon l\in L\}$ is a subgroup of $G{=}L_2$.

Exercise 9.20 With notation as above, prove that H is a maximal subgroup of G.

Definition 9.21 We say that a primitive permutation group has *special simple diagonal type* if $G = L_2$ for some almost simple group $S \leq L \leq \mathrm{Aut}(S)$, and G_ω has the form $G_\omega = \{(l,l^\alpha)\colon l \in L\}$ for some $\alpha \in \mathrm{Aut}(S)$.

For our second type, we need to recall that the socle $\mathrm{soc}(X)$ of a finite group X is the product of the minimal normal subgroups of X.

Let $W = J \wr \mathrm{Sym}_t$, where $J \leq \mathrm{Sym}(\Delta)$ is a primitive permutation group on a finite set Δ, and $t > 1$. Fix $\delta \in \Delta$, set $I := J_\delta$, and consider the naturally embedded subgroup $I^t \leq J^t \leq W$. Set

$$H := I^t \rtimes \mathrm{Sym}_t \cong I \wr \mathrm{Sym}_t \leq W, \text{ and } \Omega := H\backslash W. \qquad (9.2.1)$$

Exercise 9.22 Prove that H is a maximal subgroup of W (i.e. W acts primitively on Ω). (Hint: show that $I^t \leq H$ is maximal as a proper Sym_t-invariant subgroup of J^t.)

Definition 9.23 We will say that a primitive permutation group $G \leq \mathrm{Sym}(\Omega)$ has *special product action type* if $G \leq W := J \wr \mathrm{Sym}_t$, where:

(a) J is primitive of special simple diagonal type;
(b) G_ω has the form $G_\omega = G \cap H$, where H is as in (9.2.1) above;
(c) the projection $G \to \mathrm{Sym}_t$ has transitive image; and
(d) G contains the naturally embedded subgroup $\mathrm{soc}(J)^t \leq J^t \leq W$.

Note that a primitive permutation group of special simple diagonal type as in Definition 9.21 above has two minimal normal subgroups, each isomorphic to S. A primitive permutation group of special product action type, as in Definition 9.23, also has two minimal normal subgroups, each isomorphic to S^t (where S^2 is the socle of J). By the O'Nan–Scott theorem, these are the only examples of a primitive permutation group with more than one minimal normal subgroup. (See [9] for more details, and for proofs of the assertions made in this paragraph.)

Remark If G is a finite group, then a subgroup H of the direct product G^k of k copies of G is said to be a *diagonal subgroup* if H has the form

$$H = \{(g, g^{\alpha_2}, \ldots, g^{\alpha_k}) : g \in G\}$$

for some automorphisms $\alpha_i \in \mathrm{Aut}(G)$. Thus, in this language a primitive permutation group has simple diagonal type if there exists a finite simple group T such that $T \times T \leq G \leq \mathrm{Aut}(T) \times \mathrm{Aut}(T)$, and a point stabiliser in G intersects $T \times T$ in a diagonal subgroup.

9.2.2 Back to Equivalence Classes of Chief Factors

Now, we have already seen an example of a primitive permutation group of simple diagonal type. Namely, take $G = \mathrm{Alt}_5 \times \mathrm{Alt}_5$ to be as in Example 9.17 (so that $T = \mathrm{Alt}_5$), and take Ω to be the set of right cosets of the diagonal subgroup $\{(t,t) : t \in \mathrm{Alt}_5\}$.

For our purposes, the important thing about this group was that it gave us an example of a finite group G with G-equivalent chief factors which are not G-isomorphic. We have seen already that two abelian chief factors of G are G-isomorphic if and only if they are G-isomorphic. By a result of Jiménez-Seral and Lafuente [7, Proposition 4.1], the non-Frattini chief factors in a finite group which are G-equivalent but not G-isomorphic occur in a very similar way to Example 9.17.

Proposition 9.24 *Let G be a finite group, and let X_1/Y_1 and X_2/Y_2 be non-Frattini chief factors of G. Then X_1/Y_1 and X_2/Y_2 are G-equivalent if and only if one of the following holds:*

(i) X_1/Y_1 and X_2/Y_2 are abelian and G-isomorphic; or
(ii) G has a maximal subgroup containing $Y_1 \cap Y_2$ such that $\mathrm{core}_G(M) = Y_1 \cap Y_2$ and $G/\mathrm{core}_G(M)$ is a primitive permutation group of simple diagonal type, with minimal normal subgroups G-isomorphic to X_1/Y_1 and X_2/Y_2.

Proposition 9.24 gives us a useful way to determine the equivalence classes of chief factors in a finite group. Some important examples are as follows.

Example 9.25 Let G be a finite p-group, for p prime. Then the non-Frattini chief factors of G all occur in $G/\Phi(G)$ – an elementary abelian p-group. Thus, all non-Frattini chief factors of G are G-isomorphic to the trivial G-group $\mathbb{F}_p$.

Example 9.26 Let L be a primitive monolithic group with minimal normal subgroup N, and let $G = L_k$ be the crown-based power of L of length k (see Definition 9.11). Then $G = N^k \operatorname{diag}(L^k)$. In particular, each element of G can be written uniquely in the form $g = (l, n_2 l, \ldots, n_k l)$, for $l \in L$ and $n_i \in N$.

For $1 \leq i \leq k$, let A_i be the ith coordinate subgroup of N^k. That is,

$$A_i := \{(1, \ldots, 1, \underbrace{a_i}_{i\text{th position}}, 1 \ldots, 1) \colon a_i \in N\}.$$

We claim that A_i is G-equivalent to A_j for all i, j. Indeed, for fixed $1 \leq i, j \leq k$, define $\varphi \colon A_i \to A_j$ by

$$\varphi(1, \ldots, 1, \underbrace{a_i}_{i\text{th position}}, 1 \ldots, 1) := (1, \ldots, 1, \underbrace{a_i}_{j\text{th position}}, 1 \ldots, 1).$$

Also, define $\Phi \colon A_i \rtimes G \to A_j \rtimes G$ as follows: for a generic element

$$x := ((1, \ldots, 1, \underbrace{a_i}_{i\text{th position}}, 1 \ldots, 1), (l, n_2 l, \ldots, n_k l))$$

of the external semidirect product $A_i \rtimes G$, define

$$\Phi(x) := ((1, \ldots, 1, \underbrace{a_i n_j^{-1}}_{j\text{th position}}, 1 \ldots, 1), (l, n_2 l, \ldots, n_k l)).$$

It is routine (though non-trivial) to prove that φ and Φ are homomorphisms, and that the associated diagram from (9.1.2) commutes. Thus, all A_i and A_j are G-equivalent.

The following is an illustration of how one finds representatives for the equivalence classes of non-Frattini chief factors of G in a specific example.

Example 9.27 Consider $G = \mathrm{Sym}_4$. Since G is soluble, two chief factors A_1 and A_2 are G-equivalent if and only if they are G-isomorphic. Now, a chief series for G is

$$1 < V_4 < \mathrm{Alt}_4 < G,$$

where $V_4 = \langle (1,2)(3,4), (1,3)(2,4) \rangle$. Since G, $G/V_4 \cong \mathrm{Sym}_3$ and $G/\mathrm{Alt}_4 \cong Z_2$ all have trivial Frattini subgroups, each of the associated chief factors are non-Frattini. Furthermore, finding a set of representatives for the G-equivalence classes of Frattini chief factors for G is easy in this case, since the chief factors V_4, Alt_4/V_4 and G/Alt_4 are pairwise non-isomorphic as groups, so are certainly pairwise non-isomorphic as G-groups. Thus, $\{V_4, \mathrm{Alt}_4/V_4, \mathrm{Sym}_4/\mathrm{Alt}_4\}$ is a complete set of representatives for the G-equivalence classes of chief factors in $G = \mathrm{Sym}_4$.

Exercise 9.28 Find a set of representatives for the non-Frattini chief factors in the cases $G = \mathrm{GL}_2(3)$ and $G = Z_p \rtimes Z_{p-1}$, where p is prime and Z_{p-1} acts on Z_p as $\mathrm{Aut}(Z_p)$.

9.2.3 The Set of Crowns in a Finite Group

In this subsection, we will define the set of crowns in a finite group G. As the terminology suggests, and as the next lemma shows, crown-based powers play an important role in this definition.

Lemma 9.29 *Let G be a finite group, and let A be a non-Frattini chief factor of G. With $C_G(A)$ as defined on page 320, define L_A, the* monolithic primitive group *associated to A, by*

$$L_A := \begin{cases} A \rtimes (G/C_G(A)) & \text{if } A \text{ is abelian,} \\ G/C_G(A) & \text{otherwise.} \end{cases}$$

Then:

(i) There exists a normal subgroup N of G such that $G/N \cong L_A$.
(ii) Set $R_G(A) := \bigcap_N N$, where the intersection runs over all $N \trianglelefteq G$ such that $G/N \cong L_A$. Then $G/R_G(A) \cong (L_A)_k$, where k is the number of non-Frattini chief factors in any chief series for G which are G-equivalent to A.

Proof Since (i) is trivial in the case where A is non-abelian, we may assume that A is abelian.

Let G be a counterexample to (i) of minimal order, and let $Y \leq X$ be normal subgroups of G with $A = X/Y$. By minimality, we may assume that $Y = 1$, so that $A = X$ is a minimal normal subgroup of G.

Since A is non-Frattini, A has a complement H in G (see the discussion after the statement of Lemma 9.10). Thus, $G = A \rtimes H$. It is then clear that $N := C_H(A) \trianglelefteq G$, and $G/N \cong L_A$. This proves (i).

We will now prove (ii). To do so, we need a few standard group theoretic facts. In what follows, let G be a finite group and let N_1 and N_2 be distinct normal subgroups of G.

(1) The factor group $G/N_1 \cap N_2$ is isomorphic to a subgroup of $G_1 \times G_2$, where $G_i := G/N_i$ via the embedding $\theta \colon G/N_1 \cap N_2 \hookrightarrow G_1 \times G_2$, $(N_1 \cap N_2)g \to (N_1 g, N_2 g)$ for $g \in G$.
(2) Let $\pi_i \colon G_1 \times G_2 \to G_i$ be the canonical projection. Then $\pi_i(\theta(G/N_1 \cap N_2)) = G_i$, for $i = 1,2$ (we say that $G/N_1 \cap N_2$ projects onto both G_1 and G_2). This is clear from the definition of ρ.
(3) If E and F are finite groups and X is a subgroup of $E \times F$ projecting onto both E and F, then $X_E := X \cap (E \times 1) \trianglelefteq E$, $X_F := X \cap (1 \times F) \trianglelefteq F$, $E/X_E \cong F/X_F$ and $X/X_E X_F$ is a diagonal subgroup of $(E/X_E) \times (F/X_F)$. We leave the proofs of these assertions as an exercise for the reader.

We now prove (ii), only in the case $k = 2$ (the proof for larger k follows the same line of argument, and is left as an exercise for the reader). So let N_1 and N_2 be distinct normal subgroups of G with $G/N_1 \cong G/N_2 \cong L_A$, and $R_G(A) = N_1 \cap N_2$. Note that $N_1, N_2 > 1$, since N_1 and N_2 are distinct and have the same order.

We need to show that $G/R_G(A) \cong (L_A)_2$. Thus, by factoring out $R_G(A)$, we may assume that $R_G(A) = 1$. Suppose first that A is non-abelian. Then by Proposition 9.24(ii), and since $R_G(A) = N_1 \cap N_2 = 1$, we see that G is isomorphic to a primitive permutation group of simple diagonal type, with two minimal normal subgroups, each G-isomorphic to A. By definition of primitive groups of simple diagonal type, we deduce that $G \cong (L_A)_2$.

Suppose next that A is abelian. Since $R_G(A) = 1$, Fact (1) above implies that G embeds as a subgroup of $L_1 \times L_2 \cong (L_A)^2$, where $L_i := G/N_i \cong L_A$. To avoid unnecessary additional notation, we will omit reference to the embedding given in Fact (1) and assume, for the remainder of the proof, that G is a subgroup of $L_1 \times L_2$. By Fact (3) and the definition of the embedding $G \hookrightarrow L_1 \times L_2$, we may then assume that $G \cap (L_1 \times 1) = N_1 \times 1$; $G \cap (1 \times L_2) = 1 \times N_2$; and $G/(N_1 \times N_2)$ is isomorphic to a diagonal subgroup of $L_1/N_1 \times L_2/N_2$.

Write A_i for the unique minimal normal subgroup of L_i. We claim that $N_i = A_i$ for each i. To this end, note first that since $N_i \neq 1$, N_i contains A_i. Also, since A is abelian and A_1 and A_2 are G-equivalent, Proposition 9.24 and Remark 9.1.3 imply that $C_G(A_1) = C_G(A_2)$. On the other hand, since A_1 is the unique minimal normal subgroup of L_1 and A_1 is non-Frattini, we have $A_1 = C_{L_1}(A_1)$. Thus, the centraliser of A_1 in $G \leq L_1 \times L_2$ is precisely $(A_1 \times 1) \times (G \cap 1 \times L_2) = A_1 \times N_2$. Similarly, $C_G(A_2) = N_1 \times A_2$. Since $C_G(A_1) = C_G(A_2)$, it follows that $N_1 = A_1$ and $N_2 = A_2$, as claimed.

Thus, we have shown that G is isomorphic to a subgroup of $(L_A)^2$ containing A^2, and that, under this embedding, G/A^2 is isomorphic to a diagonal subgroup of $(L_A/A)^2$. We then see from the definition of $(L_A)_2$ that $G \cong (L_A)_2$. □

Lemma 9.29 is the key lemma in the theory of crowns, and allows us to define the following:

Definition 9.30 Let G be a finite group, and let A be a non-Frattini chief factor of G.

(a) The normal subgroup $R_G(A)$ from Lemma 9.29 is called the *A-core* of G.
(b) The subgroup $I_G(A)$ is defined so that $I_G(A)/R_G(A) = \mathrm{soc}(G/R_G(A))$ is the socle of $G/R_G(A)$. We call $I_G(A)/R_G(A)$ the *A-crown* of G.
(c) As proved in Lemma 9.29, $I_G(A)/R_G(A) \cong A^k$. We define $\delta_G(A) := k$, so that $\delta_G(A)$ is the number of non-Frattini chief factors G-equivalent to A in any chief series for G.

We can now define the *set of crowns* in a finite group G.

Definition 9.31 Let G be a finite group. The set

$$\{I_G(A)/R_G(A) \colon A \text{ a non-Frattini chief factor of } G\}$$

is called the *set of crowns for* G.

9.3 An Application of Crowns: Minimal Generator Numbers

In this section, our aim is to demonstrate one of the most useful applications of the theory of crowns to problems in finite group theory. Namely, finding the minimal number of elements required to generate a finite group G.

For a finite group G, define $d(G) := \min\{|X| \colon X \subseteq G, \langle X \rangle = G\}$ to be the minimal size of a generating set for G. Thus, if G is cyclic, for example, then $d(G) = 1$. If V is an elementary abelian group of order p^n, then G may be viewed as a vector space of dimension n over the finite prime field $\mathbb{F}_p$, and then $d(G)$ is just the $\mathbb{F}_p$-dimension of G: that is, $d(G) = n$.

The last example shows that the function d is well behaved when G is a vector space: namely, $d(H) \leq d(G)$ when H is a subgroup (i.e. subspace) of G. But this is not true in general. For example, take G to be the wreath product $R \wr S$ (see Definition 9.6) where $R \cong S \cong Z_p$ (S is viewed as $S = \langle s \rangle$, with $s = (1,2,\ldots,p) \in \mathrm{Sym}_p$ in this case). The base group $H \cong R^p$ of G is elementary abelian of order p^p, and so $d(H) = \dim_{\mathbb{F}_p}(H) = p$. However, if we set $X := \{\underbrace{(r,1\ldots,1)}_{p}, s\} \subseteq G$, where r is a generator for R, then it is easy to see that $G = \langle X \rangle$. Thus, $d(G) = 2$, since G is not cyclic.

The above example shows that, in general, there can be no bound on $d(H)$ in terms of $d(G)$ for subgroups H of a finite group G, even for finite p-groups.

Finite p-groups are, however, quite straightforward to deal with when it comes to finding $d(G)$, as the next result shows.

Proposition 9.32 *Let G be a finite group. Then $d(G) = d(G/\Phi(G))$. In particular, if G is a finite p-group, for p prime, then $d(G)$ is the dimension of the $\mathbb{F}_p$-vector space $G/\Phi(G)$.*

Proof That $d(G) \leq d(G/\Phi(G))$ follows immediately from Lemma 9.10 (since $\Phi(G)$ is the set of "non-generators" for G). On the other hand, if X is a generating set for G, and N is any normal subgroup of G, then the set $\{xN : x \in X\}$ is a generating set for G. Hence $d(G/N) \leq d(G)$. In particular, $d(G/\Phi(G)) \leq d(G)$, so $d(G) = d(G/\Phi(G))$. □

Example 9.33 Recall from Example 9.25 that a finite p-group has a unique G-equivalence class of non-Frattini chief factors, represented by the trivial $\mathbb{F}_p[G]$-module $A := \mathbb{F}_p$. Since all non-Frattini chief factors of G occur as chief factors of $G/\Phi(G)$, and all chief factors of $G/\Phi(G)$ are non-Frattini, we deduce that G has precisely $d(G)$ non-Frattini chief factors of G (all G-equivalent to A). Since G acts trivially on A, we have $L_A = A$ (where L_A is as in Lemma 9.29), and so $G/R_G(A) \cong A^{d(G)} \cong (\mathbb{F}_p)^{d(G)}$. In particular, $R_G(A) = \Phi(G)$ and $\delta_G(A) = d(G)$ in this case.

Remark During the course of the proof of Proposition 9.32, we proved that if G is a finite group and N is a normal subgroup of G, we have $d(G/N) \leq d(G)$. An often-used inductive tool (for deriving upper bounds on $d(G)$) is the (almost trivial) upper bound $d(G) \leq d(G/N) + d(N)$.

As mentioned in Example 9.33, we have $d(G) = d(G/R_G(A))$ for a finite p-group G, where A is the (up to G-equivalence) unique non-Frattini chief factor of G (here, $R_G(A) = \Phi(G)$). The next result shows that this (perhaps surprisingly) can be made more general.

Theorem 9.1 *[4, Theorems 1.4 and 2.7] Let G be a finite group with $d(G) \geqslant 3$. Then G has a non-Frattini chief factor A such that $d(G) = d(G/R_G(A))$. Moreover:*

(1) if G is abelian, then $d(G) = d(G/R_G(A)) \leq \delta_G(A) + 1$;
(2) if G is non-abelian, then $d(G) = d(G/R_G(A)) \leqslant \lceil \log_{|A|}\left(\frac{90}{54}\delta_G(A)\right) + 5/4 \rceil$.

The proof of Theorem 9.1 is beyond the scope of this course, but we refer the interested reader to [4, Theorems 1.4 and 2.7] for details.

Theorem 9.1 is an extremely useful tool for determining the minimal generator numbers in various classes of finite groups. We close the section by illustrating this with some examples.

Example 9.34 Let $G = T^k$ be the direct product of k copies of a non-abelian finite simple group T. As mentioned in Remark 9.3, we have $d(G) \leq d(G/N) + d(N)$ for any normal subgroup N of G. Hence, since every finite simple group can be generated by two elements, we have $d(G) \leq 2k$.

Let us see if we can do any better using the theory of crowns. We will assume that $d(T^k) \geqslant 3$. Then $k \geqslant 2$, since $d(T) = 2$ (as mentioned above). Now, note that G is isomorphic to the crown-based power T_k (this is, in some sense, the "trivial crown-based power associated to T"). It follows that G has a unique equivalence class of non-Frattini chief factors, isomorphic to T, by Example 9.26. Clearly, $\delta_G(T) = k$. Hence, Theorem 9.1 yields $d(G) \leq \log_{|T|}(k) + 1$ – a far tighter bound than $d(G) \leq 2k$.

Example 9.35 Let G be the wreath product $R \wr S$, where $R = Z_p$ is cyclic of prime order p, and $S = \mathrm{Alt}_s$ is the alternating group of degree $s \geqslant 5$. Consider the following subgroups of the base group R^s of G:

$$A_1 := \{(x,x,\ldots,x) : x \in R\} \text{ and } A_2 := \{(x_1,x_2,\ldots,x_s) : x_i \in R, \prod_{i=1}^{s} x_i = 1\}.$$

The subgroups A_1 and A_2 are clearly normal in G. Since $|A_1| = p$, A_1 is a minimal normal subgroup of G. Note that $A_1 \leq A_2$ if $p \mid s$, and $A_1 \nleq A_2$ otherwise. Since A_2 has order p^{s-1}, we deduce that $A_1A_2 = A_2$ has index p in R^s if $p \mid s$, and $A_1A_2 = R^s$ otherwise. It is not difficult to prove (see [8, Proposition 5.4.1]) that Alt_s acts irreducibly on the $\mathbb{F}_p[\mathrm{Alt}_s]$-module A_1A_2/A_1, and hence that A_1A_2/A_1 is a chief factor of G. Thus, since $G/R^s \cong \mathrm{Alt}_s$ is simple, we deduce that

$$1 < A_1 < A_1A_2 \leq R^s < G$$

is a chief series for G. Since $|A_1A_2/A_1| = p^{s-2}$ if $p \mid s$, and $|A_1A_2/A_1| = p^{s-1}$ otherwise, we have that G has two chief factors G-isomorphic to the trivial $\mathbb{F}_p[G]$-module $\mathbb{F}_p$ if $p \mid s$, and one chief factor G-isomorphic to $\mathbb{F}_p$ otherwise. Hence, $\delta_G(\mathbb{F}_p) \leq 2$. Furthermore, we clearly have $\delta_G(A_1A_2/A_1) = 1$, and $\delta_G(\mathrm{Alt}_s) = 1$. In fact, it is not difficult to prove that if $p \mid s$, then $A_1 \leq \Phi(G)$, so $\delta_G(A) = 1$ for all non-Frattini chief factors A of G. Hence, Theorem 9.1 yields $d(G) \leq 2$. Thus, $d(G) = 2$, since G is not cyclic.

Recommended Further Reading

A more detailed account of crowns in finite groups in given in [2, Chapter 1], which we can certainly recommend for further reading.

The theory of crowns also naturally arises in the study of the first cohomology group of a finite group: see [1] for more details.

Bibliography

[1] Aschbacher M., and Guralnick, R. M. 1984. Some applications of the first cohomology group, *J. Algebra*, **90**(2), 446–460.

[2] Ballester-Bolinches, A. and Ezquerro, L. M., Classes of finite groups, Mathematics and Its Applications (Springer), vol. 584, Springer, Dordrecht, 2006.

[3] Bidwell, J. N. S., Automorphisms of direct products of finite groups II, Arch. Mat. 91 (2008), 111–121.

[4] Dalla Volta, F. and Lucchini, A., Finite groups that need more generators than any proper quotient, J. Austral. Math. Soc., Series A, 64 (1998) 82–91.

[5] Doerk, K. and Hawkes, T., Finite soluble groups, de Gruyter, Berlin, 1992.

[6] Isaacs, I. M., Finite Group Theory, Graduate Studies in Mathematics, vol. 92, American Mathematical Society, Providence, 2008.

[7] Jiménez-Seral, P. and Lafuente, J. On complemented nonabelian chief factors of a finite group, Israel J. Math. 106 (1998), 177–188.

[8] Kleidman, P. and Liebeck, M. W., The subgroup structure of the finite classical groups, CUP, Cambridge, 1990.

[9] Liebeck, M. W., Praeger, C. E. and Saxl, J., On the O'Nan-Scott Theorem for finite primitive permutation groups, J. Austral. Maths. Soc (Series A) 44 (1988), 389–396.

10
An Introduction to Totally Disconnected Locally Compact Groups and Their Finiteness Conditions

Ilaria Castellano

10.1 Introduction

The class of locally compact groups generalises discrete and Lie groups. Locally compact groups came to light in the first half of the 20th century, and since then they have played a central role among topological groups. The 20th century has been characterised by intense activity on the structure theory of many algebraic objects, e.g., simple finite groups, fields and rings with ascending chain condition. What about the general structure of locally compact groups? A basic strategy to understand the structure of a locally compact group G is to split it into smaller groups: let G_0 be the largest connected subset of G containing the identity element, which is a closed subgroup (see Fact 10.1), and produce the short exact sequence

$$1 \to G_0 \to G \to G/G_0 \to 1,$$

where G_0 is a **connected** locally compact group and G/G_0 is a **totally disconnected** locally compact group (i.e., the connected components of G/G_0 are reduced to singletons). Therefore, G can be regarded as an extension of its connected component by the totally disconnected piece G/G_0. It follows that questions about the structure of locally compact groups may be dealt with by treating separately the cases where G is connected and where G is totally disconnected and then combining the two answers. With the solution of Hilbert's fifth problem, our understanding of connected locally compact groups has significantly increased: they can be approximated by Lie groups (see Theorem 10.8). Therefore, the contemporary structure problem on locally compact groups concerns the class of totally disconnected locally compact groups.

The first part of the notes introduces the objects under investigation: topological groups that are locally compact and totally disconnected (TDLC-groups, for short). In particular, it includes basic properties of topological groups, a proof of van Dantzig's theorem, which is the fundamental structure theorem of TDLC-groups, and several examples.

The study of this class of topological groups can be made more manageable by dividing the infinity of objects under investigation into classes of types with "similar structure:" we focus on TDLC-groups satisfying some finiteness conditions. The most common finiteness condition for (totally disconnected) locally compact groups is *compact generation* – i.e., the topological group is algebraically generated by a compact subset – which naturally generalises the notion of finite generation that has been widely (and fruitfully) used in group theory to study infinite groups. Since every TDLC-group is a directed union of compactly generated open subgroups (see Fact 10.19), one can confine the investigation to the compactly generated case without losing too much information (at least from a local perspective).

All compactly generated TDLC-groups fall in the class of automorphism groups of locally finite connected graphs (see § 10.3.3). Indeed, [1] proved that, by using van Dantzig's theorem, we can always construct a locally finite connected graph Γ on which the group G acts vertex-transitively and with compact open stabilisers, the so-called *Cayley–Abels graph of G* (see § 10.4.2). To be precise, every compactly generated TDLC-group comes with a family of Cayley–Abels graphs which satisfy the property of being quasi-isometric to each-other: the Cayley–Abels graph Γ of a compactly generated TDLC-group is unique up to quasi-isometry. Therefore, by analogy with the theory of finitely generated groups and their Cayley graphs, we can study a compactly generated TDLC-group as a geometric object and all the properties of Γ that are invariant up to quasi-isometry become *geometric group invariants*[1] of G. It leads to a new line of research in topological group theory, *geometric group theory for compactly generated TDLC-groups*, which can be traced back to the work of [31].

As a consequence, TDLC-groups need to be viewed simultaneously as geometric objects and topological objects. Therefore, the interaction between the local structure (i.e., the topological one) and the large-scale structure (i.e., the geometric one) becomes important. Since profinite groups are trivial as geometric groups and discrete groups are trivial as topological groups, it is not surprising that the profinite groups and the discrete groups constitute the atomic pieces in the theory of TDLC-groups and that we are curious to understand to what extent well-known results about the geometry of infinite discrete groups find an analogue in such a topological context.

[1] The number of ends, growth and hyperbolicity are examples of geometric group invariants of a TDLC-group G.

The final part of the chapter attempts to introduce the reader to some homological finiteness conditions for TDLC-groups that generalise compact generation in higher dimensions. Since these notes are meant to be on the non-specialist level, we only provide definitions and references, letting the reader decide how deep to dig into the subject.

The conclusive part briefly introduces the seminal work of [51, 52] which was a fundamental breakthrough in the theory of TDLC-groups after several years of stillness.

Willis' theory gave start to the research interest we now benefit from.

Pre-requisites: The reader is supposed to have mastered linear algebra, point-set topology and basic notions from group theory. These notes aim to help beginning Ph.D. students taking the first steps towards the theory of TDLC-groups. Therefore, this text is (supposed to be) accessible to non-specialists and self-contained (most of the results are proved in these notes) but, as with any advanced topic, some of the results are stated without proofs. In such a case, references are provided. All the proofs provided in these notes are rather classical and can be found in the literature.

The interested reader can also have a look at a few conference proceedings that collect part of the progress made with the general theory of TDLC-groups and include some interesting open problems; see for example [14] and [57].

Notation: We denote by

- $\mathbb{N}$ the set of natural numbers $\{0, 1, 2, ...\}$,
- $\mathbb{Z}$ the ring of rational integers and $\mathbb{Q}$ the ring of rational numbers,
- $\mathbb{R}$ the field of real numbers and $\mathbb{C}$ the field of complex numbers,
- $\mathbb{R}_+$ the subset of non-negative real numbers.
- If R is a commutative ring with unit, $R^\times$ stands for its multiplicative group of units. For example, $\mathbb{R}_+^\times$ is the group of positive real numbers.

10.2 Preliminaries on Topological Groups

A complete and detailed introduction to the theory of topological groups can be found in several textbooks; for example, [27]. For convenience, all topological spaces appearing below are assumed to satisfy the Hausdorff separation axiom.[2]

[2] Totally disconnected locally compact groups are necessarily Hausdorff.

10.2.1 Warming-up

Definition 10.1 (Topological group) A **topological group** is a group $(G,\cdot)$ which is also a topological space such that the following maps are continuous:

- the group operation

$$\_\cdot\_\colon G\times G\to G,\quad (x,y)\mapsto x\cdot y,\quad \forall x,y\in G,$$

  where $G\times G$ is endowed with the product topology;
- the inversion map

$$\_^{-1}\colon G\to G,\quad x\mapsto x^{-1},\quad \forall x\in G.$$

If the underlying group G is cyclic (resp., abelian, nilpotent, etc.), the topological group G is also called cyclic (resp. abelian, nilpotent, etc.). The additive notation $(G,+)$ can describe some topological groups but only in the case when the topological group is abelian.

Remark Similarly, one has **topological rings** and **topological fields**.[3]

Every abstract group G can be viewed as a topological group if given the discrete topology. In such a case, G is called a **discrete group** and, since these notes concern topological groups, we will often refer to (abstract) groups as discrete groups.

Exercise 10.2 Let G be a topological group. Prove that:

1. If a subgroup $H\leqslant G$ is open, then it is also closed.
2. If a subgroup contains an open set then it is open.
3. A connected subset $C\ni x$ is contained in the intersection of all clopen[4] subsets $O\ni x$ of G.
4. If G is connected, then the only open subgroup of G is G itself.
5. Every quotient map of a topological group is open.
6. For $H\leqslant G$ normal, the quotient group G/H is discrete iff H is open. In particular, the group G is discrete iff the point 1_G is isolated, i.e., the singleton $\{1_G\}$ is open.

Other (less trivial) examples of a topological group are provided by the additive group $(\mathbb{R},+)$ of the reals equipped with the usual topology, its subgroups $\mathbb{Z}$ and $\mathbb{Q}$ (with the subspace topology) and the quotient $\mathbb{T}:=\mathbb{R}/\mathbb{Z}$ (with the quotient topology). These extend to all powers $(\mathbb{R}^d,+)$, and also $(\mathbb{C},+)$, because it can be easily proved that products of topological groups are again topological

[3] By "field" we always mean "commutative field".
[4] A set which is both closed and open.

groups. Moreover, if R is a topological ring, then the ring $M(n,R)$ of all $n \times n$ matrices with entries in R is a topological ring if endowed with the product topology of $R^{n\times n}$.

Exercise 10.3 Let R be a commutative topological ring such that inversion is continuous on the set of invertible elements (for example, if R is a topological field). Prove that

1 the group $(GL(n,R),\cdot)$ of all invertible $n \times n$ matrices with entries in R is a topological group (notice that $GL(n,R)$ is a subset of $M(n,R)$ but not a subgroup; in this case Cramer's rule helps with the continuity of the inversion map);
2 $SL(n,R) = \{M \in GL(n,R) \mid \det(M) = 1\}$ is closed in $GL(n,R)$.

A topology τ on the group G such that the space (G,τ) is a topological group is called a **group topology** on G. Obviously, a topology τ on G is a group topology if, and only if, the map $G \times G \to G, \quad (x,y) \mapsto xy^{-1}$, is continuous for all $x,y \in G$. Notice that, for every $g \in G$, the **left translation** $x \mapsto gx$, the **right translation** $x \mapsto xg$, as well as the **conjugation** $x \mapsto gxg^{-1}$ are continuous; in other words, every topological group is a homogeneous topological space.

Exercise 10.4 Prove the assertions above.

As a consequence, the topology of G is determined by a neighbourhood basis[5] of the identity: a family $\{U_\alpha\}_{\alpha\in I}$ of arbitrarily small neighbourhoods of 1_G determines the family $\{gU_\alpha\}_{\alpha\in I}$ of arbitrarily small neighbourhoods of any other group element g.

Example 10.5 One can define group topologies on G by declaring well-behaved collections of subsets to be neighbourhood basis at 1_G (see [9]):

- the **profinite topology** is determined by the family of all normal subgroups of finite index of G;
- the **pro-p topology** is determined, for a prime p, by all normal subgroups of G of finite index that is a power of p;
- the **p-adic topology** is determined, for a prime p, by the family $\{U_n\}_{n\in\mathbb{N}}$ of normal subgroups of G, where U_n is generated by the powers $\{g^{p^n} \mid g \in G\}$.

Example 10.6 (Absolute values on fields) An **absolute value** on a field $\mathbb{K}$ is a function $|\_|\colon \mathbb{K} \to \mathbb{R}$ satisfying

[5] A family $\mathcal{B}$ of neighbourhoods of the point x is said to be a *neighbourhood basis* of x if for every neighbourhood U of x there exists $V \in \mathcal{B}$ contained in U.

$|1|$. $|x| \geqslant 0$ for every $x \in \mathbb{K}$, and $|x| = 0$ if and only if $x = 0$,
$|2|$. $|xy| = |x||y|$, for all $x, y \in \mathbb{K}$,
$|3|$. $|x+y| \leqslant |x| + |y|$, for all $x, y \in \mathbb{K}$.

If the absolute value $|\_|$ satisfies the stronger condition

$|\bar{3}|$. $|x+y| \leqslant \max\{|x|, |y|\}$, for all $x, y \in \mathbb{K}$,

it is said to be **non-Archimedean**, otherwise it is **Archimedean**. For example, the usual absolute value on $\mathbb{R}$ is Archimedean.

It is clear that $d(x,y) = |x-y|$ gives $\mathbb{K}$ a structure of a metric space, and the topology for which

$$B_\varepsilon(0) := \{x \in \mathbb{K} \mid |x| < \varepsilon\}, \quad \varepsilon > 0,$$

form a basis of neighbourhoods of 0 is a field topology.

Exercise 10.7 Let $\mathbb{K}$ be a field equipped with the absolute value $|\_|$. The field topology defined on $\mathbb{K}$ by $|\_|$ is discrete iff $|x| = 1$ for all $x \neq 0$.

Example 10.8 (The field of p-adic numbers) For a prime p, the **p-adic absolute value** on $\mathbb{Q}$ is defined as $|x|_p = p^{-n}$, for every $x \in \mathbb{Q}$, where n is the unique integer such that $x = p^n(\frac{a}{b})$ and neither of the integers a and b is divisible by p (with the convention, $|0|_p = 0$). It is an example of non-Archimedean absolute value. The **p-adic metric** $d_p(x,y) = |x-y|_p$ induces a field topology on $\mathbb{Q}$ which is called the **p-adic topology** of $\mathbb{Q}$; see Example 10.5. As with the topology on $\mathbb{Q}$ inherited from $\mathbb{R}$, the metric space $(\mathbb{Q}, d_p)$ is not complete (i.e., not every Cauchy sequence converges in $(\mathbb{Q}, d_p)$). Let $\mathbb{Q}_p$ denote the completion of $\mathbb{Q}$ with respect to the p-adic metric. The field $\mathbb{Q}_p$ has characteristic zero and it is called the **field of p-adic numbers**. The metric d_p, and so the p-adic topology, extends from $\mathbb{Q}$ to $\mathbb{Q}_p$ and yields a topological field; see [37, § 12.3.4].

The identity component of G. On a topological space X one defines the equivalence relation $\sim$ as follows: $x \sim y$ if there exists a connected subspace $C \subseteq X$ such that $x, y \in C$. Each equivalence class is a maximal connected subspace which is called a **connected component** of X.

Definition 10.9 Given a topological group G, the connected component containing the identity is the **identity component** of G, and it is denoted by G_0. Clearly, G_0 is the union of all connected subspaces of G containing 1_G, and the topological group G is connected if, and only if, $G = G_0$. A topological group is said to be **totally disconnected** if the identity is its own connected component, that is, $G_0 = \{1_G\}$.

Notice that, for every $g \in G$, the set $gG_0 = G_0g$ is nothing but the connected component containing g because continuous maps preserve connectedness and translations are continuous. As a consequence, one has the following important fact.

Fact 10.1 *Given a topological group, the identity component is a closed characteristic subgroup.*

Proof The subspace G_0 is closed since the closure of a connected subspace is still connected.

For every $x \in G_0$, the translate $x^{-1}G_0 \ni 1_G$ is connected. It follows that $x^{-1} \in x^{-1}G_0 \subseteq G_0$, i.e., G_0 is closed under taking inverses. On the other hand, for all $x, y \in G_0$, one has $xy \in xG_0$ but xG_0 is the connected component of x which is in the same connected component as 1_G (because $x \sim 1_G$), i.e., $xy \in xG_0 = G_0$ and G_0 is a subgroup.

Finally, G_0 is characteristic (and in particular normal) because continuous homomorphisms preserve connectedness. □

Proposition 10.2 *Let G be a topological group and let G_0 be the identity component. Then G/G_0 is a totally disconnected group.*

Proof Let $\pi\colon G \to H := G/G_0$ be the quotient map. One has to prove that $H_0 = 1$, i.e., $\pi^{-1}(H_0) = G_0$. By the maximality of G_0, it suffices to prove that $\pi^{-1}(H_0) \supseteq G_0$ is connected. To this end, suppose by contradiction $\pi^{-1}(H_0)$ is a disjoint union $C_1 \cup C_2$ of non-empty closed subsets of G. Since G_0 is connected, for every $g \in \pi^{-1}(H_0)$, either $gG_0 \subseteq C_1$ or $gG_0 \subseteq C_2$. Therefore, C_1 and C_2 are unions of G_0-cosets. Consequently, H_0 is the disjoint union of non-empty closed subsets, a contradiction. □

10.2.2 Profinite Groups

For a complete introduction to the realm of profinite groups the reader is referred to [40, 56] and [14, Chapter 3]. We recall only the definition and provide some easy examples.

A **directed poset** $(I, \preceq)$ is a set I with a binary relation $\preceq$ satisfying:

(DP1) $i \preceq i$, for $i \in I$;
(DP2) $i \preceq j$ and $j \preceq k$ imply $i \preceq k$, for $i, j, k \in I$;
(DP3) $i \preceq j$ and $j \preceq i$ imply $i = j$, for $i, j \in I$; and
(DP4) if $i, j \in I$, there exists some $k \in I$ such that $i \preceq k$ and $j \preceq k$.

An **inverse system of topological groups over** I consists of a family $\{G_i \mid i \in I\}$ of topological groups together with continuous group morphisms $\varphi_{ij}\colon G_i \to G_j$, defined whenever $j \preceq i$, such that the diagram

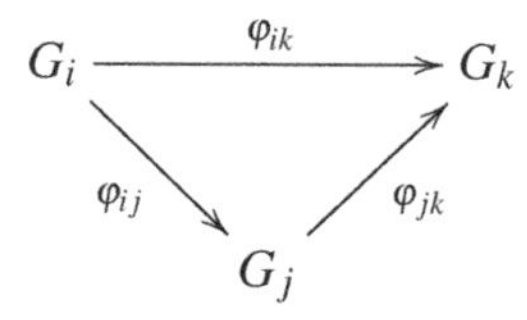

commutes whenever $k \preceq j \preceq i$. In addition, we assume that φ_{ii} is the identity morphism for every $i \in I$. A projective system of topological groups is said to be surjective if every morphism φ_{ij} is surjective. A family of continuous group morphisms $\varphi_i\colon G \to G_i$ is said to be **compatible** with the inverse system (G_i, φ_{ij}, I) if, for every $i \preceq j$, the diagram

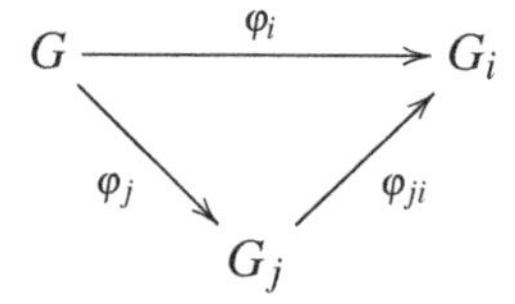

commutes. A topological group G together with a compatible family $\{\varphi_i\colon G \to G_i\}_{i\in I}$ of continuous morphisms is an **inverse limit** of the inverse system (G_i, φ_{ij}, I) if the following universal property is satisfied: for every topological group $\tilde{G}$ together with a compatible family $\{\psi_i\colon \tilde{G} \to G_i\}_{i\in I}$ of continuous group morphisms, there exists a unique continuous group morphism $\psi\colon \tilde{G} \to G$ such that the diagram

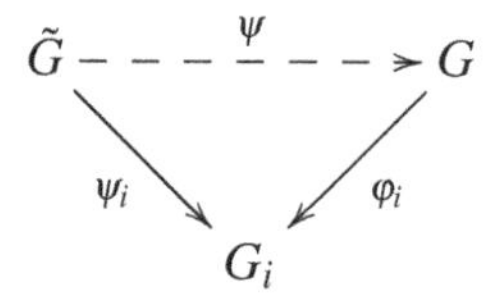

commutes for every $i \in I$. In such a case, we denote the inverse limit by

$$G = \varprojlim_{i\in I} (G_i, \varphi_{ij}),$$

and call the maps $\varphi_i\colon G \to G_i$ *projection morphisms*. If the family φ_{ij} is clear from the context, we use simply $G = \varprojlim_{i\in I} G_i$.

Proposition 10.3 ([40, Proposition 1.1.1]) *Let (G_i, φ_{ij}, I) be an inverse system of topological groups over a directed poset I. Then the following hold:*

(a) *there exists an inverse limit of the inverse system* (G_i, φ_{ij}, I)*;*
(b) *this limit is unique in the following sense: if* (G, φ_i) *and* (H, ψ_i) *are two limits of* (G_i, φ_{ij}, I)*, then there is a unique topological isomorphism* $\varphi\colon G \to H$ *such that* $\psi_i \varphi = \varphi_i$ *for each* $i \in I$*.*

In particular, the inverse limit (G, φ_i) can be constructed as follows:

- $G\colon = \{(g_i)_{i\in I} \in \prod_{i\in I} G_i \mid \varphi_{ij}(g_i) = g_j \text{ if } j \preceq i\}$;
- $\varphi_i\colon G \to G_i$ is the restriction of the projection $\prod_{i\in I} G_i \to G_i$;
- the group topology on G is the subspace topology inherited by the product topology of $\prod_{i\in I} G_i$ (and G turns out to be a closed subset).

Fact 10.4 ([40, Lemma 1.1.2]) *If* (G_i, φ_{ij}, I) *is an inverse system of topological groups, then* $\varprojlim_{i\in I} G_i$ *is a closed subgroup of the product* $\prod_{i\in I} G_i$*.*

Definition 10.10 A **profinite group** G is the inverse limit $\varprojlim_{i\in I} G_i$ of a surjective inverse system (G_i, φ_{ij}, I) of finite groups G_i, where each finite group G_i is assumed to have the discrete topology.

For a profinite group G, a neighbourhood basis at 1_G is $\{\ker(\varphi_i)\}_{i\in I}$, where $\varphi_i\colon G \to G_i$ are the canonical projection homomorphisms.

Fact 10.5 *A profinite group G is compact and totally disconnected.*

Proof It easily follows from the fact that G is a closed subset of the product of finite groups (see Fact 10.4). To see this, one needs to recall the following basic properties:

- closed subspaces and products of compact (resp. totally disconnected) spaces are compact (resp. totally disconnected);
- a finite group is a compact discrete group; in particular it is totally disconnected.

□

Example 10.11 (a) Let R be a profinite commutative ring with unit. Then the following groups (with topologies naturally induced from R) are profinite groups: $R^\times$, $GL(n,R)$ and $SL(n,R)$.

(b) Consider the natural numbers $I = \mathbb{N}$, with the usual partial ordering, and the group of integers $\mathbb{Z}$. Form the inverse system $\{\mathbb{Z}/n\mathbb{Z}, \varphi_{nm}\}$, where the map $\varphi_{nm}\colon \mathbb{Z}/n\mathbb{Z} \to \mathbb{Z}/m\mathbb{Z}$ is the natural projection for $m \leqslant n$. The inverse limit produces the profinite group $\widehat{\mathbb{Z}}$, which can be identified with the set of equivalence classes of tuples of integers

$$\{\overline{(x_1, x_2, x_3, \ldots)} \mid x_n \in \mathbb{Z}, \forall n \in \mathbb{Z}, \text{ and } x_m = x_n \bmod m \text{ whenever } m|n\}.$$

Note that $\widehat{\mathbb{Z}}$ naturally inherits a structure of profinite ring from the finite rings $\mathbb{Z}/n\mathbb{Z}$. The ring $\widehat{\mathbb{Z}}$ is the **profinite completion** of $\mathbb{Z}$.

(c) Let p be a prime and form the profinite group defined by the inverse limit over the system $\{\mathbb{Z}/p^n\mathbb{Z}, \psi_{mn}\}$ given by the canonical projections. It is the **pro-p completion** of $\mathbb{Z}$ and it is topologically isomorphic to the ring of p-adic integers $\mathbb{Z}_p$: it suffices to prove that $\mathbb{Z}_p$ is the inverse limit of its quotients $\mathbb{Z}_p/p^n\mathbb{Z}_p$ (where the family $\{p^n\mathbb{Z}_p\}_{n\in\mathbb{N}}$ is the neighbourhood basis at 0 in the group of p-adic integers) and that each $\mathbb{Z}_p/p^n\mathbb{Z}_p$ is isomorphic to the finite group $\mathbb{Z}/p^n\mathbb{Z}$.

The set of the elements of $\mathbb{Z}_p$ can be then identified with the set of all equivalence classes of sequences $(a_1, a_2, a_3, \ldots)$ of natural numbers such that $a_m = a_n (\operatorname{mod} p^m)$, whenever $m \leqslant n$.

Exercise 10.12 1 Let $\{G_i \mid i \in I\}$ be a collection of finite groups. Is the direct product a profinite group?

2 Consider the natural numbers $I = \mathbb{N}$, with the usual partial ordering, and form the constant inverse system $\{\mathbb{Z}, id\}$. Compute the limit.

3 Let G be a profinite group.

(a) A closed normal subgroup $H \leqslant G$ is open iff it has finite index.

(b) Every open subgroup H of G contains a subgroup H_G that is normal and open in G. (Hint: $H_G = \bigcap_{g\in G} gHg^{-1}$.)

10.2.3 Locally Compact Groups

An arbitrary topological space X is locally compact if every point admits a compact neighbourhood. A topological group G is **locally compact** if 1_G admits a compact neighbourhood. The additive group $(\mathbb{R}, +)$ with its usual topology is a locally compact, non-compact, abelian group. Clearly, the multiplicative group $(\mathbb{R}^\times, \cdot)$ is also locally compact (here $\mathbb{R}^\times$ carries the induced topology) and the same holds for the groups $(\mathbb{C}, +)$ and $(\mathbb{C}^\times, \cdot)$. On the other hand, $\mathbb{Q}$ is not locally compact with the topology inherited by $\mathbb{R}$ (see Proposition 10.7): local compactness is not inherited by subgroups; see Exercise 10.14.

Different examples of locally compact groups are discrete groups and profinite groups. If $\mathbb{T} = \mathbb{R}/\mathbb{Z}$ is the circle group,[6] then Tychonov's theorem yields that every power $\mathbb{T}^I$ of $\mathbb{T}$ is again compact and, in particular, locally compact. This is actually the most general example of a compact abelian group: every compact abelian group is isomorphic to a closed subgroup of a power of $\mathbb{T}$; see [23, Corollary 11.2.2].

[6] The circle group is compact because it is the image of the compact subspace $[0,1] \subset \mathbb{R}$.

Example 10.13 (Local fields) Non-discrete locally compact fields have been classified by [48]. A non-discrete locally compact field is either connected or totally disconnected. A non-discrete locally compact field is connected if and only if it is Archimedean (see Example 10.6), and then isomorphic to either $\mathbb{R}$ or $\mathbb{C}$. Non-Archimedean locally compact fields are called **local fields**.[7] For further details see [37, § 12.3.4] and references there.

Exercise 10.14 1 A closed subgroup of a locally compact group is locally compact. The closure condition is necessary, see Proposition 10.7.

2 If R is a locally compact ring and n is a natural number, then $R^{n\times n}$ is a locally compact ring.

3 Every quotient of a locally compact group is locally compact.

4 The product of a finite family of locally compact groups is locally compact (for infinite products to be locally compact the condition "all but a finite number of factors are compact" is necessary).

5 If $\mathbb{K}$ is a topological field, then $GL(n,\mathbb{K})$ is open in $\mathbb{K}^{n\times n}$. Consequently, $(GL(n,\mathbb{K}),\cdot)$ is locally compact exactly if $\mathbb{K}$ is.

6 If $\mathbb{K}$ is a locally compact field, then $(SL(n,\mathbb{K}),\cdot)$ is locally compact.

Proposition 10.6 *A locally compact countable group is discrete.*

Proof Recall that a *Baire space* is a topological space with the property that for each countable collection of open dense sets $\{U_n\}_{n\in\mathbb{N}}$ their intersection $\bigcap_{n\in\mathbb{N}} U_n$ is dense. By the Baire category theorem – see [28, Theorem A] and references there – every locally compact group is a Baire space. Let G be a non-discrete locally compact group; in particular, each singleton $\{g\}$ is closed but not open in G. If G is countable, then

$$G=\{g_1,\dots,g_n,\dots\}=\bigcup_{n\in\mathbb{N}}(G\setminus\{g_n\})$$

but $\bigcap_{n\in\mathbb{N}}(G\setminus\{g_n\})=\varnothing$, a contradiction. □

Proposition 10.7 ([27, Theorem 5.11]) *If a subgroup H of a topological group G is locally compact, then it is closed.*

Since the identity component G_0 is a closed normal subgroup of the locally compact group G (see Fact 10.1), one can form the quotient group G/G_0. By Exercise 10.14(3), G/G_0 is locally compact. Moreover, the locally compact group G/G_0 is totally disconnected by Proposition 10.2. Therefore, one has the short exact sequence

$$1\to G_0\to G\to G/G_0\to 1 \tag{10.2.1}$$

[7] Some authors define a local field to be any commutative non-discrete locally compact field.

and every locally compact group G is an extension of a TDLC-group G/G_0 by the connected locally compact group G_0, which can be approximated by Lie groups.

Theorem 10.8 *Every connected locally compact group is an inverse limit of Lie groups.*

The latter description is obtained as a consequence of the following fundamental result.

Theorem 10.9 (Gleason–Yamabe Theorem) *Let G be a locally compact group. For any open neighbourhood U of the identity there exists an open subgroup G' of G and a compact normal subgroup K of G' in U such that G'/K is isomorphic to a Lie group.*

We omit the proof of Gleason–Yamabe theorem but the reader is referred to [46, Theorem 1.1.13].

Proof of Theorem 10.8 Let G be connected and locally compact. In order to construct a projective system of Lie groups, let $\mathcal{N}$ be the set of all open neighbourhoods U of the identity 1_G. By Exercise 10.2(4), the only open subgroup G' of G is the whole group G. Therefore, for every $U \in \mathcal{N}$, there exists a compact normal subgroup K_U of G such that G/K_U is isomorphic to a Lie group. Let $\mathcal{K} = \{K_U \mid U \in \mathcal{N}\}$ be indexed by inverse inclusion. In particular, $(\mathcal{K}, \supseteq)$ is a directed poset; see [29, Glöckner's lemma, p. 148]. Whenever $M \supseteq N$ in $\mathcal{K}$, one has the continuous morphism $\varphi_{NM}\colon G/N \to G/M$ given by $gN \mapsto gM$ and $(G_N, \varphi_{NM}, \mathcal{K})$ is a surjective inverse system of Lie groups. By the universal property of the inverse limit, there exists a continuous morphism $\gamma\colon G \to \varprojlim(G_N, \varphi_{NM}, \mathcal{K})$; namely, $\gamma(g) = (gN)_{N\in\mathcal{K}}$. It is clear that the kernel of γ is given by $\bigcap\{N \mid N \in \mathcal{K}\}$ which is trivial since the elements of $\mathcal{K}$ can be arbitrarily small (see the Gleason–Yamabe theorem above). Thus one needs to prove that γ is open and surjective but it follows by [29, Theorem 1.33] and the fact that locally compact groups are always complete (see, for example, [29, Remark 1.31]). □

As [45] writes on his blog: *this theorem asserts the "mesoscopic" structure of a locally compact group (after restricting to an open subgroup G' to remove the macroscopic structure, and quotienting out by K to remove the microscopic structure) is always of Lie type.*

Remark (Hilbert's fifth problem) During the International Congress of Mathematicians (Paris, 1900) D. Hilbert presented a list of 23 open problems which turned out to be very influential for the mathematics of the 20th century. The fifth of these problems asked for a *topological description of Lie groups*: does every locally euclidean topological group admit a Lie group structure? Recall that a topological group is said to be *locally euclidean* if its identity element has a neighbourhood homeomorphic to an open subspace of $\mathbb{R}^n$. A positive solution to this problem was achieved in the early 1950s by [25], [35] and [58]. Hilbert's fifth problem motivated an enormous volume of work on locally compact groups that shed light on the structure of connected locally compact groups. An exposition on the celebrated solution of Hilbert's fifth problem can be found in [46].

Remark By Theorem 10.8, Lie group techniques may be used to analyse the structure of connected groups and their automorphisms. A canonical form for automorphisms of TDLC-groups has been developed in [51, 52]; see § 10.5.1.

10.3 Totally Disconnected Locally Compact Groups

10.3.1 van Dantzig's Theorem

By using arguments from [27], we prove van Dantzig's theorem – together with some consequences – which can be considered as the key theorem in the theory of TDLC-groups: the topology of a TDLC-group admits a well-behaved basis of identity neighbourhoods.

Theorem 10.10 (van Dantzig, 1936) *Let G be a TDLC-group. Then every neighbourhood of the identity contains a compact open subgroup.*

Proof We follow [23, Proof of Theorem 7.4.2.(b)]. By Vedenissov's Theorem (see [3, Theorem B.6.10]), there exists a neighbourhood basis $\mathcal{B}$ at the identity consisting of compact open sets. Let $K \in \mathcal{B}$. For each $x \in K$, there is an open set $U_x \ni 1_G$ with $xU_x \subseteq K$ (because left translation by x is continuous at 1_G) and an open set $V_x \ni 1_G$ with $V_xV_x \subseteq U_x$ (because the multiplication $\mu\colon G \times G \to G$ is continuous at $(1_G, 1_G)$). Since also the inversion map is continuous, the open set V_x can be chosen to be symmetric. For K is compact, there are $x_1, \dots, x_n$ such that $K \subseteq x_1V_{x_1} \cup \dots \cup x_nV_{x_n}$. Set $V = V_{x_1} \cap \dots \cap V_{x_n}$ and notice that

$$KV \subseteq \Big(\bigcup_{i=1}^{n} x_iV_{x_i}\Big)V \subseteq \bigcup_{i=1}^{n} x_iV_{x_i}V_{x_i} \subseteq \bigcup_{i=1}^{n} x_iU_{x_i} \subseteq K.$$

The inclusion $V = 1_GV \subseteq KV \subseteq K$ implies that $VV \subseteq U$, $VVV \subseteq U$, etc. Since V is symmetric, the subgroup H generated by V is given by

$$H = \bigcup_{n\in\mathbb{N}} \underbrace{V\cdots V}_{n} \subseteq K.$$

By Exercise 10.2, H is open (and so closed) and, for $H \subseteq K$, H is also compact. □

Hence, every TDLC-group contains arbitrarily small (compact) open subgroups. This is the opposite of what occurs in the connected case, where the only open subgroup is the whole group (see Exercise 10.2). This is also the opposite of what occurs in Lie Groups, which admit a neighbourhood of the identity that contains only the trivial subgroup.

10.3.2 Consequences of van Dantzig's Theorem

Here we collect a few long-known consequences of van Dantzig's theorem.

Proposition 10.11 *Given a locally compact group G, the identity component G_0 coincides with the intersection of all open subgroups of G.*

Proof To prove that G_0 is contained in the intersection it suffices to notice that every open subgroup H of G is a clopen set (see Exercise 10.2(3)).

For the reverse inclusion, we show that, for every $x \in G \setminus G_0$, there is an open subgroup H_x not containing x. By Proposition 10.2, the group G/G_0 is TDLC. Therefore, van Dantzig's theorem yields a neighbourhood basis at G_0 given by compact open subgroups of G/G_0. It follows that there is a compact open subgroup $K \subseteq G/G_0$ not containing xG_0. Given the quotient map $\pi_0 \colon G \to G/G_0$, we set $H_x = \pi_0^{-1}(K)$. □

Corollary 10.12 *If a topological group G admits a neighbourhood basis $\mathcal{B}$ at 1_G consisting of compact open subgroups, then G is TDLC.*

Proof The only part that needs some work is the total disconnectedness of G. By Proposition 10.11, $G_0 = \bigcap\{U \mid U \in \mathcal{B}\}$. But $\bigcap\{U \mid U \in \mathcal{B}\} = 1$ since we assume topological groups to be Hausdorff. □

In other words, van Dantzig's theorem characterizes TDLC-groups among topological (Hausdorff) groups: they are the ones whose compact open subgroups form a neighbourhood basis at 1_G. This property will reveal itself to be the most fruitful property of TDLC-groups.

In general, total disconnectedness is not preserved under taking quotients. Thanks to van Dantzig's theorem this is not the case for locally compact groups.

Proposition 10.13 *The quotient of a TDLC-group by a closed normal subgroup is totally disconnected.*

Proof Let G be a TDLC-group and let N be a closed normal subgroup of G. It follows from van Dantzig's theorem that the collection of all compact open subgroups of G forms a neighbourhood basis at 1_G. Since quotient maps are open, the quotient G/N admits a neighbourhood basis at N formed by compact open subgroups that are the quotients of all compact open subgroups of G. Thus G/N is TDLC by Corollary 10.12. □

Exercise 10.15 Every locally compact group G contains an open subgroup H which is compact-by-connected.

Proposition 10.14 *A compact totally disconnected group is a projective limit of finite groups. In particular, a topological group is profinite if, and only if, it is compact and totally disconnected.*

Proof Let G be a compact totally disconnected group. The set $\mathcal{O}$ of all compact open subgroups of G forms a neighbourhood basis at 1_G. Since G is compact, every subgroup $H \in \mathcal{O}$ contains a subgroup which is both open and normal in G (see Exercise 3(2)). Thus, the family $\mathcal{NO} = \{H \in \mathcal{O} \mid H \trianglelefteq G\}$ is a neighbourhood basis at 1_G. Therefore, the morphism $G \to \prod_{H \in \mathcal{NO}} G/H$ of compact groups is injective and continuous, and provides a topological isomorphism from G to a closed subgroup of the latter product of finite groups. □

10.3.3 Examples of TDLC-groups

Many examples of a TDLC-group can be produced by means of van Dantzig's theorem.

- Discrete groups and profinite groups are rather trivial examples.
- (Local fields) Let $\mathbb{K}$ be a local field; see Example 10.13. Then $\mathbb{K}$ has a unique maximal compact subring

$$\mathfrak{o}_{\mathbb{K}} = \{x \in \mathbb{K} \mid \{x^n \mid n \geqslant 1\} \text{ is relatively compact}\},$$

which has a unique maximal ideal

$$\mathfrak{p}_{\mathbb{K}} = \{x \in \mathbb{K} \mid \lim_{n \to \infty} x^n = 0\}.$$

Both $\mathfrak{o}_{\mathbb{K}}$ and $\mathfrak{p}_{\mathbb{K}}$ are compact and open in $\mathbb{K}$. The ideal $\mathfrak{p}_{\mathbb{K}}$ is principal in $\mathfrak{o}_{\mathbb{K}}$: there is $\pi \in \mathbb{K}$ such that $\mathfrak{p}_{\mathbb{K}} = (\pi)$. The nested sequence of ideals

$$(\pi) \supset \cdots \supset (\pi^n) \supset (\pi^{n+1}) \supset \cdots$$

constitutes a basis of compact open subgroups at 0 in $\mathbb{K}$.
Local fields fall into two families (see [37, § 12.3.4]):

1 the fields of p-adic numbers, $\mathbb{Q}_p$ and their finite extensions, and

2 the fields of formal Laurent series, $\mathbb{F}_q((t))$, over some finite field $\mathbb{F}_q$.

Note that $\mathbb{Q}_p$ admits $\{p^n\mathbb{Z}_p\}_{n\in\mathbb{N}}$ as a basis of compact open subgroups, where $\mathbb{Z}_p$ is the ring of p-adic integers, and $\mathbb{F}_q((t))$ has $\{t^n\mathbb{F}_q(t)\}_{n\in\mathbb{N}}$ as basis of compact open subgroups, where $\mathbb{F}_q(t)$ is the ring of formal Taylor series over $\mathbb{F}_q$.

- (Linear groups over local fields) Let $\mathbb{K}$ be a local field. For instance, the group $GL_n(\mathbb{K})$ is TDLC with the topology inherited by $\mathbb{K}^{n^2}$.
- (Lie groups over local fields) An exact analogue of Lie theory exists for analytic groups defined over local fields such as $\mathbb{Q}_p$ and $\mathbb{F}_q((t))$; see [14, § 4] and [8].
- (Automorphism groups of connected locally finite graphs) A **graph** Γ is a pair $(V\Gamma, E\Gamma)$ where $V\Gamma$ is a set and $E\Gamma$ is a collection of unordered distinct pairs of elements from $V\Gamma$. The elements of $V\Gamma$ are called **vertices** and the elements of $E\Gamma$ are called **edges**. We will need a bit of terminology for graphs: two vertices v and u are said to be adjacent, if $\{v,u\}$ is an edge in Γ; a graph is **locally finite** if each vertex v has a finite number of adjacent vertices; a **path of length** n from v to u is a sequence $(v = v_0, v_1, \ldots, v_n = u)$ of vertices, such that v_i and v_{i+1} are adjacent for $i = 0, 1, \ldots, n-1$; a graph is **connected** if for any two vertices v and u there is a path from v to u in the graph. An **automorphism** of a graph Γ is a bijection $\varphi\colon V\Gamma \to V\Gamma$ such that $\{\varphi(v), \varphi(w)\} \in E\Gamma$ if and only if $\{v,w\} \in E\Gamma$. The collection of automorphisms forms a group under composition, and it is denoted by $\mathrm{Aut}(\Gamma)$. Let Γ be a connected graph and endow $\mathrm{Aut}(\Gamma)$ with the compact-open topology via considering $V\Gamma$ to be a discrete space. Namely, a basis of this topology is given by the sets

$$\Sigma_{v,w} = \{g \in \mathrm{Aut}(\Gamma) \mid g(v_i) = w_i\},$$

where $v = (v_1, \ldots, v_n)$ and $w = (w_1, \ldots, w_n)$ range over all finite[8] tuples of vertices of Γ. In particular, two automorphisms of Γ are "close" to each other if they agree on "many" vertices.

Exercise 10.16 The compact-open topology is a group topology on $\mathrm{Aut}(\Gamma)$ and it coincides with the pointwise convergence topology.

Remark The compact-open topology on $\mathrm{Aut}(\Gamma)$ also coincides with the **permutation topology** (cf. [30]).

Let $G = \mathrm{Aut}(\Gamma)$. The identity element of G belongs to an open set $\Sigma_{a,b}$ iff $a = b$. Consequently, the compact-open topology on G has a neighbourhood basis at

[8] Notice that the length of the tuples is arbitrary.

the identity formed by the pointwise stabilisers of finite sets of vertices. Recall that, given a vertex $v \in \Gamma$, the set $G_{(v)} = \{g \in G \mid g(v) = v\}$ is a subgroup of G which is called the **stabiliser of** v. Given a finite set of vertices $\mathcal{V}$, the intersection $G_{(\mathcal{V})} = \bigcap_{v \in \mathcal{V}} G_{(v_i)}$ is the **pointwise stabiliser** of the set $\mathcal{V}$. Clearly, pointwise stabilisers of finite sets of vertices are open subgroups of G.

Proposition 10.15 *Let Γ be a connected locally finite graph. The pointwise stabilisers of finite sets of vertices are compact in the compact-open topology. In particular, $G = \mathrm{Aut}(\Gamma)$ is a TDLC-group.*

Sketch of the proof. It suffices to prove the claim for the stabiliser $G_{(v)}$ of an arbitrary vertex v. For $k > 0$, set

$$S_k(v) = \{w \in V\Gamma \mid d_\Gamma(v,w) \leqslant k\},$$

where $d_\Gamma(v,w)$ denotes the length of the shortest path from v to w. Since Γ is locally finite, every $S_k(v)$ is finite. Let $Sym(S_k(v))$ be endowed with the discrete topology. Clearly, the stabiliser $G_{(v)}$ permutes the elements in each $S_k(v)$ i.e., there exists a group homomorphism $\varphi_k \colon G_{(v)} \to Sym(S_k(v))$. The family $\{\varphi_k\}_{k>0}$ can then be used to construct an injective group homomorphism from $G_{(v)}$ to the profinite group $\prod_{k>0} Sym(S_k(v))$ which is continuous and closed. □

Remark The latter result does not state the "non-discreteness" of Aut(Γ). Indeed, it is a difficult task to determine whether a given group is discrete.

- (Neretin group of spheromorphisms of a d-regular tree) A **tree** is a connected graph without nontrivial cycles, where by nontrivial cycle we mean a path $(v_0, \ldots, v_n)$ such that $n \geqslant 1$ and $v_0 = v_n$. A vertex $v \in VT$ of degree 1 (i.e., has a unique adjacent vertex) is called a **leaf**. For $d \in \mathbb{N}$, an **infinite d-regular tree** is an infinite tree whose vertices have degree $d+1$. A **finite d-regular tree** is a finite tree whose every vertex is either a leaf or has degree $d+1$, i.e., it is an **internal vertex**. A **rooted tree** is a tree with a distinguished vertex $o \in T$, called its **root**. If the tree is rooted then d-regularity requires the root to have degree d instead of $d+1$.
  An important property of a tree T is given by the fact that there exists a unique path connecting two vertices v and u. A **ray** in T is defined to be an infinite path, i.e., a sequence $(v_0, v_1, \ldots)$ of distinct vertices of T such that the consecutive ones are adjacent. Two rays are said to be **asymptotic** if they have common tails.[9] Equivalence classes of asymptotic rays are called

[9] A tail of a sequence is a subsequence obtained after removing finitely many initial elements.

the **ends** of T. All ends of T form the set ∂T which is called the **boundary** of the tree.

The construction of $\mathcal{N}_d$: Here we follow the construction and the arguments of [14, Chapter 8]. Let T be a d-regular tree. For every finite d-regular subtree $F \subseteq T$, denote by $T \setminus F$ the (no longer connected) graph obtained by removing from T all the edges and internal vertices of F. The connected components of $T \setminus F$ are rooted d-regular trees whose roots are the leaves of F. In particular, $T \setminus F$ is a rooted d-regular forest such that $\partial(T \setminus F) = \partial T$.[10]

Let $F_1, F_2 \subseteq T$ be two finite d-regular subtrees with the same number of leaves. Each forest isomorphism $\varphi\colon T \setminus F_1 \to T \setminus F_2$ induces a homeomorphism φ_* of ∂T, called *spheromorphism* of T. Clearly, different choices of subtrees F_1, F_2 can induce the same spheromorphism. Therefore, φ is just a representative of φ_*. This is important because, for each pair of spheromorphisms φ_* and ψ_*, we find *composable* representatives: we enlarge the finite trees in the representation of the spheromorphisms in order to make the isomorphisms φ and ψ share a target forest and a source forest. This procedure shows that the spheromorphism $\psi_* \circ \varphi_*$ is well defined (in particular, it coincides with the composition in $\mathrm{Homeo}(\partial T)$). Hence, the set of all spheromorphisms of a d-regular tree is a group, which is called a **Neretin group** and denoted by $\mathcal{N}_d$. The group of spheromorphisms of a d-regular tree has been introduced by [36] by analogy with the diffeomorphism group of a circle. Roughly speaking, a spheromorphism of ∂T is a transformation induced in the boundary ∂T by a piecewise tree automorphism; indeed, spheromorphisms are also known as *almost automorphisms of T*.

Fact 10.16 *The set of all spheromorphisms is a subgroup of the homeomorphism group of the boundary ∂T. Moreover, every automorphism φ of T induces an isomorphism $T \setminus F \to T \setminus \varphi(F)$ of forests which is independent on the finite d-regular subtree F. Thus, $\mathrm{Aut}(T)$ can be regarded as a subgroup of $\mathcal{N}_d$.*

Remark Let φ_* be a spheromorphism of T_d and suppose that φ_* admits a representative $\varphi\colon T_d \setminus F \to T_d \setminus F$ that leaves the trees of $T_d \setminus F$ in place. Then φ can be extended to an automorphism of the tree T_d which belongs to

[10] Since a forest is a disjoint union of trees the notion of boundary can be easily extended to forests.

the pointwise stabiliser of the finite tree F. As a consequence, φ_* belongs to the image of $\mathrm{Aut}(T_d)$ in $\mathcal{N}_d$.

In order to topologize the group $\mathcal{N}_d$, the first attempt is to endow $\mathrm{Homeo}(\partial T)$ with the compact-open topology and then give $\mathcal{N}_d$ the subspace topology. Unfortunately, the resulting topological group is not locally compact: $\mathcal{N}_d$ is not closed in $\mathrm{Homeo}(\partial T)$ with respect to the compact-open topology (see Proposition 10.7).
Instead of restricting a topology from a larger topological group, we could try to "copy and paste around" a topology coming from an abstract subgroup which is also a topological group.

Lemma 10.17 ([14, Lemma 8.4, p. 137]) *Suppose that an abstract group G contains a topological group H as a subgroup. If, for all open subsets $U \subseteq H$ and $g, g' \in G$, the intersection $gUg' \cap H$ is open in H, then G admits a unique group topology such that the inclusion $H \to G$ is continuous and open.*

Theorem 10.18 *Neretin group $\mathcal{N}_d$ admits a unique group topology such that the natural embedding $\mathrm{Aut}(T_d) \to \mathcal{N}_d$ is continuous and open. With this topology, $\mathcal{N}_d$ is a TDLC-group.*

Proof By the lemma above, one needs to show that for every open $U \subseteq \mathrm{Aut}(T_d)$ and all $\varphi_*, \psi_* \in \mathcal{N}_d$, the subset $\mathrm{Aut}(T_d) \cap \varphi_* U \psi_*$ is open in $\mathrm{Aut}(T_d)$. A sub-basis of the compact-open topology on $\mathrm{Aut}(T_d)$ is given by vertex stabilisers and therefore one has to show the claim only for the sets in the sub-basis.
To this end, let v be a vertex of Γ. Let S be a sufficiently large sphere centred at v such that the spheromorphisms φ_* and ψ_* admit representatives $\varphi\colon T_d \setminus F_1 \to T_d \setminus S$ and $\psi\colon T_d \setminus S \to T_d \setminus F_2$. Denote by $G_{(S)}$ the pointwise stabiliser of S. Since $G_{(S)}$ is an open subgroup of $\mathrm{Aut}(T_d)$ contained in $G_{(v)}$, there exist finitely many elements $g_1, \ldots, g_n \in G_{(v)}$ such that $G_{(v)} = \bigsqcup_{i=1}^n g_i G_{(S)}$. Therefore,

$$\psi_* G_{(v)} \varphi_* = \bigsqcup_{i=1}^{n} \psi_* g_i G_{(S)} \varphi_* = \bigsqcup_{i=1}^{n} \psi_* g_i \varphi_* (\varphi_*^{-1} G_{(S)} \varphi_*).$$

By Remark 10.3.3, $\varphi_*^{-1} G_{(S)} \varphi_*$ coincides with the pointwise stabiliser of the finite tree F_1 and, therefore, it is contained in the image of $\mathrm{Aut}(T_d)$ in $\mathcal{N}_d$. It then follows that $\psi_* G_{(v)} \varphi_* \cap \mathrm{Aut}(T_d)$ is open in $\mathrm{Aut}(T_d)$ (it is the union of translates of the open subgroup $G_{(F_1)}$). □

- (Fundamental groups of graphs of profinite groups) Bass–Serre theory carries over to the realm of TDLC-groups. Let $(\mathcal{G},\Lambda)$ be a finite graph of profinite groups; see [17]. The fundamental group $\pi_1(\mathcal{G},\Lambda)$ can be endowed easily with a TDLC-group topology.
- (Topological semi-direct products) Let G and H be topological groups. Suppose that G acts on H continuously, i.e., there is a group action of G on H such that the map $\alpha\colon G\times H\to H$ defined by the action is continuous. The **topological semi-direct product** is the abstract semi-direct product $H\rtimes G$ endowed with the product topology.

  Exercise 10.17 Let G and H be TDLC-groups such that G acts continuously on H. The topological semi-direct product $H\rtimes G$ is a TDLC-group.

- (Powers of topological groups) Let G be a locally compact group. For I infinite, the power G^I fails to be locally compact as soon as G is non-compact. To deal with this issue, given a compact open subgroup U in G, one defines the **semi-restricted power**

$$G^{I,U}=\{(g_i)_{i\in I}\in G^I\mid g_i\in U \text{ for all but finitely many } i\in I\}.$$

  There is a unique group topology on $G^{I,U}$ that makes the embedding of U^I a topological isomorphism onto an open subgroup. Moreover, such a group topology is locally compact; see [20, Proposition 2.4]. TDLC-groups are full of compact open subgroups, and therefore they are amenable to this construction; in particular, semi-restricted powers of TDLC-groups are again TDLC-groups.

10.4 Finiteness Properties for TDLC-groups

10.4.1 Compact Generation and Presentation

There are several finiteness conditions that a TDLC-group can satisfy. At this early stage, we are interested in two finiteness conditions that naturally generalise the notions of finite generation and finite presentation in the context of locally compact groups.

Definition 10.18 A locally compact group G is said to be

(CG) **compactly generated** if it has a compact generating set S.

(CP) **compactly presented** if it has a presentation $\langle S\mid R\rangle$ as an abstract group with the generating set S compact in G and the relators in R of bounded length.

It is straightforward that (CP) implies (CG). The converse is not true; see for instance [21, Example 8.A.28]. The notion of compact presentation was introduced in 1964 by Kneser but it has attracted much less attention than compact generation until recently: for example, [21] proved that compact presentation is equivalent to a metric condition and [17] provided an equivalent notion of compact presentation via generalised presentations which is based on van Dantzig's theorem and the notion of a fundamental group of finite graphs of profinite groups.[11]

Example 10.19 1 Every profinite group is (trivially) compactly generated and compactly presented.

2 Every compactly generated abelian TDLC-group is topologically isomorphic to $\mathbb{Z}^n \times K$, where $n \in \mathbb{N}$ and K is a compact abelian group; see [23, Theorem 12.5.5]. In particular, it is compactly presented.
3 The field of p-adic numbers $\mathbb{Q}_p$ is not compactly generated because it is the ascending union of nested compact open subgroups, i.e., $\mathbb{Q}_p = \bigcup_{n\in\mathbb{Z}} p^n\mathbb{Z}_p$.
4 The automorphism group of a d-regular tree is compactly generated; see Corollary 10.21.
5 The special linear group $\mathrm{SL}_2(\mathbb{Q}_p)$ is compactly generated. Indeed, by Ihara's Theorem [43, p. 143, Corollary 1 to Theorem 3], we can decompose $\mathrm{SL}_2(\mathbb{Q}_p)$ into the amalgamated free product

$$\mathrm{SL}_2(\mathbb{Z}_p) *_I \mathrm{SL}_2(\mathbb{Z}_p),$$

where I is a compact open subgroup.
6 More generally, the fundamental group of a finite graph of profinite groups satisfies (CP); see [17, § 5.8].
7 The Neretin groups are compactly presented; see [12] and [32].

Fact 10.19 *Every locally compact group is a directed union of compactly generated open subgroups.*

Proof It suffices to notice that, for any $g \in G$ and any compact open neighbourhood V_g of g, the subgroup $\bigcup_{n>0}(V_g \cup V_g^{-1})^n$ is open in G and compactly generated. □

10.4.2 The Cayley–Abels Graphs

Recall that a **group** G **acts on a graph** Γ if the set of vertices $V\Gamma$ is a G-set and, for every $g \in G$, $\{gv, gw\} \in E\Gamma$ if and only if $\{v,w\} \in E\Gamma$. The group G

[11] Which plays a role in the geometric group theory of TDLC-groups that can be compared with the role played by free groups in the discrete case.

acts **vertex-transitively** on Γ if $V\Gamma$ is a transitive G-set. Given a vertex $v \in V\Gamma$, the set $G_{(v)} = \{g \in G \mid gv = v\}$ is the **vertex stabiliser** of v in G.

Definition 10.20 For a TDLC-group G, a locally finite connected graph Γ on which G acts vertex-transitively with compact open vertex stabilisers, is called a **Cayley–Abels graph of** G.

Exercise 10.21 Let Γ be a Cayley–Abels graph of G. Let $V\Gamma$ be endowed with the discrete topology. Prove that the map $G \times V\Gamma \to V\Gamma$ is continuous; that is, a TDLC-group G always acts continuously on its Cayley–Abels graphs. Moreover, prove that, as far as G is non-discrete, the action of G on Γ is never free.[12]

Proposition 10.20 *Let G be a TDLC-group. If G has a Cayley–Abels graph, then G is compactly generated.*

Proof Let Γ be a Cayley–Abels graph of G and $v \in V\Gamma$. Since Γ is locally finite, one has $\mathrm{star}(v) = \{v_1, \dots, v_n\}$. Since G acts on Γ vertex-transitively, for every $i = 1, \dots, n$, there is a $g_i \in G$ such that $v_i = g_i v$. We claim that, for every $g \in G$, there is an $h \in \langle g_1, \dots, g_n \rangle$ such that $gv = hv$. This implies that $h^{-1}g \in G_{(v)}$, which is compact and open by hypothesis. In other words, $G = \langle G_{(v)}, g_1, \dots, g_n \rangle$.

Let us prove the claim: for every $g \in G$ there is a path in Γ connecting v and gv because Γ is connected. We proceed by induction on the length of the path. For $k = 0$ there is nothing to prove. Suppose the hypothesis for k and prove it for $k+1$. For Γ is vertex-transitive, a path of length $k+1$ connecting v and gv is given by a $(k+1)$-tuple $(v, \gamma_1 v, \dots, \gamma_k v, gv)$ with $\gamma_1, \dots, \gamma_k \in G$. By the inductive hypothesis, there is $h \in \langle g_1, \dots, g_n \rangle$ such that $\gamma_k v = hv$. Therefore, the group element h^{-1} maps the edge $\{\gamma_k v, gv\}$ to the edge $\{v, h^{-1}gv\}$. In other words, $h^{-1}gv$ is adjacent to v, i.e., $h^{-1}gv = v_j = g_j v$ for some $j \in \{1, \dots, n\}$ and the claim holds. □

Corollary 10.21 *$Aut(\mathcal{T}_d)$ is compactly generated for every $d \in \mathbb{N}$.*

Now, we do the converse: we start with a compactly generated TDLC-group G and we construct a (family of) Cayley–Abels graph(s) of G. In particular, we show that, for every compact open subgroup U of G, there is a Cayley–Abels graph admitting U as stabiliser of some vertex.

Let U be a compact open subgroup of G. For every symmetric subset $S = S^{-1} \subseteq G \setminus U$ define the graph $\Gamma^G_{U,S}$ by setting

$$V\Gamma^G_{U,S} = \{gU \mid g \in G\}, \quad \text{and} \quad E\Gamma^G_{U,S} = \{\{gU, gsU\} \mid g \in G, s \in S\}.$$

Clearly, G acts transitively on the set of vertices of $\Gamma^G_{U,S}$.

[12] A groups acts *freely* on a set if point stabilisers are trivial.

Proposition 10.22 *With the above notation, the following hold:*

1 *if S is a finite set, then $\Gamma^G_{U,S}$ is locally finite;*
2 *$\Gamma^G_{U,S}$ is connected if, and only if, $G = \langle S \cup U \rangle$.*

If G is compactly generated, there exists a Cayley–Abels graph of G.

Proof 1. Since the action is vertex-transitive, it suffices to prove that the vertex U has finitely many neighbours if S is finite. Since $U = xU$, for every $x \in U$, one has that the set $\{xU, xsU\}$ is an edge for every $s \in S$, and therefore the set of all neighbours of U coincides with $\{xsU \mid x \in U, s \in S\}$. In order to determine the cardinality of such a set, one must count the number of left cosets of U that are necessary to cover each double coset UsU. But this number is finite because U is open and the double coset UsU is compact (since U is compact). Therefore, if S is finite, the graph $\Gamma^G_{U,S}$ is locally finite.

2. Suppose the graph $\Gamma^G_{U,S}$ is connected. For every $g \in G$, there is a path $p = (v_0, \ldots, v_n)$ connecting the vertex U to the vertex gU. In particular, the vertices of the path p can be written as

$$v_0 = U, \quad v_1 = u_1 s_1 U, \quad \ldots, \quad v_n = u_1 s_1 \cdots u_n s_n U,$$

where each $u_i \in U$ and each $s_j \in S$. Since $u_1 s_1 \cdots u_n s_n U = gU$, it follows that g belongs to the subgroup generated by $U \cup S$.

Conversely, suppose that $G = \langle U \cup S \rangle$. Let gU and hU be any two vertices of the graph. The group element $g^{-1}h$ can then be written as a word $u_1 s_1 \cdots u_n s_n u_{n+1}$ such that each $u_i \in U$ and each $s_j \in S$. Thus, the sequence of vertices

$$(gU, gu_1 s_1 U, gu_1 s_1 u_2 s_2 U, \cdots, gu_1 s_1 \cdots u_n s_n u_{n+1} U = hU)$$

is a path in $\Gamma^G_{U,S}$ connecting gU and hU. □

Remark The first (technical) construction of the Cayley–Abels graph is due to [1]. A less technical approach to Cayley–Abels graphs was provided in [31], where the Cayley–Abels graphs were at the time called *rough Cayley graphs.* Today the widely accepted nomenclature is "Cayley–Abels graph".

Proposition 10.23 *Let G be a compactly generated TDLC-group with Cayley–Abels graph Γ. The group homomorphism $\psi \colon G \to \mathrm{Aut}(\Gamma)$ defined by the action of G on Γ is continuous, the kernel of ψ is compact and the image of ψ is closed.*

Proof A basis of the compact-open topology of $\mathrm{Aut}(\Gamma)$ is given by the family of pointwise stabilisers in $\mathrm{Aut}(\Gamma)$ of finite sets of vertices. The pre-image of each of these sets is the intersection of finitely many open subgroups of G (that are the stabilisers $G_{(v)}$ of the vertices v in the finite set). Since each stabiliser $G_{(v)}$ is open, ψ is continuous.

The kernel of ψ is closed because it is the pre-image of the closed set $\{1\}$ and then it is compact since $\ker(\psi) \subseteq G_{(v)}$, for any $v \in V\Gamma$.

To prove that the image $\psi(G)$ is closed, it suffices to see that $\psi(G) \cap H$ is closed for every H in the basis of the compact-open topology of $\mathrm{Aut}(\Gamma)$. Since every such H is the intersection of finitely many vertex stabilisers in $\mathrm{Aut}(\Gamma)$, we only need to prove that $\psi(G) \cap H$ is closed whenever H is the stabiliser in $\mathrm{Aut}(\Gamma)$ of a single vertex v. In such a case, $\psi(G) \cap H$ coincides with $\psi(G_{(v)})$, which is compact because ψ is continuous and $G_{(v)}$ is compact. In particular, $\psi(G) \cap H$ is closed. □

Corollary 10.24 *Compactly generated TDLC-groups are second countable modulo a compact normal subgroup.*

The representation $\psi \colon G \to \mathrm{Aut}(\Gamma)$ is called the **Cayley–Abels representation of** G.

Remark One can say more on the image $\psi(G)$ of the Cayley–Abels representation: $\psi(G)$ is a cocompact subgroup of $\mathrm{Aut}(\Gamma)$, see [50, Lemma 3.12].

Exercise 10.22 Suppose G is a compactly generated TDLC-group and Γ is a Cayley–Abels graph of G. Show that a closed subgroup $K \leqslant G$ is compact if and only if, for all $v \in V\Gamma$, $Kv := \{kv \mid k \in K\}$ is finite.

The geometric structure of compactly generated TDLC-groups:

Definition 10.23 ([26]) Two metric spaces (X, d_X) and (Y, d_Y) are said to be **quasi-isometric** if there is a map $\varphi \colon X \to Y$ and constants $a \geqslant 1$ and $b \geqslant 0$ such that, for all $x_1, x_2 \in X$, one has

$$\frac{1}{a} d_X(x_1, x_2) - \frac{b}{a} \leqslant d_Y(\varphi(x_1), \varphi(x_2)) \leqslant a d_X(x_1, x_2) + ab,$$

and, for all $y \in Y$,

$$d_Y(y, \varphi(X)) \leqslant b.$$

Such a map φ is called a **quasi-isometry**. Moreover, being quasi-isometric is an equivalence relation on the class of metric spaces.

Every connected graph Γ can be regarded as a metric space: two vertices v and w are points at distance 1 if, and only if, there is an edge connecting v and w. In other words, we endow the set $V\Gamma$ with the path-length metric $d_\Gamma \colon V\Gamma \times V\Gamma \to \mathbb{N}$ defined as follows:

$$d_\Gamma(v, w) = \min\{\text{length of } p \mid p \text{ path connecting } v \text{ and } w\}, \quad v, w \in V\Gamma.$$

Theorem 10.25 ([1], [31, Theorem 2]) *Let G be a compactly generated TDLC-group. Any two Cayley–Abels graphs of G are quasi-isometric.*

The quasi-isometric invariance allows us to define *geometric invariants* of a compactly generated TDLC-group G by considering quasi-isometric invariants of a Cayley–Abels graph associated to G. For example, one can give the following definitions (that are long-known for finitely generated discrete groups) for a compactly generated TDLC-group G:

(Hyp) G is said to be **hyperbolic** if some (and hence any) Cayley–Abels graph of G is hyperbolic.

(Ends) The **number of ends** of G is defined to be the number of ends of some (and hence any) Cayley–Abels graph of G.

The class of hyperbolic TDLC-groups is a rich source of compactly presented TDLC-groups; see [21]. Indeed, geometric invariants often reflect structural properties of the group: for instance, an analogue of the famous Stallings decomposition theorem is available in the context of TDLC-groups; see [1, Struktursatz 5.7 and Korollar 5.8], [31, Theorem 1.3] and [15].

10.4.3 Finiteness Conditions in Higher Dimension

For discrete groups, finite generation is the first in two sequences of increasingly stronger properties: the homological finiteness conditions, the **types** $(\mathbf{FP}_n)_{n\in\mathbb{N}}$ over a commutative unital ring R and the homotopical finiteness conditions, the **types** $(\mathbf{F}_n)_{n\in\mathbb{N}}$. We recall the definitions here, but the reader is referred to [11, Chapter VIII] for details:

(FP_n) A discrete group G is **of type** $\mathbf{FP}_n$ ($0 \leqslant n < \infty$) over R if there is a projective resolution $\{P_i\}$ of the trivial module R over $R[G]$ such that each projective $R[G]$-module P_i is finitely generated for $i \leqslant n$. (If $R = \mathbb{Z}$, the reference to the ring R usually drops.)

(F_n) A discrete group G is **of type** $\mathbf{F}_n$ ($0 \leqslant n < \infty$) if there exists a contractible G-CW-complex with finite cell stabilisers and such that G acts on the n-skeleta with finitely many orbits. See [33, Lemma 4.1].

We will refer to the properties above as the classical finiteness properties for discrete groups. These properties are known to satisfy the following:

- A discrete group G is of type F_1 if, and only if, it is finitely generated if, and only if, it is of type FP_1 over R.
- A discrete group G is of type F_2 if, and only if, it is finitely presented but being of type FP_2 is strictly weaker than finite presentation; see [7].

- For each $n \geqslant 1$, a discrete group of type F_n is of type FP_n over R but the converse is not true (the converse becomes true if the group is supposed to be finitely presented and $R = \mathbb{Z}$).
- Being of type FP_n over R (resp. of type F_n) is a geometric property of the finitely generated group, that is, it is invariant up to quasi-isometry.

A first attempt at generalising these properties to the realm of locally compact groups is due to Abels and Tiemeyer [3]. They introduced **compactness properties** for locally compact groups – we avoid here the (very technical) definitions – which are two sequences $(\mathbf{CP}_n)_{n\geqslant 0}$ and $(\mathbf{C}_n)_{n\geqslant 0}$ of increasingly stronger properties satisfying:

- for all $n \geqslant 1$, a discrete group is of type (CP_n) (resp. C_n) if and only if it is of type FP_n (resp. F_n);
- a locally compact group is of type C_1 if, and only if, it is compactly generated if, and only if, it is of type CP_1;
- a locally compact group is of type C_2 if, and only if, it is compactly presented but being of type CP_2 is strictly weaker than compact presentation;
- for each $n \geqslant 1$, a locally compact group of type C_n is also of type CP_n but the converse is not true;
- being of type CP_n (resp. C_n) is invariant "up to compactness": the compactness properties remain unchanged by passing to a cocompact[13] subgroup or by taking the quotient by a compact normal subgroup. Such an invariance is weaker than the invariance up to quasi-isometry among the class of compactly generated locally compact groups.

For the (more amenable) class of TDLC-groups, a different approach to finiteness conditions was recently introduced and investigated in [17] and [16]:

<u>(FP_n)</u> A TDLC-group G is **of type FP_n** ($0 \leqslant n < \infty$) over R if there is a resolution $\{P_i\}$ of the trivial module R over $R[G]$ such that each P_i is a permutation $R[G]$-module[14] with compact open stabilisers and finitely many orbits for $i \leqslant n$.

<u>(F_n)</u> A TDLC-group G is **of type F_n** ($0 \leqslant n < \infty$) if there exists a contractible G-CW-complex X with compact open cell stabilisers such that G acts on the n-skeleta of X with finitely many orbits.

[13] A closed subgroup H is *cocompact* if the quotient G/H, equivalently $H\backslash G$, equipped with the quotient topology is compact.

[14] A *permutation $R[G]$-module* is a module $R[\Omega]$ freely R-generated by a G-set Ω.

These finiteness conditions for TDLC-groups satisfy the following:

- For $R = \mathbb{Q}$, permutation $\mathbb{Q}[G]$-modules with compact open stabilisers are projective objects in the category ${}_{\mathbb{Q}[G]}\mathbf{dis}$ of rational discrete $\mathbb{Q}[G]$-modules[15] (see [17]) and so we recover the homological flavour of the classical definition. In particular, whenever G is discrete, the new definition reduces to the classical notion of type FP_n over $\mathbb{Q}$.
- For all $n \geqslant 1$, a discrete group is of type F_n in the category of TDLC-groups if, and only if, it is of type F_n in the classical sense because requiring compact open discrete stabilisers reduces to finite stabilisers.
- A TDLC-group is of type F_1 if, and only if, it is compactly generated if, and only if, it is of type FP_1 over R (see [16, Proposition 3.4] and [17, Proposition 5.3]).
- A TDLC-group is of type F_2 if and only if it is compactly presented by [16, Proposition 3.4] but being of type FP_2 is strictly weaker than compact presentation.
- For each $n \geqslant 1$, a TDLC-group of type F_n is also of type FP_n over R (see [16, Fact 2.7]) but the converse is not true (the converse becomes true if the group is supposed to be compactly presented and R is replaced by $\mathbb{Z}$).
- Being of type FP_n over R (resp. F_n) is a geometric property (see [16, Corollary 5.7]).

Remark All the finiteness conditions above can be extended to $n = \infty$. [42] showed that Neretin groups are of type F_∞ which, in particular, implies the fact – observed already in [17] – that these groups are of type FP_∞ over $\mathbb{Q}$.

Example 10.24 Hyperbolic TDLC-groups are of type F_n for some finite n. It is a classical result that, for a large enough constant d, the (topological realisation of the) Rips complex $P_d(\Gamma)$ is contractible whenever Γ is a hyperbolic (locally finite) graph.

[16] showed that it is possible to introduce further two sequences (**types KP**$_n$)$_{n\geqslant 0}$ and (**types K**$_n$)$_{n\geqslant 0}$ of increasingly stronger compactness properties. The concrete motivation comes from the fact that permutation $R[G]$-modules with compact open stabilisers fail to be projective over the ring $R = \mathbb{Z}$ in the abelian category ${}_{\mathbb{Z}[G]}\mathrm{dis}$. In such a case, being of type FP_n lack of a description based on the existence of partial projective resolutions of the trivial module $\mathbb{Z}$ of finite type. The strategy to get back such a description is to embed the category ${}_{\mathbb{Z}[G]}\mathrm{dis}$ into a quasi-abelian category ${}_{\mathbb{Z}[G]}\mathrm{top}$ where the discrete

[15] An $R[G]$-module M is said to be *discrete* if the action $G \times M \to M$ is continuous when M carries the discrete topology.

permutation $\mathbb{Z}[G]$-modules with compact stabilisers go back to being projective again. The objects of ${}_{\mathbb{Z}[G]}\mathrm{top}$ are the so-called *k-$\mathbb{Z}[G]$-modules*, i.e., module objects in the category of k-spaces over the k-$\mathbb{Z}$-algebra $\mathbb{Z}[G]$. The reader is referred to [22] for the definition and the background on this category. Since all locally compact Hausdorff spaces are k-spaces, every TDLC group is automatically a k-group and one can take advantage of the homological machinery developed in [22] to define the following finiteness conditions:

(KP_n) A TDLC-group G is **of type $\mathbf{KP}_n$** ($0 \leqslant n < \infty$) if there is a projective resolution $\{P_i\}$ of the trivial module $\mathbb{Z}$ in ${}_{\mathbb{Z}[G]}\mathrm{top}$ such that each P_i is a free k-$\mathbb{Z}[G]$-module on a compact space for $i \leqslant n$.

(K_n) A TDLC-group G is **of type $\mathbf{K}_n$** ($0 \leqslant n < \infty$) if there exists a contractible G-KW-complex[16] X with cocompact n-skeleta, in the compact Hausdorff model structure on G-k-spaces.

Remark In [16] the sequence (types $\mathrm{KP}_n)_{n \geqslant 0}$ is defined over an arbitrary ring R.

The latter properties relate to types FP_n and types F_n as follows.

- A TDLC group G is of type FP_n over $\mathbb{Z}$ if, and only if, it is of type KP_n (see [16, Theorem 3.10]).
- If G has type F_n then it has type K_n (see [16, Theorem 3.23]).

Open Problem Despite the abundance of finiteness properties that are available in the TDLC context, the theory of finiteness conditions for TDLC-groups is still much less developed than the one for discrete groups. Moreover, very little is known about the relation (if one exists) among the properties of different sequences $\mathrm{CP}_n, \mathrm{FP}_n$ and KP_n (resp. $\mathrm{C}_n, \mathrm{F}_n$ and K_n). For example, is it true that type K_n implies type F_n?

Open Problem It would be relevant to find an example of a TDLC-group of type FP_2 over $\mathbb{Q}$ which is not compactly presented and is "sufficiently" non-discrete (for example, it is not quasi-isometric to a discrete group). Unfortunately, the strategy developed in [7] does not seem to have a TDLC-analogue.

Open Problem For a TDLC-group G one can introduce several homological invariants. Which homological invariants are geometric? In 1991 Gromov asked whether the cohomological dimension of discrete groups is a geometric invariant. Under additional finiteness assumptions, [24] proved that quasi-isometric groups have the same cohomological dimension and [41] removed

[16] The definition of such an object can be found in [22].

the finiteness assumptions in the case of rational cohomological dimension. This yields the natural question: is the rational discrete cohomological dimension defined in [17] a geometric invariant of a TDLC-group?

10.5 Willis' Theory of TDLC-groups: a Sketch

As mentioned at the end of the introduction, Willis' theory of the scale function was a fundamental breakthrough in the theory of TDLC-groups after several years of stillness and it motivated a vast research interest in the theory of TDLC-groups, which led to a systematic study of the class of compactly generated, topologically simple TDLC-groups that are non-discrete. In this section we confine ourselves to a brief introduction of the scale function and to some comments on (topologically) simple TDLC-groups. Nevertheless, the student that is approaching the study of TDLC-groups is encouraged to dig further into this fundamental subject.

10.5.1 Scale Function and Tidy Subgroups

Let $\alpha \in \mathrm{Aut}(G)$ and U be a compact open subgroup of G. Then

$$[\alpha(U) \colon U \cap \alpha(U)] < \infty$$

because $U \cap \alpha(U)$ is open in the compact set U. The **scale of** α is

$$s(\alpha) = \inf\{[\alpha(U) \colon U \cap \alpha(U)] \mid U \text{ compact open subgroup of } G\}.$$

A subgroup U is **tidy** for α if the infimum is attained at U. Tidy subgroups for α always exists since $s(\alpha)$ is the minimum of a subset of $\mathbb{N}$. Every tidy subgroup U can be decomposed as the product of a subgroup where α expands and a subgroup where α shrinks:

$$\text{if } U_{\pm} := \bigcap_{k>0} \alpha^{\pm k}(U), \text{ then } U = U_{+}U_{-}.$$

By construction, U_{+} and U_{-} are closed subgroups, $\alpha(U_{+}) \geqslant U_{+}$ and, similarly, $\alpha(U_{-}) \leqslant U_{-}$. Moreover, it can be shown that $s(\alpha)$ represents the factor by which α expands U_{+}, i.e., $s(\alpha) = [\alpha(U_{+}) \colon U_{+}]$.

A striking result in the theory of the scale is the existence of an algorithm, the so-called **tidying procedure**, which produces a tidy subgroup when the input is an arbitrary compact open subgroup.

Definition 10.25 The **scale function** of G is defined as the map

$$s\colon G \to \mathbb{Z}^+, \quad x \mapsto s(\alpha_x), \quad x \in G,$$

where α_x denotes the inner automorphism $y \mapsto xyx^{-1}$.

The scale function s is known to satisfy the following properties (see [51, 52]):

(s1) s is continuous if $\mathbb{Z}^+$ carries the discrete topology;
(s2) $s(x) = 1 = s(x^{-1})$ if and only if there is a compact open subgroup U such that $xUx^{-1} = U$;
(s3) $s(x^n) = s(x)^n$, for every $x \in G$ and $n \geqslant 0$;
(s4) $\Delta(x) = s(x)/s(x^{-1})$, where $\Delta\colon G \to \mathbb{Q}^+$ is the modular function;
(s5) $s(\alpha(x)) = s(x)$ for every $x \in G$ and $\alpha \in \mathrm{Aut}(G)$.

The scale function encodes structural information of the group G. For a summary on the scale function (which in particular highlights the fact that tidy subgroups for automorphisms of TDLC-groups are analogues of the Jordan canonical form of linear transformations) the reader is referred to [55] and references there.

Remark In recent years, Willis' theory has been investigated from different points of view bringing new approaches to the scale function of a TDLC-group. For example, [34] offers an interpretation of the fundamental ingredients of Willis' theory (that are tidy subgroups and scale function) in the setting of permutation group theory. Another example is given by the work initiated in [6], where Willis' topological dynamics of automorphisms has been reformulated in terms of the long-known theory of topological entropy.

10.5.2 Comments on Simple TDLC-groups

Simple groups play an important role in group theory as the "indecomposable factors". Indeed several types of simple groups have been completely classified; for instance, the simple finite groups and the simple connected Lie groups. Long-known classes of simple TDLC-groups are the class of simple Lie groups over local fields (see [10]) and a class of automorphism groups of trees (see [47]).

In the realm of simple TDLC-groups, a distinction between topological simplicity (i.e., every closed normal subgroup is trivial) and abstract simplicity (i.e., the underlying abstract group is simple) is necessary. Examples show that a topologically simple TDLC-group can fail to be abstractly simple, see [54], but no example is known of a topologically simple compactly generated TDLC-group that fails to be abstractly simple. Among compactly generated TDLC-groups, [44] showed that there are $2^{\aleph_0}$ non-isomorphic compactly generated abstractly simple TDLC-groups.

Scale function: The theory of the scale produces invariants that could be important tools in the classification. For example:

- The set of values of the scale function: if G is compactly generated, the set of prime divisors of the values of the scale is finite (see [53]); this set could distinguish between compactly generated simple TDLC-groups.
- The flat-rank: a notion of rank for TDLC-groups which is defined via a notion of distance on the space of compact open subgroups (see [5]).

Remark In some cases, the flat-rank can be related to a cohomological invariant as shown in [17].

Local-to-Global principle: Relates the global properties of a compactly generated topologically simple TDLC-group with the structural properties of its compact open subgroups. This approach was initiated by

- [54] who showed that simplicity affects the local structure of the compactly generated TDLC-group: if G is compactly generated and topologically simple, then no compact open subgroup is solvable;
- [4] who addressed the question of which profinite groups can occur as compact open subgroups of compactly generated topologically simple TDLC-groups.

Decomposition theory: Includes methods for "breaking" a given TDLC-group into smaller (simple) pieces; see [13]. This approach has been successful for several classes of groups; for example, finite groups, profinite groups and algebraic groups. Therefore, one would hope to obtain analogous results for TDLC-groups.

General decomposition results have been obtained by [38, 39] who made use of the theory of *elementary groups* introduced by [49]. Elementary groups are TDLC-groups that are built out of discrete and compact pieces.

Geometrisation: Cayley–Abels graphs allow the study of compactly generated TDLC-groups from a geometric perspective. Often a TDLC-group admits other types of geometric objects, e.g., *buildings*, that one can consider. For example, semisimple Lie groups over a local field act on *affine buildings*, and also on *spherical buildings* (e.g., the affine building of $\mathrm{SL}_2(\mathbb{Q}_p)$ is a tree and the spherical building is its boundary). Buildings are able to determine some properties of the group they are associated to. For instance, in some cases, they facilitate the computation of the Euler characteristic, which turns out to have a curious connection with the value of a zeta-function in -1 (see [19]).

Since profinite groups are trivial as geometric objects, the geometric behaviour of compactly generated TDLC-groups is often related to the geometric behaviour of discrete groups. There is an ongoing programme of studying geometric properties of TDLC-groups by analogy with discrete groups. The aim is to understand to what extent long-known results on discrete groups find an analogue in the framework of TDLC-groups; see, for example, [2, 18, 21, 31].

Bibliography

[1] Abels, H. 1973/74. Specker-Kompaktifizierungen von lokal kompakten topologischen Gruppen. *Math. Z.*, **135**, 325–361.

[2] Arora, S., Castellano, I., Corob Cook, G. and Martínez-Pedroza, E. 2021. Subgroups, hyperbolicity and cohomological dimension for totally disconnected locally compact groups. *Journal of Topology and Analysis*, 1–27.

[3] Außenhofer, L., Dikranjan, D. and Giordano Bruno, A. 2021. *Topological Groups and the Pontryagin-van Kampen Duality: An Introduction*. Vol. 83. Walter de Gruyter GmbH & Co KG.

[4] Barnea, Y., Ershov, M. and Weigel, Th. 2011. Abstract commensurators of profinite groups. *Trans. Amer. Math. Soc.*, 5381–5417.

[5] Baumgartner, U., Rémy, B. and Willis, G. A. 2007. Flat rank of automorphism groups of buildings. *Transformation Groups*, **12**(3), 413–436.

[6] Berlai, F., Dikranjan, D. and Giordano Bruno, A. 2013. Scale function vs topological entropy. *Topology Appl.*, 2314–2334.

[7] Bestvina, M. and Brady, N. 1997. Morse theory and finiteness properties of groups. *Invent. Math.*, 445–470.

[8] Bode, A. and Dupré, N. 2021. Locally analytic representations of p-adic groups. www.icms.org.uk/seminars/2020/lms-autumn-algebra-school-2020

[9] Bourbaki, N. 1998. *General topology. Chapters 5–10*. Springer-Verlag, Berlin.

[10] Bourbaki, N. 2002. *Lie groups and Lie algebras. Chapters 4–6*. Springer-Verlag, Berlin.

[11] Brown, K. S. 1994. *Cohomology of groups*. Springer-Verlag, New York.

[12] Caprace, P.-E. and De Medts, T. 2011. Simple locally compact groups acting on trees and their germs of automorphisms. *Transform. Groups*, 375–411.

[13] Caprace, P.-E. and Monod, N. 2011. Decomposing locally compact groups into simple pieces. *Math. Proc. Cambridge Philos. Soc.*, 97–128.

[14] Caprace, P.-E. and Monod, N. 2018. Future directions in locally compact groups: a tentative problem list. In: *New directions in locally compact groups*. Cambridge Univ. Press, Cambridge.

[15] Castellano, I. 2020. Rational discrete first degree cohomology for totally disconnected locally compact groups. *Math. Proc. Cambridge Philos. Soc.*, 361–377.

[16] Castellano, I. and Corob Cook, G. 2020. Finiteness properties of totally disconnected locally compact groups. *J. Algebra*, 54–97.

[17] Castellano, I. and Weigel, Th. 2016. Rational discrete cohomology for totally disconnected locally compact groups. *J. Algebra*, 101–159.

[18] Castellano, I., Marchionna, B. and Weigel, Th. 2022. A Stallings-Swan theorem for unimodular totally disconnected locally compact groups. *arXiv preprint arXiv:2201.10847*.

[19] Castellano, I., Chinello, G. and Weigel, Th. In preparation. The Hattori-Stallings rank, the Euler characteristic and the ζ-functions of a totally disconnected locally compact group.

[20] Cornulier, Y. 2019. Locally compact wreath products. *Journal of the Australian Mathematical Society*, **107**(1), 26–52.

[21] Cornulier, Y. and de la Harpe, P. 2016. *Metric geometry of locally compact groups*. EMS Tracts in Mathematics, vol. 25. European Mathematical Society (EMS), Zürich. Winner of the 2016 EMS Monograph Award.

[22] Corob Cook, G. 2017. Homotopical algebra in categories with enough projectives. *arXiv preprint arXiv:1703.00569*.

[23] Dikranjan, D. 2018. Introduction to topological groups. http://users.dimi.uniud.it/~dikran.dikranjan/ITG.pdf.

[24] Gersten, S. 1993. Quasi-isometry invariance of cohomological dimension.

[25] Gleason, A. M. 1952. Groups without small subgroups. *Annals of mathematics*, 193–212.

[26] Gromov, M. 1984. Infinite groups as geometric objects. Pages 385–392 of: *Proceedings of the International Congress of Mathematicians*, vol. 1.

[27] Hewitt, E. and Ross, K. A. 2012. *Abstract Harmonic Analysis: Volume I Structure of Topological Groups Integration Theory Group Representations*. Springer Science and Business Media.

[28] Hofmann, K. H. 1980. A note on Baire spaces and continuous lattices. *Bulletin of the Australian Mathematical Society*, **21**(2), 265–279.

[29] Hofmann, K. H. and Morris, S. A. 2007. *The Lie theory of connected pro-Lie groups: a structure theory for pro-Lie algebras, pro-Lie groups, and connected locally compact groups*. Vol. 2. European Mathematical Society.

[30] Karrass, A. and Solitar, D. 1956. Some remarks on the infinite symmetric groups. *Math. Z.*

[31] Krön, B. and Möller, R. G. 2008. Analogues of Cayley graphs for topological groups. *Math. Z.*, 637–675.

[32] Le Boudec, A. 2017. Compact presentability of tree almost automorphism groups. *Annales de l'Institut Fourier*.

[33] Luĉk, Wolfgang. 2000. The type of the classifying space for a family of subgroups. *Journal of Pure and Applied Algebra*, **149**(2), 177–203. Elsevier.

[34] Möller, R. G. 2002. Structure Theory of Totally Disconnected Locally Compact Groups via Graphs and Permutations. *Canadian Journal of Mathematics*.

[35] Montgomery, D. and Zippin, L. 1952. Small subgroups of finite-dimensional groups. *Proceedings of the National Academy of Sciences of the United States of America*, **38**(5), 440.

[36] Neretin, Y. A. 1992. On combinatorial analogs of the group of diffeomorphisms of the circle. *Izvestiya Rossiiskoi Akademii Nauk*.

[37] Palmer, T. W. 1994. *Banach Algebras and the General Theory of *-Algebras: Volume 2, *-Algebras*. Cambridge University Press.

[38] Reid, C. D. and Wesolek, P. R. 2018a. Dense normal subgroups and chief factors in locally compact groups. *Proceedings of the London Mathematical Society*.

[39] Reid, C. D. and Wesolek, P. R. 2018b. The essentially chief series of a compactly generated locally compact group. *Mathematische Annalen*.

[40] Ribes, L. and Zalesskii, P. 2000. Profinite groups. Pages 19–77 of: *Profinite Groups*. Springer.

[41] Sauer, R. 2003. Homological Invariants and Quasi-Isometry. *Geometric & Functional Analysis GAFA*, **16**, 476–515.

[42] Sauer, R. and Thumann, W. 2017. Topological models of finite type for tree almost automorphism groups. *International Mathematics Research Notices*, **2017**(23), 7292–7320.

[43] Serre, Jean-Pierre. 1980. SL 2. Pages 69–136 of: *Trees*. Springer.

[44] Smith, S. M. 2017. A product for permutation groups and topological groups. *Duke Mathematical Journal*.

[45] Tao, T. 2011. What's new. *https://terrytao.wordpress.com/2011/10/08/254a-notes-5-the-structure-of-locally-compact-groups-and-hilberts-fifth-problem/*.

[46] Tao, T. 2014. *Hilbert's fifth problem and related topics*. American Mathematical Soc.

[47] Tits, J. 1970. Sur le groupe des automorphismes d'un arbre. In: *Essays on topology and related topics*. Springer.

[48] Van Dantzig, D. 1931. *Studien over topologische algebra*. Ph.D. thesis, HJ Paris Amsterdam.

[49] Wesolek, P. R. 2015. Elementary totally disconnected locally compact groups. *Proceedings of the London Mathematical Society*, **110**(6), 1387–1434.

[50] Wesolek, P. R. 2018. *An introduction to totally disconnected locally compact groups*. Available from https://zerodimensional.group/reading_group/190227_michal_ferov.pdf

[51] Willis, G. A. 1994. The structure of totally disconnected, locally compact groups. *Mathematische Annalen*, **300**, 341–363.

[52] Willis, G. A. 2001a. Further properties of the scale function on a totally disconnected group. *Journal of Algebra*.

[53] Willis, G. A. 2001b. The number of prime factors of the scale function on a compactly generated group is finite. *Bulletin of the London Mathematical Society*, **33**(2), 168–174.

[54] Willis, G. A. 2007. Compact open subgroups in simple totally disconnected groups. *Journal of Algebra*, **312**, 405–417.

[55] Willis, G. A. 2008. *A canonical form for automorphisms of totally disconnected locally compact groups*. De Gruyter.

[56] Wilson, J. S. 1998. *Profinite groups*. Clarendon Press.

[57] Wood, D. R., De Gier, J., Praeger, C. E. and Tao, T. 2018. *2016 MATRIX Annals*. Springer.

[58] Yamabe, H. 1953. A generalization of a theorem of Gleason. *Annals of Mathematics*, 351–365.

11
Locally Analytic Representations of p-adic Groups

Andreas Bode & Nicolas Dupré

11.1 Introduction

The p-adic representation theory of p-adic groups is the subject of much ongoing research. It is primarily motivated by the desire to better understand the conjectural p-adic Langlands correspondence, but it is also an interesting branch of representation theory in its own right. The purpose of this chapter, which is based on a series of lectures given at the LMS 'Autumn Algebra School', is to give an introduction to Schneider and Teitelbaum's theory of admissible locally analytic representations to people with an algebra/representation theory background without assuming any familiarity with the Langlands program or the number theoretic motivations for it. We therefore begin by briefly recalling what the conjectured p-adic Langlands correspondence is.

Fix two primes ℓ and p. When $\ell \neq p$, there is a classical Local Langlands Correspondence, which can be roughly stated as follows: if $n \geqslant 1$ there is an injection

$$\left\{ \begin{array}{c} \text{Frobenius-semisimple, continuous} \\ \text{representations of } \mathrm{Gal}(\overline{\mathbb{Q}_p}/\mathbb{Q}_p) \text{ on} \\ n\text{-dimensional } \overline{\mathbb{Q}_\ell}\text{-vector spaces,} \\ \text{up to isomorphism} \end{array} \right\} \to \left\{ \begin{array}{c} \text{Irreducible, admissible,} \\ \text{smooth representations of} \\ \mathrm{GL}_n(\mathbb{Q}_p) \text{ on } \overline{\mathbb{Q}_\ell}\text{-vector} \\ \text{spaces, up to isomorphism} \end{array} \right\}.$$

In fact, one usually considers the larger class of Frobenius-semisimple representations of the Weil–Deligne group on the left in order to get a bijection. This correspondence can be uniquely characterised by various properties (e.g. equivalence of L- and ε-factors of pairs), and can be stated more generally by replacing $\mathbb{Q}_p$ with some fixed finite extension F of $\mathbb{Q}_p$. This correspondence was first established by Harris–Taylor [23] and Henniart [24], and more recently by Scholze [40].

When $\ell = p$ one would like to obtain an analogue of the above correspondence, often referred to as the 'p-adic Local Langlands Correspondence'. However, the situation is trickier as there are now 'more' Galois representations on the left, due to the fact that the topologies of $\mathrm{Gal}(\overline{\mathbb{Q}_p}/\mathbb{Q}_p)$ and of $\overline{\mathbb{Q}_p}$ are compatible. As a result one would need to enlarge the right-hand side, and to this end Breuil [9] proposed to consider certain (not necessarily irreducible) Banach space representations of $\mathrm{GL}_n(\mathbb{Q}_p)$ over some fixed finite extension of $\mathbb{Q}_p$. In the case $n = 2$, a correspondence was established using the fact that both sides can be classified very explicitly in terms of (φ, Γ)-modules (see [14], [15], [12], [13] amongst others). We note that this correspondence was only established for $\mathrm{GL}_2(\mathbb{Q}_p)$ and not for $\mathrm{GL}_2(F)$, where F is more generally a finite extension of $\mathbb{Q}_p$. The general situation (working with $n > 2$ or over some non-trivial finite extension of $\mathbb{Q}_p$) is still very mysterious.

In the early 2000s, Schneider and Teitelbaum started the systematic study of locally analytic representations of p-adic groups in a series of papers [34, 35, 36, 37, 38]. These locally analytic representations arise naturally in various geometric constructions, and they are related to Banach space representations by the process of taking locally analytic vectors. Part of the challenge in working with $\mathrm{GL}_n(\mathbb{Q}_p)$-representations over topological vector spaces is that one needs to find the right finiteness condition to replace the notion of admissibility that occurred in the $\ell \neq p$ case, in order to get a well-behaved theory which still retains all the important examples that arose in the Langlands program. The class of locally analytic representations that Schneider and Teitelbaum introduced to serve that purpose are called admissible, and they generalise the above notion of admissible smooth representations. The main aim of this chapter is to explain their theory in detail.

In Section 11.2, we recall all the basic definitions required in order to define locally analytic representations and we provide examples. In Section 11.3, we explain the construction of the distribution algebra and give some of its properties. In Section 11.4 we explain the notion of a Fréchet–Stein algebra and define admissible representations. We also give a short account of some further developments in this theory. Finally, to make the material more accessible we also include an appendix summarising all the notions from non-archimedean functional analysis that we require.

Acknowledgements

We would like to thank the organisers of the LMS 'Autumn Algebra School' for giving us the opportunity to give lectures there and for offering the possibility to expand our notes into this chapter. We would also like to thank Simon Wadsley for reading a draft of our notes and for giving helpful comments.

11.2 Basic Definitions

We begin by recalling the definitions of p-adic Lie groups and locally analytic representations.

11.2.1 Non-archimedean Fields

Throughout, L will denote a field.

Definition 11.1 A non-archimedean absolute value (NAAV) on L is a function $|\cdot| : L \to \mathbb{R}$ such that for all $a, b \in L$:

(i) $|a| \geqslant 0$;
(ii) $|a| = 0 \iff a = 0$;
(iii) $|a \cdot b| = |a| \cdot |b|$; and
(iv) $|a + b| \leqslant \max\{|a|, |b|\}$.

This gives a metric on L via $d(a,b) := |a - b|$, making L into a topological field. The unit ball $\mathcal{O}_L := \{a \in L : |a| \leqslant 1\}$ is a subring.

From now on, we assume that L is equipped with a NAAV and that it is *complete*, i.e. Cauchy sequences converge. It is possible to develop a whole theory of functional analysis over fields equipped with such absolute values, see the Appendix.

Remark We can more generally topologise L^n for any n by equipping it with the norm $||(a_1, \dots, a_n)|| := \max\{|a_1|, \dots, |a_n|\}$. It then becomes an L-Banach space.

Example 11.2 Let p be a prime number and let $a \in \mathbb{Q}$. Define $|a|_p := p^{-r}$ if $a = p^r \cdot \frac{m}{n}$ where $(m,p) = (n,p) = 1$. This is a NAAV on $\mathbb{Q}$ and the completion of $\mathbb{Q}$ with respect to $|\cdot|_p$ is denoted by $\mathbb{Q}_p$, the *field of p-adic numbers*, and the unit ball of $\mathbb{Q}_p$ is denoted by $\mathbb{Z}_p$, the ring of *p-adic integers*. Moreover, the NAAV on $\mathbb{Q}_p$ extends uniquely to a NAAV on L for any finite field extension $L/\mathbb{Q}_p$.

Concretely, elements of $\mathbb{Z}_p$ are 'infinite base p expansions', i.e. can be represented uniquely as a series

$$a_0 + a_1 p + a_2 p^2 + \dots + a_n p^n + \dots,$$

where $a_i \in \{0,1,\ldots,p-1\}$ for all i. We then have $\mathbb{Q}_p = \mathbb{Z}_p[1/p]$.

Going back to general L, convergence of series will be central to the basic definitions of locally analytic functions. Part (iv) in the definition of a NAAV has the following immediate and useful consequence in that regard: if (a_n) is a sequence in L, then

$$\sum_{n\geqslant 0} a_n \text{ converges } \iff a_n \to 0 \text{ as } n \to \infty.$$

In particular, as a function, a power series $f(x) = \sum_{n\geqslant 0} a_n x^n$ will converge on a ball $B_{0,\varepsilon} := \{a \in L : |a| \leqslant \varepsilon\}$ if and only if $\varepsilon^n |a_n| \to 0$ as $n \to \infty$.

11.2.2 *p*-adic Lie Groups

Given $\alpha = (\alpha_1,\ldots,\alpha_n) \in \mathbb{N}^n$, we adopt the notation $|\alpha| := \alpha_1 + \ldots + \alpha_n$ and $\underline{X}^\alpha := X_1^{\alpha_1} \cdots X_n^{\alpha_n}$.

Definition 11.3 Let V be an L-Banach space and fix $n \geqslant 1$.

(i) If $U \subseteq L^n$ is open, then a function $f : U \to V$ is *locally analytic* if for all $x_0 \in U$, there exists $\varepsilon > 0$ and a power series $F(\underline{X}) = \sum_{\alpha\in\mathbb{N}^n} v_\alpha \underline{X}^\alpha$, where $v_\alpha \in V$ and $\varepsilon^{|\alpha|} \cdot ||v_\alpha|| \to 0$ as $|\varepsilon| \to \infty$, such that for all $x \in U$ with $||x - x_0|| \leqslant \varepsilon$ we have $f(x) = F(x - x_0)$.
(ii) Given $x_0 \in L^n$ and $\varepsilon > 0$, write $B(x_0,\varepsilon) = \{x \in L^n : ||x - x_0|| < \varepsilon\}$. We say that a map $f : B(x_0,\varepsilon) \to V$ is *holomorphic* if $f(x) = F(x - x_0)$ for all $x \in B(x_0,\varepsilon)$, with F as in (i). The vector space

$$\mathcal{F}(x_0,\varepsilon,V) := \{f : B(x_0,\varepsilon) \to V | f \text{ holomorphic}\}$$

is then an L-Banach space, with norm $||\sum_{\alpha\in\mathbb{N}^n} v_\alpha \underline{X}^\alpha|| = \sup_\alpha \varepsilon^{|\alpha|} ||v_\alpha||$.

This definition can be extended to the case where V is a Hausdorff locally convex space. In that case, we say $f : U \to V$ is locally analytic if it factors as a composite $U \to W \to V$, where W is an L-Banach space, the map $U \to W$ is locally analytic as defined above and $W \to V$ is a continuous injective L-linear map.

Next we can now introduce manifolds:

Definition 11.4 Let M be a Hausdorff topological space and let $n \geqslant 1$. An *atlas of dimension n* on M is a set $\mathcal{A} = \{(U_i,\varphi_i)\}_{i\in I}$ such that

- $U_i \subset M$ is open for all $i \in I$ and $M = \bigcup_{i\in I} U_i$;
- $\varphi_i : U_i \to L^n$ is a homeomorphism onto an open subset of L^n for all $i \in I$; and

- for all $i, j \in I$, the maps

$$\varphi_i(U_i \cap U_j) \underset{\varphi_i \circ \varphi_j^{-1}}{\overset{\varphi_j \circ \varphi_i^{-1}}{\rightleftarrows}} \varphi_j(U_i \cap U_j)$$

  are locally analytic.

We say an atlas $\mathcal{A}$ is *maximal* if for any other atlas $\mathcal{B}$ such that $\mathcal{A} \cup \mathcal{B}$ is also an atlas, then $\mathcal{B} \subseteq \mathcal{A}$. We then say that M is a (locally L-analytic) *manifold of dimension n* if it is equipped with a maximal atlas, and the pairs (U_i, φ_i) are called *charts*.

Given a Hausdorff locally convex L-vector space V, we say that a map $f : M \to V$ is *locally analytic* if $f \circ \varphi^{-1} : \varphi(U) \to V$ is locally analytic for each chart (U, φ) of M. Similarly, a map between manifolds is locally analytic if it is locally analytic on the charts.

Finally we can talk about groups:

Definition 11.5 A manifold G is a *Lie group* if it is a group such that the multiplication $m : G \times G \to G$ is locally analytic.

Remark Given a Lie group G, the inversion map $g \mapsto g^{-1}$ is automatically a locally analytic isomorphism of manifolds (c.f. [33, Proposition 13.6]).

Example 11.6 The following are all examples of Lie groups:

(i) $(L^n, +)$ or $(\mathcal{O}_L^n, +)$.
(ii) $(L^\times, \cdot)$ or $(\mathcal{O}_L^\times, \cdot)$.
(iii) $(1 + p\mathbb{Z}_p, \cdot) \leqslant (\mathbb{Q}_p^\times, \cdot)$, i.e. elements of the form $1 + a_1 p + a_2 p^2 + \ldots$.
(iv) $\mathrm{GL}_n(L)$, $\mathrm{GL}_n(\mathcal{O}_L)$, $\mathrm{SL}_n(L)$, $\mathrm{SL}_n(\mathcal{O}_L)$.
(v) More generally, the L-valued points of any connected algebraic group over L. In particular, the Borel subgroup

$$B = \left\{ \begin{pmatrix} * & \cdots & * \\ & \ddots & \vdots \\ & & * \end{pmatrix} \in \mathrm{GL}_n(L) \right\}$$

and the maximal torus

$$T = \left\{ \begin{pmatrix} * & & \\ & \ddots & \\ & & * \end{pmatrix} \in \mathrm{GL}_n(L) \right\}$$

of $\mathrm{GL}_n(L)$ are Lie groups.

(vi) The *Iwahori subgroup* of $\mathrm{GL}_2(\mathbb{Z}_p)$

$$I = \left\{ \begin{pmatrix} a & b \\ c & d \end{pmatrix} \in \mathrm{GL}_2(\mathbb{Z}_p) : c \in p\mathbb{Z}_p \right\}.$$

Most of these examples are algebraic in nature, but the point of the analytic setup is that we may study a class of representations larger than the algebraic ones.

11.2.3 Locally Analytic Representations

From now on, we fix complete non-archimedean fields $L \subseteq K$ such that the NAAV on K extends the one on L, and we fix G a locally L-analytic Lie group. Note that in this setup K is an L-Banach space. We will study representations of G on K-vector spaces. We assume that V is a Hausdorff locally convex K-vector space and we write

$$C^{\mathrm{an}}(G,V) := \{f : G \to V | f \text{ locally analytic}\}.$$

Definition 11.7 A representation $\rho : G \to \mathrm{GL}(V)$ is *locally analytic* if for each $v \in V$, the map $g \mapsto \rho(g)v$ belongs to $C^{\mathrm{an}}(G,V)$.

Remark This only depends on each vector $v \in V$, so given *any* representation on V, it makes sense to consider the *locally analytic vectors*,

$$V^{\mathrm{an}} := \{v \in V : (g \mapsto \rho(g)v) \in C^{\mathrm{an}}(G,V)\},$$

a locally analytic subrepresentation.

Example 11.8 (i) If G is algebraic (e.g. $\mathrm{GL}_n(L)$) then any algebraic representation of G is locally analytic, because the orbit maps are polynomial functions on G.

(ii) If $G = (\mathbb{Z}_p, +)$, we can define a character $\chi : G \to K^\times$ as follows. Pick $z \in K^\times$ such that $|z-1| < 1$. Then, for $a \in \mathbb{Z}_p$, set

$$\chi_z(a) = z^a := \sum_{n=0}^{\infty} (z-1)^n \binom{a}{n}.$$

Here the binomial coefficient is defined as $\binom{a}{n} = \frac{a(a-1)\dots(a-n+1)}{n!} \in \mathbb{Z}_p$. It was shown by Amice [1, 16.1.15] that χ_z is locally analytic.

(iii) Let $G = \mathrm{GL}_2(\mathbb{Q}_p)$, B the Borel subgroup, T the maximal torus. Let $\chi : T \to K^\times$ be a locally analytic character. As T can be identified with a quotient of B, we may lift χ to B. Then we have the locally analytic induction

$$\mathrm{Ind}_B^G(\chi) := \{f \in C^{\mathrm{an}}(G,K) : f(gb) = \chi(b^{-1})f(g)\ \forall g \in G, b \in B\}.$$

This is a locally analytic representation of G when G acts by left translation, called a *principal series* representation.

In fact, more generally, given any locally L-analytic group G and any locally analytic subgroup P such that G/P is compact, and given any locally analytic representation V of P, the induction

$$\mathrm{Ind}_P^G(V) := \left\{f \in C^{\mathrm{an}}(G,V) : f(gb) = b^{-1} \cdot f(g)\ \forall g \in G, b \in P\right\}$$

is a locally analytic G-representation (see [22, Satz 4.1.5]).

(iv) When $\chi = \mathbf{1}$ in (iii), the corresponding principal series representations $\mathrm{Ind}_B^G(\mathbf{1})$ can be identified with the space of locally analytic functions on $\mathbb{P}^1$, because $G/B \cong \mathbb{P}^1$. Moreover, we have a natural injection $\mathbf{1}_G \to \mathrm{Ind}_B^G(\mathbf{1})$ with image the constant functions $G \to K$. The quotient $\mathrm{St} := \mathrm{Ind}_B^G(\mathbf{1})/\mathbf{1}_G$ is called the *Steinberg representation*. This also has a geometrical interpretation, namely by a theorem of Morita [26] the Steinberg representation is isomorphic to the strong dual of the space of 1-forms on the Drinfeld upper half plane $\mathbb{P}^1(\mathbb{C}_p) \setminus \mathbb{P}^1(\mathbb{Q}_p)$.

Remark Even if $G = (\mathbb{Z}_p, +)$, we can construct infinitely many irreducible, infinite dimensional, locally analytic representations. If $z \in K^\times$ as in example (ii) and z is transcendental over $\mathbb{Q}_p$, and assuming that K is the smallest complete field containing z, then Diarra [16, Théorème 5] showed that K is then an irreducible $\mathbb{Q}_p$-representation V_z of G via

$$\rho(a)v = \sum_{n=0}^{\infty} (z-1)^n \binom{a}{n} v,$$

and moreover he showed that if we instead choose $z' \in K^\times$ with $|z'| \neq |z|$, then V_z is not isomorphic to $V_{z'}$. Hence locally analytic representations are too wild to study in general. We need a nicer subclass of representations within it.

11.3 The Distribution Algebra

The analytic nature of both the groups and representations makes it hard to work with them directly. In order to study these representations more algebraically, we define an algebra $D(G,K)$ such that

$$\left\{ \begin{array}{l} \text{sufficiently nice} \\ \text{loc. an. representations} \end{array} \right\} \leftrightarrow \left\{ \begin{array}{l} \text{sufficiently nice} \\ D(G,K)\text{-modules} \end{array} \right\}.$$

Here, 'sufficiently nice' will have to be some topological properties. Later, we will see how to replace some of these topological properties with more algebraic ones.

We fix fields $\mathbb{Q}_p \subseteq L \subseteq K$ with $L/\mathbb{Q}_p$ finite, and assume further that K is *spherically complete* with respect to a NAAV extending the one on L. We refer the reader to the Appendix for the precise meaning of this, but we recall that this is automatically satisfied if K is a finite extension of $\mathbb{Q}_p$.

11.3.1 The Space of Distributions

Recall that given a locally L-analytic manifold M, we have $C^{\mathrm{an}}(M,K) = \{f : M \to K | f \text{ is locally analytic}\}$. Assume that M is strictly paracompact, i.e. any open cover can be refined to one where the opens are disjoint, and of dimension d. We point out that this is not a very restrictive condition, for instance it is always satisfied when M is a Lie group, see [33, Corollary 18.8]. In such a setting, Féaux de Lacroix [22, 2.1.10] defined a K-locally convex structure on $C^{\mathrm{an}}(M,K)$, which we recall now.

First, note that if $(U_i, \varphi_i)_{i \in I}$ are the charts, by strict paracompactness we can refine it to assume that the U_i are disjoint. Then since $f : M \to K$ is locally analytic if and only if $f \circ \varphi_i^{-1} : \varphi_i(U_i) \to K$ is locally analytic by definition, it now follows that $C^{\mathrm{an}}(M,K) = \prod_{i \in I} C^{\mathrm{an}}(U_i,K)$. Moreover, by definition of what it means for $f \circ \varphi_i^{-1}$ to be locally analytic and by strict paracompactness, for each f we can cover $\varphi_i(U_i)$ by disjoint open balls B_{ij} (which depend on f) so that f is holomorphic on each B_{ij}.

Summarising the above, given $f \in C^{\mathrm{an}}(M,K)$ there is a family of charts $(V_i, \varphi_i)_{i \in J}$ of M and real numbers $\varepsilon_i > 0$ $(i \in J)$ such that:

(i) $M = \coprod_{i \in J} V_i$;
(ii) $\varphi_i(V_i) = B(x_i, \varepsilon_i)$ for some $x_i \in L^d$; and
(iii) $f \circ \varphi_i^{-1}$ is holomorphic on $B(x_i, \varepsilon_i)$.

We then define an *index* $\mathcal{J}$ to be a family of charts $(V_i, \varphi_i)_{i \in J}$ of M satisfying (i)–(ii) above. For each index $\mathcal{J}$, we may form the product

$$\mathcal{F}_{\mathcal{J}}(M,K) := \prod_{i \in J} \mathcal{F}(x_i, \varepsilon_i, K).$$

Since each $\mathcal{F}(x_i, \varepsilon_i, K)$ is Banach by Definition 11.3, this can be given the structure of a locally convex K-vector space (see Appendix). From the above we can see that

$$C^{\text{an}}(M,K) = \varinjlim_{\mathcal{I}} \mathcal{F}_{\mathcal{I}}(M,K),$$

where the limit is over all indices $\mathcal{I}$, and it therefore has the structure of a locally convex K-vector space as an inductive limit of locally convex spaces (see Appendix).

Definition 11.9 With M as above, the *space of distributions* on M is the dual $D(M,K) := C^{\text{an}}(M,K)'$.

Lemma 11.10 *(i)* ([36, Lemma 2.1]) *If M is compact then $D(M,K)$ is a nuclear Fréchet space, and so in particular reflexive.*

(ii) ([22, 2.2.4]) *If $M = \coprod_{i \in I} M_i$, where the M_i are pairwise disjoint compact open subsets, then $D(M,K) = \bigoplus_{i \in I} D(M_i,K)$ topologically.*[1]

At first sight, it isn't a priori clear what the elements of $D(M,K)$ look like. The only easy examples are the *Dirac distributions* which are defined as follows: let $m \in M$, then we have an element $\delta_m \in D(M,K)$ given by $\delta_m(f) := f(m)$ for $f \in C^{\text{an}}(M,K)$. This gives a map $M \to D(M,K)$, $m \mapsto \delta_m$.

11.3.2 The Convolution Product

From now on, $M = G$ is a Lie group. We now sketch the construction of the product on $D(G,K)$. It can be motivated as follows:

Note that, given a finite group H and a field k, one can identify the group algebra kH with the dual vector space of the space $C(H,k)$ of all functions $H \to k$. One can define Dirac distributions δ_h for $h \in H$ as above. The group operation $H \times H \to H$ gives rise to a map $C(H,k) \to C(H \times H,k)$. Passing to duals we get a map $k(H \times H) \cong kH \otimes_k kH \to kH$, which gives a k-algebra structure to kH and which agrees with the group operation on the Dirac deltas, i.e. $\delta_h \otimes \delta_{h'} \mapsto \delta_{hh'}$.

We now replicate this construction for the distributions on G. The key fact is that there is a canonical isomorphism

$$D(G \times G,K) \cong D(G,K) \widehat{\otimes}_K D(G,K),$$

where $\widehat{\otimes}_K$ denotes the completion of the usual algebraic tensor product with respect to the so-called inductive topology – see [39, Section 12]. Also, the group multiplication $m : G \times G \to G$ induces a map $C^{\text{an}}(G,K) \to C^{\text{an}}(G \times G,K)$, $f \mapsto f \circ m$. Dually this gives a map $D(G \times G,K) \to D(G,K)$.

[1] This is useful when $M = G$ is a Lie group and the M_i are left cosets of some compact open subgroup G_0 (e.g. $G = \mathrm{GL}_n(\mathbb{Q}_p)$ and $G_0 = \mathrm{GL}_n(\mathbb{Z}_p)$).

Given $u, v \in D(G,K)$, we define their *convolution* $u * v$ to be the image of $u \otimes v$ under the composite

$$D(G,K)\widehat{\otimes}_K D(G,K) \xrightarrow{\cong} D(G \times G, K) \to D(G,K).$$

The main result is then:

Theorem 11.11 ([21, 4.4.1 & 4.4.4]) *Convolution defines a separately continuous product on $D(G,K)$ with unit δ_1. When G is compact, this makes $D(G,K)$ into a Fréchet algebra i.e. $* : D(G,K) \times D(G,K) \to D(G,K)$ is continuous.*

The main moral of the story to come is that we gain more control by working with $D(G,K)$-modules rather than locally analytic G-representations directly.

We now briefly describe more concretely the above definition of the distribution algebra when $G = \mathbb{Z}_p$. By the compact topology on $\mathbb{Z}_p$, any index is finite and consists of disjoint balls which may be shrunk so that they all have the same radius. This radius is a negative power of p, and moreover the set of open balls of radius p^{-j} in $\mathbb{Z}_p$, for a fixed $j \geqslant 1$, are in bijection with $\mathbb{Z}/p^j\mathbb{Z}$. Thus we see that

$$C^{\mathrm{an}}(\mathbb{Z}_p, K) = \varinjlim_{j \geqslant 1} \prod_{b \in \mathbb{Z}/p^j\mathbb{Z}} \mathcal{F}(b, p^{-j}, K)$$

with the obvious restriction maps in the directed system. Passing to the dual, we have

$$D(\mathbb{Z}_p, K) = \varprojlim_{j \geqslant 1} \prod_{b \in \mathbb{Z}/p^j\mathbb{Z}} \mathcal{F}(b, p^{-j}, K)'.$$

As an algebra, $D(\mathbb{Z}_p, K)$ also has a geometric description. Indeed, recall the locally analytic characters from Example 11.8(ii): given any $z \in K$ with $|1-z|<1$, there is a locally analytic character χ_z of $\mathbb{Z}_p$. Thus, given any distribution $\lambda \in D(\mathbb{Z}_p, K)$, we may 'evaluate' λ at z by evaluating λ at $\chi_z \in C^{\mathrm{an}}(\mathbb{Z}_p, K)$. This associates to λ a function called its 'Fourier transform'. Using Fourier theory, Amice showed that this explicitly realises the distribution algebra as analytic functions on an open ball:

Theorem 11.12 ([2, 1.3 & 2.3.4]) *Let $X = \{a \in K : |a| < 1\}$ be the open unit ball. Then $D(\mathbb{Z}_p, K)$ is canonically isomorphic to the ring of rigid analytic functions on X.*

11.3.3 A More Explicit Description of the Distribution Algebra

So far the only elements of $D(G,K)$ that we have come across are the Dirac delta distributions δ_g for $g \in G$,

$$\delta_g : C^{\mathrm{an}}(G,K) \to K,$$
$$f \mapsto f(g).$$

The definition of the convolution product yields immediately the following lemma.

Lemma 11.13 *The map $g \mapsto \delta_g$ is a continuous map of monoids $G \to D(G,K)$, i.e. $\delta_{gh} = \delta_g \cdot \delta_h$ for any $g,h \in G$.*

In particular, there is a natural algebra morphism $K[G] \to D(G,K)$. If G is a compact, we can even go further: define the completed group algebra (or Iwasawa algebra) by

$$K[[G]] := (\varprojlim \mathbb{Z}_p[G/N]) \otimes_{\mathbb{Z}_p} K,$$

where the inverse limit is taken over all open normal subgroups N of G (note that G/N is then a finite group by compactness of G).

In fact (compare e.g. the introduction of [38]), $K[[G]]$ is the dual of all *continuous* functions $G \to K$, so our morphism actually extends naturally to an algebra morphism $\theta : K[[G]] \to D(G,K)$.

Theorem 11.14 ([38, Theorem 5.2]) *Let G_0 denote G, but regarded as a Lie group over $\mathbb{Q}_p$ rather than L. The morphism $L[[G]] \to K[[G]] \to D(G_0,K)$ is faithfully flat.*

In other words, we can study the ('more classical') $L[[G]]$-modules by passing to $D(G_0,K)$-modules (applying $D(G_0,K) \otimes_{L[[G]]} -$) without losing any information.

Proposition 11.15 ([36, Lemma 3.1]) *The Dirac distributions δ_g span a dense subspace of $D(G,K)$.*

Proof We only sketch the idea here. Let $H \leqslant G$ be a compact open subgroup. Then the coset decomposition of Lemma 11.10(ii) yields

$$D(G,K) = \bigoplus_{g \in G/H} \delta_g * D(H,K).$$

Using Lemma 11.13, we can thus reduce to the case where G itself is compact. In particular, $D(G,K)$ is a reflexive Fréchet space (Lemma 11.10(i)), so that $D(G,K)' \cong C^{\mathrm{an}}(G,K)$. Since an element f of $C^{\mathrm{an}}(G,K)$ is zero if and only if

$f(g) = \delta_g(f)$ is zero for all g, the result then follows from the Hahn–Banach theorem as given in Theorem 11.35 in the Appendix (as there is no non-zero functional on $D(G,K)$ vanishing on the closure of the K-span of all Dirac distributions). □

We thus often think of $D(G,K)$ as an 'analytic' group algebra. A special property of $D(G,K)$ is that apart from the distributions induced by the group G, it also contains distributions induced by the Lie algebra:

Let $\mathfrak{g}$ be the Lie algebra of G, e.g. $G = \mathrm{SL}_2(\mathbb{Q}_p)$, $\mathfrak{g} = \mathfrak{sl}_2(\mathbb{Q}_p)$. Write $\mathfrak{g}_K = \mathfrak{g} \otimes K$. If $x \in \mathfrak{g}$, we can form the distribution $\mathrm{dist}(x)$ by

$$\mathrm{dist}(x)(f) = \frac{\mathrm{d}}{\mathrm{d}t}(f(\exp(tx)))|_{t=0}.$$

This gives a linear map $\mathrm{dist} : \mathfrak{g} \to D(G,K)$, sending $[x,y]$ to the commutator $\mathrm{dist}(x)\mathrm{dist}(y) - \mathrm{dist}(y)\mathrm{dist}(x)$ – so we obtain an algebra morphism $d : U(\mathfrak{g}_K) \to D(G,K)$.

Lemma 11.16 ([22, Korollar 4.7.4]) *The map d is injective. The closure of $U(\mathfrak{g}_K)$ in $D(G,K)$ is a Fréchet algebra which we denote by $\widehat{U(\mathfrak{g}_K)}$.*

At first, the object $\widehat{U(\mathfrak{g}_K)}$ might seem strange, but its elements are actually very concrete. If $x_1, \ldots, x_d$ is an ordered K-basis of $\mathfrak{g}_K$, then the Poincaré–Birkhoff–Witt theorem states that $U(\mathfrak{g}_K)$ admits a K-basis of the form

$$x^\alpha = x_1^{\alpha_1} x_2^{\alpha_2} \ldots x_d^{\alpha_d},$$

where $\alpha = (\alpha_1, \ldots, \alpha_d) \in \mathbb{N}_0^d$. Now [36, Lemma 2.4] shows that an arbitrary element of $\widehat{U(\mathfrak{g}_K)}$ can be written uniquely as

$$\sum_{\alpha \in \mathbb{N}_0^d} \lambda_\alpha x^\alpha, \; \lambda_\alpha \in K, \; \pi^{-|\alpha| n} \lambda_\alpha \to 0 \text{ as } |\alpha| \to \infty \; \forall n,$$

where $\pi \in K$ is any non-zero element with $|\pi| < 1$.

It is worth contemplating this particular convergence condition for a while, as it appears quite naturally in several places in p-adic representation theory and rigid analytic geometry.

Recall from subsection 11.2.1 that a power series $\sum \lambda_\alpha X^\alpha$ converges on a ball of radius ε if and only if $\varepsilon^{|\alpha|} |\lambda_\alpha| \to 0$. We should therefore think of the convergence condition in the description of $\widehat{U(\mathfrak{g}_K)}$ as requiring an infinite radius of convergence – i.e. elements of $\widehat{U(\mathfrak{g}_K)}$ 'look like' analytic functions on the dual vector space $(\mathfrak{g}_K)^*$. Just as $U(\mathfrak{g}_K)$ is a non-commutative deformation of the ring of polynomial functions on $(\mathfrak{g}_K)^*$, as expressed in the Poincaré–Birkhoff–Witt isomorphism $\mathrm{gr}\, U(\mathfrak{g}_K) \cong \mathrm{Sym}(\mathfrak{g}_K)$, we should think of $\widehat{U(\mathfrak{g}_K)}$ as a non-commutative version of the ring of (globally) analytic functions on $(\mathfrak{g}_K)^*$.

In summary, $D(G,K)$ is a topological group algebra which is sufficiently thickened to also incorporate the infinitesimal information, present in the form of the Lie algebra.

We now turn to the study of $D(G,K)$-modules and their role in locally analytic representation theory. Just as with the usual group algebra, a locally analytic G-representation carries a natural $D(G,K)$-module structure (this is actually a bit subtle to show, see [36, section 3] for details). It turns out, however, that it is more useful to dualize this operation to get a better handle on the topology.

Theorem 11.17 ([36, Corollary 3.3]) *There is an anti-equivalence of categories*

$$\{\textit{locally analytic } G\textit{-representations on spaces of compact type}\}$$
$$\downarrow$$
$$\{\textit{sep. continuous } D(G,K)\textit{-modules in nuclear Fréchet spaces}\}$$

given by sending V to its strong dual V'.

Remark We refer to the appendix for the definitions of 'compact type' and 'nuclear', and only comment on their function in this result: compact type is a property that ensures that V is reflexive, i.e. $(V')' \cong V$, and nuclear Fréchet spaces are those spaces which are dual to spaces of compact type. The conditions are therefore necessary in the theorem above to ensure that the duality functor is an anti-equivalence on the underlying topological vector spaces.

11.4 Fréchet–Stein Algebras

We saw above how we can think of locally analytic G-representations (of compact type) as certain topological modules over the distribution algebra $D(G,K)$. The problem persists however that these are topological modules, and doing algebra with topological objects is quite difficult. For instance, the category of topological $D(G,K)$-modules knows no 'isomorphism theorem': if $f: M \to N$ is a morphism, then $M/\mathrm{ker} f$ need not be isomorphic to the image of f, as the quotient topology on $M/\mathrm{ker} f$ need not agree with the subspace topology on $\mathrm{im} f$. In particular, we are dealing with categories that are not abelian.

11.4.1 Toy Model: Noetherian Banach Algebras and Finitely Generated Modules

Let A be a Noetherian Banach K-algebra, i.e. it is a Noetherian K-algebra which is complete with respect to some (submultiplicative) norm. The category of normed A-modules (or of Banach modules if we insist on completeness) displays the same problems alluded to above. But here there is an excellent remedy.

Theorem 11.18 (see [7, sections 3.7.2, 3.7.3]) *Any (abstract) finitely generated A-module can be endowed with a canonical Banach norm such that any A-module map between finitely generated modules is continuous. These norms are compatible with the formation of submodules, quotients and direct sums.*

More abstractly: there is a fully faithful functor from (abstract!) finitely generated A-modules to the category of Banach A-modules, exhibiting the former as an (abelian!) subcategory of the latter.

Proof (Rough sketch.) If M is a finitely generated A-module, there exists some surjection $A^r \to M$. We can check that this endows M with a Banach norm by checking that every submodule of A^r is closed. This quotient norm then has the property that $M \to N$ is continuous if and only if the composition $A^r \to M \to N$ is. But if N is another finitely generated A-module endowed with such a norm, then any A-linear map $A^r \to N$ is a sum of action maps and hence continuous. This shows that any A-module map $M \to N$ is automatically continuous. In particular (taking $M = N$), Banach norms arising from a different generating set give rise to an equivalent norm. Now check that any submodule of a finitely generated A-module is a closed subspace with respect to this norm, by reducing to the case of A^r. □

Remark This applies e.g. to the Tate algebra

$$K\langle x\rangle = \left\{ \sum_{i\in\mathbb{N}_0} a_i x^i : |a_i| \to 0 \right\}$$

of analytic functions on the closed unit disk, ensuring that p-adic analytic geometry (and its theory of coherent modules) is well behaved.

11.4.2 Fréchet–Stein Algebras and Coadmissible Modules

It turns out that $D(G,K)$ is hardly ever Noetherian Banach – recall for example from Theorem 11.12 that $D(\mathbb{Z}_p, K)$ is the ring of analytic functions on the *open* unit disk. As the open disk is the union of countably many closed disks (with radius approaching 1), $D(\mathbb{Z}_p,K)$ is not Noetherian Banach, but rather an inverse limit of Noetherian Banach algebras.

Definition 11.19 ([38, section 3]) Let A be a Fréchet K-algebra. We say that A is a *Fréchet–Stein algebra* if A can be written as a countable inverse limit $A = \varprojlim A_n$, where each A_n is a Noetherian Banach K-algebra such that $A_{n+1} \to A_n$ has dense image and turns A_n into a flat A_{n+1}-module on both sides.

An A-module M is called *coadmissible* if $M = \varprojlim M_n$, where M_n is a finitely generated A_n-module, and the natural morphism $A_n \otimes_{A_{n+1}} M_{n+1} \to M_n$ is an isomorphism.

Example 11.20 As before, let $\pi \in K$ be non-zero with $|\pi| < 1$. Let

$$A_n = K\langle \pi^n x\rangle = \left\{\sum a_i x^i : \pi^{-in} a_i \to 0\right\}$$

be the ring of analytic function on a closed disk of radius $|\pi|^{-n}$. Then $A = \varprojlim A_n$ is the ring of analytic functions on $X = \cup \mathrm{Sp} A_n$, the 'affine line'. Then A is a Fréchet–Stein algebra, and coadmissible A-modules are precisely the global sections of coherent $\mathcal{O}_X$-modules.

By exactly the same argument, $D(\mathbb{Z}_p,K)$ is a Fréchet–Stein algebra, and coadmissible $D(\mathbb{Z}_p,K)$-modules are given as global sections of coherent modules on the open unit disk.

Just as in these examples, the M_n can in general be recovered from M, which allows us to go back and forth between M and its 'Noetherian levels'.

Lemma 11.21 ([38, Corollary 3.1]) *If M is a coadmissible A-module, then the natural morphism $A_n \otimes_A M \to M_n$ is an isomorphism.*

Proof (Sketch.) By our density assumption, $A_n \otimes_A M \to M_n$ has dense image. It follows from our toy model that any (automatically finitely generated) A_n-submodule of M_n is closed, so the morphism is in fact surjective. For injectivity, suppose that $\sum_{i=1}^k b_i \otimes x_i \in A_n \otimes_A M$ is in the kernel. We now consider the morphisms $\varphi_m : A_m^k \to M_m$ sending $(a_1, \dots, a_k)$ to $\sum a_i \overline{x_i}$, where $\overline{x_i}$ denotes the image of x_i in M_m. By construction, $\varprojlim \ker \varphi_m$ is a coadmissible A-module, and applying the same surjectivity argument as above, this time to $\varprojlim \ker \varphi_m$, we can compute that $\sum b_i \otimes x_i = 0$. This proves injectivity. □

We now extend the ideas from our toy model to the Fréchet–Stein setting.

Lemma 11.22 ([38, paragraph after Lemma 3.6]) *Any coadmissible A-module can be endowed with a canonical Fréchet topology such that any A-module* *morphism between coadmissible A-modules is continuous. Kernel and cokernel of any morphism between coadmissible A-modules are also coadmissible.*

Proof Equip each M_n with its canonical Banach norm and take the limit. Any morphism $M \to N$ then gives rise to A_n-module morphisms $M_n \to N_n$ by Lemma 11.21, and these are continuous by our toy model. Thus $M \to N$ is continuous by definition of the inverse limit topology. It is easy to verify that kernels and cokernels are given as the inverse limit of the corresponding kernels and cokernels on the Noetherian level. □

Proposition 11.23 ([38, Lemma 3.6]) *Let A be a Fréchet–Stein algebra and let M be a coadmissible A-module. Let $N \leqslant M$ be a submodule. The following are equivalent:*

(i) N is coadmissible.
(ii) N is closed with respect to the canonical topology on M.
(iii) M/N is coadmissible.

Corollary 11.24 *The category of coadmissible A-modules is an abelian category and contains all finitely presented A-modules.*

In other words, the theory of coadmissible modules over Fréchet–Stein algebras follows the same philosophy as finitely generated modules over Noetherian Banach algebras did in our toy model: they provide us with an abelian category of 'topological' modules which can be manipulated purely algebraically.

11.4.3 Distribution Algebras of Compact Groups Are Fréchet–Stein

It thus remains to show that distribution algebras are indeed Fréchet–Stein algebras (provided that G is compact). The proof of this result, given in [38], is quite involved, but we can at least indicate the main ideas. We have already discussed the example of $G = \mathbb{Z}_p$ above, which is not difficult to generalize to $G = \mathbb{Z}_p^d$. We then generalize further to the case where G is still structurally similar to $\mathbb{Z}_p^d$ – namely, that G is uniform pro-p. Since any compact p-adic Lie group contains such a group as an open subgroup, we can then conclude with the usual coset decomposition from Lemma 11.10(ii).

Instead of giving a formal definition, let us treat uniform pro-p groups as a black box and discuss which properties might help us to generalize from $\mathbb{Z}_p^d$ to a more general class of groups.

Definition 11.25 Let G be a locally $\mathbb{Q}_p$-analytic group which is pro-p. We say that G admits an *ordered basis* if there exists an ordered set of topological generators $h_1, \dots, h_d$ such that the map

$$\mathbb{Z}_p^d \to G,$$
$$(x_1, \dots, x_d) \mapsto h_1^{x_1} \dots h_d^{x_d}$$

is a homeomorphism.

In particular, an ordered basis yields a global chart for the group G, giving us isomorphisms of topological vector spaces $C^{\mathrm{an}}(\mathbb{Z}_p^d, K) \cong C^{\mathrm{an}}(G, K)$ and hence $D(\mathbb{Z}_p^d, K) \cong D(G, K)$. Note, however, that this tells us nothing about the algebra structure on $D(G, K)$: if the h_i do not commute, then we obtain a noncommutative convolution product which cannot be read off from the chart. In other words, we can use this description to bring elements of $D(G, K)$ into a standard form, but the question remains how to multiply two such expressions.

On the other hand, if the h_i do commute, then the above chart is a group isomorphism and we have indeed reduced to the case of $G = \mathbb{Z}_p^d$. The crucial idea is therefore to consider groups admitting an ordered basis which is commutative 'up to higher order terms'.

Lemma 11.26 ([17, Theorem 8.18, Theorem 4.9, Lemma 4.10]) *Let G be a uniform pro-p group. Then G is locally $\mathbb{Q}_p$-analytic and admits an ordered basis $\mathrm{h}_1, \dots, \mathrm{h}_d$ such that each commutator $h_i h_j h_i^{-1} h_j^{-1}$ is a pth power, i.e. there exists $g \in G$ such that $h_i h_j h_i^{-1} h_j^{-1} = g^p$.*

With this result in place, one can employ graded methods to show (after a significant amount of work) the following.

Proposition 11.27 ([38, Theorem 4.10]) *If G is a uniform pro-p group, viewed as a locally $\mathbb{Q}_p$-analytic group, then $D(G, K)$ is a Fréchet–Stein algebra.*

Lemma 11.28 ([17, Corollary 8.34]) *Every compact locally $\mathbb{Q}_p$-analytic group contains an open normal subgroup which is uniform pro-p.*

Theorem 11.29 ([38, Theorem 5.1], [30, Theorem 2.3]) *Let G be a compact locally L-analytic group and let $\mathfrak{g}$ be its Lie algebra. Then $D(G, K)$ and $\widehat{U(\mathfrak{g}_K)}$ are Fréchet–Stein algebras.*

Proof (Sketch.) For $\widehat{U(\mathfrak{g}_K)}$, note the similarity to the example of analytic functions discussed before. See [30, Theorem 2.3], which reformulates [25, Theorem 1.4.2]. A more general argument can be found in [5, Theorem 6.7].

Let G_0 be G, viewed as a locally $\mathbb{Q}_p$-analytic group, and let H be an open, normal, uniform pro-p subgroup. By Proposition 11.27, $D(H, K)$ is Fréchet–Stein, so we can write $D(H, K) \cong \varprojlim D(H, K)_n$ for some suitable Noetherian

Banach algebras. By Lemma 11.10.(ii),

$$D(G_0,K) \cong \oplus_{g\in G_0/H}\delta_g * D(H,K),$$

i.e. $D(G_0,K)$ is a free $D(H,K)$-module of finite rank (by compactness). It is then not too difficult to check that $D(G_0,K)$ is also Fréchet–Stein – it is trivially a coadmissible $D(H,K)$-module, and the only thing one needs to verify is that the algebra structure on $D(G_0,K)$ extends to an algebra structure on

$$D(G_0,K)_n := D(H,K)_n \otimes_{D(H,K)} D(G_0,K) \cong \oplus\delta_g * D(H,K)_n,$$

as $D(G_0,K)_n$ then inherits all desired properties from $D(H,K)_n$. Thus $D(G_0,K)$ is Fréchet–Stein.

Finally, embedding locally $\mathbb{Q}_p$-analytic functions $G \to K$ into the space of locally L-analytic functions $G \to K$ yields dually a surjection $D(G_0,K) \to D(G,K)$, exhibiting $D(G,K)$ as a topological quotient algebra of $D(G_0,K)$. We can thus deduce that $D(G,K)$ is Fréchet–Stein. □

It is worth pointing out a feature which is quite common in this theory: the general case is often much less accessible than the case '$L = \mathbb{Q}_p$', so that we are forced to take a detour via G_0 and descend along $D(G_0,K) \to D(G,K)$. This also explains the restrictions in the faithful flatness result Theorem 11.14: while some properties, like being Fréchet–Stein, are preserved when taking the (Fréchet) quotient, this certainly need not be the case for the property of being faithfully flat. We can prove the result for G_0 (again making use of uniform pro-p subgroups), but a priori this does not tell us anything about the general case.

Let G be a compact locally L-analytic group. By the above, it now makes sense to talk about coadmissible $D(G,K)$-modules. We note that coadmissible modules are indeed contained in the category appearing in the equivalence of Theorem 11.17.

Lemma 11.30 ([38, Lemma 6.1]) *Let G be compact. Then any coadmissible $D(G,K)$-module, endowed with its canonical Fréchet structure, is a nuclear Fréchet space.*

This justifies the following definition.

Definition 11.31 Let G be a locally L-analytic group, and let V be a locally analytic G-representation of compact type. We say that V is *admissible* if V' is a coadmissible $D(H,K)$-module for some (equivalently, for any) compact subgroup $H \leqslant G$.

We have thus reached our goal. We have the following commutative diagram (for G compact), where the horizontal arrows are (anti-)equivalences of categories and the vertical arrows are fully faithful embeddings.

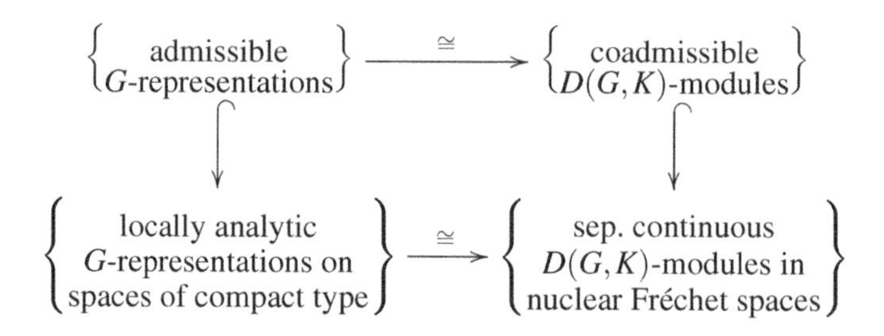

As mentioned before, the crucial feature is that the category of coadmissible $D(G,K)$-modules is an abelian category which embeds naturally into the category of abstract $D(G,K)$-modules. In this way, admissible representations can now be studied purely algebraically.

11.4.4 Some Further Directions

We give here a (non-exhaustive) list of further developments in the field:

(i) As mentioned in the introduction, the p-adic local Langlands conjecture relates n-dimensional Galois representations to certain unitary Banach representations of $\mathrm{GL}_n(\mathbb{Q}_p)$. Often it is helpful to restrict a unitary representation to its locally analytic vectors (analogously to the smooth vectors for complex representations). From this viewpoint, the faithful flatness result Theorem 11.14 ensures that the functor of taking locally analytic vectors is exact and does not annihilate any non-zero representations (see [38, Theorem 7.1]).

(ii) These lectures should already have illustrated that some ideas from p-adic geometry arise quite naturally in the study of locally analytic representations. It is therefore not surprising that a number of distinctly geometric tools have been developed, often influenced by complex geometric representation theory. For instance, if $G = \mathbb{G}(L)$ for some split reductive algebraic group $\mathbb{G}$, one can establish an equivalence of categories between coadmissible $D(G,K)$-modules with trivial infinitesimal central character and a certain class of G-equivariant p-adic $\mathcal{D}$-modules on the p-adic analytic flag variety, analogously to Beilinson–Bernstein theory over the complex numbers – see e.g. [4], [3], [29].

There is also a growing body of literature on particular representations coming from geometry as certain cohomology groups (e.g. the cohomology of Drinfeld's upper half plane and its tower of coverings), often motivated by the Langlands programme – see e.g. [27], [20], [18].

(iii) One phenomenon which we have exploited regularly is the description of $D(\mathbb{Z}_p, K)$ as the ring of functions on the open unit disk, where we could view the open unit disk as the set of all locally analytic characters of $\mathbb{Z}_p$. This observation is also quite helpful in the theory of (φ, Γ)-modules (and hence in Colmez's work on the pLLC for $\mathrm{GL}_2(\mathbb{Q}_p)$). Schneider and others have generalised this approach to consider 'character varieties' for the ring of integers of more general finite field extensions of $\mathbb{Q}_p$ (see [34], [6]).

(iv) The study of coadmissible $\widehat{U(\mathfrak{g}_K)}$-modules can be developed along similar lines as the representation theory of complex Lie algebras. For instance, the centre of $\widehat{U(\mathfrak{g}_K)}$ can be described via an extension of the usual Harish-Chandra isomorphism, c.f. [25, Theorem 2.1.6]. In fact, if G is the group of L-rational points of a connected, split reductive algebraic group over L and if the field K is discretely valued, the Harish-Chandra isomorphism extends to the distribution algebra $D(G, K)$, c.f. [25, Theorem 2.4.2]. There is also an analogue of the BGG category $\mathcal{O}$ for $\widehat{U(\mathfrak{g}_K)}$, see [31].

(v) For more details on locally analytic induction, see the discussion of principal series representations of GL_2 in [36]. We also mention that Orlik–Strauch [28] give a functor which constructs a locally analytic G-representation out of a P-equivariant $\widehat{U(\mathfrak{g}_K)}$-module in the BGG category $\mathcal{O}$ and a smooth P-representation (P being a parabolic subgroup of a split reductive group G).

(vi) As one might imagine, there are many links between the p-adic theory and the mod p theory: for example, if V is a unitary Banach representation of G over $\mathbb{Q}_p$, then its unit ball is a representation over $\mathbb{Z}_p$ and we can consider its reduction mod p. There is a mod p Langlands conjecture which is being developed in tandem with the p-adic one, see e.g. [11], [10].

11.5 Appendix: Some Non-archimedean Functional Analysis

Throughout we made use of various notions appearing in non-archimedean functional analysis. Here we collect the various facts and definitions required. A main reference for this material is e.g. [32]. Throughout this appendix, K is a field equipped with a NAAV and is complete.

11.5.1 Locally Convex Spaces

Definition 11.32 A (non-archimedean) *semi-norm* on a K-vector space V is a function $q : V \to \mathbb{R}_{\geqslant 0}$ satisfying:

(i) $q(av) = |a|\,q(v)$ for all $v \in V$ and $a \in K$; and
(ii) $q(v+w) \leqslant \max(q(v), q(w))$ for all $v, w \in V$.

Condition (ii) is called the *strong triangle inequality*. Note that (i) implies in particular that $q(0) = 0$ and (i) and (ii) imply that the set $\{v \in V : q(v) = 0\}$ is a vector subspace of V. If furthermore the condition

(iii) $q(v) = 0$ if and only if $v = 0$

is satisfied then we say that q is a *norm*. When q is a norm, it is conventional to denote it by $||\cdot||$.

Given a family of semi-norms $(q_i)_{i \in I}$ on V, we may assign to V the coarsest topology such that $q_i : V \to \mathbb{R}$ is continuous for all $i \in I$ and all translation maps $v + \cdot : V \to V$, for $v \in V$, are continuous. In such a case we call V a *locally convex space*. If the topology is defined by a single (semi-)norm q, we call V a *(semi-)normed space*.

We note here that when a locally convex space is finite dimensional, it has nice properties:

Proposition 11.33 ([32, Proposition 4.13]) *Every Hausdorff locally convex topology on a finite dimensional vector space $V = K^n$ $(n \geqslant 1)$ is equivalent to the one defined by the norm $||(a_1, \ldots, a_n)|| = \max_{1 \leqslant i \leqslant n} |a_i|$.*

There is another equivalent definition of a locally convex topology (see [32, Proposition 4.4]). Given a K-vector space V, a *lattice* in V is an $\mathcal{O}_K$-submodule M of V such that the canonical map $M \otimes_{\mathcal{O}_K} K \to V$ is an isomorphism. Then V is locally convex if and only if there exists a non-empty family of lattices $\{M_i\}_{i \in I}$ of V with the properties

(i) for any $i \in I$ and $a \in K^\times$ there exists $j \in I$ such that $M_j \subseteq aM_i$; and
(ii) for any two $i, j \in I$, there exists $k \in I$ such that $M_k \subseteq M_i \cap M_j$.

A basis for the topology on V is then given by the subsets of the form $v + M_i$ for $v \in V$ and $i \in I$.

We now describe explicitly various constructions with locally convex spaces that we use in this chapter.

- Suppose $\{V_i\}_{i\in I}$ is an inductive system of locally convex K-vector spaces. Then we can equip $\varinjlim_{i\in I} V_i$ with the final locally convex topology with respect to the maps $\varphi_j : V_j \to \varinjlim_{i\in I} V_i$. Explicitly, it is defined by the family of all lattices $M \subseteq \varinjlim_{i\in I} V_i$ such that $\varphi_j^{-1}(M) \subseteq V_j$ is open for all $j \in I$.
- Suppose $\{V_i\}_{i\in I}$ is a family of locally convex K-vector spaces. Then we may equip $\bigoplus_{i\in I} V_i$ with the final locally convex topology with respect to the inclusions $V_j \to \bigoplus_{i\in I} V_i$ as above.
- Suppose again that $\{V_i\}_{i\in I}$ is a family of locally convex K-vector spaces. Then we may define a locally convex topology on the direct product $\prod_{i\in I} V_i$ this time using seminorms as follows: if for each $j \in I, (q_{jk})_{k\in J}$ denotes the family of seminorms defining the topology on V_j, then we give $\prod_{i\in I} V_i$ the locally convex topology defined by the family of seminorms

$$\prod_{i\in I} V_i \to V_j \xrightarrow{q_{jk}} \mathbb{R}_{\geqslant 0}$$

for $j \in I$ and $k \in J$.
- Given a locally convex K-vector space V, we let

$$V' = \{\varphi \in V^* : \varphi \text{ is continuous}\}$$

be the continuous dual of V. We define a subset $B \subseteq V$ to be *bounded* if for any open lattice $M \subseteq V$, there exists $a \in K$ such that $B \subseteq aM$.[2] Given a bounded $B \subseteq V$, we may define a seminorm p_B on V' by

$$p_B(\varphi) := \sup_{b\in B} |\varphi(b)|.$$

The seminorms $\{p_B : B \text{ bounded in } V\}$ define a locally convex topology on V' called the *strong topology*. When V is normed, this topology is just the normed topology on V' given by the operator norm

$$||\varphi|| := \sup_{0\neq v\in V} \frac{||\varphi(v)||}{||v||}.$$

We will always assume in this chapter that our duals are equipped with the strong topology. We next turn to the Hahn–Banach theorem. For this we need the notion of a spherically complete base field.

Definition 11.34 We say that our field K is *spherically complete* if for any decreasing sequence $B_1 \supset B_2 \supset \cdots$ of balls in K, the intersection $\bigcap_{i\in\mathbb{N}} B_i$ is non-empty.

[2] When V is normed this is equivalent to saying $||B||$ is bounded in $\mathbb{R}$.

As an example, any finite extension of $\mathbb{Q}_p$ is spherically complete because these fields are locally compact. But not all complete non-archimedean fields are spherically complete. For instance, one may consider the algebraic closure $\overline{\mathbb{Q}_p}$ of $\mathbb{Q}_p$. There is a unique extension of the p-adic absolute value to it, but it is no longer complete. Its completion, denoted by $\mathbb{C}_p$, is a complete non-archimedean field and is not spherically complete. See [32, Section I.1] for more details.

This condition is required in order to get that the dual V' of a locally convex vector space is non-zero, which is a consequence of the following Hahn–Banach theorem:

Theorem 11.35 ([32, Proposition 9.2]) *Suppose that K is spherically complete. Let V be a K-vector space, q a seminorm on V and $V_0 \leqslant V$ a vector subspace. Then for any linear form $f_0 : V_0 \to K$ such that $|f_0(v)| \leqslant q(v)$ for all $v \in V_0$, there exists a linear form $f : V \to K$ such that $f|_{V_0} = f_0$ and $|f(v)| \leqslant q(v)$ for all $v \in V$.*

It follows as a corollary of this theorem that any continuous seminorm on a vector subspace V_0 of a locally convex space V extends to a continuous seminorm on V.

11.5.2 Fréchet and Banach spaces

If $||\cdot||$ is a norm on V, the normed space topology on V as we defined it is the usual metric topology coming from $||\cdot||$.

Definition 11.36 A normed space that is complete, meaning that all Cauchy sequences converge, is called a *Banach space*. More generally, a locally convex space that is metrizable and complete is called a *Fréchet space*.

Of course, any Banach space is in particular a Fréchet space. Also, K is itself a Banach space with the absolute value as a norm, and more generally K^n is Banach with the topology given in Proposition 11.33.

In general, a Fréchet space V can be described as follows. Because V is metrizable, there is an increasing countable family of seminorms $q_1 \leqslant q_2 \leqslant \cdots$ defining the topology on V (see [32, Proposition 8.1]). Then for each $i \geqslant 1$, we denote by V_i the metric space completion of $V/\{v \in V : q_i(v) = 0\}$, which is a Banach space with the norm induced by q_i. There is moreover a continuous linear map $V_{i+1} \to V_i$. Then V is canonically isomorphic to the projective limit $\varprojlim V_i$. So one may describe Fréchet spaces as being the countable projective limits of Banach spaces.

We now turn to the notion of a compact type space:

Definition 11.37 Let V and W be Hausdorff locally convex K-vector spaces.

(i) A subset $B \subseteq V$ is called *compactoid* if for any open lattice $M \subseteq V$, there are finitely many vectors $v_1, \ldots, v_n \in V$ such that $B \subseteq M + \mathcal{O}_K v_1 + \ldots + \mathcal{O}_K v_n$.
(ii) A bounded $\mathcal{O}_K$-submodule $B \subseteq V$ is called *c-compact* if it is compactoid and complete.
(iii) A continuous linear map $f : V \to W$ is called *compact* if there is an open lattice $M \subseteq V$ such that the closure of $f(M)$ in W is bounded and c-compact.
(iv) V is said to be of *compact type* if it is the locally convex inductive limit of a sequence

$$V_1 \xrightarrow{\iota_1} V_2 \xrightarrow{\iota_2} V_3 \xrightarrow{\iota_3} \cdots$$

where each V_i is a Banach space and each map ι_i is an injective compact map.

The key property of these spaces is the following:

Theorem 11.38 ([32, Proposition 16.10]) *Any locally convex K-vector space of compact type $V = \varinjlim_{n \geqslant 1} V_n$ is reflexive, i.e. the evaluation map $V \to (V')'$ is an isomorphism, its strong dual V' is Fréchet and there is a canonical isomorphism $V' \cong \varprojlim_{n \geqslant 1} V_i'$.*

There is a dual notion:

Definition 11.39 A Fréchet space V is called *nuclear* if $V \cong \varprojlim_{n \geqslant 1} V_n$ is a countable projective limit of Banach spaces, where the transition maps $V_{n+1} \to V_n$ are all compact.

There is more generally a notion of nuclear locally convex space, see [32, Section 19]. The key fact we'll need is the following:

Theorem 11.40 ([36, Theorem 1.3]) *A Fréchet space V is the strong dual of a locally convex K-vector space of compact type if and only if V is nuclear.*

Bibliography

[1] Amice, Y. 1964. Interpolation p-adique. *Bull. Soc. Math. France*, **92**, 117–180.
[2] Amice, Y. Duals. *Proc. Conf. on p-Adic Analysis*, Nijmegen 1978, 1–15.
[3] Ardakov, K. Equivariant $\mathcal{D}$-modules on rigid analytic spaces. arXiv 1708.07475. To appear in *Astérisque*.

[4] Ardakov, K., Wadsley, S. On irreducible representations of compact p-adic analytic groups. *Ann. of Math.* (2) 178 (2013), no. 3, 453–557.

[5] Ardakov, K., Wadsley, S. $\mathcal{D}$-modules on rigid analytic spaces I. arXiv 1501.02215.

[6] Berger, L., Schneider, P., Xie, B. Rigid character groups, Lubin–Tate theory, and (φ,Γ)-modules. *Mem. Amer. Math. Soc.* 263 (2020), no. 1275.

[7] Bosch, S., U. Güntzer, Remmert, R. Non-Archimedean analysis. A systematic approach to rigid analytic geometry. *Grundlehren der Mathematischen Wissenschaften* 261, Springer-Verlag, Berlin, 1984.

[8] Breuil, C. Sur quelques représentations modulaires et p-adiques de $\mathrm{GL}_2(\mathbb{Q}_p)$. I. *Compositio Math.* 138 (2003), no. 2, 165–188.

[9] Breuil, C. The emerging p-adic Langlands program. *Proceedings of I.C.M. 2010*, Vol II, (2010), 203–230.

[10] Breuil, C., Herzig, F., Hu, Y., Morra, S., Schraen, B. Conjectures and results on modular representations of $\mathrm{GL}_2(K)$ for a p-adic field K. arXiv 2102.06188.

[11] Breuil, C., Paskunas, V. Towards a Modulo p Langlands correspondence for GL_2. *Mem. Amer. Math. Soc.* 216 (2012), no. 1016.

[12] Breuil, C., Schneider, P. First steps towards p-adic Langlands functoriality. *J. Reine Angew. Math.* 610 (2007), 149–180.

[13] Caraiani, A., Emerton, M., Gee, T., Geraghty, D., Paskunas, V., Shin, S.W. Patching and the p-adic local Langlands correspondence. *Camb. J. Math.* 4 (2016), no. 2, 197–287.

[14] Colmez, P. Représentations de $\mathrm{GL}2\mathbb{Q}_p$ et (φ,Γ)-modules. *Astérisque* No. 330 (2010), 281–509.

[15] Colmez, P., Dospinescu, G., Paskunas, V. The p-adic local Langlands correspondence for $\mathrm{GL}2\mathbb{Q}_p$. *Camb. J. Math.* 2 (2014), no. 1, 1–47.

[16] Diarra, B. Sur quelques représentations p-adiques de $\mathbb{Z}_p$. *Indagationes Math.* 41 (1979), 481–493.

[17] Dixon, J.D., du Sautoy, M.P.F., Mann, A., Segal, D. Analytic pro-p groups. *LMS lecture note series* 157, Cambridge University Press, Cambridge, 1991.

[18] Dospinescu, G., A.-C. Le Bras. Revêtements du demi-plan de Drinfeld et correspondance de Langlands p-adique. *Ann. of Math.* (2) 186 (2017), no. 2, 312–411.

[19] Emerton, M. Locally analytic vectors in representations of locally p-adic analytic groups. *Mem. Amer. Math. Soc.* 248 (2017), no. 1175.

[20] Emerton, M. Completed cohomology and the p-adic Langlands program. *Proceedings of the International Congress of Mathematicians – Seoul 2014* Vol. II, 319–342. Kyung Moon Sa, Seoul, 2014.

[21] Féaux de Lacroix, C.T. p-adische Distributionen. *Diplomarbeit*, Kóln, 1993.

[22] Féaux de Lacroix, C.T. Einige Resultate über die topologischen Darstellungen p-adischer Liegruppen auf unendlich dimensionalen Vektorräumen über einem p-adischen Körper. Thesis. Köln 1997, *Shriftenreihe Math. Inst. Univ. Münster*, 3. Serie, Heft 23 (1999), 1–111.

[23] Harris, M., Taylor, R. The geometry and cohomology of some simple Shimura varieties. *Annals of Mathematics Studies, vol. 151*. Princeton University Press, Princeton, NJ, (2001).

[24] Henniart, G. Une preuve simple des conjectures de Langlands pour $\mathrm{GL}(n)$ sur un corps p-adique. *Invent. Math.* 113(2) (2000), 439–455.

[25] Kohlhaase, J. Invariant distributions on p-adic analytic groups. *Duke Math. J.* 137, no. 1 (2007), 19–62.

[26] Morita, Y. Analytic representations of SL_2 over a p-adic number field, II. Automorphic forms of several variables (Katata, 1983), 282–297, *Progr. Math.*, 46, Birkhäuser Boston, Boston, MA (1984).

[27] Orlik, S. Equivariant vector bundles on Drinfeld's upper half space. *Invent. Math.* 172 (2008), no. 3, 585–656.

[28] Orlik, S., Strauch, M. On Jordan–Hölder series of some locally analytic representations. *J. Amer. Math. Soc.* 28 (2015), no. 1, 99–157.

[29] Patel, D., Schmidt, T., Strauch, M. Locally analytic representations of $\mathrm{GL}(2,L)$ via semistable models of $\mathbb{P}^1$. *J. Inst. Math. Jussieu* 18 (2019), no. 1, 125–187.

[30] Schmidt, T. Analytic vectors in continuous p-adic representations. *Comp. Math.* 145 (2009), 247–270.

[31] Schmidt, T. Verma modules over p-adic Arens–Michael envelopes of reductive Lie algebras. *J. Algebra* 390 (2013), 160–180.

[32] Schneider, P. Nonarchimedean Functional Analysis. *Springer Monographs in Mathematics*, Springer-Verlag, Berlin, 2002.

[33] Schneider, P. p-adic Lie groups. *Grundlehren der Mathematischen Wissenschaften*, vol. 344. Springer, Heidelberg, (2011).

[34] Schneider, P., Teitelbaum, J. p-adic Fourier theory. *Documenta Math.* 6 (2001), 447–481.

[35] Schneider, P., Teitelbaum, J. $U(\mathfrak{g})$-finite locally analytic representations. *Representation Theory* 5 (2001), 111–128.

[36] Schneider, P., Teitelbaum, J. Locally analytic distributions and p-adic representation theory, with applications to GL_2. *J. AMS* 15 (2002), 443–468.

[37] Schneider, P., Teitelbaum, J. Banach space representations and Iwasawa theory. *Israel J. Math.* 127 (2002), 359–380.

[38] Schneider, P., Teitelbaum, J. Algebras of p-adic distributions and admissible representations. *Invent. Math.* 153 (2003), 145–196.

[39] Schneider, P., Teitelbaum, J. Continuous and locally analytic representation theory. Lectures at Hangzhou (2004), available at https://ivv5hpp.uni-muenster.de/u/pschnei/publ/lectnotes/hangzhou.pdf.

[40] Scholze, P. The Langlands Correspondence for GL_n over p-adic fields. *Invent. Math.* 192(3) (2013), 663–715.